# 1st International Conference on Engineering and Applied Science (ICEAS 2019)



## Development of Engineering and Science Towards Revolution 4.0

Journal of Physics: Conference Series Volume 1381

Madiun, Indonesia
21 August 2019

**Editors:**

**Andista Candra Yusro**
**Robbi Rahim**

ISBN: 978-1-7138-1137-4
ISSN: 1742-6588

**Printed from e-media with permission by:**

Curran Associates, Inc.
57 Morehouse Lane
Red Hook, NY 12571

**Some format issues inherent in the e-media version may also appear in this print version.**



Institute of Physics
Dirac House, Temple Back
Bristol BS1 6BE UK

Phone: 44 1 17 929 7481
Fax: 44 1 17 920 0979

techtracking@iop.org

**Additional copies of this publication are available from:**

Curran Associates, Inc.
57 Morehouse Lane
Red Hook, NY 12571 USA
Phone: 845-758-0400
Fax: 845-758-2633
Email: curran@proceedings.com
Web: www.proceedings.com

# TABLE OF CONTENTS

## CHEMISTRY, BIOLOGY, MATHEMATICS



## COMPUTER SCIENCE

## EDUCATION



## ENGINEERING

## INFORMATION TECHNOLOGY

Journal of Physics: Conference Series **1381** (2019) 011001 doi:10.1088/1742-6596/1381/1/011001

# Preface

The Faculty of Engineering, Universitas PGRI, Madiun organized the 1st International Conference on Engineering and Applied Science (ICEAS 2019) on 21st August 2019 in Madiun, East Java, Indonesia. The ICEAS 2019 aims to exchange knowledge and research finding among academicians, researchers, profesionals, policy makers, and postgraduate students.

The awarness of research publication by lecturers,students, teachers, and practitioners in the Madiun area is still minimal. Because of that educational institution must be able to provide motivation and space for researcher to disseminate their research and accomodate the result of research that has been done. International Conference on Engineering and Applied Science or we call it ICEAS, is the first International Conference held by Engineering Faculty UNIPMA. This faculty have 5 departments, there are Informatics Engineering, Information System, Chemical Engineering, Industrial Engineering, and the last but not least is Electrical Engineering. This 5 departments are going through together to held this conference and brings the theme "Development of Engineering and Science Towards Revolution 4.0" .

ICEAS 2019 was attended by 158 participants, and a total of 70 papers were presented and discussed. The papers were authored by researchers from Timor Leste, Japan, and Indonesia. All papers have been scrutinized by a panel of reviewers who provide critical comments and corrections, and thereafter contributed to the improvement of the quality of the papers. Based on the reviewer's reports, 70 papers were selected and eligible to be published in the proceeding

We sincerely express our gratitude to the international/national advisory committee, presenters, organizing committee members, session chairs, the Dean and all members of the Faculty of Egnineering Universitas PGRI Madiun, student volunteers, participants, contributors and all the members ICEAS 2019. Last but not the least, we are thankful to IOP JPCS for producing the proceeding.

The Editors:

Andista Candra Yusro
Robbi Rahim

Chairman of the ICEAS 2019:

Wildanul Isnaini, S.T., M.Sc

# Peer review statement

All papers published in this volume of *Journal of Physics: Conference Series* have been peer reviewed through processes administered by the proceedings Editors. Reviews were conducted by expert referees to the professional and scientific standards expected of a proceedings journal published by IOP Publishing.

# Study of fasting effect in holy Al-Qur'an on testosterone hormones in mice using Elisa method

**N Marfu'ah[1*], N M Kusumaningtyas[1], A Y Damayanti[2], A Fadholah[1]**

[1]Pharmacy Department, Health Science Faculty, University of Darusssalam Gontor, Indonesia
[2]Nutrition Department, Health Science Faculty, University of Darusssalam Gontor, Indonesia

**Abstract.** Fasting is one of the routine worship performed by a Muslim. It has many benefits including controlling lust (Hadith of Bukhari and Muslim). Lust of an organism is controlled by the testosterone hormone. Fasting has the possibility to control the levels of testosterone in the blood so it does not exceed normal levels that be able to manage the organism's libido. The method that is used to determine hormone levels in the blood is the ELISA method (enzyme-linked immunosorbent assay). This study aims to determine the fasting effect in the Qur'an on testosterone levels in mice using the ELISA method. This study utilized 5 treatments (n=3). The types of treatments were not fasting (control), fasting in the middle of the Hijri month for 3 days; fasting on Mondays and Thursdays; Dawood fasting, and fasting every day for 36 days. Based on the type of fasting performed, fasting on Monday and Thursdays generated the highest increase in hormone levels followed by Dawood fasting then fasting every day for 36 daysIncreased levels of testosterone did not reach excessive levels; thus it can be concluded that fasting can curb lush in the organism that was accordance with Hadith of Bukhari and Muslim.

## 1. Introduction

The millennial generation, or commonly called Y Generation which is always keeping up the current development although sometimes is not accompanied by their psychological development, especially in terms of maintaining the desire to act immorally. Lust or libido is controlled by the hormone testosterone[1]. One way to control these hormone levels is by fasting as written in Hadith of Bukhari and Muslim, Prophet Rasulullah said "O young men, whoever of you has a family should be married because he can control his eyes and keep his genitals. Anyone who has not been able to fast because he can maintenance his lust". Based on the hadith, it is explained that for young people who can not get married should fast because fasting can hold lust in the normal condition[2].

Health problems that occur in humans can be treated using 2 ways, namely pharmacological therapy (using drugs) and non-pharmacological therapy (without drugs). Fasting is one way that can be used for non-pharmacological therapy for abnormalities in the testosterone levels in the blood. The influence of pharmacological and non-pharmacological therapies before being carried out in humans (clinical trials), must be done in experimental animals (preclinical testing) to ensure safety and side effects. This study is still in the preclinical testing phase using mice in the form of mice. Mice are normally used as experimental animals because these animals belong to the class of mammals so the reproductive, respiratory, and circulatory systems resemble human[3]. In addition, mice have a relatively short life cycle, many children per birth, variations in characteristics are high, easy to handle, anatomical characteristics and physiology are well characterized[4].

The hormone can be analyzed by various methods such as radioimmunoassay (RIA), enzyme immunosorbent assay (EIA), and enzyme-linked immunosorbent assay (ELISA). The advantages of the ELISA method compared the other two methods are simple, easy, economical and have a high enough level of sensitivity so it is suitable for research to test the types of hormones that are not too specific as in this study[5].

Fasting is one of the worship carried out by a Muslim and is done by refraining from eating, drinking, and things that cancel fasting from the dawn until the sunset. In Islam, there are several kinds of fasting such as obligatory fasting, namely fasting every day on the month of Ramadan (Surat al-Baqoroh: 183) and fasting of the Sunnah such as fasting on Mondays and Thursdays, Dawood fasting (Hadith of Bukhari and Muslim), and fasting 3 days in the middle of the month. Fasting has several benefits, such as the study which shows that fasting on Monday can increase body temperature[6]. Explains that Dawood's and Monday-Sunnah fasting can increase achievement motivation in students[7]. Dawood fasting can also stabilize human emotions[8]. Fasting on Monday can reduce anxiety levels in humans[9]. Thursday's fasting can reduce blood sugar levels[10]. Based on the results of the above research, the study of the effects of fasting in the Qur'an on testosterone levels in mice using the ELISA method has never existed, so this research has the urgensy to finish it.

## 2. Materials and Methods

The materials used in this research were a plastic tub, wire mesh cover, plastic bottle for drinking of mice with the volume of 50 ml, digital scales (Acis, max: 500 g, d: 1 g) to balance the mice weight, syringe of 1 mL for taking blood samples of mice, Eppendorf tube of 1,5 ml to storage the blood sample, freezer (-20 °C) to maintain the temperature of blood sample, centrifuge (80-1 Table Top Low Speed Centrifuge), ELISA (Zenix-320 Microplate Reader) blood samples for analysis the testosterone hormones, ELISA kit (DRG, Germany), wood shavings as mats cages, tap water for drinking of mice, pellets type 511 broilers produced by PT. Japfa Comfeed Indonesia Tbk. for food of mice and. The experimental animals used in this study were male Balb / C strain mice with an initial weight of 23-26 grams and an age range of 8-10 weeks obtained from the Farma Veterinary Center Surabaya, East Java, Indonesia.

The research began by adapting mice to the treatment site for 1 week. Mice were placed in 30 cages (each cage containing 1 mouse) made of plastic tub surrounded by wire mesh, and given a wood shaving which was replaced every 4 days. Mice were given feed in the form of pellets and tap water which were given in *ad libitum*. Mice were weighed every day using digital scales to determine the body weight before being treated. Mice were set fasting (not fed and not given a drink) starting at 4 AM until 6 PM for 36 days.

The mice that have been fasted then were taken the blood through the heart ± 1 ml and put it in an Eppendorf bottle. Then, the collected blood was centrifuged at 3,000 rpm for 20 minutes to get the plasma. The plasma obtained was stored in a freezer with a temperature of -20°C[11]. The blood plasma sent to the Physiology Laboratory, Faculty of Veterinary Medicine, Gajah Mada University to calculate testosterone levels using the ELISA method. Data were analyzed using one-way ANOVA with a significance level of 5%. The statistical program used for analysis is the SPSS 16 program.

## 3. Results and Discussions

### *3.1Effect of fasting on body weight*

The body weight of an organism can be influenced by the amount of food intake in the body, including mice. Fasting is done by refraining of eating and drinking from dawn to sunrise. Mice given fasting treatment will affect their weight. Data on the average body weight of mice were presented in Table 1. Based on the results of the analysis indicate that the value of the average body weight of mice is 0.317 which means that the average body weight of mice between treatment groups was not statistically significantly different (P>0.05). Nevertheless, the average body weight of mice in treatment group B decreased by 1.57% when compared to the control group. However, the average body weight of mice in the treatment group E, C, and D increased when compared to the control group, which amounted to 1.47%; 1.36% and 1.23%.

**Table 1.** Average body weigth of mice

| Treatment | n (the total of mice) | Body weight mice (gram) ± SD |
|---|---|---|
| A | 6 | 26,92 ± 1,38 [a] |
| B | 6 | 25,35 ± 2,42 [a] |
| C | 6 | 28,28 ± 3,42 [a] |
| D | 6 | 28,14 ± 3,96 [a] |
| E | 6 | 28,33 ± 2,16 [a] |

Information:
a = no significant difference at the level of 5%
n = number of experimental animals
SD = Standard Deviation
A = not fasting (control)
B = fasting in the middle of the Hijri month for 3 days
C = fasting on Mondays and Thursdays
D = Dawood fasting
E = fasting every day for 36 days

According to previous studies, fasting could cause weight loss so fasting is one way that can be done if you want to avoid a state of obesity or excessive weight[12–14]. In this study, the statement is contrary because fasting carried out in mice actually increases weight. Although the increase in body weight of mice does not cause obesity but makes mice having the ideal body weight which is close to the range of 30 grams. This fact is in accordance with the statement that the body weight of adult male mice ranges from 20-40 grams. 30 grams of body weight is the ideal body weight because it is not close to thin (20 grams) and also not close to obesity (40 grams)[15].

The increase in body weight of mice in this study was caused by observations that fasted mice had improved appetite. This was indicated by the mice who were fasted to always eat food when they are given food after fasting, is in the afternoon until the morning before they have fasted again. It is very different from the control group, which is not fasting mice. They tend to waste food given and actually their appetite decreases. One of the benefits of fasting is that it can maintain the digestive organs to work properly. If the digestive organs work well, the absorption of food will run well and the metabolic processes in the body will run well. This will cause the body to be healthy, be able to carry out activities well and also have an ideal body weight as the results shown in this study[16].

*3.2Effect of fasting on testosterone hormone levels*

The hormone testosterone is formed from cholesterol originating from food intake so that there is a possible correlation between the treatment of differences in fasting with testosterone levels. Data on the average testosterone levels of mice were presented in Table 2. Based on the results of the analysis indicate that the value of the significant average testosterone of mice was0.308 which means that the average testosterone hormone mice between treatment groups were not statistically significantly different ($P > 0.05$). Nevertheless, the average testosterone hormone in mice in the C, D, and E treatment groups rose by 73.51%; 53.21% and 28.86% when compared to the control group. However, the average testosterone level of mice in treatment group B decreased compared to the control group, which was 14.74%.

**Table 2.** Average testosterone hormone levels of mice

| Treatment | n (the total of mice) | Testosterone hormone levels of mice (ng/ml) ± SD |
|---|---|---|
| A | 6 | 0,2423 ± 1,139 [a] |
| B | 6 | 0,2066 ± 0,099 [a] |
| C | 6 | 0,9146 ± 0,346 [a] |
| D | 6 | 0,5179 ± 0,658 [a] |
| E | 6 | 0,3406 ± 0,296 [a] |

Information:
a = no significant difference at the level of 5%
n = number of experimental animals
SD = Standard Deviation
A = not fasting (control)
B = fasting in the middle of the Hijri month for 3 days
C = fasting on Mondays and Thursdays
D = Dawood fasting
E = fasting every day for 36 days

Based on the results of this study, testosterone levels in the treatment group namely fasted mice showed an increase compared to the control group of mice that did not fast. The hormone testosterone is a hormone produced by Leydig cells in the testicles. The process of synthesis of this hormone is stimulated by the hormones GnRH and LH. GnRH produced by the hypothalamus stimulates the anterior pituitary to produce LH. LH stimulates Leydig cells to produce testosterone. The testosterone hormone has two roles, namely affecting spermatogenesis in the testes and going into the blood vessels and then circulating throughout the body to control secondary sex in male organisms. The process of forming this hormone is strongly influenced by the presence of energy from respiration (burning food using oxygen). In the discussion above, it was explained that fasting can improve appetite, improve the performance of digestive organs, and metabolic processes can run well so that food absorption runs optimally. Food that is sufficiently available will produce very enough energy, one of which is for the formation of the hormone testosterone[17]. In addition, explains that the hormone testosterone is formed from cholesterol which initially in the form of LDL in the blood and LDL is also derived from food intake that enters the body of the organism[18].

The results of ANOVA analysis from this study indicated that fasting could increase testosterone levels even though statistically it was not significantly different. The normal testosterone level in adult male mice is 3.06 ng/mL[3]. The testosterone hormone in addition to controlling the process of spermatogenesis can also control lust or libido in organisms. If the hormone levels rise above the normal limit, it will cause lust also to increase excessively. Based on the results of this study, the level of male mice testosterone hormone fasted rose from the control group but did not exceed the normal limit of more than 3.06 ng/mL[1].

Based on the type of fasting performed, fasting on Mondays and Thursdays had the highest effect on the increase in testosterone, then followed by Dawood fasting and fasting every day for 36 days. Whereas the middle fasting month of the Hijri actually decreases testosterone levels compared to controls. This shows that the type of fasting performed to control testosterone levels should not be too much, such as fasting every day or not too little, such as fasting 3 days in the middle of the Hijri month. Dawood fasting is still not optimal for controlling testosterone levels because the distance between fasting with one another is too close, which is only 1 day. The best type of fasting to control testosterone levels is seen on fasting on Mondays and Thursdays. This is because the distance between fasting with one another is not too close and not too long. The process of forming testosterone from cholesterol is probably best to occur with a distance of 2 to 3 days as in the treatment of fasting on Mondays and Thursday. The biosynthesis process of testosterone hormones occurs from cholesterol in the mitochondria of testis cell. It consists of many steps, including transferring cholesterol from mitochondria membrane pass through StAR protein. After that, cholesterol will be changed to be testosterone hormones using enzymes including 3β-hydroxysteroid dehydrogenase, 17α-hydroxylase/ C17-20-lyase, and 17β-hydroxysteroid dehydrogenase, as shown in Figure 1 below[19].

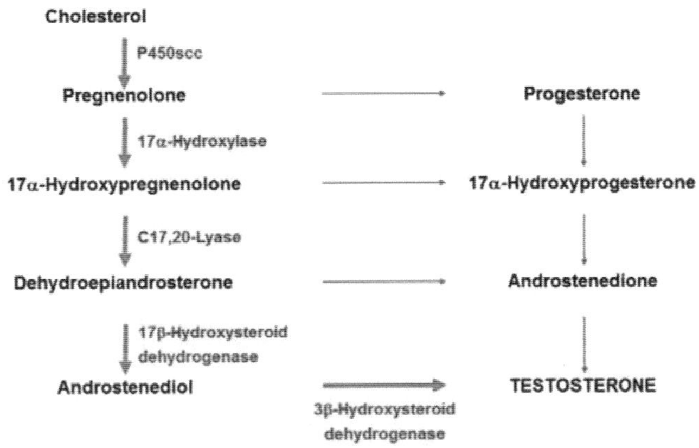


**Figure 1.** Biosynthetic pathway of testicular Testosterone synthesis

The method used to measure testosterone levels in the blood of mice in this study used the ELISA method. ELISA (enzyme-linked immunosorbent assay) is one technique designed to detect and quantify peptides, proteins, antibodies, and hormones on a plate-based. This polystyrene plate in ELISA will bind antibodies and proteins, making this technique easy to design and do. The working principle of ELISA is employing antibodies associated with enzymes complexed with antigens that are immobilized on a solid surface. While the enzyme activity on the substrate. It can be used as a method of detection in this technique. The necessary procedures in ELISA are four as in Figure 2 below (direct, indirect, sandwich, or competitive). ELISA is directly carried out by direct immobilization of the antigen by absorption on the plate. Whereas ELISA is not immediate, immobilization of antigen is not directly absorbed in the plate but on antibodies on the plate. In sandwich ELISA, antigens can interact with 2 types of antibodies so this ELISA technique tends to be specific to antigens. It has at least 2 antigenic sides (interactions with antibodies) or multivalent antigens such as polysaccharides or proteins. While Competitive ELISA is done by adding competitors to the micro titer holes on the plate[5,20].

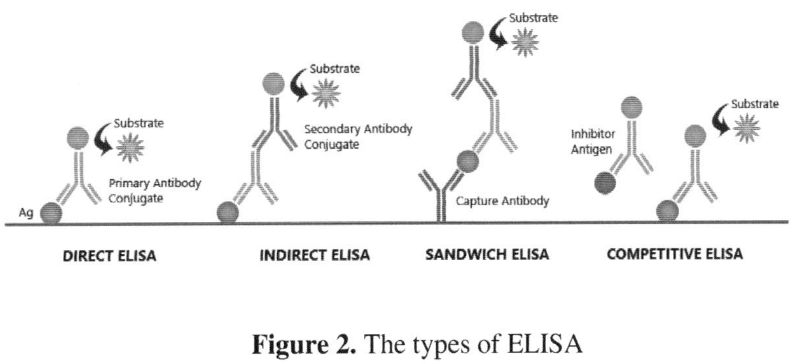


**Figure 2.** The types of ELISA

## 4. Conclusions

The results showed that fasting in the holy of Al-qur'an caused an increase in testosterone levels in mice. Although statistically the increase was not significantly different (P value> 0.05). Based on the type of fasting performed, fasting on Monday Thursday caused the highest increase in hormone levels followed by fasting every day for 36 days. Whereas the middle fasting month of the Hijri actually decreases testosterone levels compared to controls. Increased levels of testosterone did not reach excessive levels.

Thus, it could be concluded that fasting has ablity to curb lust in the organism in accordance with the word of Rosululloh in Hadith of Bukhari and Muslim.

## 5. Acknowledgement

The authors are grateful to Ministry of research, technology and higher education who gave the beginner lecturer research grant. And, the authors thanks to the Department of Pharmacy, University of Darussalam Gontor has given permission to do reseach in the laboratory.

## References

[1] Muryanti Y, Suharno and Sagi M 2006 Kadar testosteron serum darah dan kualitas spermatozoa mencit (Mus musculus L.) setelah diberi ekstrak biji saga (Abrus precatorius L.) *Sains dan Sibernatika* **19** 209–22

[2] Al-Asqalani A-H I H 2016 *Bulughul Maram* (Jakarta: Ummul Quran)

[3] Ngatijan 1991 *Petunjuk Laboratorium: Metode Laboratorium dalam Toksikologi* (Yogyakarta: PAU Bioteknologi UGM)

[4] Priyambodo S 2003 *Pengendalian Hama Tikus Terpadu* (Jakarta: Penebar Swadaya)

[5] Boster Biological Technology 2019 *ELISA Handbook Principle, Troubleshooting, Sample Preparation and Assay Protocols*

[6] Purwaningsih Y 2015 *Pengaruh Puasa Senin-Kamis terhadap Suhu Tubuh Basal Santri Pondok Pesantren Nurul Umah Putri Kota Gede Yogyakarta sebagai Alternatif Sumber Belajar Biologi Sma Kelas XI* (UIN Sunan Kalijaga)

[7] Kurniawati A 2010 *Pengaruh Aktivitas Puasa Sunnah Dawud dan Senin-Kamis terhadap Motivasi Berprestasi pada Siswa Kelas XI MAN Temanggung Tahun Ajaran 2009/2010* (Sekolah Tinggi Agama Islam negeri Salatiga)

[8] Fathonah A Y N 2011 *Pelaksanaan Puasa Daud dan Hubunganya dengan Kestabilan Emosi (Santriwati PP. Al – Fitroh Jejeran Wonokromo Plered Bantul Yogyakarta) (Yogyakarta: Universitas Islam Negeri Sunan Kalijaga)* (Universitas Islam Negeri Sunan Kalijaga Yogyakarta)

[9] Irchamni A 2011 *Pengaruh Intensitas Melakukan Puasa Senin Kamis terhadap Penurunan Tingkat Kecemasan Santri dalam Menghafal Nadham Alfiyah di Madrasah Diniyah Tsanawiyah "Mamba'ul Huda" Talok Woh Mojongawen Blora* (Semarang: Institut Agama Islam Negeri Walisongo)

[10] Saputra A B W 2016 *Pengaruh Puasa Senin dan Kamis terhadap Kadar Gula Darah pada Penderita Diabetes Melitus Tipe 2 di Dukuh Kasihan, Bantul, Yogyakarta* (Universitas Muhammadiyah Yogyakarta)

[11] Akmal M and Mahdi C 2010 *pada Tikus Akibat Paparan Ekstrak Air Biji Pinang ( The Increased of Testosterone Concentration of Rat)* vol 11

[12] Kurniasani A 2017 *Perbandingan Lemak Tubuh Sebelum dan Saat Berpuasa pada Dewasa Muda* (Institut Pertanian Bogor)

[13] Laksmi D H 2017 *Efikasi Puasa Ramadan terhadap Penurunan Berat Badan dan Indeks Massa Tubuh pada Dewasa Muda* (Institut Pertanian Bogor)

[14] Munigar M 2013 Puasa Ramadhan dan Obesitas *J. Heal. Qual.* **4** 47–53

[15] Kusumawati D 2004 *Bersahabat dengan hewan coba* (Yogyakarta: University Press UGM)

[16] Danusukarno S 1989 *Tanya Jawab Masalah Kesehatan Sehari-hari* (BPK Gunung Mulia)

[17] Wistuba J 2007 Mammalian Spermatogenesis *Adv. Genome Biol.* **4** 249–304

[18] Gill G 1987 *Biosynthesis, secretion, and metabolism of hormones. In: Endocrinology and Metabolism* (New York: McGraw-Hill)

[19] McEwan J and Brinkmann A O 2016 *Androgen Physiology: Receptor and Metabolic Disorders* ed Endotext (South Dartmouth: South Dartmouth)

[20] Biovendor 2017 *Mouse/Rat Testosterone ELISA* (Czech Republic: BioVendor Research and Diagnostic Products)

Journal of Physics: Conference Series **1381** (2019) 012002 doi:10.1088/1742-6596/1381/1/012002

# Utilization of Sap from Part of Kepok Banana Tree (Musa Mcuninata Balbisianacolla) with Variation of Extraction Solutions as Textile Dyes

**W Nuriana***, M **Winarni, Suryono**

Universitas Merdeka Madiun, Serayu Street 79 Madiun, East Java, Indonesia, 63133

nuriana@unmer-madiun.ac.id*

**Abstract**. The purpose of this study was to obtain tannin natural dyes from the sap of the fronds / stalks of kepok banana leaves, namely the percentage of yield, color intensity, time and good solvents for taking the most dyes in the sap of the banana tree. The extraction of dyes in various parts of banana trees was carried out by means of an extraction process using soxchlet, with a medium of distilled water, methanol, ethanol and ethylacetate. The raw material was sliced, dried to + 6.7% moisture content, ground powder made from size 60-80 mesh. The powder was weighed 50 grams and extracted with each medium solution in 500ml at each evaporation temperature, the results were evaporated to dryness. Test the tannin color intensity with a spectrophotometer. The results showed that the optimal amount of sap yield on the banana midrib was 16,248%, the color intensity of tannin obtained was 0.218 in methanol solvent products. The solvent which can provide the optimal yield in extracting tannins from banana leaf midrib was methanol. The time of the extraction process to the maximum out tannin was 4-4.5 hours

## 1. Introduction

The development of the textile industry and textile products in Indonesia is progressing rapidly, also with additional materials related to industrial products, namely dyes. The development of the above synthetic dyestuff solution resulting from inorganic chemical reactions can cause quite serious problems for the environment, namely the wastewater produced. Liquid waste and the textile industry contain relatively high BOD, COD, color suspended solids and can also contain heavy metals that interfere with water bodies (Paryanto, 2015). Most textile dyes on the market are carcinogenic / cancer-causing (Arifin, 2009).

Banana tree sap when it comes to clothes is difficult to remove even though washed over and over again, with the nature of banana sap is very soluble when attached to the fabric (Endang, 2010).Banana sap in the health sector is useful in healing wounds due to scratches on the surface of the skin (Riyani, 2015; Kwartiningsih, 2010; Paryanto, 2015).

Banana sap contains tannins, which are natural brown coloring pigments. Brown color is widely used in batik coloring. Many grow various types of banana plants that spread throughout the Indonesian archipelago. Most natural dyes are obtained from plant parts, for example in roots, tubers, tree bark, leaf stalks, fruit stalks, fruit peels, fruit skins, leaves, fruits and seeds.

## 2. Research methods

Journal of Physics: Conference Series **1381** (2019) 012002 doi:10.1088/1742-6596/1381/1/012002

The material used is part of the tree, namely the stem, stem and fruit stalk on the Kepok banana plant. The extraction of dyes in the banana tree is carried out by the extraction process using a soxchlet tool. Solvents for the extraction process are aquades, methanol, ethanol and ethyl acetate.

The raw material for tree parts, fronds and tangakai bananas, sorted in slices, dried in the sun for ± 5 days, blended to powder. ± 12 grams are taken to put in filter paper to be placed in Soxchlet and extracted in parallel with distilled water, methanol, ethanol and ethyl acetate. Heating was carried out at 100°C for water solvents, 64°C for methanol, 78.4°C for ethanol, and 77°C, for etylacetate solvents with time for 4-4.5 hours respectively.

The product of extraction is in the form of viscous liquid, then the evaporation process is carried out by ovenring 4-6 hours at 105°C. Absorbance testing with UV VIS spectrophotometry by dissolving the dyes produced with distilled water.

The weight of the powder product is weighed for each solvent, then tested:

1. Soaking weight in each solvent and soaking weight.
2. Absorption of dyes in various parts of a banana tree.

## 3. Result and Discussions

The powder material to be processed was measured by the water content of stem powder, stem, stem of the kapok banana and the extraction process was carried out using soxchlet with aqueous media of distilled water, methanol, ethanol, ethyl acetate for 4-4.5 hours. The tannin dye extracted in the form of a thick brown liquid, then concentrated by evaporating water with an oven at a temperature of 105oC for 8-10 hours. The results of the extraction of dyes obtained for all raw materials, tested the average water content of 0.0012%.

Absorbance was measured by diluting the results of the dyes extracted from 1gram in distilled water, then inserting a UV VIS spectrophotometer at a wavelength of 570 nm. The yield and absorbance obtained in various parts of the Kepok banana tree and various solvents can be seen in Table 1 below:

**Table 1.** Yield and Absorption of dyes produced by the soxchlet extraction process various parts of the kepok banana tree (stem, midrib, leaf, stalk) on the media solvents, aquades, methanol, ethanol, ethyl acetate

| Solvent media | Stems | | midribs | | fruit stalks | |
|---|---|---|---|---|---|---|
| | Yields(%) | Absorbance | Yields (%) | Absorbance | Yields (%) | Absorbance |
| Aquades | 1,9069 | 0,024 | 4,9034 | 0,199 | 1,4243 | 0,159 |
| Metanol | 5,6083 | 0,106 | 16,2481 | 0,218 | 7,4106 | 0,194 |
| Etanol | 2,5317 | 0,023 | 13,8665 | 0,083 | 0,3501 | 0,014 |
| Etil Asetat | 1,0231 | 9,012 | 1,4954 | 0,030 | 0,4345 | 0,005 |

Quantitative yield results of dyestuff results obtained by the highest weight of extract containing tannin dyes are banana fronds with methanol as a solvent. This is due to the fact that the methanol solvent is a polar solvent and does not have a high boiling point so it does not damage the components to be extracted when the desired compound is susceptible to high temperatures. Although distilled water is more polar compared to methanol. This is in accordance with research Riyani (2015), which carried out the extraction process on ambon banana stems obtained the highest concentration of flavonoid substances of 8.4301 mg / l in methanol solvents.

This shows that although aquades are more polar than methanol, ethanol and ethyl acetate, methanol solvents produce more tannins than aquades because methanol has a lower boiling point.

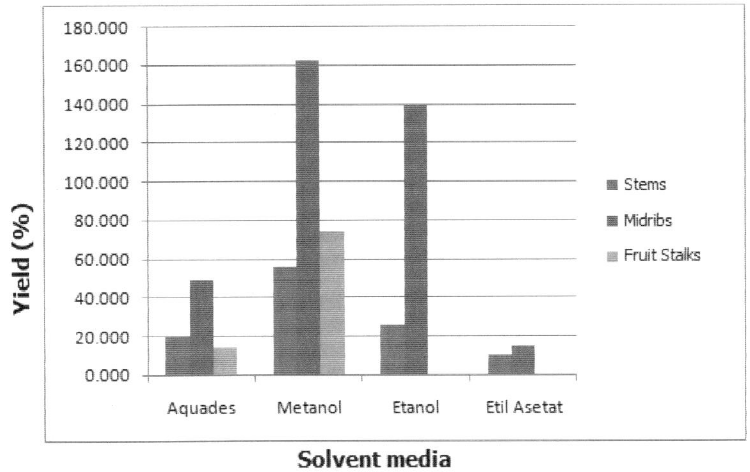


**Figure 1.** Graph of rendement of natural dyes extracted from stems, midribs, fruit stalks banana kepok with various solvents aquades, methanol, ethanol, ethyl acetate

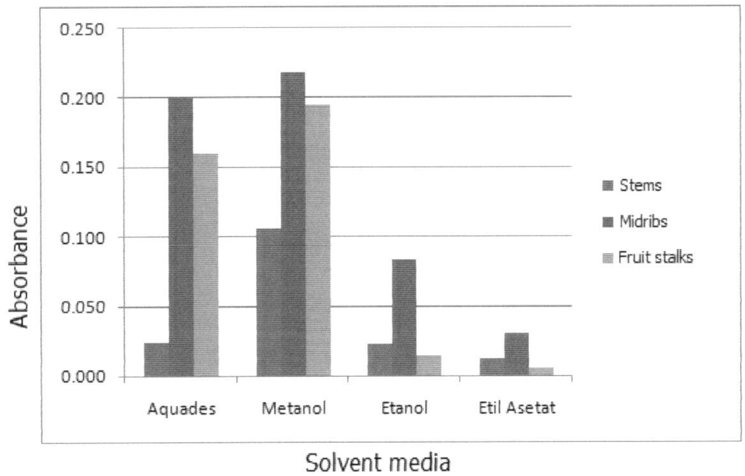


Figure 2. Graph of absorbance of natural dyes extracted from stems, midribs, fruit stalks banana kepok with various solvents aquades, methanol, ethanol, ethyl acetate

Calculating the color intensity resulting from extraction with soxchlet was tested with a UV VIS spectrophotometer, wavelength of 570 nm. The 0.5 gram dye powder was dissolved in 100 ml of distilled water and then absorbed on the spectrophotometer.

In the graph above the higher the absorbance price, the higher the color intensity. The highest absorbance price is in the original dyestuff from Kepok banana fronds with methanol extraction media producing an older color than the banana tree and fruit stalks. This is in accordance with the study of Kwartiningsih, (2010) the highest absorbance results in dyes from the Kepok banana tree compared to ambon bananas and plantains by extraction using distilled water.

Extraction time of 4-4.5 hours, it is expected that tannin dyes will be extracted entirely. In the study of Paryanto (2015), extraction of banana stems with distilled water and ethanol in the variation of the extraction process time of 20-120 minutes, the results of the extract of dyes obtained were less than the maximum.

## 4. Conclusions

Natural brown tannin dyes obtained from the extraction process using soxchlet from the part of the Kepok banana tree (stem, stem, fruit stalk). The time and temperature of the extraction process using soxchlet is 4-4.5 hours and the boiling point temperature of methanol is 64°C. The solvent that can produce the highest dye is methanol, and the portion of the banana tree that produces tannin dyes is the banana leaf midrib. The highest tannin dye composition in leaf fronds was 16.248%, and the intensity / absorbance of tannin dyes obtained 0.218 in banana leaf fronds with methanol extraction solvent.

## Acknowledgement

The researcher thanks the honorable:

1. DPRM Ristek Dikti which has provided grant funds through the Higher Education Applied Research (PTUPT) scheme;
2. Chancellor of Merdeka Madiun University who has provided facilities to be able to propose a grant for Higher Education Applied Research schemes (PTUPT);
3. Principal of SMK Negeri 3 Madiun who provides facilities for laboratory research.

## References

[1] Anonim, 2007, *Utilization of natural dyes for textile and weaving materials* www.gemaindustrikecil.com Arthazone. 2007. " *Chlorophyll Plant Substances That Have Many Health Benefits*" www.arthazone.com

[2] Azhar M, Afandy, Siti N. dan Anang W. M. D., 2017. *Purple Sweet Potato (Ipomoea Batatas L.) Extraction Using Variation of Solvents and Their Utilization as Acid-Base Indicators*, J. Akad. Kim. 6(2): 79-85, May 2017 ISSN 2302-6030(p), 2477-5185 (e)

[3] Eka, Wahyu, 2014, JURNAL REVIEW: *Purple Sweet Potato Co-pigmentation (Ipomoea Batatas var. Ayamurasaki) with Na-Casein and Whey protein co-pigmentation and its Stability to Heating. FTP* : Brawijaya University of Malang.

[4] Fitrihana N., 2007," *Exploration Techniques of Natural Dyes from Plants Around Us For Textile Material Dyeing*" www.batikindonesia.com

[5] Mahfudz, 1996, *Teak Glance*. Research Center for Biotech and Forest Plant Breeding, Yogyakarta.

[6] Moerdoko W. 1975. *Evaluation of Textile Chemistry Section* ",*Textile Technology Institute*, Bandung.

[7] Koswara, Sutrisno, 2014, *Tuber Processing Technology, Part 5: Sweet Potato Processing. IPB : Bogor* (http://scafast.ipb.ac.id) accessed April 5, 2014.

[8] Kwartiningsih, E., 2009, *Mangosteen Skin Natural Textile Dyes*. Chemical Engineering Department, Engineering Faculty: UNS

[9] Kwartiningsih, E., Atika A., Sri B., Aryo N., Fina Rahmawati, 2010, *Utilization of Various Types of Gum and Parts of Banana Trees as Natural Dyes in Textile*. Equilibrium Journal, Vol. 9, No. 1, Page. 5-10.

[10] Marpaung, Monika, 2011, *Study on the Use of Calcium Clordia (CaCl2) on the Stored Age of Purple Sweet Potatoes (Ipomoea batatas L.)*. Faculty of Agricultural Technology: PadangAndalas University

[11] Rahayu, Paramitha, 2012, *Acceptability and Nutritional Content of Supplement Made from Purple Sweet Potatoes*. Faculty of Engineering : Unnes

[12] Paryanto, Hans A. P., 2015, *Dyes from Banana Leaf Stalks* (musa ssp.) J E kuilibrium, vol.14, no. 2, hal.39-43

[13] Riyani A., Adawah R., 2015, *Flavonoid Extraction Soxchlettation Method From Ambon Banana Tree Trunk (Musa paradisacea var, sapientum) with various Types of Solvents, Proceedings of the National Symposium on Innovation and Science Learning* (SNIPS), ISBN, 978-502-19655-8-0, Bandung Indonesia

[14] Sulastri, 2013, *Antioxidant Activity of Ethanol Extract of Purple Sweet Potato (Ipomeabatatas L.) Leaves from Aceh Besar Saree Cultivation*, Chemistry Education Study Program, FKIP Unsyiah, Syiah Kuala University

[15] Winarti S., 2008, *Purple Sweet Potato (Ipomoea batatas L.,) Color Stability and Stability as Natural Dyes, Department of Food Technology*, Faculty of Industrial Technology: UPN East Java.

[16] Zulfa L. , Sri K. , Mas'ud E., 2014, Natural Dyes from teak leaves. Journal of Industria Vol 3 No 1 Hal 62 – 72

# Analysis of Differential Calculus in Economics

**R Marsitin**

Universitas Kanjuruhan Malang, Malang, Indonesia

mars_retno@unikama.ac.id

**Abstract**. The differential is one of the mathematical material in calculus which is loaded with counts. Differential counts can be applied in economics for profit optimization. This study aims to analyze differential calculus in economics. This research is a descriptive qualitative study. This research analyzes the profit optimization in the entrepreneurial world with the second differential formula in calculus. Data analysis is to describe the results of the analysis of the second differential formula with economics in optimizing profits. Calculation of calculus uses the second differential of the mathematical model with the provision that the second differential result is negative which is smaller zero. The results of the analysis show that the second differential calculus smaller than zero, so it can be concluded that to obtain profit optimization in the economy it can be applied using the second differential calculus.

## 1. Introduction

The differential is one of the calculus material in mathematics that studies a derivative function. In determining the derivative of a function, the function must be differentiable. The derivative process is called differentiation so that the derivative is closely related to the differential. In the basic theorem of calculus states that differentiation is a reversal of integration. Differentials in calculus describe changes in a function with a small change in the function variable. In addition, differential calculus discusses solving problems related to a change that is the change in the form of independent variables and dependent variables. These variables are formed in a mathematical model. Mathematical models related to differential can be applied in various fields, for example in the fields of physics, biology, economics. Modern economics leads to the emergence of fundamental concepts and methods in economic theory, which allows the use of differential and integral calculus to describe economic phenomena, effects, and processes [1–3]. Calculus is becoming more accessible in the business because it uses calculus as a tool, a constructive calculus course in meeting economic needs [4,5]. Differential calculus in economics is used in simplifying economic problems including optimizing profits, balancing supply and demand prices. Problems in the economy can be simplified by limiting the number of variables by assuming that other variables remain constant. The results of previous studies [6] states that the buyer's economic order quantity and the vendor's optimal number of deliveries are derived by setting the first derivatives to zero and solving the simultaneous equations. In addition, the results of previous studies [7] states that such derivatives also are the key to the formulation of subproblems determining the response of a problem's solution when the data values on which the problem depends are perturbed. [8] states that the calculus theory and many economic models are dynamic models to relate those two subjects to motivate the use of the calculus of variations and optimal control problems on time scales in the study of economic models.

Regarding the descriptions that have been described, it is very interesting to discuss research related to calculus derivatives in economics and objectives of the research was to analyze derivative calculus in economics.

1.1. Calculus

Calculus material, for the concept and differential definition of calculus includes the understanding of functions, limits and derivatives [9], namely:

*1.1.1. Function.*

The concept of function is one of the most important concepts in mathematics, especially calculus.
A function $f$ is a correspondence rule that connects each $x$ object in a set, called the origin (domain), with a single value $f(x)$ from a second set. The set of values obtained is thus called the function result range.

*1.1.2. The limit.*

The concept of limits can be understood through the question, what happens to the function $f(x)$ when $x$ approaches a constant c.
Intuitive definition of limits:
To state that $\lim_{x \to c} f(x) = L$, means that when $x$ is close but differs from $c$ then $f\ (x)$ is close to $L$.

*1.1.3. Derivative Function.*

Derivatives are neutral words in mathematical terms. Derivatives as keywords in calculus and additional words to functions and limits.
Definition: The derivative function f is another function $f\,'(read\ "f\ accent")$ whose value at any number $c$ is:

$$f'(c) = \lim_{h \to 0} \frac{f(c+h) - f(c)}{h} \ or\ f'(x) = \lim_{h \to 0} \frac{f(x+h) - f(x)}{h}$$

(provided that this derivative exists and not $\infty\ atau - \infty$)
If this derivative does exist, it is said that $f$ is differentiated in $c$. The search for derivatives is called differentiation and the part associated with derivatives is called the differentiation calculus. The differentiation operation takes a function f and produces a new function that is $f'$ (first derivative). If $f'$ is differentiated, it still produces other functions that are stated with $f''$ (second derivative). If $f''$ is differentiated again, it will still produce other functions called $f'''$ (third derivative), then the fourth derivative is declared $f''''$, the fifth derivative is expressed $f'''''$and so on [10,11].

1.2. Derivative calculus

Calculus material, for the concept and differential definition of second derivative calculus relating to economics, namely:

*1.2.1. Maximum and Minimum.*

Definition:
Suppose that a function *f (x)* is given with the area of origin $S$ and contains the point $c$, it is said that: $f(c)$ is the maximum value of $f$ on $S\ if\ f(c) \geq f(x)$ for all $x$ in $S$, $f(c)$ is the minimum value $f$ on $S$ if $f(c) \leq f(x)$ for all $x$ in $S$, $f(c)$ is the extreme value of $f$ on S if it is a maximum or minimum value and the function we want to maximize or minimize is an objective function
Theorem $A$ (the maximum-minimum theorem of existence): if $f$ is continuous at a closed interval $[a, b]$ then f has a maximum value and a minimum value. Note the key words in theorem $A$: f is required to be continuous and the set $S$ is required to be a closed interval.

Theorem $B$ (critical point theorem): if $f$ is defined in the interval $I$ containing point $c$. If $f(c)$ is an extreme value, $c$ must be a critical point. In other words, $c$ is one of: endpoint of $I$, stationary point of $f$ and that is the point where $f'(c) = 0$ or singular point of $f$ and that is the point where $f'(c)$ does not exist.

*1.2.2. Monotonic and concave.*
Theorem $A$ (monotonic theorem): suppose that there are continuous $f$ intervals of $I$ and are defined at each inner point of $I$ then: if $f'(x) > 0$ for all $I$-points, then f rises at $I$ and if $f'(x) < 0$ for all points in $I$ then f falls at $I$.
Theorem $B$ (concave theorem): suppose f is defined twice at the open interval $I$ then: if $f''(x) > 0$ for all x in I then f concaves upward on $I$ and if $f''(x) < 0$ for all x in I then f concaves downward on $I$.

*1.2.3. Maximum and Minimum with the First Derivative Test and the Second Derivative Test.*
Theorem $A$ (first derivative test): if $f$ be continuous at an open interval (a, b) containing a critical point $c$: if $f'(x) > 0$ for all $x$ in $(a, c)$ *and* $f'(x) < 0$ for all $x$ in $(c, b)$ then $f(c)$ is the local maximum value of $f$, if $f'(x) < 0$ for all $x$ in $(a, c)$ *and* $f'(x) > 0$ for all $x$ in $(c, b)$ then $f(c)$ is the local minimum value $f$, if $f'(x)$ is marked the same on all parties of $c$, then $f(c)$ is not an extreme local value of $f$.
Theorem $B$ (second derivative test): suppose $f'$ and $f''$ exist at each interval point $(a, b)$ containing $c$, and for example $f'(c) = 0$: if $f''(c) < 0$, then $f(c)$ is the local maximum value of $f$ and if $f''(c) > 0$, then $f(c)$ is the local minimum value $f$.

1.3. Differential calculus in economics
In the economic field, the calculation uses the second derivative of calculus to obtain profit optimization. There are three functions in the economy, namely: the cost function of each production $(C)$, the amount of production function $(Q)$, the revenue function $(R)$ and the profit function $(\pi)$. If all product units are sold, the relationship of the functions is: $\pi = R - C$. If $(\pi'' < 0)$ then in the economy the profit optimization is obtained. The profit function π is analyzed using the second derivative of π so that in the economy the profit optimization is obtained: if $\pi'' < 0$, then π is maximum, profit optimization is obtained, if $\pi'' > 0$, then π is minimum, profit (loss) minimization is obtained and if $\pi'' = 0$ then π is a turning point
The result of the calculation on the second differential calculus was negative, which was smaller than zero $\pi'' < 0$, so in the economy the optimization of profits were obtained [12].

## 2. Method
This research was a qualitative research. The research was done by analyzing differential calculus in economics on food entrepreneurship. Differential calculus in economics is related to differential calculations applied in profit optimization. Differential calculus in economics is through analyzing the differential calculus formula with its application in obtaining the results of calculations on the second differential calculus is negative ie smaller than zero $(\pi'' < 0)$ by determining the cost function of each production $(C)$, the demand function of each production $(P)$, the sum of production $(Q)$, the revenue function $(R)$ and the profit function $(\pi)$ [13].

## 3. Result and Discussion
The results of research on food entrepreneurship obtained data on the cost function of each production $(C)$, the demand function of each production $(P)$, the number of production functions $(Q)$, the revenue function $(R)$ and the profit function $(\pi)$ obtained the following data.
Fixed costs in production amounted to $Rp$ 33,250,000 with details:

Journal of Physics: Conference Series **1381** (2019) 012003 doi:10.1088/1742-6596/1381/1/012003

Monthly salary for employees = $Rp$ 15,750,000, equipment maintenance = $Rp$ 1,500,000, tax = $Rp$ 13,000,000, promotional advertising = $Rp$ 500,000, electricity, and water and telephone = $Rp$ 2,500,000
Expenses in production with details:
Fried chicken = $Rp$ 10,500,000, grilled chicken = $Rp$13,250,000, crispy chicken = $Rp$ 12,500,000, roasted chicken = $Rp$ 12,750,000, spicy chicken = $Rp$ 11,000,000, chicken roll = $Rp$ 12,250,000 chicken steak = $Rp$ 11,750,000
Production expenses in 2017 and 2018 on the Surabaya Indonesia crispy chicken entrepreneur with details in table 1:

**Table 1.** Production Results

| Description | Production results in 2017 | | Production results in 2018 | |
|---|---|---|---|---|
| | Volume | Unit price (Rp) | Volume | Unit price (Rp) |
| Fried chicken | 6700 | 9,500 | 5400 | 10,500 |
| Grilled chicken | 6730 | 11,000 | 4730 | 12,500 |
| Crispy chicken | 6500 | 9,500 | 4900 | 10,500 |
| Roasted chicken | 5250 | 10,500 | 3250 | 12,000 |
| Spicy chicken | 4300 | 8,500 | 3000 | 9,500 |
| Chicken roll | 5600 | 9,000 | 4300 | 10,200 |
| Chicken steak | 4550 | 10,500 | 3300 | 11,500 |

The average cost of each production variable can be obtained from the description of the table above, namely: fried chicken = $Rp$ 1,944, grilled chicken = $Rp$ 2,801, crispy chicken= $Rp$ 2,551, roasted chicken = $Rp$ 3,923, spicy chicken = $Rp$ 3,666, chicken roll = $Rp$ 2,849, chicken steak = $Rp$ 3,561
Profit optimization analysis using differential calculus that is looking for the cost function of each production ($C$), the number of production ($Q$), the revenue function ($R$) and the profit function ($\pi$) in each production variable. The results in table 2:

**Table 2.** Cost Functions, Demand Functions and Revenue Functions

| Description | Cost function($C$) | Demand function ($P$) | Revenue Function ($R$) |
|---|---|---|---|
| Fried chicken | $33{,}2500{,}00 + 1{,}944Q$ | $-0.77Q + 14{,}654$ | $-0{,}77Q^2 + 14{,}654Q$ |
| Grilled chicken | $33{,}2500{,}00 + 2{,}801Q$ | $-0.75Q + 24{,}688$ | $-0.75Q^2 + 24{,}688Q$ |
| Crispy chicken | $33{,}2500{,}00 + 2{,}551Q$ | $-0.625Q + 13{,}563$ | $-0.625Q^2 + 13{,}563Q$ |
| Roasted chicken | $33{,}2500{,}00 + 3{,}923Q$ | $-0.75Q + 14{,}438$ | $-0.75Q^2 + 14{,}438Q$ |
| Spicy chicken | $33{,}2500{,}00 + 3{,}667Q$ | $-0.77Q + 11{,}808$ | $-0.77Q^2 + 11{,}808Q$ |
| Chicken roll | $33{,}2500{,}00 + 2{,}849Q$ | $-0.92Q + 14{,}169$ | $-0.92Q^2 + 14{,}169Q$ |
| Chicken steak | $33{,}500{,}00 + 3{,}561Q$ | $-0.83Q + 14{,}250$ | $-0.83Q^2 + 14{,}250Q$ |

The description in table 2, the form of economic equations in the cost function (C), the demand function (P) and the opinion function (R) is obtained from the fixed costs of production, expenditure costs, production volume and unit price of the variable production. In the cost function, demand function, revenue function and profit function, we can get the function of production amount (Q) and profit function ($\pi$) in table 3:

**Table 3.** Function of Total Production and Profit Function

| Description | Profit Function ($\pi$) | Profit Optimization ($\pi''$) |
|---|---|---|
| Fried chicken | $\pi = -0.77Q^2 + 12{,}710Q - 33{,}2500{,}00$ | $\pi' = -1.54Q + 12{,}710$<br>$\pi'' = -1.54$ |

| | | |
|---|---|---|
| Grilled chicken | $\pi = -0.75Q^2 + 21{,}887Q - 33{,}2500{,}00$ | $\pi' = -1.5Q + 21{,}887$<br>$\pi'' = -1.5$ |
| Crispy chicken | $\pi = -0.625Q^2 + 11{,}012Q - 33{,}2500{,}00$ | $\pi' = -1.25Q + 11{,}012$<br>$\pi'' = -1.25$ |
| Roasted chicken | $\pi = -0.75Q^2 + 10{,}515Q - 33{,}2500{,}00$ | $\pi' = -1.5Q + 10{,}515$<br>$\pi'' = -1.5$ |
| Spicy chicken | $\pi = -0.77Q^2 + 8{,}141Q - 33{,}2500{,}00$ | $\pi' = -1.54Q + 8{,}141$<br>$\pi'' = -1.54$ |
| Chicken roll | $\pi = -0.92Q^2 + 11{,}320Q - 33{,}2500{,}00$ | $\pi' = -1{,}84Q + 11{,}320$<br>$\pi'' = -1{,}84$ |
| Chicken steak | $\pi = -0.83Q^2 + 10{,}689Q - 33{,}2500{,}00$ | $\pi' = -1.66Q + 10{,}689$<br>$\pi'' = -1.66$ |

In table 3, the results are fried chicken with $\pi'' = -1.54$, grilled chicken $\pi'' = -1.5$, crispy chicken $\pi'' = -1.25$, roasted chicken $\pi'' = -1.5$, spicy chicken $\pi'' = -1.54$, chicken roll $\pi'' = -1{,}84$, chicken steak $\pi'' = -1.66$. This can be said to be a second differential calculus with $\pi'' < 0$. The total production function $(Q)$ shows the optimization of production in each production variable and to obtain profit optimization by using the second differential calculus in $(\pi)$ and $(\pi'' < 0)$. This is in accordance with [13] the opinion which states that when the optimization problem is represented by its objective function, the first and second-order conditions for optimality can be expressed in terms of the first and second epi-derivatives of that function. The statement is related to the opinion of [7] who states that optimal production lot size are by using the differential calculus on the production inventory cost function with the need to prove optimality first.
Differential calculus in economics to obtain profit optimization in determining the cost function of each production (C), the demand function of each production (P), the total of production functions (Q), the revenue function (R) and the profit function (π) with completion stages, namely: a) to determine the demand function, the equation of the line is used; b) the line equation in the economic equation with $Q_1$ is the initial demand function, $Q_2$ is the final demand function, $P_1$ is the initial demand function, $P_2$ is the initial demand function so that the equation is obtained: $\frac{P-P_1}{P_2-P_1} = \frac{Q-Q_1}{Q_2-Q_1}$; c) the revenue function with $R = P \cdot Q$ and the profit function with $\pi = R - C$. This is in accordance with [14] the opinion which states the quantity of economic production using differential calculus and solving economic equations in two simultaneous equations (derived from assigning two first partial derivatives to zero) with the need to prove the optimality of conditions with second-order derivatives). In addition, [5] the basic equation of equilibrium theory is supply equals demand. Since we are in a situation of several markets, there are several variables in this equation. [15] Financial Derivatives remain the only ones that can attract people outside the mathematics and physics community because they explain how and why practical financial problems in economics.

## 4. Conclusion

Research conclusions from the results exposure and discussion in economics regarding optimal profit that the second calculus differential calculation can be used in economics to get profit optimization in entrepreneurship. The optimization of production in each production variable and to obtain profit optimization in economics using the second differential calculus in $(\pi)$ and $\pi'' < 0$. Optimize the advantages of using differential calculus by making economic equations in mathematical models, solving mathematical models and distinguishing first and then differentiating the second with negative second differential results in production variables.

**References**

[1] Tarasov V E 2019 On History of Mathematical Economics: Application of Fractional Calculus *Mathematics* **7** 509
[2] Wilmott P, Howison S and Dewynne J 2012 *The Mathematics of Financial Derivatives*
[3] Sent E-M 2018 Enginering Dynamic Economics *Hist. Polit. Econ.* **29** 41–62
[4] Verzosa D 2015 Reading the World with Calculus *PRIMUS* **25** 349–68
[5] Sharma S 2017 New Advancements in the Applications of Fractional Calculus in Science and Engineering *Int. J. Trend Sci. Res. Dev.* **Volume-1** 471–6
[6] Yang P C and Wee H M 2002 The economic lot size of the integrated vendor-buyer inventory system derived without derivatives *Optim. Control Appl. Methods* **23** 163–9
[7] Leung K N F 2009 Optimal production lot sizing with backlogging, random defective rate, and rework derived without derivatives *Proc. Inst. Mech. Eng. Part B J. Eng. Manuf.* **223** 1081–4
[8] Guzowska M, Malinowska A B and Sidi Ammi M R 2015 Calculus of variations on time scales: applications to economic models *Adv. Differ. Equations* **2015** 1–15
[9] FAA 2008 Chapter 1 - Mathematics *Aviation Maintenance Technician Handbook* pp 1–30
[10] Fripp A L, Fripp J B and Fripp M L 2003 Chapter 6 - Calculus *Just-In-Time Math for Engineers* pp 125–76
[11] Smale S 2011 global analysis in economic theory *The New Palgrave Dictionary of Economics* p 1
[12] Davis H T and Allen R G D 2006 Mathematical Analysis for Economists *Natl. Math. Mag.* **14** 60
[13] Poliquin R A and Rockafellar R T 1993 A Calculus of EPI-Derivatives Applicable to Optimization *Can. J. Math.* **45** 879–96
[14] Cárdenas-Barrón L E 2001 The economic production quantity (EPQ) with shortage derived algebraically *Int. J. Prod. Econ.* **70** 289–92
[15] Hirsa A and Neftci S N 2014 *An Introduction to the Mathematics of Financial Derivatives: Third Edition*

Journal of Physics: Conference Series **1381** (2019) 012004 doi:10.1088/1742-6596/1381/1/012004

# Growth and yield of Shallot (*Allium cepa* L.) in respons of organic fertilizers and *Trichoderma asperellum*

**Parwi*, U Isnatin, M Hamawi, and U Etica**
Agrotechnology Department, Universitas of Darussalam Gontor, Indonesia
* p.parwi@yahoo.com

**Abstract**. Increased shallot yield with aplication of organic fertilizer. Cajeput waste as a source of organic fertilizer is very abundant in Ponorogo. Research aimed to study the aplication of organic fertilizers and *Trichoderma asperellum* to incerase growth and yield of shallot. The experiment was laid-out in factorial randomized completely design with 3 replications and eight treatment combinations. The first factors is organic fertilizer ( without organic fertilizer; 6 ton ha$^{-1}$ of cajeput waste compost, 3 ton ha$^{-1}$ of cajeput waste compost + 3 ton ha$^{-1}$ of rabbit manure, 6 ton ha$^{-1}$ of rabbit manure). The second factors is *Trichoderma asperellum* (without *T. asperellum* and 1 L ha$^{-1}$ of *T. asperellum*). The results of this study showed that treatmen of 3 ton ha$^{-1}$ of cajeput waste compost + 3 ton ha$^{-1}$ of rabbit manure without *Trichoderma asperellum* increased plant hight and leaf dry weight of shallot by 25% and 135,66%. The application of combination rabbit manure and without *T. asperellum* increased shallot tillering number by 9,67 tillering. The application of 3 ton ha$^{-1}$ of cajeput waste compost + 3 ton ha$^{-1}$ of rabbit manure increased shallot yield and diameter of shallot bulbs.

## 1. Introduction

Shallot production in Indonesia from 2016 to 2020 extimated to increase [1]. Chemical fertilizers are generally used excessively by farmers to increase the yield of shallot. Excessive use of chemicl fertilizers exceeds the recommendations causing accumulation of fertilizer mineral salts which causes soil degradation. Significant soil degradation affects soil structure and nutrient uptake [2]. That it has an impact on soil health and shallot productifity in the next growing season.

Long-term use of chemical fertilizers and pesticides will reduce the potential for land production and increase the toxic effects on crop yields [3]. Reducing the dependence of chemical inputs in agriculture is very important with increasing the utilization of organic fertilizers and beneficial microorganisms. The use of beneficial microorganisms that are applied as biofertilizers and/or biocontrol agents is an important tool for controlling plant diseases and reduction of chemical fertilizers [4]. Fungi as biofertilizers have the potential to significantly increase shallot yield [5].

The application of organic fertilizers 20 ton ha$^{-1}$ can reduce 50% the dose of anorganic fertilizers [6] Cajeput (*Melaleuca leucadendron*) waste has the potential to be used as raw material for organic fertilizer. A Significant increase in soybean production with application of cajeput waste compost [7]. The amount of nutrients in the cajeput waste compost is small so it needs to be combined with other sources of organic matter such as rabbit manure. Rabbit manure can increase yield of barley (*Hordeum vulgare*) [8]. The organic manure from rabbit manure mixed with rice straw helps seed growth [9]. There are many reports that *Trichoderma sp*. as biofertilizer increases crop yields such as wheat (*Triticum aestivum*)[10], Rice (*Oriza sativa* L.) [11], Sugarcane (*Saccharum officinarum* L.) [12], Mustard (*Brassica rapa* L.) and Tomato (*Solanum lycopersicon* Mill.) [13].

This research was undertaken to determine the effect of cajeput waste compost as organic fertilizer and *T. asperellum* as biofertilizer on growth and yield of shallot. The application of organic fertilizer from cajeput waste compost and *T. Asperellum* was expected to provide information in increasing the effectiveness of shallot fertilization

## 2. Materials and Methods

The experiment was conducted in Greenhouse, Agrotechnology Department, University of Darussalam Gontor, Ponorogo, Indonesia from Mei 2019 to July 2019.

### *2.1 Material*

The tool used in research are : polybag, calipers, analytic scales, centimeter rulers, scissors, oven. The materials used in the study are : cajeput waste compost, rabbit manure, *Trichoderma asperellum*, soil, NPK fertilizer (15-15-15), Bauji variety of shallot.

### *2.2 Research Experimental Design*

The experiment was laid-out in factorial randomized completely design with 3 replications and eight treatment combinations. The first factors is organic fertilizer with 4 kinds of organic fertilizer (K0 = without organic fertilizer; K1 = 6 ton $ha^{-1}$ of cajeput waste compost; K2 = 3 ton $ha^{-1}$ of cajeput waste compost + 3 ton $ha^{-1}$ of rabbit manure; K3 = 6 ton $ha^{-1}$ of rabbit manure). The second factors is *Trichoderma asperellum* with 2 levels (T0 = without *T. asperellum*; T1 = 1 L $ha^{-1}$ of *T. asperellum*).

### *2.3 Methodology*

2.3.1 *Prosedure.* Soil as a growing medium was put into polybags as much as 3 kg per polybag. The fresh bulb was hand planted at polybags with one bulb per polybag. The polybags was irrigated every day. NPK fertilizer (15-15-15) of 150 kg $ha^{-1}$ was applied at two days before planting. Bulbs are harvested at 85 days after planting. In organic fertilizer application, we applied 6 ton $ha^{-1}$ of cajeput waste compost (K1); 3 ton $ha^{-1}$ of cajeput waste compost + 3 ton $ha^{-1}$ of rabbit manure (K2); and 6 ton $ha^{-1}$ of rabbit manure (K3) 7 days before planting by mixing organic fertilizer with soil. In *T. Asperellum* application, we applied 1 L $ha^{-1}$ at 1 day after planting,

2.3.2 *Observations.* The data of growth parameters recorded include : plant hight, leaves dry weight per plant, and number of tillers per plant. Plant hight is measured when the shallot crops are in a state of maximum height. That is at the age of one month after planting by extending from ground level to tip of longest leaf (when held vertically). Leaves dry weight was weighed leaf from leaf until leaf sheath after harvesting. The leaf was dried in oven then weigh it. The number of tillers were counted at one month after planting.
The date of yield parameters recorded consist of wet weight of bulbs per plant and bulbs diameter per plant. The bulb separated from each polybag was weighed of bulb yield per plant. The bulb after harvest was measured bulb diameter using calipers.
2.3.3 *Data Analysis.* Obsevational data were statically analyzed using the analysis of variance procedure to assess the differences of treatmen. Test the significant differences among treatment means using Least Significance Different Test.

## 3. Results

This study evaluated the effect of cajeput waste compost as organic fertilizer and *T. asperellum* as biofertilifer on growth and yield of shallot. According to table 1, there was a significant difference among treatment combinations on observations of plant height, dry weight of leaves, and number of tillers. Treatment combination of cajeput waste compost + rabbit manure without *T. Asperellum* was produced plant hight of 40 cm that heighest and significantly different than the other treatment combinations. Treatment combination of cajeput waste compost + rabbit manure without *T. Asperellum* was produced dry weight of leaves of 3.37 g that heighest and no significantly different than the other treatment combinations as treatment combination of *T. Asperellum* without organic fertilizer (2.67 g), treatment combination of cajeput waste compost with T. Asperellum (2.48 g), and treatment combination of rabbit manure without *T. Asperellum* (2.38 g). The highest number of tillers was treatment combination of rabbit manure without T. Asperellum (9.67 tillers), however not significanly different from treatmen combinaton of cajeput waste compost without *T. Asperellum* ( 8.67 tillers).

**Table 1**. Effect of treatment combinations on growth performance of shallot in term of measured values and standart errors

| Treatment Combinations | Plant Height (cm) | Dray Weight of Leaves (g) | Number of Tillers (tillers) |
|---|---|---|---|
| K0T0 | 32.00 ±1.00 a | 1.43±0.96 a | 6.75±0.90 a |
| K0T1 | 35.67±2.51 b | 2.67±0.34 bc | 7.25±0.25 a |
| K1T0 | 32.00±1.00 a | 2.05±0.59 ab | 8.67±1.04 bc |
| K1T1 | 37.00±1.00 b | 2.48±0.65 abc | 7.00±0.25 a |
| K2T0 | 40.00±1.00 c | 3.37±1.42 c | 7.83±1.25 ab |
| K2T1 | 36.00±2.00 b | 2.30±0.50 a | 6.50±0.91 a |
| K3T0 | 30.33±4.72 a | 2.38±0.40 abc | 9.67±0.52 c |
| K3T1 | 37.00±6.56 b | 1.64±0.30 ab | 7.00±0.25 a |

Data followed by the same letter in the same colum were not significantly different based on the 5% LSD test. Ko = without organic fertilizer, K1 = 6 ton $ha^{-1}$ of cajeput waste compost, K2 = 3 ton $ha^{-1}$ of cajeput waste compost + 3 ton $ha^{-1}$ of rabbit manure, K3 = 6 ton $ha^{-1}$ of rabbit manure, T0 = without *T. Asperellum*, T1 = with *T. Asperellum*

Figure 1. show that treatment combinations organic fertilizer with *T. Asperellum* were no significantly difference on yield performance of shallot in term of measured wet weight of bulbs and bulbs diameter. However treatment of organic fertilizer was on yield performance of shallot in term of measured wet weight of bulbs and bulbs diameter. Rabbit manure treatmen produced highest of wet weight of bulb, and not significantly difference with treatment of cajeput waste compost + rabbit manure. However, treatment combination of cajeput waste compost + rabbit manure produced hieghest of bulbs diameter.

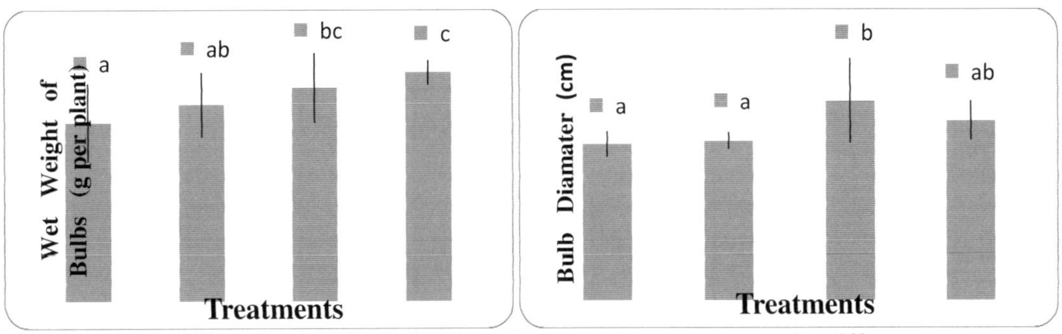


**Figure 1**. Wet Weight of Bulbs and Bulbs Diameter of Shallot under different treatment

## 4. Discussion

Understanding how shallot responds to the organic fertilizer from cajeput waste compost and *T. asperellum* is essential for further effectiveness of fertilizer on shallot. The results of experiments showed that organic fertilizer from cajeput waste compost + rabbit manure without *T. asperellum* were significanly able to improve growth in shallot through plant height, and dry weight of leaves. This treatment increased plant height by 8.11 % until 25.00 % compared to all treatment combinations. This is different from the results of the study reported that treatmen of Organic fertilizer + *Trichoderma spp.* increased plant height on red onion [14].

The highest number of tillers was found in treatment combination of rabbit manure without *T. Asperellum* (9.67 tillers per plant). Rabbit urine treatment incresed number of tillers on shallot [15]. Treatment of rabbit menure fertilizer incresed number of branches per plant on Periwinkle (*Catharanthus roseus* L.) [16] and help seed growth [8].

The application of organic fertilizer was a significantly difference at wet weight of bulbs and bulb diameter. The application of Rabbit manure produced wet weight of bulbs and bulbs diameter on shallot which not significantly difference with treatment of cajeput waste compost + rabbit manure. There are many reported that cuttle dung, agriculture wast, compost of mushroom growing media, and granural organic fertilizer can incresed plant height, bulb diameter, and bulb weight on shallot [17] [18] [19]. Generally hight bulb yield on onion may be due to the increase in photosynthesis process rate and the assimilation of such products in plant tissu [20]. A number of organic fertilizers in the root zone can increase nutrient availability in soil solution that support plant growth, so increasing yield of shallot [21].

Treatment of *T. Asperellum* can't increase wet weight of bulbs. This is different from the results of research which explains that onion with Tricoderma treatment can increase wet weight of bulbs compared with control [22]. Seed Treatments with Trichoderma increase shallot yield on sandy coastal [23]. *Tricoderma spp.* have the ability to synthesize cellulose enzymes [24] so there is a possibility that the *T. Asperellum* is applied to function as decomposers of cajeput waste that has not been completely decomposed. Trichoderma can be used as biological fertilizer [9], organic decomposers [24], biocontrol agents [25].

## 5. Conclusion

In conclusion, the results of this study showed that treatmen of 3 ton ha$^{-1}$ of cajeput waste compost + 3 ton ha$^{-1}$ of rabbit manure without *T. asperellum* increased plant hight and leaf dry weight of onion by 25% and 135,66%. The application of combination rabbit manure and without *T. asperellum* increased shallot tillering number by 9,67 tillering. The application of 3 ton ha$^{-1}$ of cajeput waste compost + 3 ton ha$^{-1}$ of rabbit manure increased shallot yiled and diameter of shallot bulbs.

## Acknowledgments

University of Darussalam Gontor has funded this program at 2019.

## References

[1] Nuryati L and Warianto B 2016 Pus. Data dan Sist. Inf. Pertanian Kementerian Pertanian p 43

[2]. Massah J and Azadegan B 2016 *Agricultural Mechanization In Asia, Africa, And Latin America* **47(1)** 44-50

[3]. Kumari K A, Kumar K N R and Narasimha-Rao CH 2014 *Journal of Chemical and Pharmaceutical Sciences*, JCHPS Special Issue **3** 150-151

[4] Russo A, Carrozza G P, Vettori L and Felici C 2012 Plant Beneficial Microbes and Their Application in Plant Biotechnology, Innovations in Biotechnology Dr. Eddy C. Agbo (Ed.) ISBN: 978-953-51-0096-6, InTech, pp. 68

[5] Laila A, Trisnaningrum N and Hamawi M 2019 *IOP Conf. Series: Earth and Enviromental Science* **292**

[6] Ramadhan A F N and Sumarni T 2018 *Jurnal Produksi Tanaman* **6 (5)** 815-822

[7] Isnatin U, Parwi and Mulyanto T 2017 *Gontor AGROTECH Science Journal* **3(2)** 119-130

[8] Islas-Valdez S, Constantino C A L, Betran-Hernandez R I, Gomez-Mercado R and Jimenez A 2017 *Environmental Science and Pollutan Research* **24 (33)** 25731-25740

[9] Li-li B, Tie-jun Y, Bin W, Lin B, De-gui T and Xiang-chao F 2013 *J. Agr. Sci. Tech.* **15** 1069-1081

[10] Mahato S, Bhuju S, and Shertha J 2018 *Malaysian Journal of Sustainable Agriculture* **2 (2)** 01-05
[11]. Khan H I 2018 *Rice Science* **25(6)** 357–362
[12] Srivastava S N, Singh V and Awasthi S K 2006 *Sugar Tech* **8 (2&3)** 166 -169
[13] Haque Md M, Ilias G N M and Molla A H *2012 A Scientific Journal of Krishi Foundation, The Agriculturists* **10 (2)** 109 – 119
[14] Galindez J L, Porciuncula F- L, Pascua M P, Claus S M and Lopez L L M A 2016 *Journal of Agricultural Science and Technology* B **6** 10-17
[15] Simamora A L B, Simanungkalit T and Ginting J 2014 *Jurnal Online Agroekoteknologi* **2 (2)** 533- 546
[16] Hassan E A 2012 *Plants Australian Journal of Basic and Applied Sciences* **6 (13)** 443-453
[17] Ali M, Khan N, Khan A, Ullah R, Naeem A, Khan M W, Khan K, Farooq S and Rauf K 2018 *Pure Appl. Biol.* **7 (3)** 1161-1170
[18] Brotodjojo R R R and Arbiwati D 2017 *Int'l Journal of Advances in Agricultural & Environmental Engg.* (IJAAEE) **4 (1)** 89-92
[19] Prusty M, Mishra N, Kar D S and Pal S 2019 *International Journal of Agriculture Science* **11 (4)** 7910-7912
[20] Gadelrab H M and Elamin SM *2013 Journal of Science and Technology* **14** 61-68
[21] Aisha A H, Rizk F A, Shaheen A M and Abdel-Mouty M M 2007 *Research Journal of Agriculture and Biological Sciences* **3 (5)** 380-388
[22] Garcia J G O, Belmont R M, Monroy M R, Trujillo J S R, Rodriguez R S and Jimenez G S 2016 *Scientia Horticulturae* **195** 8-16
[23] Darsan S, Sulistyaningsih E and Wibowo A 2016 *Ilmu Pertanian (Agricultural Science)* **1 (3)** 094-099
[24] Pandey S, Srivastawa M, Shohid M, Kumar V, Singh A, Trivedi S and Srivastava Y K 2015 *Journal of Data Mining in Genomics and Proteomics* **6 (2)**

Journal of Physics: Conference Series **1381** (2019) 012005 doi:10.1088/1742-6596/1381/1/012005

# Drought resistance selection in ponorogo local rice (*oryza sativa l.*) Varieties

**K Jadid, L D Cahyanti, Muhammad, H Setyaningrum, N Trisnaningrum***

Agotechnology Department, University of Darussalam Gontor, Ponorogo, Indonesia

*niken.trisnaningrum@unida.gontor.ac.id

**Abstract**. Drought is one of the main problems for the development of rice plants. The aims of this study were selecting several Ponorogo local rice cultivars that have resistance to drought stress based on germination ability in various PEG 6000 (Polyethylene Glycol 6000) concentration treatment, root penetration ability to wax layer (consisted of 60% wax and 40% vaseline) and chlorophyll content during drought stress. The germination test with PEG 6000 used a completely randomized design (CRD) factorial pattern consisting of 2 (two) treatment factors, namely cultivar and PEG concentration with 3 replications. The first treatment factor was a variety of Ponorogo local rice cultivars consisting of 5 cultivars, namely: Legowo, Mlarak 1, Malihan Ketan, Philips and Edok. The second treatment factor was PEG concentration, consisting of 4 (four) levels, namely: 0 (Control), 15, 20 and 25%. Rice seeds that passed the selection using PEG 6000 in the germination phase, were used in the root penetration ability test. Data were analyze dusing variance analysis with 95% confidence level and continued with Duncan multiple distance test at 5% level. The results of the study revealed that Legowo cultivars showed the best germination in 25% PEG treatment with 15% germination percentage, 0,5 cm seminal root length, 0,1cm shoot length, seminal root length ratio per shoot length 4,2 and vigor index 0,1. At 20% PEG treatment, Edok cultivars gave the best results with 85% germination percentage, 1,7 cm seminal root length, 0,3 cm shoot length, seminal root length ratio per-shoot length 5,4 and vigor index 1,7. The highest number of roots that can penetrate the wax layer was produced by Edok cultivars averaging 19,3 strands with an average length of 25,5 cm.

## 1. Introduction

In order to survive in drought condition, plant have a series of mechanisms to deal with drought in terms of morphology, physiology, biochemistry, cellular and molecular[6][19]. These mechanisms include drought escape or escaping, drought evoidance, drought tolerance, and drought recovery mechanisms[5]. The mechanisms of plant resistance to drought stress appear in root and leaf characters, the ability to adjust osmotic pressure, water potential, ABA[12] content, and cell membrane stability. These characters can be used as indicators to evaluate the drought resistance in plants[19].

The development of drought resistant rice variety is one option to overcome the problem to defeat unpredicted dry season[24]. Local rice varieties are potential genetic resources to obtain the types of drought resistance rice varieties. Ponorogo is one of rice crop production center in Indonesia. Ponorogo has several local rice varieties that has been cultivated for generations, but has not been studied extensively to determine its resistance to drought[21]. Research to get drought tolerant rice types has been carried out with various experimental techniques including in vitro selection with Polyethylene glycol (PEG)[14][20], root permeability testing[22][15], and analysis of chlorophyll content[23].

Polyethylene glycol (PEG) is often used as a selection agent to identify drought-tolerant genotypes in the germination phase because of its ability to reduce the water potential in the germination media[14][20] as a simulation of hardness and thickness of the soil layer which is equivalent to 12 bar[22][16]. Root penetrability is a response to plant resistance to morphological drought.

The aim of the research was to select local Ponorogo rice cultivars through PEG treatment.

## 2. Material and Methods

The research was conducted at Genetic and Biotechnology Laboratory, University of Darussalam Gontor, from July until October 2018.

### *2.1. Material*

The tools used in the study include: petri dishes, hot plates and magnetic stirers, glassware, pellets, pipettetes, rulers and meters, scissors, analytical scales, pots (from 7.5cm diameter PVC pipes), electric grinders, mesh nets. The research materials used are: PEG6000, paper paper, distilled water, rice seeds, paraffin, vaseline, black tape / aquatic tape. Local rice seeds used are seeds taken directly on land through selection based on morphological differences consisting of five accessions, namely: Legowo, Mlarak 1, Malihan Ketan, Philips and Edok.

### *2.2 Research experimental design*

The study was an experimental study using a completely randomized design (CRD) factorial pattern consisting of 2 (two) treatment factors, namely cultivar and PEG concentrations with 3 replications. The first treatment factor was the type of local Ponorogo rice cultivar consisting of 5 cultivars, namely: Legowo, Mlarak 1, Malihan Ketan, Philips and Edok. The second treatment factor is the PEG concentration, consisting of 4 levels, namely: 0% (Control), 15%, 20%, and 25%.

### *2.3 Methodology*

*2.3.1 PEG treatment.* PEG treatment was done based on the described method by Lestari and Mariska (2006). The PEG concentrations were 0% (Controls), 15%, 20% and 25%. Each petridish in each treatment was filled with 20 seeds and as a control was used sterile water. The germinated seeds were planted on individual pots and root penetration test was conducted.

*2.3.2 Root penetration test.* The shoot that grew from PEG test were planted on individual pots which given parrafin and candle mixture (60% : 40%) as the base of the pots.

*2.3.3 Data analysis*. The data were analyzed using analysis of variance (ANOVA) with a confidence level of 95% and followed by a double distance test with a level of 5%. To help with data analysis, SPSS version 24 and Microsoft Excel software were used.

## 3. Result

### *3.1 Germination rate*

Table 1 showed that the percentage of germination was only significantly different in the 20% PEG treatment, whereas in the control treatment, 15%, and 25% were not significantly different between cultivars. The highest percentage of germination in the control treatment was produced by cultivars Legowo (98.3%) and Mlarak 1 (98.3%). While the 15% PEG treatment, the highest percentage of germination was shown by Legowo cultivar (100%) and was not significantly different from other cultivars. In the PEG treatment, the 20% concentration level of cultivars that gave the highest percentage of germinated seeds, Edok cultivar (85%), was not significantly different from Legowo (80%) and significantly different from Mlarak (56.7%), Malihan Ketan (35%) and Philips (75%). Whereas in the 25% PEG treatment, the highest percentage of germination was produced by Legowo cultivar (15%) and was not significantly different from other cultivars.

Table 1. Germination rate of five Ponorogo local rice varieties in PEG treatment

| Variety | PEG TREATMENT | | | |
|---|---|---|---|---|
| | 0 % | 15 % | 20 % | 25 % |
| Legowo | 98,3±2,9 ab | 100,0±0,0 a | 80,0±5,0 bc | 15,0±5,0 f |
| Mlarak 1 | 98,3±2,9 ab | 96,7±2,9 ab | 56,7±22,5 d | 3,3±2,9 f |
| Malihan Ketan | 96,7±2,9 ab | 91,7±2,9 abc | 35,0±20,0 e | 1,7±2,9 f |
| Philips | 95,0±5,0 ab | 95,0±5,0 ab | 75,0±21,8 c | 8,3±7,6 f |
| Edok | 96,7±5,8 ab | 98,3±2,9 ab | 85,0±0,0 abc | 11,7±12,6 f |

*3.2 Ratio of seminal root length and shoot length*

The results of the ratio of seminal root length per shoot length in table 2 shows that the Legowo cultivar gives the largest ratio of root length per shoot length in control treatments, 15 and 25%, although it is not significantly different from other cultivars in the same treatment. While the highest ratio of PEG 15% was produced by Mlarak 1 cultivar and not significantly different from other cultivars. In the 20% PEG treatment, Edok cultivar gave the biggest ratio (5.4), not significantly different from Legowo (4.5), Mlarak1 (3.8) and Malihan Ketan (3.6) but significantly different compared to Philips cultivar. (3.0). Whereas the highest ratio of PEG 25% was produced by cultivars Legowo (4.2), not significantly different from Mlarak 1 (2.0) and Edok (2.1) and significantly different from cultivars Malihan Ketan (1.0) and Philips (0.8).

Table 2. Ratio of seminal root length and shoot length of five Ponorogo local rice cultivars in PEG treatment

| Cultivar | PEG TREATMENT | | | |
|---|---|---|---|---|
| | 0 % | 15 % | 20 % | 25 % |
| Legowo | 2,1±0,1 cdef | 3,1±0,2 a-f | 4,5±0,3 ab | 4,2±2,5 abc |
| Mlarak 1 | 1,1±0,2 ef | 3,2±0,1 a-f | 3,8±0,8 abcd | 2,0±2,0 cdef |
| Malihan Ketan | 1,4±0,0 ef | 2,4±0,3 b-f | 3,6±0,5 a-e | 1,0±1,7 f |
| Philips | 1,1±0,1 ef | 3,0±0,0 a-f | 3,0±0,7 b-f | 0,8±1,4 f |
| Edok | 1,5±0,1 def | 3,1±0,5 a-f | 5,4±0,7 a | 2,1±3,6 cdef |

*3.3 Seed index vigor*

In the control treatment the best seed vigor index was owned by Legowo cultivars, Mlarak 1 and Edok, significantly different from Malihan Ketan and Philips cultivars, as shown at Table 3. At the 15% PEG concentration, the best seed vigor index was owned by the Legowo cultivar, not significantly different from the Philips and Edok cultivars, and significantly different from the Mlarak 1 and Malihan Ketan cultivars. While in the 20% PEG treatment, the best seed vigor index was owned by Edok cultivars and significantly different from Malihan Ketan. Whereas in the 25% PEG treatment, the seed vigor index produced was not significantly different between cultivars. Based on the seed vigor index, it is known that the 20% PEG concentration is the best concentration for early screening of drought resistance in the germination phase.

Table 3. Seed index vigor of five Ponorogo local cultivar in PEG treatment

| Cultivar | PEG TREATMENT | | | |
|---|---|---|---|---|
| | Kontrol | 15 | 20 | 25 |
| Legowo | 6,9±0,7 a | 4,2±0,4 c | 1,6±0,1 e | 0,1±0,0 f |
| Mlarak 1 | 6,9±0,9 a | 2,9±0,6 d | 0,8±0,6 ef | 0,0±0,0 f |
| Malihan Ketan | 5,8±0,4 b | 1,8±0,6 e | 0,2±0,2 f | 0,0±0,0 f |
| Philips | 5,2±0,5 b | 3,5±0,5 cd | 1,2±0,8 e | 0,1±0,1 f |
| Edok | 6,8±1,2 a | 3,6±0,6 cd | 1,7±0,3 e | 0,1±0,1 f |

*3.4 Root penetration test through waxy layer*

Observation of root permeability in the Legowo cultivar, showed that of the 20 plants tested that succeeded in penetrating the waxy layer as many as 19 plants. One plant that cannot penetrate the waxy layer is a plant whose seeds are germinated using 25% PEG. The highest average number of roots that can penetrate the waxy layer in the Legowo cultivar is found in plants from which the seeds are germinated with a 25% PEG. While the highest average root length that penetrates the waxy layer is a plant whose seeds are germinated using 20% PEG (Table 4).

In Mlarak 1 cultivar, 17 plants tested were all able to penetrate the waxy layer. The highest average number of roots that penetrate the waxy layer is found in plants germinated with 25% PEG. While the highest average root length that penetrates the waxy layer is a plant whose seeds are germinated using 20% PEG (Table 4).

Based on the data in table 4 Malihan Ketan cultivar shows that of the 16 plants tested were able to penetrate the waxy layer as many as 15 plants. One plant that cannot penetrate the waxy layer is a plant whose seeds are germinated at a 20% PEG solution concentration. In Malihan Ketan cultivar, plants whose seeds were germinated with 25% PEG gave the highest average number of roots while the highest average root length came from plants whose seeds were germinated with 20% PEG.

To known the ability of Philips cultivar roots to penetrate the waxy layer, Table 4 showed that from 20 plants tested, 19 plants can penetrate the waxy layer. While only 1 plant did not see its root penetrating the waxy layer, the plant whose seeds were germinated in 25% PEG solution. The highest average number of roots was shown in plants where the seed origin was germinated with 25% PEG and the highest average root length was produced by plants where the seeds were germinated using PEG20%.

While the data of Edok cultivar rice roots showed that of 20 plants tested that could penetrate the waxy layer of 18 plants, while the roots did not penetrate the waxy layer of 2 plants (each seed was added to PEG 20 and 25%). The highest average number of roots was indicated by plants germinated in PEG25% solution. While the highest average root length is produced by plants where the seeds are germinated using PEG15%. Table 4 shows that Edok cultivars have the best root permeability to the waxy coating. Edok cultivars produced the most number of roots penetrating the waxy layer, 23.6 cm (from the seed of the control treatment), 16.6 cm (PEG 15%), 14.8 cm (PEG20%), and 22.3 cm (PEG25%) with root length 26.4cm (Control), 29.5cm (PEG15%), 23.6cm (PEG20%) and 22.6cm (PEG 25%).

Table 4. The average of root number and root length that penetrated the waxy layer

| Cultivar | Treatment | Tested plants | Penetrate root plants | The number of root average | The length of root average |
|---|---|---|---|---|---|
| Legowo | Kontrol | 5 | 5 | 14,2 | 22,7 |
| | PEG 15% | 5 | 5 | 11,2 | 26,1 |
| | PEG 20% | 5 | 5 | 11,8 | 27,4 |
| | PEG 25% | 5 | 4 | 13,8 | 20 |
| Mlarak 1 | Kontrol | 5 | 5 | 14 | 25,8 |
| | PEG 15% | 5 | 5 | 8,8 | 19,6 |
| | PEG 20% | 5 | 5 | 8,8 | 20,7 |
| | PEG 25% | 2 | 2 | 17,5 | 20,1 |
| Malihan Ketan | Kontrol | 5 | 5 | 15,8 | 26,6 |
| | PEG 15% | 5 | 5 | 8,2 | 28,5 |
| | PEG 20% | 5 | 4 | 10,3 | 35,4 |
| | PEG 25% | 1 | 1 | 16 | 31,9 |
| Philips | Kontrol | 5 | 5 | 12 | 24 |
| | PEG 15% | 5 | 5 | 11 | 22,9 |
| | PEG 20% | 5 | 5 | 6,6 | 25,6 |
| | PEG 25% | 5 | 4 | 17,3 | 22,3 |
| Edok | Kontrol | 5 | 5 | 23,6 | 26,4 |
| | PEG 15% | 5 | 5 | 16,6 | 29,5 |
| | PEG 20% | 5 | 4 | 14,8 | 23,6 |
| | PEG 25% | 5 | 4 | 22,3 | 22,6 |

## 4. Discussion

### *4.1 The effect of PEG treatment to seed germination*

The results of the study noted that an increase in PEG concentration resulted in a decrease in the percentage of germination. The percentage of germination in PEG 0 MPa is greater than -0.25 and -0.5 MPa, even in the treatment of PEG-0.75 and -1 MPa, no sprouts were seen[2][2]. Moreover, the local Tanangge upland rice germinated with PEG 6000 solution at 0 bar pressure produced the best percentage of normal germination compared to pressures of -3 and -4bar but did not differ from the treatment of giving solution at pressures -1 and -2 bar[4].

The percentage of germination decreasing occurs due to disruption of the process of imbibition and tissue hydration on rice seeds. Low water potential due to increase osmotic media levels results in plant cells not being able to take water from the medium or the amount of water absorbed by seeds decreases[2]. In addition, the germinability and viability of different seeds also affects the decrease in the percentage of germination[7].

Drought stress conditions due to PEG treatment during germination in this study also showed inhibition of seminal root growth and shoots. The Edok cultivar produced the longest seminal root at PEG20% treatment of 1.7 cm, followed by the Legowo cultivar of 1.6 cm. Whereas in the 25% PEG treatment, the longest seminal root produced by the Legowo cultivar was 0.5 cm and was followed by the Edok cultivar 0.4 cm. While the longest shoots in PEG 20 and 25% treatments were shown by Legowo cultivars, which were 0.4 cm and 0.1 cm, not significantly different from other cultivars. The increasing concentration of PEG to the media make the ratio of root length to shoot length increases. An increase in the ratio of root length to shoot length indicates tolerance to rice dryness with a decrease in shoot length. This happens because the growth of leaves is more reduced than the growth of roots[3], translocation of photosynthesis results into the root system and the presence of hormonal signals induced in roots in response to water shortages[4].

From the description above, it was concluded that the treatment of PEG 6000 in the germination phase of several local Ponorogo rice cultivars after observing on the 7th day showed a different resistance response between cultivars towards drought conditions. Based on all the observed germination parameters, it was found that the Legowo and Edok cultivars could germinate better, mean that they could be said to be more resistant to drought compared to cultivars Mlarak 1 and Philip which were somewhat resistant and Malihan Ketan which was more sensitive to drought. Several factors that can determine the ability of germination seeds, including the differences in seed size, thickness of seed coat and seed vigor[9].

Plants that have drought resistance are able to mobilize food reserves in seeds for root and plumular growth, even in conditions of water deficit. Tolerant plants respond to water deficits by optimizing physiological processes in critical phases so that plants can grow well [1]. When water stresses occur when the seeds germinate, the metabolism of the seeds is disrupted due to insufficient water needed. Therefore, only seeds that are resistant to drought are able to germinate[11][11].

*4.2 The ability of Ponorogo local rice cultivars in root penetration through waxy layer*

Water absorption and transport by the roots of rice plants is very important during drought conditions[10]. The ability to penetrate roots in hard soil layers (compact) is recognized as the most appropriate way in the characterization of plants for genetic improvement of plants to drought[10]. The use of a waxy combination of a paraffin and vaselin mixture has been carried out at IRRI to study the ability of roots to penetrate hard layers[19]. The hard layer is simulated from a mixture of paraffin (60%) and vaselin (40%) which are known to be equivalent to 12bar hardness[22][17].

The results of the root permeability test showed that seeds germinated with 25% PEG produced the highest average number of roots that could penetrate the waxy layer. Whereas the highest average root length comes from plants whose seeds are germinated using a 20% PEG concentration. From the observation of root permeability it is known that the Edok cultivar shows the best response. The Edok cultivar produces the highest number of roots that can penetrate the waxy layer, which is an average of 19.3 strands with an average length of 25.5 cm. Roots that are able to penetrate the basic layer of the pot form waxy layers, more than 10 cm long and relatively large in number of roots, are expected to be more tolerant of drought[8].

The form of morphological adaptation of plants to avoid drought stress is to produce longer roots to look for water sources that are relatively far from the ground surface[18]. Dense, deep roots with high root permeability will increase water absorption from the soil[13][13]. In relation to roots, genotypes of plants that have drought resistance have characteristics, including: (1) being able to develop their root systems when water is still available before experiencing drought stress so that plants can extract water from the inner soil, (2) modifying the root system so as to be able to extract water from the innermost layer of water under water stress conditions[8].

## 5. Conclusions

The conclusion of this research were :

1. Cultivar Legowo shows the best germination in drought stress induced by PEG 6000 in the control treatment, 15 and 25%. While Edok cultivars gave the best results in the treatment of PEG20%.

2. The highest number of roots that can penetrate the waxy layer (a mixture of 60% paraffin and 40% vaseline) is produced by the Edok cultivar, which is an average of 19.3 strands with an average length of 25.5 cm.

**References**

[1] Afa, L. A., B. S. Purwoko, A. Junaedi, O. Haridjaja, dan I. S. Dewi. 2013 Deteksi Dini Toleransi Padi Hibrida terhadap Kekeringan menggunakan PEG 6000 *J. Agron. Indonesia* **41** 9-15.
[2] Ballo, M., Nio, S.A., Dingse, P., dan Feky, R.M. 201. Respon Morfologis Beberapa Kultivar Padi (*Oryzasativa*L.) terhadap Kekeringan pada Fase Perkecambahan *Jurnal bioslogos* **2** 88-95.
[3] Blum, A. 2005 Drought Resistance, Water-Use Efficiency, and Yield Potential—Are They Compatible, Dissonant, or Mutually Exclusive? *Australian Journal of Agriculture Research* **56** 1159- 1168. doi: 10.1071/AR05069.
[4] Daksa,W.R., A.Ete dan Andrianton. 2014 Identifikasi Toleransi Kekeringan Padi Gogo Lokal Tanangge pada Berbagai Larutan PEG. *e-J. Agrotekbis* **2** 114-120.
[5] Effendi, Y. 2008 Kajian Resistensi Beberapa Kultivar Padi Gogo (*Oriza sativa* L.) terhadap Cekaman Kekeringan *Tesis Program Pascasarjana. Program Studi Agronomi* (Universitas Sebelas Maret, Surakarta)
[6] Fang, Y.and L. Xiong. 2015 General Mechanisms of Drought Response and Their Application in Drought Resistance Improvement in Plants *Cell.Mol.LifeSci* **72** 673-689. DOI10.1007/s00018-014-1767- 0.
[7] Gardner, F. P., R. B. Pearce, and R. L. Mitchell. 1991 Fisiologi Tanaman Budidaya (UI-Press, Jakarta)
[8] Lestari, E.G. 2006 Mekanisme Toleransi dan Metode Seleksi Tumbuhan yang Tahan Terhadap Cekaman Kekeringan. Tinjauan Ulang (Review) *Berita Biologi* **8(3)** 215-222.
[9] Lestari, E.G., dan I. Mariska. 2006 Identifikasi Somaklon Padi Gajahmungkur, Towuti, dan IR64 Tahan Kekeringan Menggunakan *Polyethylene Glycol Bul. Agron* **34(2)** 71-78.
[10] Lestari, E. G., E. Guhardja, S. Harran, dan I. Mariska. 2006 Uji Toleransi Kekeringan Padi Kultivar Gajah Mungkur, Towuti, dan IR64 Menggunakan Uji Daya Tembus Akar *Inovasi Teknologi Padi Menuju Swasembada Beras Berkelanjutan. BB Padi* 749-761.http://digilib.litbang.pertanian.go.id/repository/repository/artikel/0/0/2006/0/5426.
[11] Lestari, E. G., I. Mariska, D. Sukmadjaja, dan D. Suardi. 2004 Seleksi *in Vitro* dan Identifikasi Tanaman Padi Kultivar Gajahmungkur, Towuti, dan IR64 yang Tahan Kekeringan (Prosiding) *Kumpulan Makalah Seminar hasil Penelitian BB-Biogen* 170-173
[12] Li, J., Y.Li, Z. Yin, J. Jiang, M. Zhang, X. Guo, Z. Ye, Y.Zhao, H. Xiong, Z. Zhang, Y.Shao, C. Jhiang, H. Zhang, G. An, N-C. Paek, J. Ali, and Z. Li. 2017 *OsASR5* Enhances Drought Tolerance through a Stomatal Closure Pathway Associated with ABA and $H_2O_2$ Signalling in Rice. *Plant Biotechnology Journal* **15** 183–196. DOI: 10.1111/pbi.12601.
[13] Mira, U. Widyastuti dan E.D. Mustikarini. 2011 Pengaruh Cekaman Kekeringan Terhadap Pertumbuhan dan Produksi Padi Lokal Bangka di Media Sandy Clay Pasca Penambangan Timah. Enviagro *Jurnal Pertanian dan Lingkungan* **3(1)** 1-43.
[14] Yunita, R. 2009 Pemanfaatan Variasi Somaklonal dan Seleksi i*n Vitro* dalam Perakitan Tanaman Toleran Cekaman Kekeringan *Jurnal Litbang Pertanian* **28(4)** 142-148.
[15] Suardi, D. 2000 Kajian Metode Skrining Padi Tahan Kekeringan *Buletin AgroBio* **3(2)** 67-73.
[16] Suardi,D., E.Lubis, dan S. Moeljopawiro. 2001 Daya Tembus Akar Galur Persilangan $BC_2F_2$ Varietas Padi Unggul *Prosiding Seminar Hasil Penelitian Rintisan dan Bioteknologi Tanaman*
[17] Suardi, D. 2002 Perakaran Padi Dalam Hubunganya dengan Toleransi Tanaman Terhadap Kekeringan dan Hasil *Jurnal Litbang Pertanian* **2(3)** 100-108.
[18] Sujinah dan Jamil, A. 2016 Mekanisme Respon Tanaman Padi terhadap Cekaman Kekeringan dan Varietas Toleran *Iptek Tanaman Pangan* **11(1)** 1-8.

[19] Suwarno, P.M., D. Wirnas, dan A. Junaedi. 2016 Kendali Genetik Toleransi Kekeringan pada Padi Sawah (*Oryza sativa* L.) *J.Agron.Indonesia* **44(2)**119-125.

[20] Swain, P., M. Anumalla, S. Prusty, B.C. Marndi and G.J.N. Rao. 2014 Characterization of Some Indian Native Land Race Rice Accessions for Drought Tolerance at Seedling Stage *Australian Journal of Crop Science (AJCS)* **3(8)** 324-331.

[21] Trisnaningrum, N., and A. Laila. 2017 Morphological Variation of Early Days to Flowering on Lokal Rice Accessions Collected from Ponorogo *Jurnal Teknosains* **7(1)** 59-66. DOI 10.22146/ teknosains.31134.

[22] Yu, L.X., J.D. Ray, J.C. O'Toole, and H.T. Nguyen. 1995 Use of Wax- Petrolatum Layers for Screening Rice Root Penetration *Crop. Sci.* **35** 684-687.

[23] Zain, N.A.M., M.R. Ismail, A. Puteh, M. Mahmood, and M.R. Islam. 2014 Impact of Cyclic Water Stress on Growth, Physiological Responses and Yield of Rice (*Oryza sativa* L.) Grown in Tropical Environment *Ciência Rural* **44(12)** 2136-2141. http://dx.doi.org/10.1590/0103-8478cr20131154

[24] Zhang, F., F. Cui, L. Zhang, X. Wen, X. Luo, Y.Zhou, X. Li, Y. Wan, J. Zhang, and J. Xie. 2014 Development and Identification of a Introgression Line with Strong Drought Resistance at Stage Derived from *Oriza sativa* L. Mating with *Oriza rufipogon* Griff. *Euphytica* Online 24 April 2014. DOI: 10.1007/s10681-014- 1121-5.

Journal of Physics: Conference Series **1381** (2019) 012006 doi:10.1088/1742-6596/1381/1/012006

# Designing a building automation system with open protocol communication and intelligent electronic devices

**S. Suhanto[1], F Faizah[1], K. Kustori[1]**

[1]Airport Electrical Department, Aviation Polytechnic of Surabaya
73 Jemur Andayani 1 Street, Surabaya East Java

Email: nadiafather@gmail.com

**Abstract**. Building Automation System (BAS) is a microprocessor-based control network setup for controlling a building's technical service systems such as air conditioning, ventilation, lighting, and hydraulics. BAS main microcontroller automatically handles all electronics in the building. To control the liked devices to suit the needs of the buildings' occupants, the controller needs to be smart or intelligent. The system will be very flexible if the communication network between control and device used an open communication protocol. In this paper, we present a miniature BAS solution implemented using intelligent electronic devices (IED) as controlling devices. The system is built using 8-bit microcontrollers, sensors, and actuators. There are three IED in use, each handling lighting, air conditioning, and water pumps respectively. Communication between the devices and the human-machine interface (HMI) used the free and open Modbus TCP protocol.

## 1. Introduction

The main principles of smart building development are for increasing the tenants' comfort value and reducing its maintenance costs. Building energy-efficient structures are a response to increasing human need for electronics to fulfill their needs. Efforts in smart building development are made possible by developments in information technology and automatic control engineering. More and more electronics integrated into common home and office tools pose more problems, such as increasing electricity costs.

Common structures and buildings consume about 40% of the national energy usage, and contribute about the same percentages to greenhouse gases, and also contribute to 70% of national electricity usage. The government has put efforts into developing its electricity infrastructure to match the increasing demand, but it is not enough. This pushes electricity costs up, increasing with the increase in coal, gas, or oil prices, due to the energy plants' general usage of coal, gas or oil as its main fuel.

To reduce energy usage, the government has required governmental offices to use green building standards in planning for new buildings, with old buildings being retrofitted to help conserve energy. Energy conservation involves efficient and rational energy usage without reducing energy usage towards for important purposes. This energy conservation effort is applied in all utilization steps, starting from energy source utilization to last-mile utilization, using energy-efficient technologies, and campaigning for an energy-efficient lifestyle.

A part of energy conservation efforts is automatic control and oversight of electricity usage in a building, commonly defined as a Building Automation System (BAS). BAS are expected to be able to optimize environmental and security concerns by economical means. This aspect is reached by using computers integrated into a distributed control system. This integration is easier to achieve on buildings that are still in the planning stage, as opposed to older buildings that did not adopt the smart building concept on their planning step. The difficulties that arose from integrating BAS into older buildings range from difficulties in integrating sensors and actuators into existing equipment, and insufficient integration between the BAS communication network and existing communication network. Thus, it is important for a BAS to have an open and free communication protocol. In this paper, we propose a miniature BAS using an open and free communication protocol, with the protocol being Modbus TCP. The distributed controllers used in this research were 8-bit microcontrollers using fuzzy control.

## 2. Basic Theory

The following subsections cover the hardware and software used in this research.

### *2.1. BuildingAutomation System (BAS)*

The purpose of modern automation controllers is to achieve automatic control of all networked facilities. Distributed control systems integrated into electronic equipment work to oversee and control mechanical, security, fire prevention, lighting, HVAC, humidity control, and ventilation systems in buildings.

BAS's main functions are to simultaneously control and oversee systems in a given building. A building controlled by a BAS is called a smart building. Commercial and historically important buildings use proven protocols such as BACnet and Modbus. Almost all green buildings are designed to accommodate BAS for energy, air, and water conservation. A unique function of BAS is responding to electrical demand accrued from active equipment, such as ventilation and humidity monitoring for a tighter building. Green building also use as much low-power DC devices as possible, usually integrated with power over Ethernet so as to allow control by BAS through it. A typical BAS usually has the following components :

*1. Controller.* Controllers used in a BAS usually consisted of one or more Programmable Logic Controllers (PLC), using specific programming. These are used to control equipment used in a building.

*2. Occupancy Sensor.* Occupancy is usually based on the daily schedule of the building. Override switches and sensors may be used to monitor occupancy in the internal areas of the building.

*3. Lighting.* Lighting may be turned off or on using BAS based on the time of the day, using timers and sensors. An example would be turning off a room's lights half an hour after the last person left the room.

*4. Air Handler.* Air handlers are used to control the air's entrance and exit to the building. This adjustment is done to keep the air quality in the building according to accepted human standards.

*5. Central Plant.* A central plant supplies the air handler with the needed air.

*6. Alarms and Security.* Many BAS solutions include alarm capabilities. If an alarm is detected, the system may be programmed to alert the authorities. The security system may also be integrated into the automation system. If there is a present occupancy sensor, it may also perform as a thief alarm.

*7. Topology.* A building, automation network consists of primary and secondary buses, consisting of controllers, input/output, and a user interface in the form of a human interface device. The primary and secondary bus may be in the form of fiber optic, Ethernet, ARCNET, RS-232, RS-485, or wireless network.

### *2.2. Modbus Protocol*

The Modbus protocol was a data communication protocol using the master-slave method. In this protocol, there are only one master and one or more slaves bound together into a network. Communicating using the Modbus protocol starts from a query from the master node, with the slave nodes responding by sending data or acting out the command given by the master node. The master node may only communicate once at any given moment. The slave nodes may only communicate if there is a query from the master node and may not communicate with other slave nodes. When sending to the slave nodes, the master uses 2 addressing modes as follows. Unicast mode, the master node sends a query to a single slave node, who after receiving and processing the query sends a response to the master node. Broadcast mode, the master node sends a query to all linked slave nodes. In this addressing mode, the slave nodes did not send responses to the master nodes. The Modbus protocol forms a message format for querying and responding between master and slaves. The query-response cycle is shown in Figure 1.

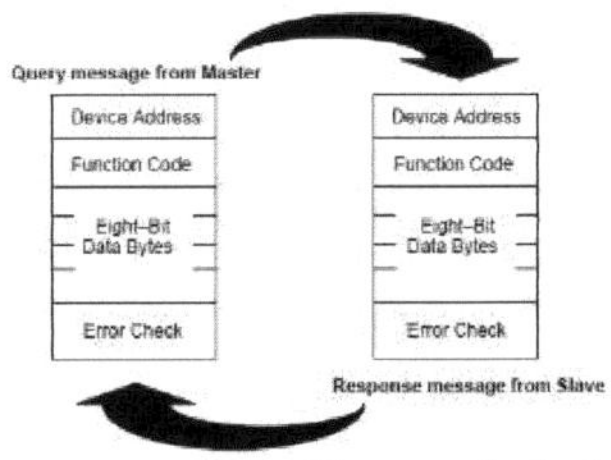


**Figure 1**. Query-response sending cycle in Modbus protocol

## 3. System Design

The plan for building automation system involves both hardware and software.

### *3.1. System Block Diagram*

The BAS proposed in this research used the block diagram shown in Figure 2.

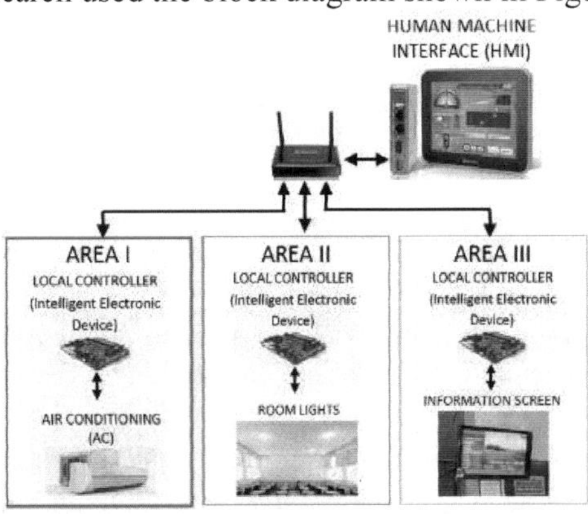


**Figure 2**. Building Automation System Block Diagram.

Areas I, II and III represent 3 different rooms in different places. Each room has a piece of different electronic equipment and a local controller for all the equipment in the room. The local controllers use fuzzy logic in controlling the devices, thus the system may be called intelligent electronic devices (IED). Communication between HMI and local control uses the Modbus TCP protocol. Using this protocol allows the system to be integrated with PLC, SmartRelay or other kinds of the controller using the same protocol and physical layer.

The BAS proposed in this paper used a computer-based HMI, using EasyBuilder Pro from WINTEK Electronics.

### *3.2. Intelligent Electronic Devices (IED) Design*

Intelligent electronic devices (IED) are an integrated solution between microcontrollers, solid state relays (SSR), sensors and signal conditioners. The block diagram for each IED used are shown in Figure 3.

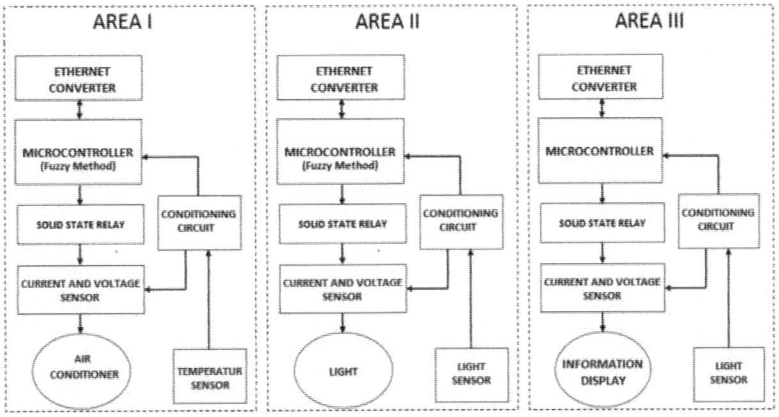


**Figure 3**. Block diagram of IED for each area

Each system used the same architecture, with differing sensor packs. The system used the closed loop concept, and has 2 main parts, that of hardware and software

*1. Current and voltage sensor design*

The current and voltage sensor is used to measure the grid voltage and the current used by the electronic device. A voltage sensor uses an isolator transformer, while a current sensor uses a solid state transformer, specifically ACS71205. The output of the sensors are still using AC current, with the magnitude of the current or voltage being proportional to the signal's amplitude. To change the current type to DC and to fit with the ADC standard, the signal is conditioned with a peak-to-peak detector as shown in Figure 4.

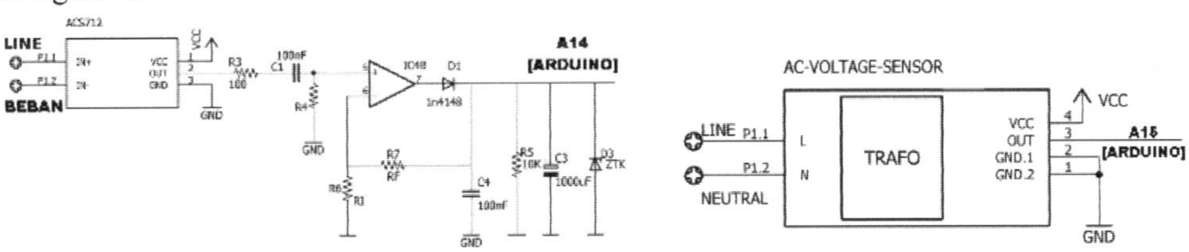


(a) (b)

Figure 4. a) Current sensor and conditioning circuit, b) voltage sensor

*2. Temperature sensor and conditioning circuit design*

In air conditioning devices, the controlled variable is that of the room's temperature. Thus, the sensor used measures temperature. The sensor used in this system uses IC LM35, a solid state sensor with 10 mV/$^0$C of output resolution. The sensor's output signal is conditioned using a signal conditioning circuit so as to fit with the ADC standard. The circuit used has an op-amp design with a non-inverting amplifier configuration and 5x amplification. Figure 5 shows a signal conditioning circuit for sensor output.

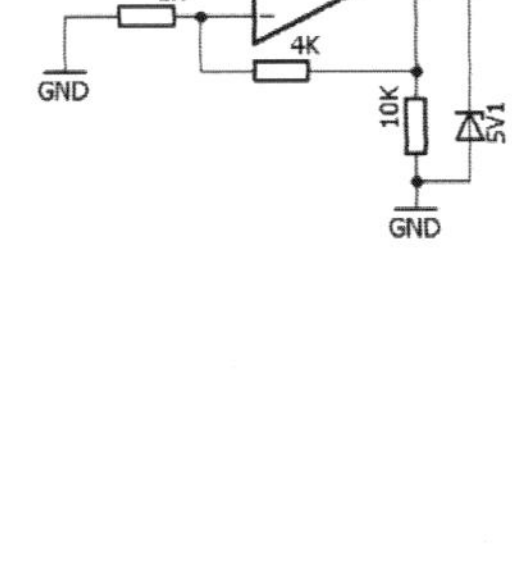

Journal of Physics: Conference Series **1381** (2019) 012006 doi:10.1088/1742-6596/1381/1/012006

**Figure 5**. Signal conditioning circuit for the temperature sensor

*3. Light and Motion Sensor Design*

A light sensor is used to measure the intensity of light in a room. The data is then used by the controller to control the number of lights to be turned on for optimal lighting. This sensor used a Light Dynamic Resistor (LDR). LDR is also used to sense the condition of an information display from its given light. Figure 6a shows a light sensor circuit with LDR, using the voltage divider design. A motion sensor may also be added to an air conditioning system to detect human presence in the room. The sensor used is a Passive Infrared (PIR) sensor. Figure 6b shows a motion sensor using PIR.

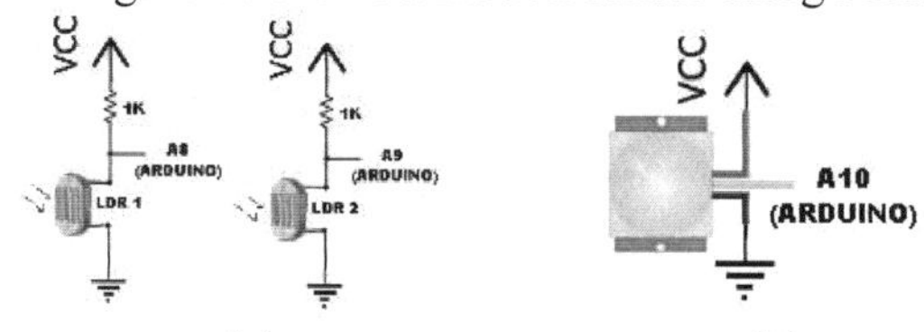


(a) (b)

**Figure 6**. a) Light sensor circuit with LDR, b) Motion sensor circuit with PIR

*4. Microcontroller System Design*

An IED's microcontroller functions as the main controller, and has a hand in reading sensor data, processing control algorithms and controlling devices through solid-state relays. An integration scheme between microcontrollers and other components are shown in Figure 7 and the addresses are compiled in Table 1.

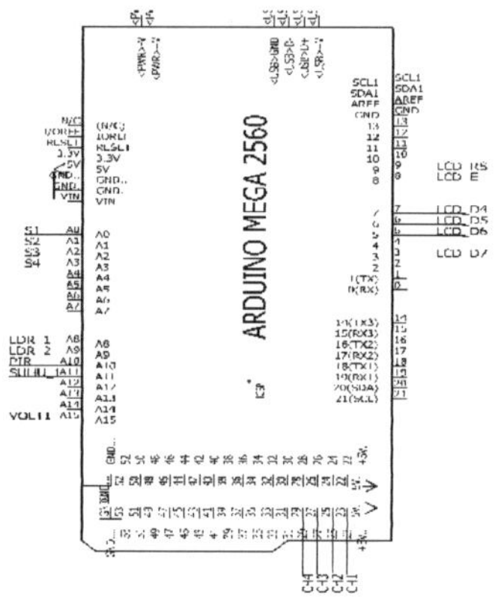


**Figure 7**. Microcontroller integration with sensors and actuators

**Table 1**. Arduino pin integration address with supporting components

| Supporting Components | Arduino Pins |
|---|---|
| LCD 16x2 | 9,8,7,6,5,3 |
| Current Sensor (ACS) | A14 |
| Voltage Sensor | A15 |
| LDR1,LDR2 Sensor | A8,A9 |

| | |
|---|---|
| PIR Sensor | A10 |
| SSR | 23,25,27,29 |
| Ethernet Shield | MOSI,MISO,SCK |

*5. Solid State Relay Driver Design*

A solid state switch (SSR) is an electronic switch using a semiconductor material as in a transistor. This switch type has advantages in not causing a fire or spark, compact design and not needing a large current to activate. Figure 8 shows a basic solid-state relay design.

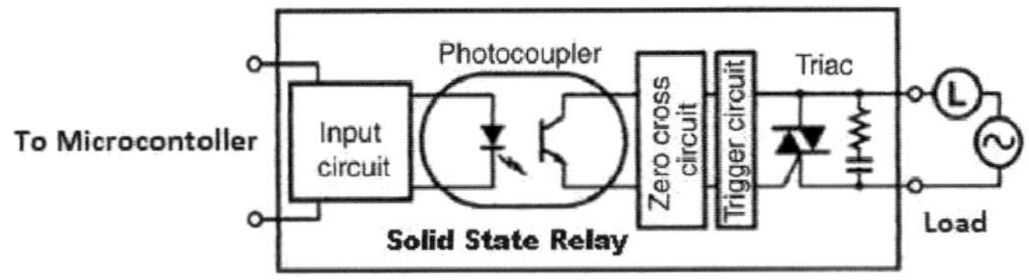


**Figure 8**. SSR internal circuit

The basic components of an SSR are optocoupler and TRIAC. The amount of load that can be handled by the SSR depends on the TRIAC's specifications, while the trigger voltage stays between 3-30 VDC.

*3.3. Software Design*

The BAS proposed in this research has two different software, which is a control algorithm in IED and HMI in the server.

*1. IED Software Design*

IED's algorithm in this system uses fuzzy lookup, which is suitable for implementation in a low-speed microcontroller due to not needing a high amount of computation. Membership function for lighting control is shown in Figure 9.

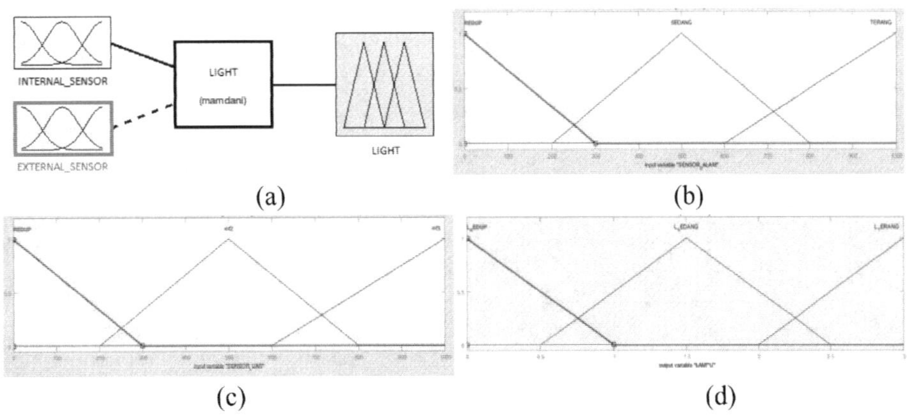


**Figure 9**. Membership function for lighting control, with a) fuzzy system diagram, b) sensor input in the room, c) inner sensor input, d) output control.

*2.HMI Software Design*

The HMI system in our proposed BAS uses a touchscreen-based minicomputer with the EasyBuilder Pro software installed. The HMI's display design has four main screens, which is the main display, the office, the classroom, and the meeting room. The main screen and the classroom screen is shown in Figures 10a and 10b

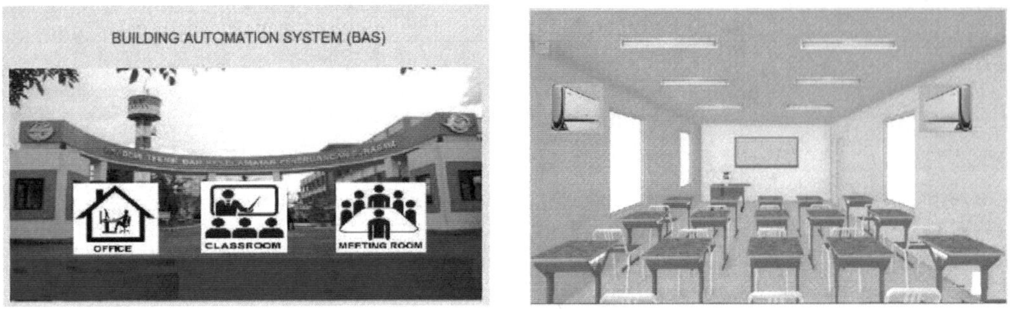


(a) (b)

**Figure 10**. HMI design for a) Main screen, b) Classroom screen

The HMI's hardware aspect has 192.168.1.8 as its IP address. HMI communicates with the microcontroller using the Modbus TCP protocol, so as to be able to link with all the devices in the network as long as they also used the same protocol. There are three microcontrollers in a network with the HMI, with each unit's address being 192.168.1.2, 192.168.1.3, and 192.168.1.4.

## 4. Testing and Analysis

System testing aims to find the performance characteristics of the system. The testing involves control and monitoring testing for each microcontroller as the local controller for each room.

### *4.1. Classroom Lighting Control Test*

Classroom testing stands for lighting control testing. This testing aims to find out the capabilities of the controller to adjust the lighting for adequate reading brightness. By dividing the activation of the lamps to different switches, the appropriate amount of light is turned on for adequate brightness is made possible. The testing is done by shutting the room to adjust for outside lighting. The results of the testing are shown in Table 2..

**Table 2**. Room lighting control test

| NO | HMI Control | Room Intensity Level | Outer Intensity Level | Turned on Lights | Average Power (Watt) |
|---|---|---|---|---|---|
| 1 | ON | 710 | 100 | 4 | 320 |
| 2 | ON | 690 | 300 | 3 | 240 |
| 3 | ON | 675 | 400 | 2 | 160 |
| 4 | ON | 640 | 700 | 1 | 80 |
| 5 | OFF | 500 | 700 | 0 | 0 |

The testing results show that the lighting control system was able to respond to natural lighting changes by turning on light until optimal brightness for reading is reached.

### *4.2. Air Conditioning Control Test*

The air conditioning control test is done by controlling the AC's activation to reach the setpoint temperature. The setpoint value depends on the presence of humans in the room, using the motion sensor. The results of the testing are shown in Table 3.

**Table 3**. Air conditioning control test

| NO | HMI Control | PIR Sensor | AC status | Temp.setpoint (⁰C) | Room temp. (⁰C) | Average Power (Watt) |
|---|---|---|---|---|---|---|
| 1 | OFF | Not detected | OFF | - | - | 0 |
| 2 | ON | Not detected | ON | 30 | 30 | 200 |
| 3 | ON | Detected | ON | 25 | 25 | 500 |
| 4 | ON | Not detected | ON | 30 | 30 | 230 |
| 5 | OFF | Detected | OFF | - | - | 0 |

The testing results show that the controller is able to change the setpoint temperature depending on the presence in the room. This behavior aims to reduce electricity usage. The setpoint value when unoccupied and at work, hours are set at 300 C, due to this value being the maximum temperature of the AC unit and still being better for power usage than just turning the AC off. Rapidly turning the AC on and off may damage the unit and accrue higher power penalties while starting the unit again.

### *4.3. Information Board Control Testing*

The information board control testing aims to find out how much the system is able to know the status of the display and adjust its brightness. The results of the testing are shown in Table 4..

**Table 4**. Information board control testing

| NO | HMI Control | Display Status | Sensor Status |
|---|---|---|---|
| 1 | OFF | OFF | OFF |
| 2 | ON | ON | ON |
| 3 | ON | ON | ON |
| 4 | OFF | OFF | OFF |
| 5 | ON | ON | ON |

The testing result shows that the system is able to monitor the condition of the display and control it from the control room through the HMI.

## 5. Conclusion

The conclusion that may be drawn from our research are as follows :

1. The building automation system allows for lower energy consumption in rooms
2. The room lighting control system was able to control the room's lighting to suit the needs of the occupants.
3. The air conditioning control system works to suit the current needs in the room

## References

[1] Jati L and Gata Grace 2011 Webcam Monitoring Ruangan Menggunakan Sensor Gerak Pir (Passive Infra Red) *BIT*

[2] Gifson A, Utara P and Selatan J 2009 Dengan Sensor Passive Infrared Berbasis Mikrokontroler At89S52 *Telkomnika* **7** 201–6

[3] Inc P 2012 Passive Infrared Sensor (#555-28027) 1–5

[4] Wilson M T 2004 Technology Advancement in Intelligent Buildings: a Through Preplanning Process Pertaining To Long-Term Maintainability

[5] Marinakis V, Doukas H, Karakosta C and Psarras J 2013 An integrated system for buildings' energy-efficient automation: Application in the tertiary sector *Appl. Energy* **101** 6–14

Journal of Physics: Conference Series **1381** (2019) 012007 doi:10.1088/1742-6596/1381/1/012007

# Cluster analysis of lombok island local buffalo (*Bubalus bubalis*) based on Principle Component Analysis (PCA)

**A Sukri[1*], T L Hajiriah[1], H Jannah[1], Andika[2], M Lukitasari[3]**

[1] Department of Biology Education, IKIP Mataram, Indonesia
[2]Dinas PPKP Kecamatan Kayangan, NTB, Indonesia
[3]Department of Biolgy, Unipma, Indonesia
*Email : akhmadsukri@ikipmataram.ac.id

**Abstract**— The Preservation of buffalo cattle in North Lombok and Central Lombok regencies can be done through inventory activities and identification of cattle performance through analysis of morphological characters in order to determine the kinship relationships of buffalo. This study examines the grouping of local buffalo in Central Lombok and North Lombok based on morphological characters. The sample of this study consisted of 16 male and female buffaloes taken from both regions. *One way anova* analysis results on eight morphological characters of buffalo which include (1) body length, (2) shoulder height, (3) hip height, (4) chest circumference, (5) chest width, (6) head length, (7) head height, and (8) head width shows no difference in mean values of each character in the two observation areas. The results of PCA analysis revealed a map of buffalo morphology distribution which was divided into two groups, i.e. group 1 representing North Lombok buffalo located in quadrant III, while group 2 represented morphological mixtures of North Lombok and Central Lombok buffalo located in quadrant I, II, and IV. The results of the cluster analysis show that some buffalos from North Lombok has similarities with Central Lombok buffalo based on the eight morphological characters.

## 1. Introduction

North Lombok Regency is one of the regions in West Nusa Tenggara Province which has a considerable potential in agricultural sector. These potentials include agriculture, plantation, forestry, livestock and fisheries [1]. In the field of animal husbandry, the district of North Lombok has large and small livestock resources that continue to be developed. One of the big livestock that becomes potential is buffalo cattle. Based on 2014 buffalo population data, North Lombok Regency ranks fourth in the number of buffalo populations from Lombok region after Central Lombok, West Lombok and East Lombok with a population of 442 buffaloes [2].

In general, the number of buffalo populations in North Lombok Regency is still relatively small compared to other regions in Lombok Island. Although it has not been able to provide a high enough contribution to the economy of North Lombok, the potential of the livestock sector needs to be continuously developed [1]. The development of the livestock sector, especially buffaloes in North Lombok Regency, is supported by land resources, socio-cultural conditions of the community, and the bio-physical condition of buffaloes. Effective land area that can be used as a cultivation area including buffalo livestock is 21,858 ha or 27% of the total area of North Lombok Regency [1]. Whereas in terms of socio-culture, the ownership of buffalo cattle has a special prestige for some people in North Lombok Regency such as in Bayan District. Some people of Bayan still hold ancestral traditions known as *wetu telu*. People who hold fast to *Wetu Telu* culture has many traditional ceremonies, sometimes in certain traditional ceremonies such as circumcisions, slaughtering buffalo cattle is a must. In such circumstances, the existence of buffaloes has a very important meaning.

The advantages of bio-physical buffalo support the development of buffalo cattle in North Lombok Regency. Buffalo has the ability to survive well on dry land or land by utilizing low-quality feed such as dry grass. Moreover, buffalo has high enough capacity to deal with pressures and extreme environmental changes [3]. The advantages of bio-physical are very much needed in the development of buffalo cattle in North Lombok Regency, because most of the potential area for livestock development is on dry land which includes three fields, i.e. dry land garden of 16,720 Ha (80.21%), farmland public relations amounting to 4,105 ha (19.69%), and 20 meadow land of Ha (0.10%) [1].

Unlike the North Lombok District, the buffalo population in Central Lombok Regency ranks the top of the population of buffaloes from the Lombok region. Recorded the number of buffalo population in 2014 was 19,293 [2]. The use of buffalo livestock in Central Lombok Regency is used to increase community income, and for consumption needs. Recorded in 2008 as many as 2,403 buffaloes were cut for consumption needs [4]. The geographical conditions of buffalo grazing land in Central Lombok and North Lombok districts are almost the same. Based on the results of observations, buffalo grazing land in Central Lombok Regency, especially in the southern part, has dry conditions as well as most buffalo cattle areas in North Lombok Regency.

According to Muthalib [5], the development and preservation of buffalo in West Nusa Tenggara (NTB) Province including North Lombok and Central Lombok Districts can be done through inventory activities, and identification of buffalo cattle performance in NTB. One of the way that can be done is analyzing morphological characters in order to determine the kinship relationship of buffalo. This is important as basic information in the development of buffalo cattle in NTB Province, especially in North Lombok and Central Lombok Districts. This study examines clustering of local buffalo in Central Lombok and North Lombok based on morphological characters.

## 2. Methodology

### 2.1 *Sample and Research Location*

The research sample consisted of 16 male and female buffaloes consisting of 8 buffaloes from Ungga Village, Praya Barat Daya Subdistrict, Central Lombok Regency, and 8 buffaloes taken from Dasan Glumpang Hamlet, Akar-Akar Village, Bayan District, North Lombok Regency. Sampling was done purposively by considering that several buffaloes were still wild, making it difficult to measure morphological parameters.

### 2.2 *Morphological character*

Morphological characters measured in this study include (1) body length, (2) shoulder height, (3) hip height, (4) chest circumference, (5) chest width, (6) head length, (7) head height, and (8) head width. Measurement of buffalo's morphological characters uses measuring tape with accuracy of 0.5 cm by adopting the method of measuring body parameters (Sarbaini, 2004) and cow head (Asoen and Anggraeni, 2011) in Agung et al [6] as follows:

a. Body length (cm), measured straight distance from the shoulder protrusion or tuberculum laterale humeri to the sitting bone or ischii tuber.
b. Shoulder height (cm), was measured from the highest point of the shoulder past to the back, then straight to the ground.
c. Hip height (cm), was measured perpendicular to the first sacrum bone to the ground.
d. Chest circumference (cm), was measured around the chest cavity behind the shoulder or behind the elbow of the front leg perpendicular to the axis of the body.

e. Chest width (cm), was measured from the largest distance of the chest right and left in the position of measurement of chest circumference.
f. Head length (cm), was measured from the tip of the skull to the tip of the nose.
g. Head height (cm), was measured from the tip of the skull to the lower jaw
h. Head width (cm), was measured from the left temple to the right temple.

### *2.3 Data Analysis*

After performing a morphometric analysis of 8 morphological characters including (1) body length, (2) shoulder height, (3) hip height, (4) chest circumference, (5) chest width, (6) head length, (7) head height, and (8) head width, then the next step is to do a *one way anoya* test to determine the difference in mean values of each morphological character for each district. To find out the grouping of buffalo based on their morphological characters, a PCA (Principle Componen Analysis) analysis was carried out and strengthened by cluster analysis by using the Euclidian distance method. PCA and cluster analysis are assisted using a PAST computer program.

## 3. Result and Discussion

### *3.1 Measurement of the Character of Buffalo Morphology*

The measurement results of morphological characters include (1) body length, (2) shoulder height, (3) hip height, (4) chest circumference, (5) chest width, (6) head length, (7) head height, and (8) the width of the head is shown in Table 1.1 below.

**Table 3.1**. Results of Measurement of Buffalo Morphological Characteristics

| Buffalo | Body Length | Shoulder height | Hip height | chest circumference, | Chest width | Head length | Head height | Head Width |
|---|---|---|---|---|---|---|---|---|
| LT1 | 106 | 126 | 120 | 170 | 40 | 44 | 28 | 25 |
| LT2 | 122 | 117 | 132 | 170 | 42 | 45 | 27 | 24 |
| LT3 | 124 | 131 | 128 | 168 | 40 | 44 | 31 | 24 |
| LT4 | 108 | 112 | 116 | 165 | 34 | 29 | 26 | 16 |
| LT5 | 115 | 114 | 117 | 170 | 42 | 43 | 32 | 18 |
| LT6 | 132 | 125 | 122 | 170 | 40 | 44 | 31 | 20 |
| LT7 | 141 | 117 | 112 | 165 | 36 | 42 | 28 | 18 |
| LT8 | 129 | 118 | 117 | 170 | 40 | 45 | 27 | 17 |
| LU1 | 139 | 123 | 121 | 170 | 47 | 44 | 41 | 22 |
| LU2 | 127 | 126 | 123 | 160 | 37 | 46 | 27 | 21 |
| LU3 | 127 | 124 | 123 | 170 | 44 | 41 | 29 | 20 |
| LU4 | 115 | 113 | 119 | 150 | 30 | 38 | 23 | 19 |
| LU5 | 121 | 114 | 119 | 141 | 27 | 35 | 21 | 22 |
| LU6 | 117 | 123 | 132 | 166 | 33 | 42 | 25 | 23 |
| LU7 | 132 | 129 | 123 | 170 | 45 | 44 | 41 | 23 |
| LU8 | 112 | 112 | 119 | 147 | 31 | 38 | 22 | 19 |
| **Average** | **122.94** | **120.25** | **121.44** | **163.88** | **38.00** | **41.50** | **28.69** | **20.69** |

Information:
LT: Central Lombok
LU: North Lombok

The results obtained in Table 1.1 show a slight difference with the results obtained by Anggaraeni et al [7], who found that the average body length, shoulder height, hip height, chest circumference, and chest width of male buffalo in NTB respectively were 120. 3 cm; 121.8 cm; 119.4 cm; 164.5 cm; and 57.0 cm. Whereas in this study, various results were obtained compared to the results of the study. The average body length in this study amounted to 122, 94 cm, greater than the results of Anggraeni et al [7], which is 120.3 cm, but smaller than the results of the study of Sitorus and Anggraeni [8], which amounted to 129.50 cm. Characteristics of shoulder height, hip height, chest circumference, and chest width obtained in this study (120.25 cm; 121.44 cm; 163.88 cm; and 38.00 cm) were slightly lower than the results of the measurement of buffalo swamp by Sitorus and Anggraeni [8] in Sumatra, which are 126.38 cm respectively; 125.56 cm; 182.16 cm; and 38.72 cm.

In this study, the measurement of buffalo's morphological characters was not differentiated by sex, so the measurement results obtained were a mixture of morphometric measurements from male and female individuals. In contrast to previous studies conducted by Sitorus and Anggraeni [8], and Anggaraeni et al [7] which distinguishes buffalo morphometric size based on sex, so that the size can be distinguished between male and female. However, based on the results of a study by Anggaraeni et al [7], it is known that in general there is no significant difference between male and female body sizes. The results obtained in this study reinforce the assumptions of Anggaraeni et al [7]. Based on the results of *one way anova* analysis of eight buffalo's morphological characters, it was found that there was no difference in the average value of eight buffalo morphological characters in North Lombok and Central Lombok regencies, both male and female individuals (Table 1.2). These results indicate the uniformity of buffalo body size in the two observation areas. This is presumably because the local buffalo which is the object of research has the same type, namely the type of mud buffalo (swamp buffalo). In addition, the uniformity of body measurements illustrates the high morphometric similarity of a group and the low variation in body measurements that make up body shape conformation [8].

**Table 3.2**. Results of One Way Anova Analysis of Each Morphological Character

| No | Morphological Character | Sig. |
|---|---|---|
| 1 | Body Length | .766 |
| 2 | Shoulders height | .881 |
| 3 | Hip height | .513 |
| 4 | Chest circle | .050 |
| 5 | Chest width | .402 |
| 6 | Head Length | .671 |
| 7 | Head Height | .967 |
| 8 | Head width | .539 |

### 3.2 *Result of PCA Analysis*

To find out the grouping of buffalo based on eight morphological characters from the two observation districts, namely North Lombok Regency and Central Lombok Regency, an analysis of PCA (Principle Componen Analysis) was carried out. The results of the PCA analysis is shown in Figure 3.1 below.

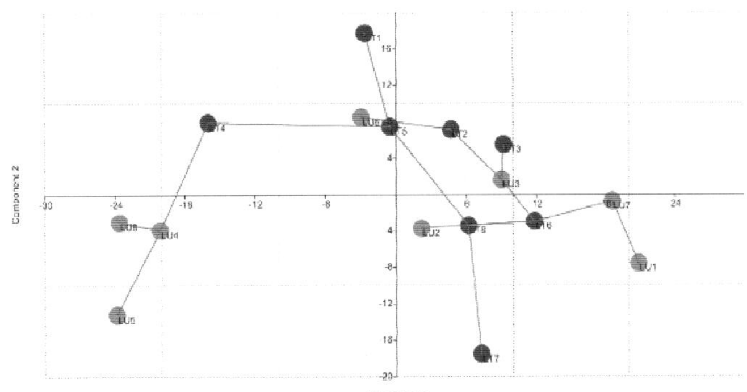


Figure 3.1 Results of PCA Analysis 16 Buffalo Individuals from Two Observation Areas (blue: individual buffalo in Central Lombok, red: individual buffalo in North Lombok)

The results of PCA analysis in Figure 3.1 reveal the spread of the morphometric linear size of buffalo bodies in North Lombok and Central Lombok. Based on Figure 3.1 it is known that the morphological distribution map of buffaloes is divided into two groups, namely group 1 which represents only North Lombok buffalo located in quadrant III, while group 2 represents buffalo's morphological mixing in North Lombok and Central Lombok which is located in quadrant I, II, and IV. These results reinforce the results of previous studies conducted by Anggaraeni et al [7] which found a grouping of buffalo morphology distribution specifically from the West Nusa Tenggara region compared to swamp buffalo from North Sumatra, Central Java, South Kalimantan, South Sulawesi, and Banten. However, this research needs to be strengthened again by increasing the number of samples and the number of districts so that they can represent the entire NTB region.

### *3.3 Result of Cluster Analysis*

The results obtained from PCA analysis need to be strengthened by analyzing the cluster of individual buffalo from the two observation regions. Cluster analysis was carried out using the euclidian distance method which was assisted by the PAST program. The results of cluster analysis is shown in the following Figure 3.2.

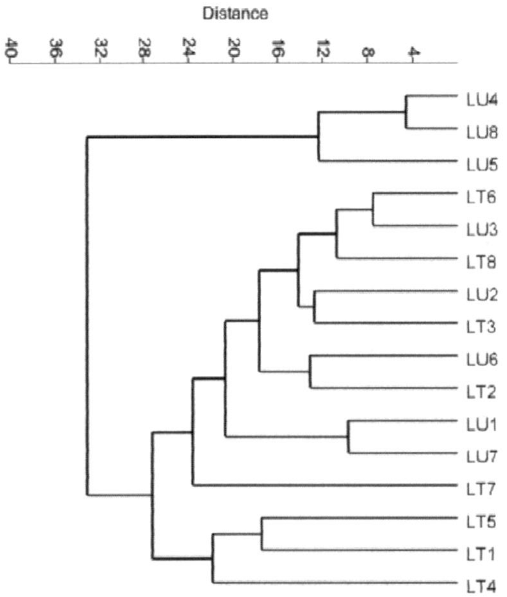


Figure 3.2 Results of Analysis of Individual Buffalo Clusters from Two Districts, namely Central Lombok Regency (LT) and North Lombok (LU)

Based on the analysis result of buffalo individual clusters from two different districts in Figure 1.2, it is known that some buffalo individuals from Central Lombok and North Lombok form separate groups as in individuals LT2 with LU6, and clusters LT6, LU3, and LT8. These results indicate that there are similarities between individual buffaloes which are grouped based on eight morphological characters. Phylogenetic analysis of local buffalo in NTB, especially in Central Lombok Regency, has been carried out by Sukri [9] who carried out phylogenetic tree reconstruction based on microsatellite DNA markers in two different buffalo populations. The results of the study show that genetic mixing occurred between the two buffalo populations studied. It is believed that it is caused by the spreading of buffalo individuals in each population is still open. Although the results of this study indicate the grouping of several individuals in North Lombok buffalo with Central Lombok in one cluster, but it cannot be ascertained whether genetic mixing has occurred or not between the two buffalo populations. Because naturally, the possibility of breeding between the two populations cannot be done because of the large number of separators between the two populations.

The results obtained in this study only describe the kinship between two buffalo populations which are seen based on their morphological characters. It is very possible if buffalo individuals in North Lombok Regency have morphological characters that are similar or similar to buffalo individuals in Central Lombok Regency, this could be due to environmental influences. According to Surahman [10], there are some weaknesses of using morphological markers in carrying out reconstruction of phylogenetic trees that is highly influenced by the environment and often cannot distinguish genotypes between individuals. To complement this study, it is expected that in future studies, phylogenetic analysis or kinship of buffalo in West Nusa Tenggara is carried out by using DNA markers in order to obtain more stable and better results.

## 4. Conclussion

Based on the results of *one way Anova* analysis of eight buffalo's morphological characters which include (1) body length, (2) shoulder height, (3) hip height, (4) chest circumference, (5) chest width, (6) head length, (7) head height, and (8) head widt, it is known that there is no difference in the mean values of each character in the two observation areas, i.e. North Lombok District and Central Lombok Regency. The results of PCA analysis revealed a map of buffalo's morphology distribution which was divided into two groups, i.e. group 1 which representing only North Lombok buffalo located in quadrant III, while group 2 represented a mixture of buffalo morphology in North Lombok and Central Lombok located in quadrant I, II, and IV. The results of the cluster analysis show that some individuals from North Lombok have similarities with Central Lombok buffalo based on eight morphological characters.

## 5. References

[1] Lombokutarakab.go.id. (2016). Kabupaten Lombok Utara., *http://www.lombokutara*.go.id/v1/images/doc/2014/gambaranekonomi.pdf.

[2] Disnakkeswan Prov NTB. (2016). Populasi Ternak. *http://www.disnakkeswan.ntbprov.go.id/informasi-2/populasi-ternak/*.

[3] Hilmiati, N. (2008). Potensi Kerbau Sebagai Substitusi Daging Sapi di Nusa Tenggara Barat, http://www.ntb.litbang.deptan.go.id/index.php?.

[4] Badan Pusat Statistik Lombok Tengah. (2010). Lombok Tengah Dalam Angka 2008/2009. *http://www.lomboktengahkab.go.id/download/tani1.pdf.*

[5] Muthalib, H.A. (2005). Potensi Sumberdaya Ternak Kerbau di Nusa Tenggara Barat. *Lokakarya Nasional Usaha Ternak Kerbau Mendukung Kecukupan Daging Sapi:* Dinas Peternakan Propinsi Nusa Tenggara Barat.

[6] Agung, PP.,Ridwan, M.,Handrie, Indriawati, Saputra, F.,Supraptono, & Erinaldi. (2014). Profil Morfologi dan Pendugaan Jarak Genetik Sapi Simmental Hasil Persilangan. *JITV, 19* (2), 112-122.

[7] Anggraeni, A.,Sumantri, C.,Praharani, L.,Dudi.,& Andreas, E. (2011). Estimasi Jarak Genetik Kerbau Rawa Lokal Melalui Pendekatan Analisis Morfologi. *JITV, 16(3),* 199-210.

[8] Sitorus, A.J, & Anggraeni, A. (2008). Karakterisasi Morfologi dan Estimasi Jarak Genetik Kerbau Rawa, Sungai (Murrah) dan Silangannya di Sumatera Utara. *http://www.peternakan.litbang.pertanian.go.id/fullteks/lokakarya/lkbo08-5.pdf.*

[9] Sukri, A. (2014). Analisis Filogenetik Kerbau Lokal Lombok Tengah (*Bubalus bubalis*) Berdasarkan Penanda DNA Mikrosatelit. *Florea, 2*(1), 52-55.

[10] Surahman, M. (2002). Peta genetika tanaman, prinsip dan aplikasinya. *Bul.Agron, 30*(1), 27-30.

Journal of Physics: Conference Series **1381** (2019) 012008 doi:10.1088/1742-6596/1381/1/012008

# Transactional database design information system web-based tracer study integrated telegram bot

**S Sucipto[1*], N C Resti[1], T Andriyanto[1], J Karaman[2], R S Qamaria[3]**
[1]Departement of Information System, Universitas Nusantara PGRI Kediri, Kediri, Indonesia
[2]Departement of Informatics, Universitas Muhammadiyah Ponorogo, Ponorogo, Indonesia
[3]Institut Agama Islam Negeri Kediri, Kediri, Indonesia

*Corresponding author's email: sucipto@unpkediri.ac.id

**Abstract**. A database is a software that is used to store data. Data stored is generally in the form of text. Database functions are not only for storing data but also used to speed up access to information systems. Optimal database management can increase access to information systems. One management that can improve performance is the appropriate relational data design. Relational is a relationship between tables. The design of the relational database must pay attention to the selection of data types, data type values, and constraints that will be selected. This study will examine the optimization of the MariaDB database on information systems in tracer studies. Database design will accommodate data from two application sources, namely web-based applications and telegram bots. Using two paths to the database to make it easier for users to register via a telegram bot, users can then access the tracer study questionnaire on web-based applications. The MariaDB database performance test shows that the highest performance average query is 6501 microsecond.

## 1. Background

Educational institutions must ensure the quality of education. In the current era, vocational secondary education is a favorite compared to secondary schools. It appears that the government fully supports the implementation of vocational secondary schools by providing various supporting facilities [1]. Evaluation is necessary to see the extent to which the quality of quality improvement for vocational school students is improved. One tool to measure the quality of graduates through tracer study. Tracer study as a means to ensure the quality of education by the needs of educational institutions [2]. In the Industrial Revolution 4.0 era, tracer study must follow information technology. Tracer study can be more effective if you use the application. Various applications can refer to online systems, such as web-based applications [3–5]. In addition to other alternative web applications, you can use social media networking. Chat based social media network. The types of chat applications include WhatsApp, BBM, Line, Telegram, and many others.

Integration of information systems with chat applications has been implemented, such as credit bill applications, PLN electricity, tickets. This integration is done to link many users who like the social media chat service a lot, one of them is a telegram. Telegram is a popular chat application that functions for integration. The user used telegram widely because of the open source and open source API support. One of the telegram features used for integration is Bot. Bot is a third party application that runs inside the Telegram Application [6]. A combination of telegram bots with various useful applications as messenger assistants for users [7]. The creation of a tracer study application needs to be considered the data storage design. Data storage of Information data system software is a database. The database is a collection of data in an application that is connected that describes the data design to meet the information needs of an organization [8]. Database applications include MariaDB, PostgreSQL, Oracle, MongoDB, and others. In this study, the researcher uses MariaDB because it adapts to the server used in the research institution. MariaDB is one of a group of DBMS (Database Management systems). Based on several studies, using MariaDB is quite capable of being applied to various applications[9,10].

Design is an essential factor in making information systems. The optimal database design information system will speed up application access. Some of the researches on information system design are carried out by Olalere Modupeola E in 2018 entitled Design, Implementation and Evaluation of a Web-Based Physical Fitness Consultation System. The study designed web-based applications consisting of PHP, HTML, and Javascript programming with MySQL databases. The use of the application shows a tremendous increase in the existing system in terms of cost-effectiveness [11]. Other research was conducted by Satria Abadi in 2018 entitled Design of online transaction models on traditional industries to increase turnover and benefits. The study discusses the design of application development models using PHP web programming and MySQL databases. The application developed to increase trust based on app access accuracy [12]. The next study was carried out by Muhamad Irfan Kurniawan in 2018 entitled Internet of Things: Raspberry Pi and Telegram Messenger-based Home Security Systems. The research combines devices with Telegram Bot for notification purposes. The use of telegram makes it easy to deliver information about home security [13].

The design of the tracer study application uses two input data lines using two applications, namely telegram and web-based applications. The purpose of the research is to design an optimal database design so that it can overcome input from the two applications.

## 2. Literature Review

The database used in this study is the MariaDB database. MariaDB is a Database Management system originally known as MySQL. Many developer communities developed MariaDB in the database field, which previously also contributed to the MySQL database. MySQL was acquired by Oracle, causing MySQL to become a proprietary licensed product. The acquisition carried out by Oracle caused the development of MySQL not to be free caused the MySQL developer community to build MariaDB [14]. Even though MySQL changed to MariaDB, but still able to maintain compatibility and API. New products from MariaDB are XtraDB and Aria, which are new storage machines. The new Machine feature is used to support transactional and non-transactional databases.

The MariaDB database engine consists of MyISAM and InnoDB. Each table has a different machine. InnoDB supports relationships between tables or is called a transactional database. MyISAM is a non-transactional type that can use three access priorities., including LOW_PRIORITY, DELAYED, and HIGH_PRIORITY. The MariaDB server must decide which statement will be processed first. Statements that change data (INSERT, UPDATE, and DELETE) are a priority compared to the SELECT statement [14].

PHP is an open source web programming language that is server-side. PHP can be part of the HTML programming language. PHP has connection capabilities with various types of databases. PHP is the language of interpreters that has a condition in the declaration of code commands. PHP can be used to update databases, create databases, and work on mathematical calculations. PHP released the latest version, namely version 7. Some functions including Linked List, Stacks, Queues, Trees, Heaps, Graph, Sorting, Searching, Dynamic Programming and Others, Numbers, and Maths - PHP Big Integer Implementation - Prime number generation – Sieve [15].

Telegram is a free and non-profit social media multiplatform instant messaging sender media application. The multiplatform telegram that works on various mobile devices or computers devices includes Android, iOS, Windows, Linux. Telegram can be used to send messages and exchange photos, videos, stickers, audio, and other types of files. Telegram also provides optional encrypted end-to-end message sending [16].

One of the telegram features is Bot. Telegram bots are the easiest modern bots to make compared to similar bots. Telegram bot supports a variety of programming languages, one of which is PHP. Bot API is an official bot provided by the developer telegram, which is a separate entity. The main function of the Bot is to help provide information. Bots help with Work and Daily Activities including calculators that calculate difficult jobs, unit conversions, calculate exchange rates, check certain status (expired, domain, ping, traceroute on computer networks), check the expedition receipts of JNE, Tiki, POS, check train schedules, check the price of goods, compare. Check automatic e-mail [17].

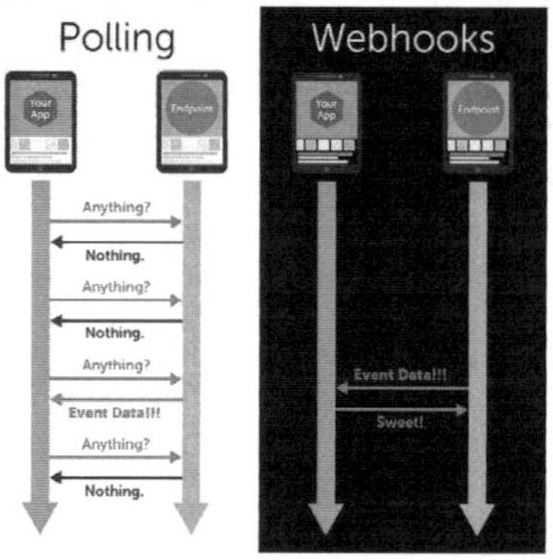


**Figure 1.** API Bot Method

There are two of the most popular bot management methods, as in Figure 1, namely long-polling and webhooks. Bot management function to share data efficiently between applications and application users. Both methods have advantages and disadvantages. Long-Polling is the default Telegram method that is portable, which means that the method can run on various devices, including computers, HP, IoT devices, and routers. This method is not required to use an online server. The disadvantages of this method include the process of reading it for quite a long time. The update process is not in real time. Webhook has the advantage of the process of reading chat faster than long-polling. This method can be run full 24 hours because it is run on an online server that has an SSL certificate. This method can update scripts in real time [6].

## 3. Research Method

The methodology in this study is to conduct a theory review of 5 pioneering SQL groups in database programming and previous research studies with literature studies explained in the previous chapter. The research method uses the waterfall model. This model approaches systematically and sequentially. The stages of this model, as illustrated in Figure 2, must step by step in each process [18].

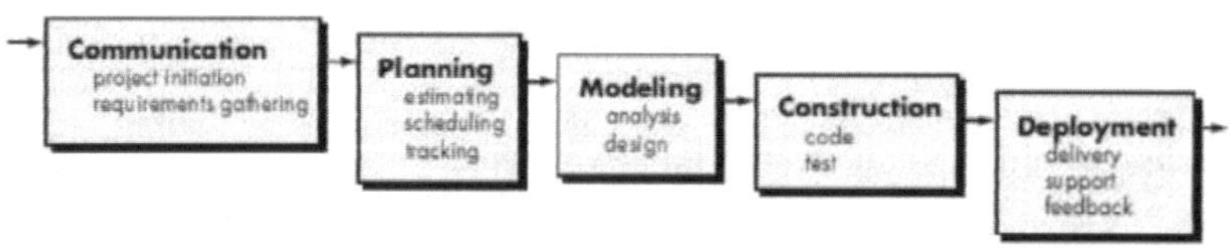


**Figure 2.** Waterfall Model

There are five stages in the model carried out in this study. At the first stage, there is Communication that is carried out before starting technical work done based on the goal of improving the performance of information systems in tracer studies. The Communication Phase has been carried out with stakeholders. Stakeholders in this study included several vocational schools and provincial service branches in Kediri. The next stage is Planning. At the Planning stage, schedule planning is carried out based on the tasks to be carried out until the testing phase. After that, it goes to the Modelling stage. In the Modelling stage, the database model adjusts the design according to the planning stage. The model selection is adjusted to the database engine in this study using the MariaDB database.

At the Construction stage, code writing is carried out, including SQL DDL (Data Definition Language), and DML (Data Manipulation Language) commands. The choice of DDL and DML requests refers to the need for a tracer study. Then at the Deployment stage, the implementation of the three final coding steps of the DCL (Data Control Language), Transaction and Select steps are carried out. At this stage, an evaluation of the cost of load code will be carried out based on monitoring using the micro time command [19]. The micro time function is one of the PHP tools used to restore the current Unix timestamp with units of microseconds [19].

## 4. Discussion and Result

The researcher uses the waterfall model to conduct this research. The telegram bot application and information systems use a database design together for the case of information systems tracer studies. Database optimization aims to smooth access to information systems. The Communication stage has been carried out by retrieving information from research objects and central studies based on book literature and previous research. The research planning phase runs for one semester, which is about six months.

The needs of the tracer conduct modeling stages of the information system database of tracer study. The database used is MariaDB using the MyISAM and InnoDB machines. The database design of tracer study will contain some information including:

a. Student Data
b. Tracer Study Data
c. Message Data
d. Announcement Data
e. Login Data

Data allocation must refer to the results of the first stage, namely the communication stage. This stage will transform into a database. Table 1 is a description of the needs of a database information system of the tracer study table.

**Table 1.** Detailed Information on Tracer Study Data

| No | Data | Information |
|---|---|---|
| 1 | Student | Information is used to hold student identity data. |
| 2 | Tracer Study | Information is used to hold data on tracer study works and lectures |
| 3 | Message | Information is used to accommodate messages from alumni |
| 4 | Announcement | Information is used for job vacancies and reunion information boards |
| 5 | Login | Information is used for alumni login status |

Based on the requirements in table 1, figure 3 is the result of the design of the tracer study system flow. The stages of construction adapt to the design of drawing three. Database design follows the system flow in Figure 3. In Figure 3, there are two actors. The actor consists of administrators and alumni. In the telegram system, both actors will use a telegram application to access tracer study information. Access to information on bot telegram-based is limited, including registration, user information, and job information. The telegram bot access method uses the webhook method. This method does not overload the web server. The bot method will make it easy for users to access information system tracer studies [20]. Bot models are arranged based on chat-bots with "/" fundamental command interactions [21].

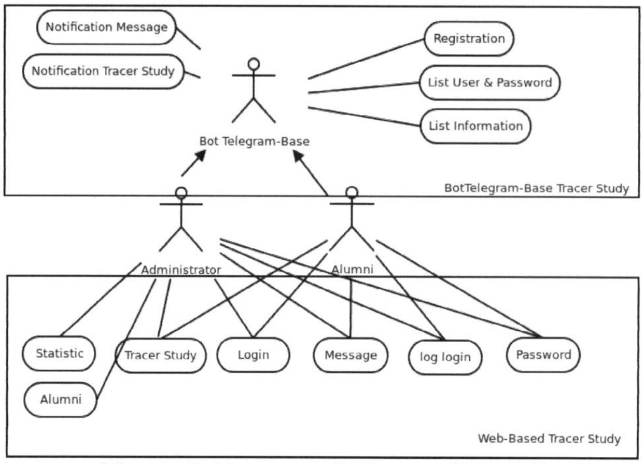


**Figure 3.** Use case of Diagram Tracer Study

The primary function of the web-based method in Figure 3 is as access to the charging tracker study. A long enough filling form is suitable for the web-based method. The registration module and user password list module are not available on the web-based method; the module can only be accessed by bot-telegram according to the telegram user id.

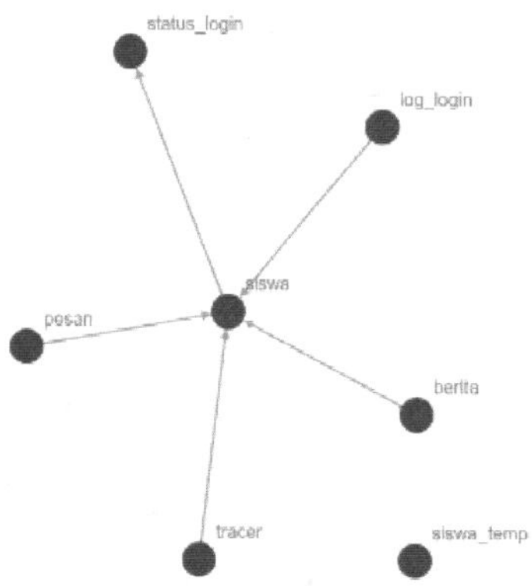


**Figure 4.** Database Relation

Figure 4 explains the information system database design. the accuracy of the relationship quality is measured to obtain data quality correctly so that it can work for data optimization in the table [22]. There are seven tables in the information system. Seven tables include the following:

a. Student table
b. Table of students temp
c. Tracer table
d. Message table
e. News table
f. Status_login table
g. Log_login table

The table in figure 3 consists of two different machines. The machines used are InnoDB and MyISAM. Only the student_temp table uses the MyISAM engine. The student_temp table functions as a temporary table of student tables when uploading data on a web-based system. In addition to the student_temp table, all other tables use the InnoDB engine which functions as a database relocation in the MariaDB database. The use of the MyISAM type machine can optimize the database with a non-relational model [12]. The use of the MyISAM engine can determine the DML priority, namely the command LOW_PRIORITY, DELAYED, and HIGH_PRIORITY [14].

The detailed structure of the database data type is adjusted to the needs of the data record — relations based on key consistency in the table.

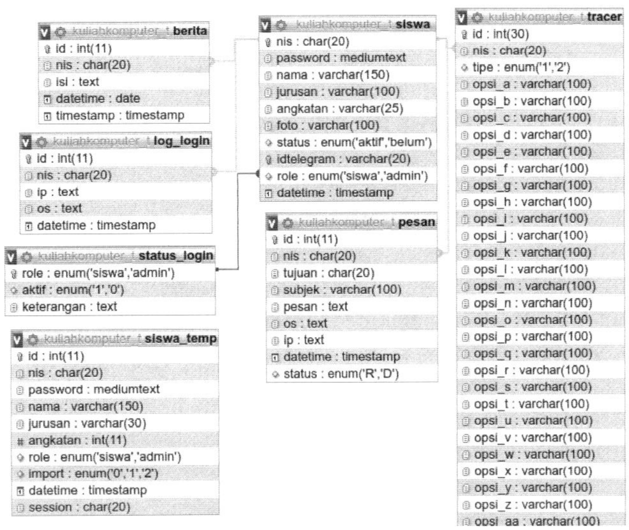


**Figure 5.** Structure Database

In the Deployment stage, Figure 5 shows the table design. Web-based applications and bot-based telegram applications will access the implementation of table designs in Figure 4. The key to being able to be integrated into the telegram bot application is in the student table, namely the telegram entity. The ID telegram entity is used to access information only accessible to certain alumni. The Log Table only works when used in the web-based method. Its function is to find out the activities of alumni who access information systems. Display bot interface on the telegram, as shown in figure 6.

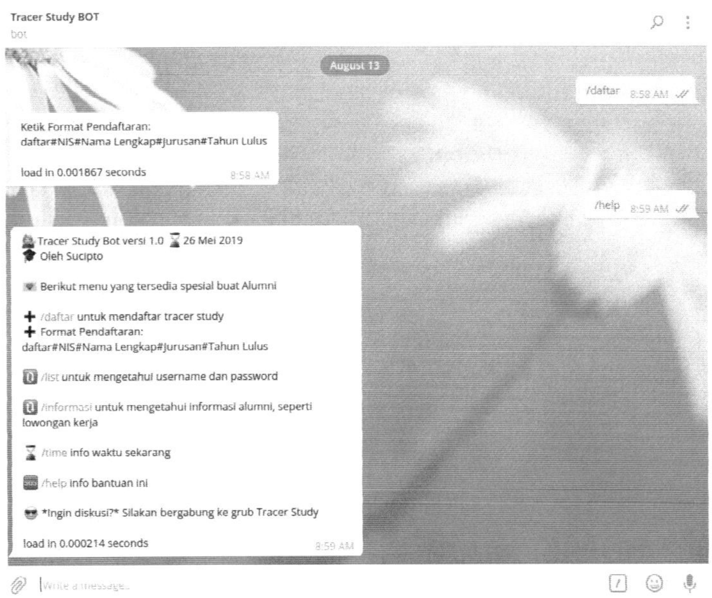


**Figure 6.** Tracer Study Bot

Figure 5 has various menus to make it easy for alumni to access information. Each list is connected to the MySQL database using the webhook model. Integration of bot display in Figure 5 with table design in Figure 4. The database table is made based on the design in figure 4, then tested using the micro time function. The basic commands for micro time are as follows:

*$ time_start = micro time (true);*
*asleep (100);*
*$ time_end = micro time (true);*
*$ time = $ time_end - $ time_start;*

Microtime functions are available in telegram applications and web-based applications. This function is used to determine the access speed of the SQL query and PHP load coding. The following is a graph based on the results of micro time testing:

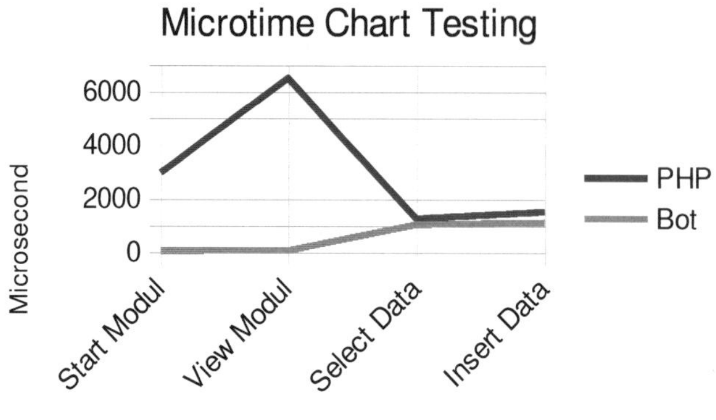


**Figure 7**. Chart Testing

Figure 6 shows the results of the access speed test performance. There are four tests performed. The initial test, namely "start module," is the module that runs at the beginning of the test. Start page on a bot with search bot then type command/start. Start a page on the web by accessing the website. The visit of "view module" is done when after the initial /start command on the bot then tests the bot with the / help command or after successfully logging in the web application. The test results of the two module start modules and module views show the results of the difference quite far between the two applications in microsecond units. The web application loads the database and web coding but on bots without database access.

Testing the next two modules is to select data and data insert. The test involves access to the database. Access the "select data" module on the bot with commands/information and / list. Access the module on the web application by clicking on the menu in the application. Access the last test module; the "data insert" module. In the bot application with the command "register" and on the web application by filling in the message and filling out the questionnaire.

The test results on the first two modules, namely the module start module and view module show a considerable time difference. At the start of the module, the difference is 2911 microseconds; the module view is 6449 microseconds. The results of testing the final two modules show a fairly thin difference in the "select data" module with a difference of 222 microseconds and the "view data" module with a difference of 394.

## 5. Conclusion

Based on the test results, the highest query average results in both applications amounted to 6501 microseconds on web-based applications. These results are still in access fast enough to be accessed by 30 users simultaneously. Information about tracker studies is fairly quickly accessed in the bot application because the telegram application directly accesses requests. Placement of information access and alumni registration on the bot application is entirely appropriate because access is quite fast compared to web applications. The placement of the questionnaire is placed on the web application because many constraints are difficult to apply to the bot application. Based on the results of testing and access to query design applications have provided satisfactory results.

## 6. Reference

[1] Kantor Staf Presiden 2016 Pemerintah Perbanyak SMK dan Tingkatkan Kompetensi Pelaku Pendidikan Kejuruan

[2] Tefera G 2019 A Tracer Study on (2011 – 2013) Debre Berhan University Graduates Integration in to the World of Work *Int. J. Second. Educ.* **6** 37–45

[3] Wicaksono G W, Hartanto A and Azhar Y 2017 SISTEM INFORMASI KARIR ALUMNI DAN TRACER STUDY STUDI KASUS PADA PROGRAM STUDI TEKNIK INFORMATIKA UNIVERSITAS MUHAMMADIYAH MALANG *Seminar Nasional Teknologi dan Rekayasa (SENTRA) 2017* (Malang: UNIVERSITAS MUHAMMADIYAH

MALANG) pp 1–12
[4] Nugroho E C and Nugroho I 2018 Sistem Pusat Karir Dan Tracer Study Perguruan Tinggi *Ijns.org Indones. J. Netw. Secur.* **7** 1–5
[5] Diana E and As'ad 2017 Analisis Dan Perancangan Sistem Informasi Tracer Study Berbasis Web *MEDIASISFO* **11** 817–29
[6] Telegram 2019 Bots: An introduction for developers
[7] Sajad F, Hossein Heidari T and Azizeh C 2019 TELEGRAM : AN INSTANT MESSAGING APPLICATION TO ASSIST DISTANCE LANGUAGE LEARNING ( App Review ) *Teach. English with Technol.* **19** 132–47
[8] Connolly T M and Begg C E 2010 *Database systems: a practical approach to design implementation and management* (America: Pearson Education)
[9] Sucipto S 2017 Perancangan Active Database System pada Sistem Informasi Pelayanan Harga Pasar *J. INTENSIF* **1** 37–45
[10] Sucipto, Indriati R and Hariawaan F B 2017 DESAIN DATABASE UNTUK OPTIMALISASI SISTEM PREDIKSI TRANSAKSI PENJUALAN *JIPI (Jurnal Ilm. Penelit. dan Pembelajaran Inform.* **2** 88–93
[11] Modupeola E O, Ishak R S, Mayowa A A, Olalekan J O and Sunday A A 2018 Design, Implementation and Evaluation of a Web-Based Physical Fitness Teleconsultation System *J. Inf. Technol. Softw. Eng.* **8** 1–4
[12] Abadi S, Huda M, Hehsan A, Marzuki Mohamad A, Basiron B, Suhaila Ihwani S, Azmi Jasmi K, Safar J, Kilani Mohamed A, Hassan Wan Embong W, Shakib Mohd Noor S, Brahmono B, Maseleno A, Nabila Fauzi A, Aminudin N and Gumanti M 2018 Design of online transaction model on traditional industry in order to increase turnover and benefits *Int. J. Eng. Technol.* **7** 231–7
[13] KURNIAWAN M I, SUNARYA U, TULLOH R and TULLOH R 2018 Internet of Things : Sistem Keamanan Rumah berbasis Raspberry Pi dan Telegram Messenger *ELKOMIKA J. Tek. Energi Elektr. Tek. Telekomun. Tek. Elektron.* **6** 1–15
[14] Dyer R J 2015 *Learning MySQL and MariaDB: Heading in the Right Direction with MySQL and MariaDB* (Gravenstein Highway North: O'Reilly Media, Inc.)
[15] Rahman M 2017 *PHP 7 Data Structures and Algorithms: Implement linked lists, stack, and queues using PHP* (Birmingham: Packt Publishing)
[16] Telegram 2019 What is Telegram?
[17] Hasan 2016 *Membuat Sendiri BOT TELEGRAM dari PHP* (Lumajang: IDT (Indonesian Digital Teacher) Group)
[18] Pressman R S 2010 *Software Enggineering: A Practitioner's Approach (7th Edition)* (New York: McGraw-Hill)
[19] Padilla A and Hawkins T 2010 *Pro PHP application performance : tuning PHP Web projects for maximum performance* (New York: Apress)
[20] R V, B B, S A and M D 2019 AI Based Student Bot for Academic Information System using Machine Learning *Int. J. Sci. Res. Comput. Sci. Eng. Inf. Technol.* **5** 590–6
[21] Wadanka K V, Waghulde R Y and Taru U 2018 Chatbot : An Application of AI *Int. J. Res. Eng. Sci. Manag.* **1** 139–41
[22] Sucipto, Kusrini and Taufiq E L 2016 Classification method of multi-class on C4.5 algorithm for fish diseases *Proceeding - 2016 2nd International Conference on Science in Information Technology, ICSITech 2016: Information Science for Green Society and Environment* (Balikpapan: Institute of Electrical and Electronics Engineers Inc.) pp 5–9

## Acknowledgments

The author would like to thank the Directorate of Research and Community Service (DRPM) of the Indonesian Ministry of Research and Technology who has provided financial support for this research through a beginner lecturer research scheme (PDP).

# Identification of ElectroEncephaloGraph signals using sampling technique and K - nearest neighbor

H Hindarto[1*], A Muntasa[2], A Efiyanti[1]

[1,1]Universitas Muhammadiyah Sidoarjo, Indonesia
[2]Universitas Trunojoyo Madura, Indonesia

hindarto@umsida.ac.id*

**Abstract**. Electroencephalograph is a device that can help humans observe and analyze the results of electrical waves produced by neurons in the brain. The results of reading the tool are called Electroencephalogram (EEG), besides being able to help diagnose physician for medical therapy, it is also developed for the Brain-Computer Interfacing (BCI) application. BCI is a method that allows humans to be able to control an external system without direct contact with the system. Research on communication between humans to control external equipment has been widely investigated, including research on brain activity to control a cursor on a computer screen. This study focuses on feature extraction for ElectroEncephaloGraph (EEG) signals using the sampling technique. *K - Nearest Neighbor is used as a classification of EEG signals to determine whether the cursor moves up or down*. The data used are EEG data originating from the 2003 BCI competition (BCI 2003 Competition). Decision making is done to classify the cursor movement up and down the cursor movement. The research data uses 250 EEG signal file training data and 50 from EEG signal file testing data, so that the whole becomes 300 EEG signal data files. The best results with K = 3 values obtained for the classification of EEG signals using K-NN are 76% of the signal data tested.

## 1. Introduction

The activity of recording electricity in the brain for electro physiological monitoring is called Electroencephalograph (EEG). Electrode placement is located along the scalp. The EEG measures voltage fluctuations that result from ion currents in brain neurons. In the clinical context, EEG refers to the recording of spontaneous electrical activity of the brain over a period of time, as recorded from several electrodes placed on the scalp [1]. Diagnostic applications generally focus on the potential associated with EEG spectral content.

Placing electrodes on the scalp follows a predetermined system, namely the 10-20 system. The right and good electrode placement is the main requirement to get good and reliable EEG recordings. Besides that the cleanliness of the scalp, the condition of the electrode, the EEG machine and the compliance of the subjects during recording are also very influential to get good results. Hans Berger stated that the human brain has continuous electrical activity that can be recorded. Brain activity can be possible to send commands to electronic equipment with the help of the Brain Computer Interface (BCI) [2]. BCI can be used for mental activities such as thinking moving hands, thinking moving arms and others. This work will produce different EEG signals [3]. The results of these different EEG signals will produce different actions for information for BCI. An evaluation of EEG signals in conducting movement activities in recent years was conducted by researchers [4]. However, to improve the performance of EEG signals in detecting the characteristics of these signals is a big bustle for researchers. For the process of taking feature extraction and classification is an important role in recognizing EEG signals.

Many are used in finding features for EEG signals, including using the Fast Fourier Transform (FFT) method [5][6]. Taking feature extraction with the Fast Fourier Transform by taking the value from the spectral using the Welch method. The weakness of this FFT method is that it only takes frequency information while time information is not taken. Research on how to combine information from frequency and time has been investigated in this study [7]. One other study is to take the Autoregressive (AR) method [8]. In addition, in taking the features used by using the AR or multivariate autoregressive (MVAR) model [9]. Wang et al also analyzed the combined time-frequency for EEG signal components [10]. The method to find the features of an EEG signal is to use Wavelet transforms by utilizing the wavelet transform coefficient [11]. In research using wavelet transform mechanisms for the production of EEG signals is rather complicated, making it difficult to obtain accurate transcendent information. Other research by taking signal features [12]. Using the Simple Random Sampling (SRS) method for extracting EEG signal features of epilepsy patients has been carried out in this study and the results of the LS-SVM classification obtained were 80.31% for training data and 80.05% for testing data [13].

Based on the above background, this research is designed as follows, the second part explains the EEG signal materials and methods used in the search for EEG features, the third part explains the results of feature extraction and the classification process of EEG signals, and the fourth section describes the conclusions of this study.

## 2. Materials And Methods

### *2.1. material*

This study takes data from the 2003 BCI competition uploaded by Dr. Birbaumer and his team at Tuebingen University, Germany (Blankertz 2004) [17]. The recording process is taken from a healthy subject by attaching 6 electrode sensors that are attached to the scalp. Recorded using 256 Hz sampling frequency and recording time in a period of 3.5 seconds. The process of recording the subject is asked to imagine the movement of the curve upwards and move the cursor down on a monitor screen. Subjects in the recording process receive visual feedback from SCP. This experiment consisted of 268 training data and 293 trial data.

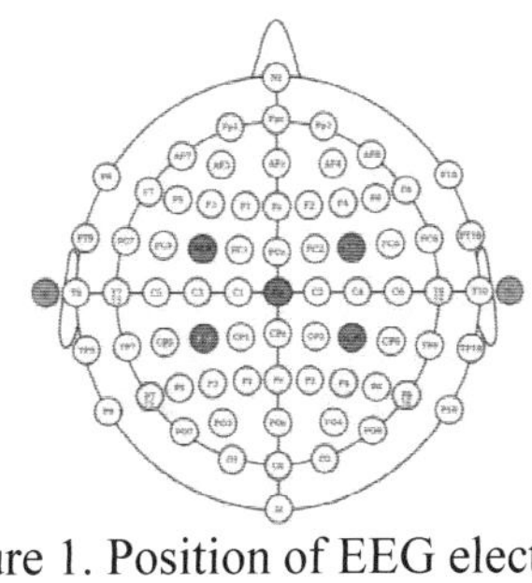

Figure 1. Position of EEG electrodes

Figure 1 shows the position of the electrode attached to the head. There are 6 sensors used, namely Cz for channel 1, A2-Cz for channel 2, FC3 for channel 3, CP3 for channel 4, FC4 for channel 5 and PC4 for channel 6 [17].

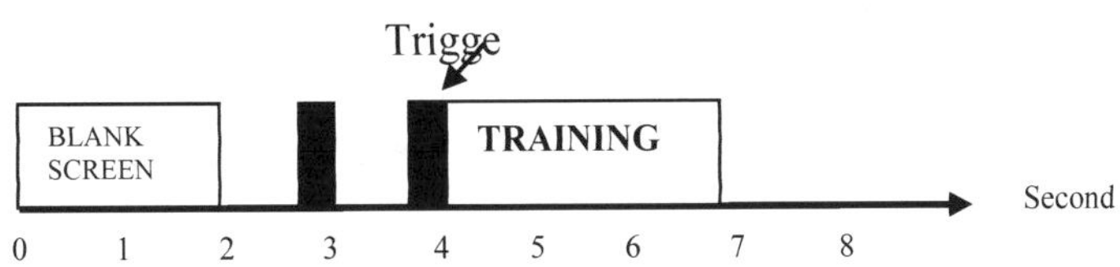


Figure 2. Timing pattern

This research takes one channel from the 6 channels installed, namely channel 3, each channel has two different classes. The data used for each class were 688 data for training and 293 data for trial data. The data used in this study are 250 training data and 50 trial data .

*2.2. Sampling technique*

In this study extraction The characteristics used are the Sampling Technique method. The EEG signal is divided into four sub-EEG signals. Of the four sub-EEG signals determined the maximum value, the minimum value, the average value and the standard deviation value to be extracted from the characteristics of the EEG signal as shown in figure 3.

.

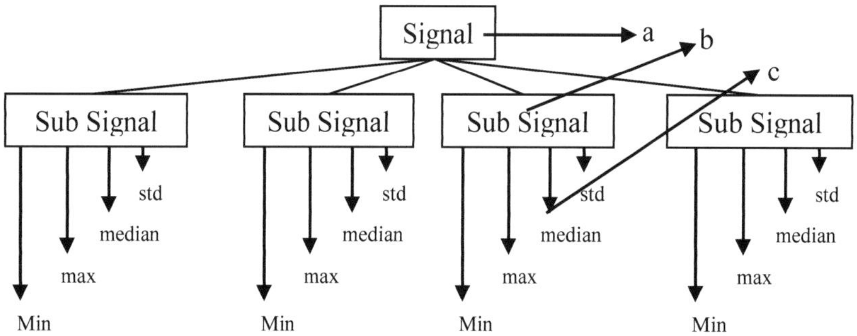


Figure 3. a. Signal sample, b. sub signal samples and c. Feature Selection

There are 300 EEG signal data files taken in this study. One EEG signal file data has 896 data points. One EEG signal is divided into four sub-signals. So that each sub signal has 214 points of data. The EEG sub signal data is searched for the minimum value, maximum value, average value and standard deviation value. There are 20 points of data obtained from 4 x 4 sub-signals.

*2.3 K-Nearest Neighbor Classification*

K-NN is a simple machine learning algorithm. This is only based on the idea that an object that is 'close' to each other will also have similar characteristics. This means that if we know the characteristics of one object, we can also predict other objects based on their closest neighbors. K-NN is an advanced improvisation of the Nearest Neighbor classification technique. This is based on the idea that each new example can be classified by the majority vote of a neighbor k, where k is a positive integer, and usually with a small number [14]. The K-NN classification algorithm predicts the sample test category according to the training sample k which is the closest neighbor to the test sample, and enters into the category that has the largest probability category [15].

In pattern recognition, the KNN algorithm is a method used to classify objects based on the closest training example in the feature space. KNN is a type of insurance based learning, or lazy learning where this function is only approached locally and all calculations are deferred to classification [16]. The K-NN classification method has several stages, the first being the k value which is the number of closest neighbors that will determine which new query goes to which class is determined. The second stage, k the nearest neighbor is searched by calculating the distance of the query point with the training point. The third stage, after knowing the distance of each training point with the query point, then see the smallest value. The fourth stage takes the smallest value, then see the class. The class that is the most is the class of the new query. Near or far points with neighbors can be calculated using the Euclidean distance. *Euclidean distance is represented as follows :*

$$j(a,b) = \sqrt{\sum_{k=1}^{k_n}(a_k - b_k)^2} \tag{1}$$

*J (a, b) is the distance between point a which is the point known to its class and b is a new point.* The distance between the new point and the training points is calculated and taken by the nearest point. New points are predicted to enter the class with the most classifications of these points.

## 3. Results And Discussion

Results of experiments conducted by researchers by taking data from the 2003 BCI competition data set Ia. First of the 6 channel electrodes / channels are taken just one channel in the sampling process. One signal from a set of signals that are joined in one channel is divided into 4 sub-signals, then from each sub-signal the search for the maximum value, minimum value, average value and standard deviation values are searched. This value will be used as a feature extraction for the identification. process

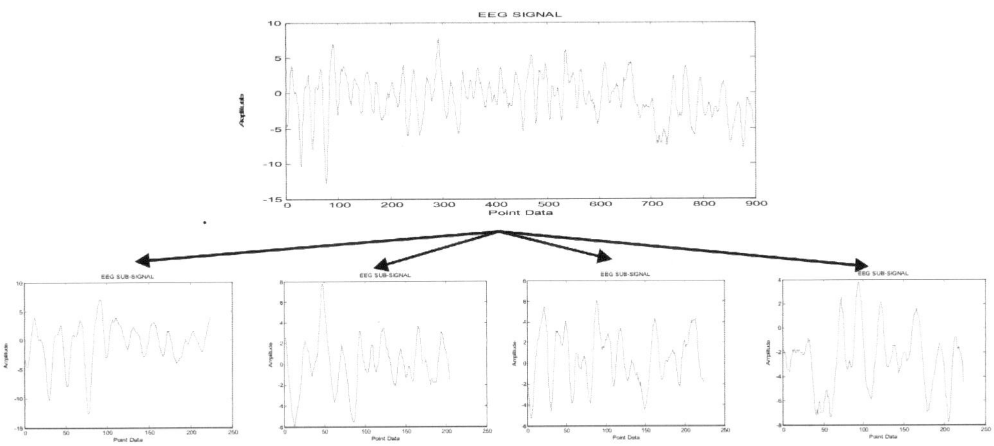


Figure 4. Results of the Sampling Technique process on EEG signals

Figure 4 is an EEG recording and the division process into four sub-signals. The four sub-signals are taken the maximum value, the minimum value, the average value and the standard deviation value for each sub-signal.

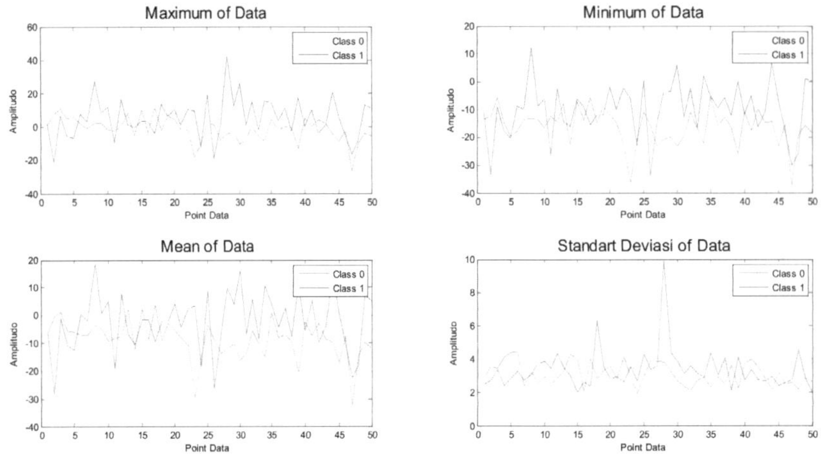


Figure 5. Comparison of class 0 and class 1 to the maximum value, minimum value, average value and standard deviation value of each sub signal

In Figure 5 shows that for values of maximum values, minimum values, average values and standard deviation values for each sub signal. Each signal for the cursor movement up and down the cursor has a difference value. With differences in values that are not the same, it shows that the level of classification by taking the maximum value, minimum value, average value and standard deviation value is quite good. Classification using K-NN is implemented using the maximum value feature, minimum value, average value and standard deviation value from the sampling technique as input. In this study, the training set numbered 250 EEG signal data and data of test is 50 EEG signal data. In table 1 is the distribution of sample classes in the training and data of test collection. Data obtained from different subjects in the training process is a way to improve the ability of K-NN. To train K-NN, use training data sets, while to verify the accuracy and effectiveness of K-NN test data to detect cursor movements up and down.

Table 1. Sampling of class distribution on training data and data of test

| Class | Training set | Test set |
|---|---|---|
| Up cursor | 125 | 50 |
| Down cursor | 125 | |

The test will be carried out with a variation of the K value in the KNN system and variations in training data and test data used. The accuracy in question is the accuracy of the classification results carried out by the program with the results of manual classification. The level of accuracy can be formulated with equation (2).

$$Level\ of\ accuracy = \frac{The\ number\ of\ EEG\ signals\ is\ correctly\ classified}{Total\ amount\ of\ test\ data}\ x\ 100\% \qquad (2)$$

In this test variations in the number of K are performed on the KNN function. k is the number of closest neighbors. The K values tested are one, three, five, seven, and nine. Odd values are chosen to avoid similarities in proximity to two different classes. Because KNN will classify based on the most class voting. From this test the average accuracy value for each k value is determined.
Classification Testing with k Value = 1

The classification results with K = 1 are shown in Table 2. The test results with K = 1 obtained an average accuracy of KNN classification of 54.4%. With the lowest accuracy 48% and the highest accuracy 64%.

Table 2. Results of testing k = 1 classification

| Test | Amount of training data | Amount of test data | Right | wrong | accuracy |
|---|---|---|---|---|---|
| 1 | 50 | 50 | 13 | 12 | 52% |
| 2 | 50 | 50 | 13 | 11 | 52% |
| 3 | 50 | 50 | 14 | 11 | 56% |
| 4 | 50 | 50 | 16 | 9 | 64% |
| 5 | 50 | 50 | 12 | 13 | 48% |
| Average | | | | | 54,4% |

Classification Test with k = 3

The classification results with K = 3 values are shown in Table 3. The test results with K = 3 obtained an average accuracy of 64%. The highest level of accuracy is found in the fifth test of 76%. The lowest level of accuracy is found in the third test of 56%.

Table 3. Results of testing k = 3 classification

| Test | Amount of training data | Amount of test data | Right | wrong | accuracy |
|---|---|---|---|---|---|
| 1 | 50 | 50 | 14 | 11 | 56% |
| 2 | 50 | 50 | 14 | 11 | 56% |
| 3 | 50 | 50 | 16 | 9 | 64% |
| 4 | 50 | 50 | 19 | 6 | 76% |
| 5 | 50 | 50 | 17 | 8 | 68% |
| Average | | | | | 64% |

Classification Test with k = 5

The classification results with K = 3 values are shown in Table 4. The test results with K=5 obtained an average accuracy of 54,4%. The highest level of accuracy is found in the fifth test of 64%. The lowest level of accuracy is found in the first and second test of 44%.

Table 4. Results of testing k = 5 classification

| Test | Amount of training data | Amount of test data | Right | wrong | accuracy |
|---|---|---|---|---|---|
| 1 | 50 | 50 | 11 | 14 | 44% |
| 2 | 50 | 50 | 11 | 14 | 44% |
| 3 | 50 | 50 | 15 | 10 | 60% |
| 4 | 50 | 50 | 16 | 9 | 64% |
| 5 | 50 | 50 | 15 | 10 | 60% |
| Average | | | | | 54,4% |

Classification Test with k = 7

The classification results with k = 7 are shown in Table 5. Testing with k = 7 produces an average accuracy of 60.8%. The highest accuracy results are found in the fifth test with a value of 68% accuracy. The lowest accuracy of 52% is in the third test.

Table 5. Results of testing k = 7 classification

| Test | Amount of training data | Amount of test data | Right | wrong | accuracy |
|---|---|---|---|---|---|
| 1 | 50 | 50 | 15 | 14 | 60% |
| 2 | 50 | 50 | 15 | 14 | 60% |
| 3 | 50 | 50 | 13 | 10 | 52% |
| 4 | 50 | 50 | 16 | 9 | 64% |
| 5 | 50 | 50 | 17 | 10 | 68% |
| Average | | | | | 60,8% |

Classification Test with k = 9

The results of classification testing with k = 9 values are shown in Table 6. Classification results with a value of k = 9 indicate an average accuracy value of ten testing times of 62.4% Of the five tests there were three tests which produced 64% accuracy, namely in the second, fourth and fifth tests. The lowest test value of 60% is found in the first and third tests.

Table 6. Results of testing k = 9 classification

| Test | Amount of training data | Amount of test data | Right | wrong | accuracy |
|---|---|---|---|---|---|
| 1 | 50 | 50 | 15 | 14 | 60% |
| 2 | 50 | 50 | 16 | 14 | 64% |
| 3 | 50 | 50 | 15 | 10 | 60% |
| 4 | 50 | 50 | 16 | 9 | 64% |
| 5 | 50 | 50 | 16 | 10 | 64% |
| Average | | | | | 62,4% |

Comparison of Average Accuracy Values

All the tests with variations in the K value the highest accuracy value was found in the value of k = 3 with an average level of accuracy of 64%. The lowest accuracy value is found in tests with k = 1 and k = 5 with an average level of accuracy of 54.4%. Overall the value of accuracy has a value close to 60%. Classification of KNN with k = 1 is very susceptible to noise this results in a low level of accuracy. With the above description, the value of k = 3 is the most optimal k value.

## 4. Conclusion

This study can be concluded that the proposed KNN classification method can classify EEG signals with an accuracy of 76% at the best neighbour of value k = 3. The feature extracted for EEG signal classification uses the sampling technique.

## 5. Acknowledgement

The author thanks the Chairman of the Universitas Muhammadiyah Sidoarjo for providing time for research that the researchers are working on and the Directorate of Research and Community Service, Directorate General of Research Strengthening, and Development of the Republic of Indonesia Ministry of Research, Technology and Higher Education which supports funding for this research.

## References

[1] Lux H D 1981 Generalized Tonic-Clonic Activity 24–8
[2] Wolpaw J R, Birbaumer N, Heetderks W J, McFarland D J, Peckham P H, Schalk G, Donchin E, Quatrano L A, Robinson C J and Vaughan T M 2000 Brain-computer interface technology: A review of the first international meeting *IEEE Trans. Rehabil. Eng.* **8** 164–73

[3] Termal K, Menggunakan I and Neighbor K 2018 No Title
[4] Millán J 2003 Adaptive brain interfaces for communication and control *Proc. 10th Int. Conf. ...* **113** 767–91
[5] ABDUL-BAREY RAOUF SULEIMAN T A-H F F e a t u r e s E x t r a c t i o n T e c h n i q u e s o f E E G S i g n a l s f o r B C I A p p l i c a t i o n s T o k a A b d u l h a m e e d F a t e h i A b d u l - B a r r y R a o u f S u l e i m a n
[6] Jaswal R 2016 Brain Wave Classification and Feature Extraction of EEG Signal by Using FFT on Lab View *Int. Res. J. Eng. Technol.* 1208–12
[7] Mensh B D, Werfel J and Seung H S 2004 BCI competition 2003 - Data set Ia: Combining gamma-band power with slow cortical potentials to improve single-trial classification of electrolencephallographic signals *IEEE Trans. Biomed. Eng.* **51** 1052–6
[8] McFarland D. J, Miner L. A, Vaughan T. M and Wolpaw J. R 2000 Mu and Beta Rhythm Topographies During Motor Imagery and Actual Movements *Brain Topogr.* **12** 177–86
[9] Stolz E 1995 Multivariate autoregressive models for the classification of spontaneous electroencephalogram during mental tasks
[10] Wang T, Deng J and He B 2004 Classifying EEG-based motor imagery tasks by means of time-frequency synthesized spatial patterns *Clin. Neurophysiol.* **115** 2744–53
[11] Adeli H, Zhou Z and Dadmehr N 2003 Analysis of EEG records in an epileptic patient using wavelet transform *J. Neurosci. Methods* **123** 69–87
[12] Der M D Van 1995 *Signal sampling techniques for data acquisition in process control Signal Sampling Techniques for Data Acquisition in Process Control Marten Derk van der Laan*
[13] Li Y and Wen P 2009 Classification of EEG Signals Using Sampling Techniques and Least Square Support Vector Machines Classification of EEG Signals Using Sampling Techniques and Least Square Support Vector Machines
[14] Koutroumbas K and Kalouptsidis N 2002 Nearest neighbor pattern classification neural networks 2911–5
[15] Suguna N and Thanushkodi K 2010 An Improved k-Nearest Neighbor Classification Using Genetic Algorithm **7** 7–10
[16] Imandoust S B and Bolandraftar M 2017 Application of K-Nearest Neighbor ( KNN ) Approach for Predicting Economic Events : Theoretical Background

Journal of Physics: Conference Series **1381** (2019) 012010 doi:10.1088/1742-6596/1381/1/012010

# Structural Performance Monitoring System Using Android RViSITS For High Rise Building

**W Riyanto*, D Irawan, T J W Adi, D Iranata, A Rizki Amalia[5,e)]**

Department of Civil Engineering, Faculty of Civil, Environmental and Geo Engineering, Institut Teknologi Sepuluh Nopember, Surabaya, Indonesia

wahyuriyanto303@gmail.com

**Abstract.** In 2017, Indonesia's National Earthquake Study Center (PuSGeN) issued an updated Indonesian sources and potentials of earthquake map which shows the possibility of new earthquake hazards potentials. This triggers the assessment of building's vulnerability towards that new earthquake hazard potential. The assessment of building vulnerability towards earthquakes is divided into two types; quick check using Rapid Visual Screening (RVS) and thorough check. In its development, RVS method was chosen because it does not require a lot of personnel, reduce time for survey, and requires lower research costs. However, RVS method that uses manual form is considered less effective for buildings with a large numbers. Nevertheless, with the existence of RViSITS android application which has developed by ITS's RVS Team based on FEMA 154, the assessment of building vulnerability and the result can be done and accessed by all sectors quickly and easily. This article aimed to assess the vulnerability level of three high-rise buildings (8-15 floors) in East Java using RViSITS android application by comparing the result of numerical analysis based on earthquake regulations applicable in Indonesia according to SNI 1726:2012. The result of analysis using RViSITS showed that the level of vulnerability of three high-rise buildings studied is "qualified" or earthquake safe. Based on numerical analysis, it shows that the control result of base shear, mass participations, structure periods and driff control are appropriate as required by SNI 1726:2012. Generally, RViSITS can be used to map the vulnerability of high-rise buildings throughout the region in Indonesia.

## 1. Introduction

An earthquake is a threat to human being. It can cause injury or death for victims as well as being a threat to the destruction of standing buildings [1]. In reality, no one can prevent the happening of earthquakes but the impact of an earthquake disaster can be minimized through the right steps and methods [2]. There are many buildings in Indonesia are not designed considering to the existing earthquake regulations. In addition, each buildings is not equipped with data about the vulnerability which is cause a lot of damage to buildings during earthquake happens [3]. In 2007, Indonesia's National Earthquake Study Center (PuSGen) has issued an update on seismic maps. It has led to the emergence of sources and potential for new earthquake hazards in Indonesia. One of them is the emergence of Surabaya and Waru faults that cross Surabaya city [4]. Thus, it is necessary to make an early effort to minimize the impact of earthquake. One step to anticipate the impact of earthquake is to analyze the vulnerability of a building.

There are two types for examination of building's vulnerability towards earthquake namely a thorough check and quick check using Rapid Visual Screening (RVS). Some things that must be considered on a thorough check are involve specialist in the field of structure, takes longer time and also more expensive cost. However, quick check using RVS does not involve a lot of personnel in research, saving time and relatively cheap [3,5]. In examining building vulnerability, researchers have combined several methods such as visual examination, probability and numerical analysis [1,3,5–12].

In its development, RVS method is widely applied to assess building vulnerability since its application is quite simple and fast [3,5]. RVS method is most popular method used because it works fast, does not require a lot of personnel or structural expert, and cheaper cost. The use of RVS is intended to determine two possibilities of a building that is ensuring the level of building vulnerability in a safe condition or as a recommendation for the percentage of building vulnerability to be followed up further [13].

However, RVS method is considered less effective and efficient for buildings with large numbers. Since, its implementation requires a long time and a lot of time also needed to process all field data from surveyors into digital data in the database, data are still vulnerable to errors in verification or calculation phase. In addition, if there are changes or errors in some of field data, surveyors have to change with new form because the old form cannot be used for several surveys [14]. Furthermore, with RViSITS android application developed by RVS Team of Institut Teknologi Sepuluh Nopember, Surabaya, Indonesia based on FEMA 154, the building vulnerability assessment can be done simpler and faster since the surveyor only need a smartphone without using manual form as many buildings as surveyed. Besides, in this application equipped with features connected to camera so that it is possible to document the object under study during the survey. This application is also connected to Global Positioning System (GPS) to determine the coordinates of building's location automatically, all field data from surveyors will be stored automatically after saving and uploading to the database, and the result can be accessed by public, government and private sector quickly and easily [3].

Therefore, the main purpose of this research is to implement RViSITS android application to assess the vulnerability towards three high-rise buildings (8-15 floors) in East Java. Then, it is compare to numerical analysis result based on earthquake regulations applied in Indonesia according to SNI 1726: 2012 using SAP 2000 which is use to verify the accuracy of the application.

## 2. Material and Method

### *2.1Rapid Visual Screening/RVS Method*

Quick check using rapid visual screening (RVS) method was initially developed by Applied Technology Council (ATC) in late 1980s to evaluate the percentage of building vulnerability due to potential earthquake [15]. RVS is one method to evaluate the safety of the building structures easily, quickly and cheaply due to an earthquake. This method can be completed in a fairly simple way by surrounding the building from the outside of the buildings and determining whether the structure requires further examination [3].

RVS method has been widely applied by several researchers to evaluate the vulnerability of seismic buildings in many places [1,3,5–12]. In this research, RVS method is applied on different types of buildings such as hospitals and various types of brick buildings, etc. Although RVS method can be done quickly, but there are some disadvantages including its implementation requires a long time and a lot of time also needed to process all field data from surveyors into digital data in the database, data are still vulnerable to errors in verification or calculation phase. In addition, if there are changes or errors in some of field data, surveyors have to change with new form because the old form cannot be used for several surveys, so RVS method with manual form is less effective and efficient [3]. In figure 1, it is a manual form that is commonly used to assess buildings vulnerability using RVS method.

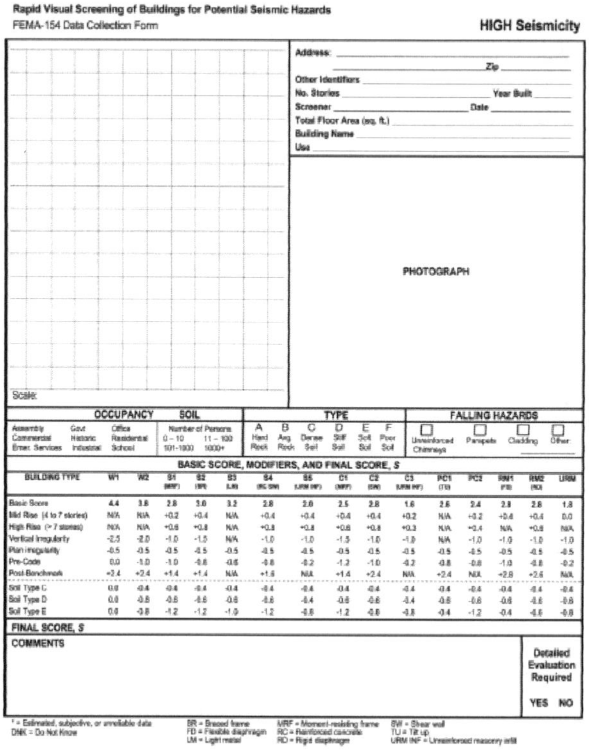

**Rapid Visual Screening of Buildings for Potential Seismic Hazards**
FEMA-154 Data Collection Form

**HIGH Seismicity**

Address: ____
____ Zip ____
Other Identifiers ____
No. Stories ____ Year Built ____
Screener ____ Date ____
Total Floor Area (sq. ft.) ____
Building Name ____
Use ____

PHOTOGRAPH

Scale:

| OCCUPANCY | | | SOIL | | TYPE | | | | | | FALLING HAZARDS | | | |
|---|---|---|---|---|---|---|---|---|---|---|---|---|---|---|
| Assembly | Govt | Office | Number of Persons | | A | B | C | D | E | F | ☐ | ☐ | ☐ | ☐ |
| Commercial | Historic | Residential | 0 – 10 | 11 – 100 | Hard Rock | Avg Rock | Dense Soil | Stiff Soil | Soft Soil | Poor Soil | Unreinforced Chimneys | Parapets | Cladding | Other: |
| Emer. Services | Industrial | School | 101-1000 | 1000+ | | | | | | | | | | |

**BASIC SCORE, MODIFIERS, AND FINAL SCORE, S**

| BUILDING TYPE | W1 | W2 | S1 (MRF) | S2 (BR) | S3 (LM) | S4 (RC SW) | S5 (URM INF) | C1 (MRF) | C2 (SW) | C3 (URM INF) | PC1 (TU) | PC2 | RM1 (FD) | RM2 (RD) | URM |
|---|---|---|---|---|---|---|---|---|---|---|---|---|---|---|---|
| Basic Score | 4.4 | 3.8 | 2.8 | 3.0 | 3.2 | 2.8 | 2.0 | 2.5 | 2.8 | 1.6 | 2.6 | 2.4 | 2.8 | 2.8 | 1.8 |
| Mid Rise (4 to 7 stories) | N/A | N/A | +0.2 | +0.4 | N/A | +0.4 | +0.4 | +0.4 | +0.4 | +0.2 | N/A | +0.2 | +0.4 | +0.4 | 0.0 |
| High Rise (> 7 stories) | N/A | N/A | +0.6 | +0.8 | N/A | +0.8 | +0.8 | +0.6 | +0.8 | +0.3 | N/A | +0.4 | N/A | +0.6 | N/A |
| Vertical Irregularity | -2.5 | -2.0 | -1.0 | -1.5 | N/A | -1.0 | -1.0 | -1.5 | -1.0 | -1.0 | N/A | -1.0 | -1.0 | -1.0 | -1.0 |
| Plan irregularity | -0.5 | -0.5 | -0.5 | -0.5 | -0.5 | -0.5 | -0.5 | -0.5 | -0.5 | -0.5 | -0.5 | -0.5 | -0.5 | -0.5 | -0.5 |
| Pre-Code | 0.0 | -1.0 | -1.0 | -0.8 | -0.6 | -0.8 | -0.2 | -1.2 | -1.0 | -0.2 | -0.8 | -0.8 | -1.0 | -0.8 | -0.2 |
| Post-Benchmark | +2.4 | +2.4 | +1.4 | +1.4 | N/A | +1.6 | N/A | +1.4 | +2.4 | N/A | +2.4 | N/A | +2.8 | +2.6 | N/A |
| Soil Type C | 0.0 | -0.4 | -0.4 | -0.4 | -0.4 | -0.4 | -0.4 | -0.4 | -0.4 | -0.4 | -0.4 | -0.4 | -0.4 | -0.4 | -0.4 |
| Soil Type D | 0.0 | -0.8 | -0.6 | -0.6 | -0.6 | -0.6 | -0.4 | -0.6 | -0.6 | -0.4 | -0.6 | -0.6 | -0.6 | -0.6 | -0.8 |
| Soil Type E | 0.0 | -0.8 | -1.2 | -1.2 | -1.0 | -1.2 | -0.8 | -1.2 | -0.8 | -0.8 | -0.4 | -1.2 | -0.4 | -0.6 | -0.8 |
| FINAL SCORE, S | | | | | | | | | | | | | | | |

COMMENTS

Detailed Evaluation Required
YES NO

* = Estimated, subjective, or unreliable data
DNK = Do Not Know
BR = Braced frame
FD = Flexible diaphragm
LM = Light metal
MRF = Moment-resisting frame
RC = Reinforced concrete
RD = Rigid diaphragm
SW = Shear wall
TU = Tilt up
URM INF = Unreinforced masonry infill

**Figure 1.** Manual Form of RVS Method According to FEMA 154

*2.2RViSITS Method*

RViSITS is an android application developed based on FEMA 154. This application serves to simplify the process of checking building vulnerabilities sent via the internet. Then, there is a server as the center for collecting and processing the data. From this data, it is possible to map vulnerability and can be used to detect affected areas which can be accessed by various sectors such as government, private sectors, and public. The development of RViSITS android application is made by adapting to the rules at FEMA 154 with conditions and rules that have been adjusted to existing regulations in Indonesia [3].

The step that must be considered before conducting research is to prepare in advance the existing data of the building to be reviewed. Data needed include years of building construction, address, number of floors, building function, soil data to determine the type of soil at the location being reviewed, and asbuild drawing as a reference for numerical modeling [3,14].

Afterward, there are 8 further steps that will be conducted after obtaining the existing data [3,15]: 1) update the data and information regarding with the considerate building based on the existing data obtained and take picture of the building; 2) walk around the building for taking picture and observe the building condition; 3) determine the building function and the floor number; 4) determine the land type; 5) determine the type of non structural elements which has potential to fall down by the earthquake; 6) identify the building type in B score and choose the appropriate one; 7) identify and choose by click on the list menu in M score in the RViSITS (e.g: the time of building set up, number of floor, and land type); 8) the final score on the S score are obtained based on B score, M score, and being modify as stated in the step 7, thus, the last result will appear automatically in RViSITS program. However, if the deep evaluation is needed, the menu “insert comment” can be applied.

*2.3Research Method*

The primary aim of this research is to evaluate susceptible building structure against three high rise building. These three building has 8th up to 15th floors in the zone D and E using the RViSITS android program. Further, this research is divided into three steps. The first step is collecting the data and conducting preliminary study deal with the building evaluation which has potential to fall cause by earthquake. The preliminary study can be described as collecting references as well as relevant regulation used in this research. Then, the second step is surveying the location, collecting others existing data as like land, shop drawing data and information relate to building. Moreover, the third step is conducting evaluation using RViSITS android program. Then, result verification numerically is obtained using SAP 2000 to control the performance of these three high rise buildings according to SNI 1726:2012. At last, numerically result analysis using SAP 2000 being compared to evaluate the RViSITS's accurate performance [3].
The research steps using RViSITS android program can be seen in the figure 2.

(a) (b) (c) (d)

(e) (f) (g) (h)

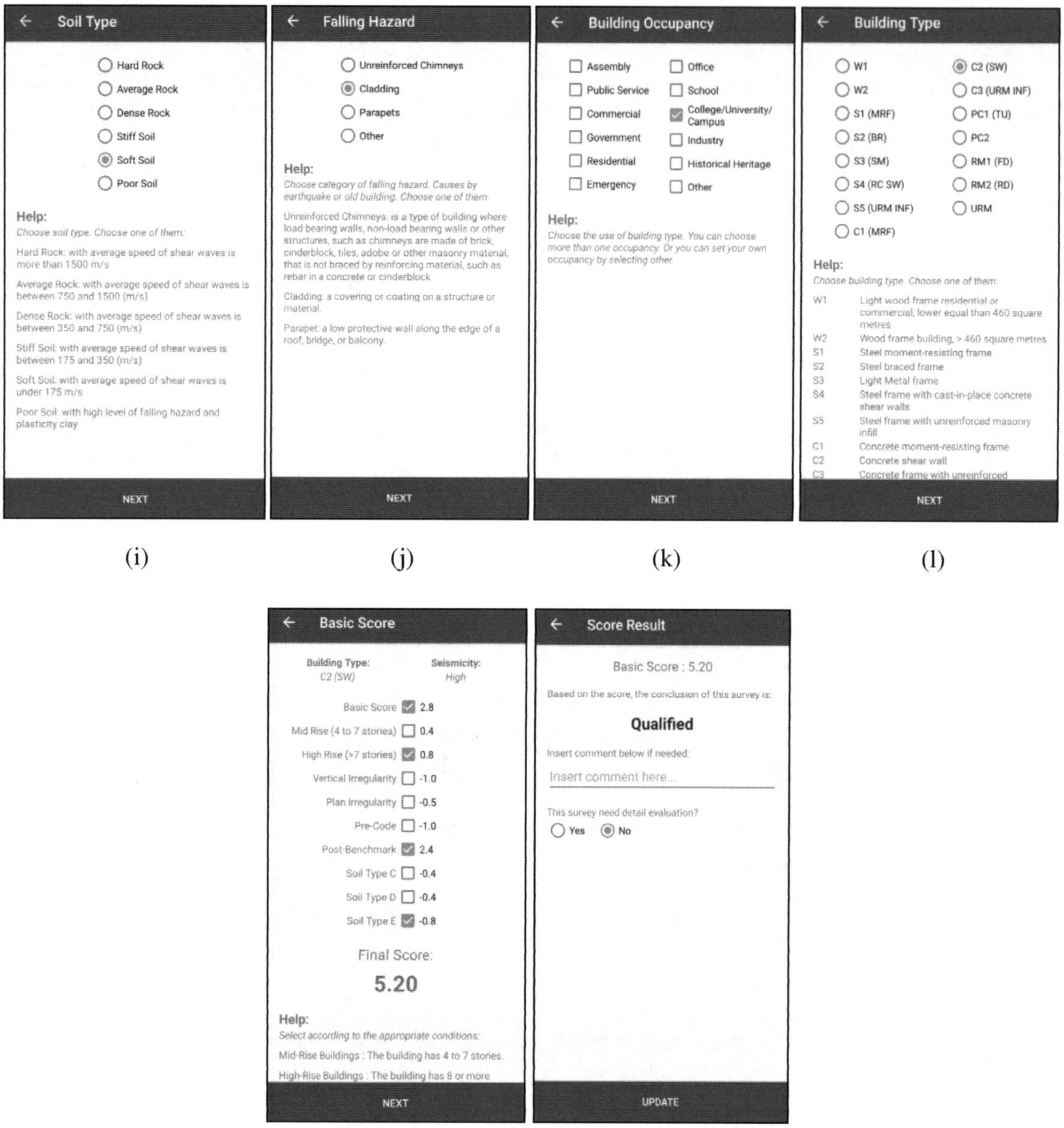


**Figure 2.** The Steps in Using RViSITS Androrid Program (a) to (n)

The research factor in FEMA 154 for every building structure is analyzed to obtain the most significant factors which has influenced on RVS score. There are three scores existed in RVS including basic score (B), modifiers score (M), and final score (S). Basic score (B) is determined based on building structure type and the earthquake potential in the considerate area. The kind of building based on the RVS method is divided into 15 kinds, namely [15,16]: 1) W1 (wood frame with acreage < 5000 $ft^2$); 2) W2 (wood frame with acreage > 5000 $ft^2$); 3) S1 (steel moment frames); 4) S2 (steel bracing frame); 5) S3 (light metal); 6) S4 (steel frame with concrete shear wall); 7) S5 (steel frame with unreinforced masonry infill); 8) C1 (Concrete moment-resisting frame); 9) C2 (concrete shear wall); 10) C3 (concrete frame with unreinforced masonry infill); 11) PC1 (tilt-up structure); 12) PC2 (precast concrete frame); 13) RM1 (Reinforced masonry buildings with flexible floor and roof diaphragms); 14) RM2 (Reinforced masonry buildings with rigid floor and roof diaphragms); 15) URM (Unreinforced Masonry).

Modifiers Score (M) is influenced by several factors such as the type of building (mid-rise building and high rise building), shape of building (vertical irregularity, and plan irregularity) pre-code, post-benchmark, and land type (type C, D, and E). The last score (S) is obtained based on the basic score (B) and modifiers score (M) chosen. Moreover, the last score usually range from 0 up to 7, in which the S score is higher, hence, the structure is better. Based on the recent seismic design criteria, final score (S) is allowed if the score is no more than 2. The building with the S score is 2 or less should be evaluate by structural expert in the earthquake planning [3,15].

## 3. Result and Discussion

*3.1Analysis Using RViSITS Android Program*

Based on the result analysis using RViSITS android is described in the figure 3.

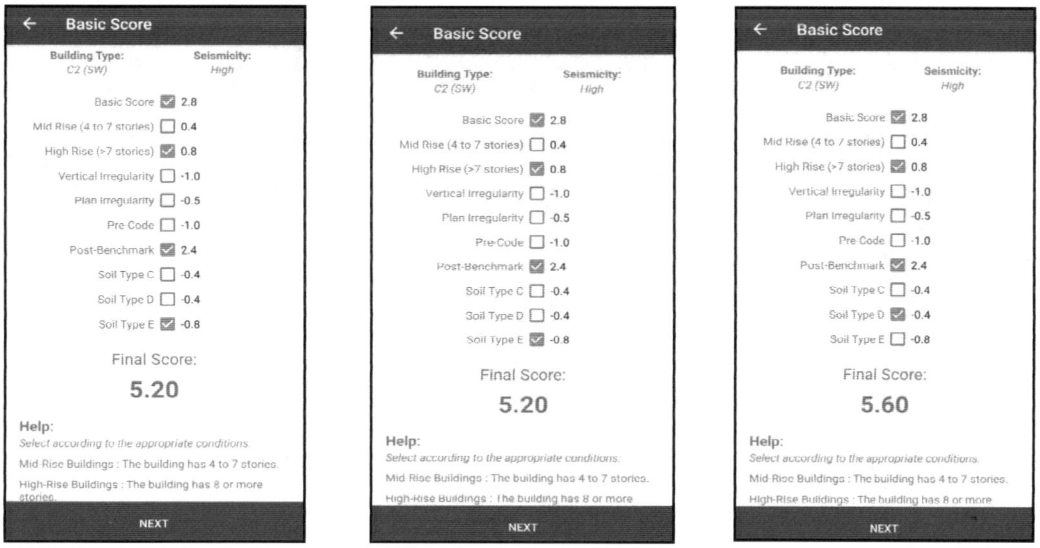


(a) (b) (c)

**Figure 3.** Output Basic Score of RViSITS Android Program a) Marine Science Faculty, Universitas Airlangga; b) At-Tauhid Building, Universitas Muhammadiyah Surabaya; c) Trauma Center & Intensive Care Building,dr. Soedono Hospital

From the figure 3, the basic score of each building shows that: (1) the score of Marine Science Faculty Building, Universitas Airlangga is 5.2; (2) the score of At-Tauhid building, Universitas Muhammadiyah Surabaya is 5.2; (3) the score of Trauma Center & Intensive Care Building, dr. Soedono Hospital, Madiun is 5.6. Based on the basic score, the three buildings have score higher than 2, hence, the susceptible of building considered to be safe against the earthquake pressure as regulated in FEMA 154 [3,15].

*3.2Numerical Analysis Using SAP 2000*

The numerical analysis is aimed to verify the analysis using RViSITS android program. The case study used involving three high rise building with the number of floor is $8^{th}$ up to $15^{th}$ in the East Java province. These three buildings are: 1) Marine Science Faculty Building, Universitas Airlangga; 2) At-Tauhid building Universitas Muhammadiyah Surabaya; 3) Trauma Center & Intensive Care Building, dr. Soedono Hospital, Madiun.

The analysis of the three buildings was obtained based on numerical analysis using SAP 2000. This analysis was used to control whether the building structure had fulfilled the earthquake regulation requirements as stated in SNI 1726:2012, and also fulfill the requirement of reinforced concrete building's structure planning as stated in SNI 2874:2013. Figure 4 presents the model of three considerate buildings using SAP 2000. Base shear control from the three buildings have score higher than 85% which means that the building has fulfill basic sliding control. In the mass participation control, the result has fulfilled the requirement given by SNI 1726:2012, in which the score of these buildings are more than 90%. Afterward, period control structure from the three buildings has exceeded the permission period. Then, in driff control, the deviation of x and y of the three buildings has not exceeded the permission limit. The three buildings categorize as new building which is firstly set up in 2003 namely Marine Science Faculty, Universitas Airlangga. Therefore, the building planning structure has considered the earthquake regulation in Indonesia. It can be seen according to the several conducted control (base shear, mass participation, structure and driff period). However, the three building are considered to be safe as required by SNI 1726:2012.

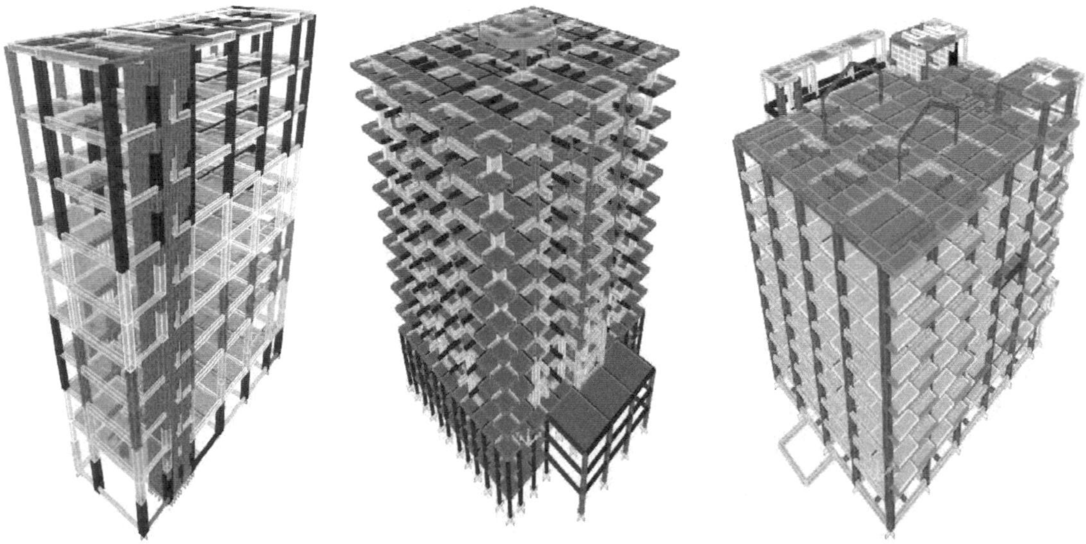

(a) (b) (c)

**Figure 4.** Numerical Modelling Using SAP 2000 a) Marine Science Faculty Building, Universitas Airlangga; b) At-Tauhid Building, Universitas Muhammadiyah Surabaya; c) Trauma Center & Intensive Care Building, dr. Soedono Hospital.

The result of numerical analysis according to SNI 1726:2012 compared with RViSITS android program analysis are present in the table 1.

**Table 1.** The Comparison Between RViSITS Analysis and Numerical Analysis as Required by SNI 1726:2012

| Name of Building | RViSITS Score | Final Score | SNI 1726:2012 |
|---|---|---|---|
| Marine Science Faculty Building | 5.2 > 2 | Qualified | Qualified |
| AT-Tauhid Building | 5.2 > 2 | Qualified | Qualified |
| Trauma Center & Intensive Care | 5.6 > 2 | Qualified | Qualified |

## 4. Conclusions

Based on the analysis that has been adjusting to the renewal of Indonesian earthquake source map 2017, the three buildings including 1) Marine Science Faculty Building, Universitas Airlangga; 2) At-Tauhid Building, Universitas Muhammadiyah Surabaya, and 3) Trauma Center building, dr. Soedono Hospital are categorized as high earthquake zone based on FEMA 154. The final score in the RViSITS android program shows that the index higher than 2, while based on numerical analysis using SAP 2000 is obtained the base shear control, mass participation, structure period, and driff control has fulfill the requirement of SNI 1726:2012. Hence, the three buildings are qualified as safe building from earthquake. In this study, it should be known that Marine Science Faculty, Airlangga University is the oldest building which set up in 2013. Meanwhile, another two buildings are being set up in 2016. However, these three building still considered as new buildings so that the planning of building structure has considered earthquake regulation in Indonesia. Furthermore, the result analysis using RViSITS shows the same result with numerical analysis. Generally, RViSITS android program can be used to map the vulnerability of high rise buildings throughout the region in Indonesia.

## References

[1] Perrone D, Aiello M A, Pecce M and Rossi F 2015 Rapid visual screening for seismic evaluation of RC hospital buildings *Structures* **3** 57–70

[2] Ningthoujam M C and Nanda R P 2018 Rapid visual screening procedure of existing building based on statistical analysis *Int. J. Disaster Risk Reduct.* **28** 720–30

[3] Wahyuni E, Iranata D, Suswanto B, Nurcahyo C B and Sutrisno W 2018 Assessment of vulnerable buildings due to earthquake loading using Rapid Visual Screening smartphone application *Int. J. Adv. Sci. Eng. Inf. Technol.* **8** 567–72

[4] Pusat Studi Gempa Nasional 2017 *Peta Sumber dan Bahaya Gempa Indonesia Tahun 2017* (Jakarta, ID)

[5] Yadollahi M, Adnan A and Zin R 2012 Seismic vulnerability functional method for rapid visual screening of existing buildings *Arch. Civ. Eng.* **58** 363–77

[6] Achs G and Adam C 2012 Rapid seismic evaluation of historic brick-masonry buildings in Vienna (Austria) based on visual screening *Bull. Earthq. Eng.* **10** 1833–56

[7] Achs G and Adam C 2012 A Rapid-Visual-Screening Methodology for the Seismic Vulnerability Assessment of Historic Brick-Masonry Buildings in Vienna *15th World Conf. Earthq. Eng.* **154**

[8] Sarmah T and Das S 2018 Earthquake Vulnerability Assessment for RCC Buildings of Guwahati City using Rapid Visual Screening *Procedia Eng.* **212** 214–21

[9] Hancilar U, Çaktö E, Erdik M, Franco G E and Deodatis G 2014 Earthquake vulnerability of school buildings: Probabilistic structural fragility analyses *Soil Dyn. Earthq. Eng.* **67** 169–78

[10] Srikanth T, Kumar R P, Singh A P, Rastogi B K and Kumar S 2010 Earthquake vulnerability assessment of existing buildings in Gandhidham and Adipur Cities Kachchh, Gujarat (India) *Eur. J. Sci. Res.* **41** 336–53

[11] Formisano A and Marzo A 2017 Simplified and refined methods for seismic vulnerability assessment and retrofitting of an Italian cultural heritage masonry building *Comput. Struct.* **180** 13–26

[12] Karapetrou S T, Fotopoulou S D and Pitilakis K D 2017 Seismic Vulnerability of RC Buildings under the Effect of Aging *Procedia Environ. Sci.* **38** 461–8

[13] Sadat M R, Huq M S and Ansary M A 2010 Seismic vulnerability assessment of buildings of Dhaka city *J. Civ. Eng.* **38** 159–72

[14] RVS Research Team ITS 2016 *Training Module - Rapid visual screening of Building using Android Based Apps. RViSITS* ed R R T ITS (Surabaya, ID)

[15] Rojahn C, Scawthorn C, Anagnos T, Wong K and Poland C D 2002 *Rapid visual screening of buildings for potential seismic hazards, Handbook, FEMA 154* (Washington, DC: Applied Technology Council)

[16] Mohamad I I, Yunus M Z M and Harith N S H 2019 Assessment of building vulnerability by integrating rapid visual screening and geographic information system : A case study of Ranau township *IOP Conf. Series: Materials Science and Engineering* (IOP Publishing) pp 1–8

Journal of Physics: Conference Series **1381** (2019) 012011 doi:10.1088/1742-6596/1381/1/012011

# Study of Fabric Movement and Influenced of Drape Measurement Based on Image Processing Using MATLAB

D Sukendar, T Totong, V G V Putra

Politeknik STTT, Bandung, Indonesia Jawa Barat, Indonesia. 40272
*dkendar@stttekstil.ac.id

**Abstract.** Generally, Drape meter for assessment of fabric drape is used in textile industry. In this paper, Image processing technique for measuring drape profile of apparel fabric has been studied and developed using MATLAB as well as modeled the fabric movement influenced of the fabric strength fort a certain mass. A modified method based on conventional drape meter was researched to determine drape parameter. Drape coefficient was selected to compare the conventional result (a standard device used in industry based on SNI ISO 08- 1511- 2004) and image processing result. In this research, a very good correlation has been found both conventional and image processing technique.

## 1. Introduction

In textile industry, yarn properties and fabric characteristics are some interested topics researched and studied by some researchers both experimentally and theoretically [1—11]. In textile, specially in fabric characteristic, drape is an important factor affecting the dynamic functionality and aesthetic of fabric characteristic. According to Kenkare and Plumlee [9], drape can be well-defined as the degree to which a fabric can bend when it is permitted to hang under its weight and without the influence of external forces. Commonly, drape ability can be defined as a phenomena of fabric deformation which is caused by the influence of fabric's force or weight of fabric caused by the gravitation force, when the fabric is suspended down without the influence of external forces.

In the study performed by Lojen and Jevsnik[10], fabric drape is not an independent fabric property. It depends on fabric parameter such as structure, yarn type, fibre content and also its finishing treatment. The drape ability of a fabric can be tested objectively and subjectively. Subjective evaluation has a lot of weakness. Drape of fabric was tested subjectively by panel of judges. Some disagreements in the test result by the panel of judges resulted in development of quantitative methods for evaluating drape. Quantitative methods to measure the drape coefficient has been developed and studied by many researchers[9-10], as subjective evaluation by panel of judges depends on the person evaluating, whereas objective evaluation of a fabric can be tested by drape coefficient which is formulated and calculated as shown below:

$$DC = \frac{S_p - \pi l^2}{\pi R^2 - \pi l^2} 100\% = \frac{\pi l^2 sin^2\theta - \pi l^2}{\pi R^2 - \pi l^2} 100\% \qquad ...(1)$$

Where $DC$ is defined as drape coefficient; $S_p$ the area under drape sample ($S_p = \pi l^2 sin^2\theta$); $\theta$ the angle of fabric drape; $\pi l^2$ the area of horizontal disc; $\pi R^2$ the area of sample before draping. Using Eq. (1), it can be calculated drape coefficient of fabric. Drape coefficient is still used for explaining drape and valid as the primary attribute for explaining the extent to which a fabric deformation under its own weight.

In this research, we developed and designed by applying image processing technique and Eq.(1) for measuring drape coefficient. In this study, the area under drape sample can be calculated by using MATLAB as image processing software. In this process, an image of each drape sample was captured by the WebCam then MATLAB software as image processing software was used to measure the area under drape sample automatically based on our listing program. In this study we also derive the equation to determine the relationship of area of drape and also the fabric strength.

## 2 Measuring Drape Using Image Processing

Recently, some computer vision techniques for analyzing drape sample have been made and introduced by some researcher. In this research we used MATLAB to analyze the drape area for an arbitrary area of drape. Generally, it is difficult to measure an exact area of an arbitrary area (area of drape) as shown in **Figure.1.** According to Solomon[7], image processing can be used to measure an arbitrary area and it provides us to overcome the problem.

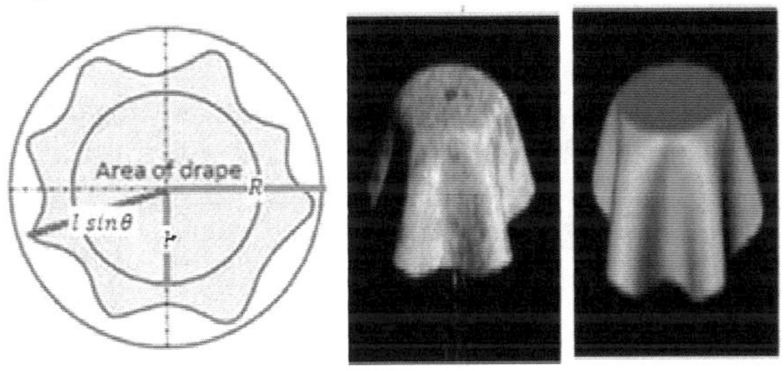


**Figure 1.** Arbitrary area (area under drape sample)

In this research, we have designed and developed a method to measure an arbitrary area as well as to determine the fabric strength. For first investigation of our aims to measure an arbitrary area, we compared a certain area (known area) to our result using image processing by MATLAB. The comparison of evaluation the certain area in pixel square both experimentally and image processing technique can be shown in **Table-1.** In this evaluation, we have made four images with difference area but having same shape.

**Table 1 −comparison of certain area in pixel square both experimental and image processing**

| No. Object | Area by image processing ($px^2$) | Area by experimental ($px^2$) |
|---|---|---|
| 1 | 1080 | 1074,2±10,29 |
| 2 | 1560 | 1589,62±12,52 |
| 3 | 2110 | 2205,06±14,74 |
| 4 | 2770 | 2826±16,69 |

**Table-1** presents the comparison of certain area both experimentally and image processing technique. In this result, the measured area, both experimentally and image processing technique, has a close relation of value. We also investigated the correlation both experimentally and image processing technique using regression test, **Figure.2** shows the program to measure the arbitrary area of the drape fabric.

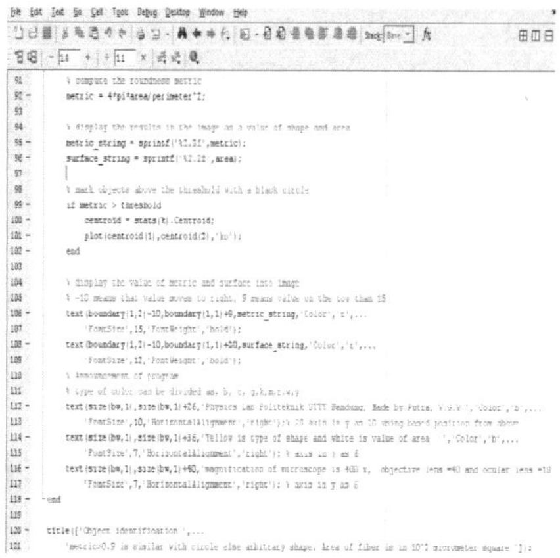

**Figure 2.** the program to measure the arbitrary area

Based on our first investigation measuring the arbitrary area by our program, we can use and implement our program to measure the drape with arbitrary area using image processing technique. In this method, a circular fabric is allowed to drape put on the circular disk with certain diameter (based on SNI ISO 08- 1511- 2004). In this modified process, the drape measurement (**Figure.3**) consists of two circular disks with different shape, a digital camera to capture the drape image of fabric and a computer to analyze the image and to measure drape coefficient. The camera was hung with a certain distance of height from the fabric and it captured the fabric image.

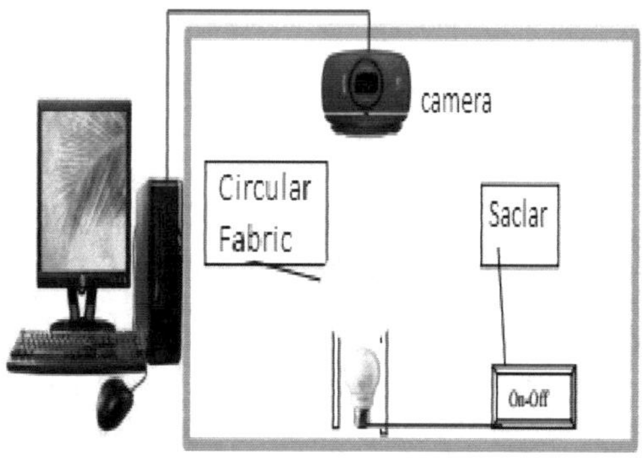


**Figure 3.** Drape measurement

In this research, we have tested some sample of fabrics (with same characteristic) both conventional method ( a standard device used in industry based on SNI ISO 08- 1511- 2004) and image processing method using MATLAB. After placed the sample of fabrics to fall inside the box (**Figure.4**), the image of the draped fabric was captured using digital camera run by the user and transferred to the computer to be analyzed and to be measured drape coefficient (DC). In this image processing technique, the picture was measured into a binary image after being processing. The threshold value can be controlled by the user to convert grey color into binary image (black and white). Using black and white image, the arbitrary area of drape can be measured using our listing program in MATLAB.

**Figure 4.** Placing the sample of fabrics to fall

The system of image processing can be shown in **Figure.5** below:

INPUT COLOR IMAGE

⇩

GRAY IMAGE CONVERSION

⇩

NOISE REMOVAL & FILTERING

⇩

BINARY IMAGE CONVERSION

⇩

THRESHOLDING

**Figure 5.** The system of image processing

## 3 Modeling of Fabric Movement influenced of Drape Measurement

To show the influence of fabric movement to the drape measurement, it can be pretended that the movement of fabric can be shown as Figure. 6.

Journal of Physics: Conference Series **1381** (2019) 012011 doi:10.1088/1742-6596/1381/1/012011

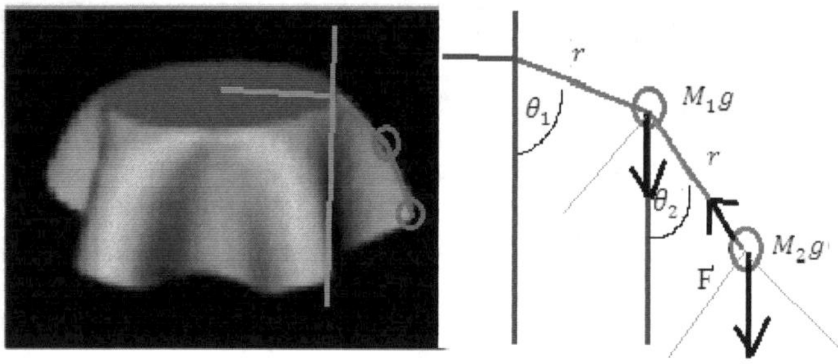


**Figure 6.** The Movement of Fabric as a function of angle

Taking Cartesian coordinate (x,y) in the vertical plane as shown in Figure.6, we have, for the mass $M_1$ can be written as shown below:

$$Y_1 = -r\ cos\theta_1 \quad ...(2)$$

$$X_1 = r\ sin\theta_1 \quad ...(3)$$

And for Mass $M_2$, we have the coordinates as written below

$$X_2 = r\ sin\theta_1 + r\ sin\theta_2 \quad ...(4)$$

$$Y_2 = -r\ cos\theta_1 - r\ cos\theta_2 \quad ...(5)$$

If we look for each speed of the masses , hence we can get for Mass $M_1$

$$\frac{dY_1}{dt} = \frac{d}{dt}(-r\ cos\theta_1) = -r\frac{d}{dt}(cos\theta_1) = -r\frac{dcos\theta_1}{d\theta_1}\frac{d\theta_1}{dt} = r\dot{\theta}_1 sin\theta_1 \quad ...(6)$$

$$\frac{dX_1}{dt} = \frac{d}{dt}(r\ sin\theta_1) = r\frac{d}{dt}(sin\theta_1) = r\frac{dsin\theta_1}{d\theta_1}\frac{d\theta_1}{dt} = r\dot{\theta}_1 cos\theta_1 \quad ...(7)$$

For mass $M_2$, we can get the equation below

$$\frac{dX_2}{dt} = \frac{d}{dt}(r\ sin\theta_1 + r\ sin\theta_2) = r\dot{\theta}_1 cos\theta_1 + r\dot{\theta}_2 cos\theta_2 \quad ...(8)$$

$$\frac{d}{dt}Y_2 = \frac{d}{dt}(-r\ cos\theta_1 - r\ cos\theta_2) = r\dot{\theta}_1 sin\theta_1 + r\dot{\theta}_2 sin\theta_2 \quad ...(9)$$

Hence the total kinetic energy and potential energy both vertical and horizontal force of the system can be written as determined below

$$T = \frac{1}{2}M_1\{(r\dot{\theta}_1 sin\theta_1)^2 + \left(r\dot{\theta}_1 cos\theta_1)^2\right\} + \frac{1}{2}M_2\{(r\dot{\theta}_1 cos\theta_1 + r\dot{\theta}_2 cos\theta_2)^2 + (r\dot{\theta}_1 cos\theta_1 + r\dot{\theta}_2 cos\theta_2)^2 + \frac{1}{2}I_{M_1}\dot{\theta}_1^{\ 2} + \frac{1}{2}I_{M_2}\dot{\theta}_2^{\ 2} \quad ...(10)$$

$$V = -M_1 gr\ cos\theta_1 - M_2 g(r\ cos\theta_1 + r\ cos\theta_2) \quad ...(11)$$

and the Lagrangian is

$$L=\frac{1}{2}M_1\{(r\dot{\theta}_1 sin\theta_1)^2+\left(r\dot{\theta}_1 cos\theta_1)^2\right\}+\frac{1}{2}M_2\{(r\dot{\theta}_1 cos\theta_1+r\dot{\theta}_2 cos\theta_2)^2+(r\dot{\theta}_1 cos\theta_1 + r\dot{\theta}_2 cos\theta_2)^2+\frac{1}{2}I_{M_1}\dot{\theta}_1^{\ 2}+\frac{1}{2}I_{M_2}\dot{\theta}_2^{\ 2}+M_1 gr\ cos\theta_1 + M_2 g(r\ cos\theta_1+r\ cos\theta_2) \quad ...(12)$$

$$L=\frac{1}{2}M_1(r\dot{\theta}_1)^2+\frac{1}{2}M_2\{(r\dot{\theta}_1 cos\theta_1)^2+2r^2\dot{\theta}_1\dot{\theta}_2 cos\theta_1 cos\theta_2+(r\dot{\theta}_2 cos\theta_2)^2+(r\dot{\theta}_1 sin\theta_1)^2 + 2r^2\dot{\theta}_1\dot{\theta}_2 sin\theta_1 sin\theta_2+\left(r\dot{\theta}_2 sin\theta_2)^2\right\}+\frac{1}{2}I_{M_1}\dot{\theta}_1^{\ 2}+\frac{1}{2}I_{M_2}\dot{\theta}_2^{\ 2} + M_1 gr\ cos\theta_1+M_2 g(r\ cos\theta_1+r\ cos\theta_2) \quad ...(13)$$

$$L=\frac{1}{2}M_1(r\dot{\theta}_1)^2+\frac{1}{2}M_2\{(r\dot{\theta}_1)^2+2r^2\dot{\theta}_1\dot{\theta}_2(cos\theta_1 cos\theta_2+sin\theta_1 sin\theta_2)+\left(r\dot{\theta}_2)^2\right\} +\frac{1}{2}I_{M_1}\dot{\theta}_1^{\ 2}+\frac{1}{2}I_{M_2}\dot{\theta}_2^{\ 2}+M_1 gr\ cos\theta_1+M_2 g(r\ cos\theta_1+r\ cos\theta_2) \quad ...(14)$$

$$L=\frac{1}{2}M_1(r\dot{\theta}_1)^2+\frac{1}{2}M_2\{(r\dot{\theta}_1)^2+2r^2\dot{\theta}_1\dot{\theta}_2 cos(\theta_1-\theta_2)+\left(r\dot{\theta}_2)^2\right\}+\frac{1}{2}I_{M_1}\dot{\theta}_1^{\ 2}+\frac{1}{2}I_{M_2}\dot{\theta}_2^{\ 2} + M_1 gr\ cos\theta_1+M_2 g(r\ cos\theta_1+r\ cos\theta_2) \quad ...(15)$$

Consider Lagrange's equation $\frac{d}{dt}\left(\frac{dL}{d\dot{q}}\right)=\frac{dL}{dq}$, therefore we have the motion for $\theta_1$

$$\frac{d}{dt}\left(\frac{dL}{d\dot{\theta}_1}\right)=\frac{dL}{d\theta_1}+F_{ext} \quad ...(16)$$

$$\frac{d}{dt}\{M_1 r^2\dot{\theta}_1+M_2[(r^2\dot{\theta}_1)+r^2\dot{\theta}_2 cos(\theta_1-\theta_2)]+I_{M_1}\dot{\theta_1}\} =-M_1 gr\ sin\theta_1-M_2 gr\ sin\theta_1+F_1 r \quad ...(17)$$

And the motion for $\theta_2$, we have

$$\frac{d}{dt}\left(\frac{dL}{d\dot{\theta}_2}\right)=\frac{dL}{d\theta_2} \quad ...(18)$$

$$\frac{d}{dt}\{M_2(r^2\dot{\theta}_2 cos(\theta_1-\theta_2)+r^2\dot{\theta}_2)+I_{M_2}\dot{\theta}_2\}=-M_2 gr\ sin\theta_2+F_2 r \quad ...(19)$$

For a special case when the acceleration and the speed of the fabric are kept constant, therefore we get from Eq.(20) the force of the fabric which can be concluded as:

$$-M_2 gr\ sin\theta_2+F_2 r=0 \quad ...(20)$$

$$M_2 gr\ sin\theta_2=F_2 r \quad ...(21)$$

$$sin\theta_2=tan\theta_2=\frac{F_2}{M_2 g} \quad ...(22)$$

and for mass $M_1$, we get

$$M_1 g\ sin\theta_1+M_2 g\ sin\theta_1=F_1 \quad ...(23)$$

$$sin\theta_1 \approx tan\theta_1 = \frac{F_1}{(M_2 + M_1)g} \quad ...(24)$$

For a small angle both $\theta_1$ and $\theta_2$, hence we can get

$$tan\theta_2 = \theta_2 \approx \frac{F_2}{M_2 g} \quad ...(25)$$

$$tan\theta_1 = \theta_1 \approx \frac{F_1}{(M_2 + M_1)g} = \frac{F_1}{M_1\left(\frac{M_2}{M_1} + 1\right)g} \quad ...(26)$$

For a special case, when the mass of $M_2$ is lower than $M_1$, hence we get

$$\theta_1 \approx \frac{F_1}{M_1\left(1 + \frac{M_2}{M_1}\right)g} \approx \frac{F_1}{M_1 g} \quad ...(27)$$

$$DC = \frac{\pi l^2 sin^2\theta - \pi l^2}{\pi R^2 - \pi l^2}100\% \approx \frac{\pi l^2\left(\frac{F_1}{M_1 g}\right)^2 - \pi l^2}{\pi R^2 - \pi l^2}100\% \quad ...(28)$$

As shown in Eq. (29) we can look that drape is defined as the amount to which a fabric will deform when it is allowed to hang under its own weight and without the influence of external forces. Commonly, drape ability can be defined as a phenomena of fabric deformation which is caused by the influence of fabric's force or weight of fabric caused by the gravitation force, when the fabric is suspended down without the influence of external forces. Based on the calculation of the fabric motion it can be determined that the higher the angle of fabric, the higher is the area under drape sample as well as the drape coefficient DC

## 4 Results and Discussions

Based on the calculation of the fabric motion, we can analyze from Eq.(23) and Eq.(25) that the value of the angle of drape is influenced by the mass of fabric and also the yarn strength *F*. the higher the angle of fabric, the higher is the strength of the fabric for a constant mass as well as drape coefficient, DC,. It has been found that the drape coefficient can be related with Eq.(1), Eq.(26) and Eq.(28). Based on theoretically approach, we can show the formula to show the relationship of angle of fabric, drape coefficient as well as the strength of fabric as stated in Eq.(29). Based on the Eq.(29) we can assume that drape ability can be defined as a phenomena of fabric deformation which is caused by the influence of fabric's force, $F_1$, or weight of fabric, $M_1 g$. Based on the validation of the experiment, we can show the drape coefficient both using image processing and also drape tester. Comparison the average of drape coefficient both conventional and image processing method can be shown in **Table-2**

**Table 2-Drape Coefficient of some fabrics**

| Sample | % Drape By Image Processing | % Conventional Drape tester |
|---|---|---|
| 1 | 35,44 | 34,18 |
| 2 | 34,18 | 36,13 |
| 3 | 34,41 | 35,09 |
| 4 | 34,87 | 34,39 |
| 5 | 34,77 | 35,57 |

The face, back and overall average drape coefficient of five samples, with same characteristic, tested with conventional and image processing method can be shown in **Figure 6**

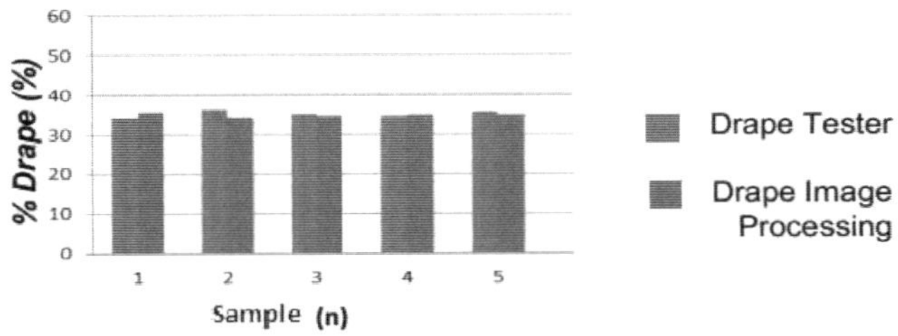


**Figure 7.** Drape coefficient of five samples tested with conventional and image processing

The difference of drape measurement of two methods can be shown in **Figure 6** using homogeneity analysis, the sig value has been found more than 0,05.It was found by t-test that there's no difference both standard drape tester and drape measurement using image processing technique. Both methods can be used to measure drape coefficient of fabric. We also conclude that from Eq. (26) and Eq. (28) the angle of $\theta_1$ and $\theta_2$ are influenced by the fabric strength $(F_1)$ for a special case when the mass of $M_2$ is lower than mass $M_1$. Based on our calculation and also looked Eq. (1) and Eq. (29) it can be assumed that the more the angle, the higher is the length as well as the higher the fabric strength, the higher the area of the drape or the lower is the drape coefficient.

## Conclusions

We have found the new equation to determine the drape coefficient related by the angle and fabric strength and we also have used Image processing technique for measuring drape profile of apparel fabric which has been studied and developed using MATLAB as an image processing software. In this research, a very good correlation has been found both conventional and image processing technique to determine the drape coefficient.

## References

[1] Putra VGV, Maruto, G & Rosyid, M.F.,2017, *Indian Journal of Fibre and Textile Research,* **42**. pp. 359-363.
[2] Putra, V G V, Rosyid M F and Maruto, G., 2016, , *Global Journal of Pure and Applied Mathematics*, **12**.,No.1 pp. 405-412.
[3] Herawati RM Fauzi I. and Putra V G V., 2015, *Applied Mechanics and Materials,* **780**, pp. 69-74.
[4] Lawrence, 2003, *Fundamentals of Spun Yarn Technology*, (New York : CRC Press)
[5] Backer, Hearle & Grosberg, 1969, *Structural Mechanics of Fibres, Yarns and Fabrics*, (New York : Wiley-Interscience)
[6] Hearle, J.W.S. & Gupta, B.S., 1965, *Textile Research Journal*, **35** pp. 788-795.
[7] Kenkare N & Plumlee T, 2005, *J Textile Apparel Tech Management*, **4**, pp 1.
[8] Lojen DZ & Jevsnik S, 2007, *FIBRES & TEXTILES Eastern Europe*, **15**, pp 39.
[9] Trommer, G., 1995, *Rotor Spinning*, (Frankfurt.: Deutscherfachverlag, )
[10] Zeidman, Shawney and Herington, 2003, *Indian Journal of Fibre and Textile Research*, **28**., pp. 123-133.
[11] Putra, V G V,Rosyid, M F. &Maruto, G.,2016, *Global Journal of Pure and Applied Mathematics*, **12**. pp. 405-412.

Journal of Physics: Conference Series **1381** (2019) 012012 doi:10.1088/1742-6596/1381/1/012012

# The design of the brantas river ecological monitoring is real time with OpenCV

**R H Irawan*, R A Ramadhani, R Helilintar, D Trianggoro**

Prodi Teknik Informatika, Universitas Nusantara PGRI Kediri, Kediri, JATIM, Indonesia

*spidole.tech@gmail.com

**Abstract**. Brantas River is a river that is used by the people of East Java for drinking water sources, irrigation, regulating water discharge, and tourist attractions. However, the bad habits of the people around the Brantas River now have damaged the ecology of the Brantas River, one of these habits is throwing garbage. To protect the ecology of the Brantas river, a system is needed to monitor the activities of people who dispose of garbage in the Brantas river in real time using Open CV. One of our efforts is to establish a Closed Circuit Television system to monitor the activities of residents in the Brantas river area which is commonly used by people to dispose of garbage. The system applied is used to provide moral punishment to people who throw garbage. The way this system works is to record an area that is often used by people to dispose of garbage, if people find it throwing garbage, the admin system will take pictures and crop, then the photo is processed with Open CV. The introduction of people who throw garbage is done from the front view. From the results of the trial as many as 125 times, the system can recognize faces correctly as many as 117 times (93%).

## 1. The first section in your paper

Brantas River is a river that is used by East Java people for drinking water sources, irrigation, regulating water debit, and tourist attractions. At present the state of the Brantas River is very concerned about the large number of people who dispose of garbage and family waste and industrial waste into the Brantas river which results in COD concentrations reaching above 50 mg / l, Phenol often appears with concentrations ranging from 0 and 0.039 mg / l [1].

The habit of people who dispose of garbage in the Brantas river, which causes pollution, and the ecological damage of the Brantas river is getting worse. The community usually throws household waste into the river; most of the waste disposed of in the waste is baby diapers and plastic waste[2].

Some efforts have been made by the East Java government to protect the ecology of the Brantas river, but this method is less effective. The most appropriate way is to provide direct supervision of the point, which is usually a place to dispose of garbage[3][4]. In this study, a system that was able to monitor the Brantas river in real time was made using Open CV[5][6].

One of our efforts to protect the ecology of the Brantas river is to install CCTV at a certain point in the Brantas river flow, which is often used by people to dispose of garbage. With this system, we hope that the ecology of the Brantas river can be saved.

The way this tool works is to record citizen activities in real time; if there are people who throw garbage, the admin system will reprimand the person through loudspeakers. To find out the name of the

Journal of Physics: Conference Series 1381 (2019) 012012 doi:10.1088/1742-6596/1381/1/012012

garbage disposal system, use the Open CV to identify the faces of the people throwing garbage. Face data is taken from the population ID card data around the Brantas river flow.

This system mengidetisikasi face of people who throw garbage from the front side. From the results of the trial as many as 125 times, the system can recognize someone's face correctly as many as 117 times (93%)

## 2. Method

In making the Brantas River ecological monitoring system in real time with OpenCV, several steps are needed. The following is the flowchart.

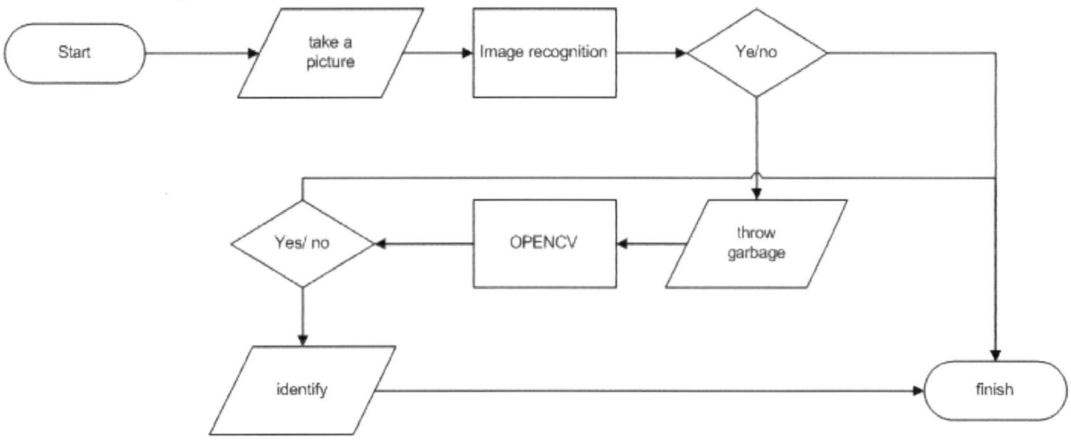


**Figure 1.** The flowchart design of ecological monitoring of the river basin is a real-time with OpenCV

### 2.1. *Take a picture*

This section is the foremost part; this section consists of CCTV cameras that are used to monitor the point that is usually used by people to dispose of garbage. Monitoring of these points is carried out every day.

### 2.2. *Image recognition*

The images obtained are then filtered by the admin; in this section, the system still uses the admin, because it is difficult to recognize movements when removing garbage. If the system finds that people who are at the point of disposing of the trash are dumping the garbage. The system will proceed to the next process, which is processing the image so that it can be identified.

### 2.3. *Identify*

After the image can be recognized by Open CV. The system will automatically give a warning through the existing speaker. Besides that, the environmental service can give a warning. Because the planning of the face database is taken from the KTP data, so that the environmental service can give a warning easily.

## 3. Result and Discussion

The Brantas River is a river that flows through a city that has developed in East Java such as Malang, Blitar, Tulunagagung, Kediri, Jombang, and Mojokerto. The awareness of the population of this city about the importance of the branches sting is still very lacking. It is vital to have an ecological monitoring system with the aim of protecting the ecology of the Brantas river from pollution. Here is a series of systems.

3.1. *The point used to dispose of garbage in the river*
At a certain point, the community often throws away garbage, so that garbage piles up, so that it covers up the small water source in the river basin and causes water pollution. This condition is very worrying considering that the Brantas River is a source of life for the people of East Java. Here is a photo of Brantas river pollution.

**Figure 2.** Pollution on the Brantas river
Source: https://nationalgeographic.grid.id

3.2. *Use of OpenCV*
Every picture taken will be recognized with OpenCV. OpenCV is a library that is used to process images. Using this library, computer-intended researchers can recognize faces precisely the same as human abilities. Shooting is done with CCTV cameras, and the results of the images are processed with OpenCV

*3.2.1.* Data Set, face data used as a benchmark for detecting residents who dispose of trash around the Brantas river is taken from ID cards. The purpose of taking photos of ID cards is to make it easier to find information about people who throw garbage in that place. Next, the face data is nailed to the database.

**Table 1. List of recognizable facial names**

| NUMER | Name |
|---|---|
| 1 | Risky Aswi |
| 2 | Didin |
| 3 | M. Agus |
| 4 | Yudi Sudarsono |
| 5 | Siti Munawaroh |

*3.2.2.* Detection results, after data, is obtained, then the photo I got is matched with the existing data set. At this stage, the system can detect faces properly. At this stage, the photo system taken is still just front. The following are the results of image detection performed.

**Figure 3.** Results of face recognition with OpenCV

The face recognition of each person is 25 times, while the faces of the people who are used as data sets are currently five people. So the system introduces faces to 125 times. From these results, images that can be recognized are 117 images. So the system's success rate is 93%. Which causes the system not to recognize the face because there is less good lighting, the slope of the face, and the distance of the face with the camera.

## 4. Conclusion

The ecological system of branches river monitoring is very necessary because the Brantas river is a source of life for the East Java community. The monitoring system needs to be installed at the point that the community uses to dispose of the garbage. The workings of this system are taking pictures (recording all activities in the river branches), identifying the faces of the people who are throwing away garbage. Currently, the system can only recognize faces from the front. There needs to be the development from the side and back. The researcher conducted a trial system 125 times, and the recognition was as successful as 117 times. It can be concluded that the level of success of this system is 93%.

## References

[1] Yudo S and Said N I 2019 Kondisi Kualitas Air Sungai Surabaya Studi Kasus: Peningkatan Kualitas Air Baku PDAM Surabaya *J. Teknol. Lingkung.* **20** 19

[2] Lubis M F and Muis A 2016 Fast shadow removal algorithm for river garbage pollution monitoring system *14th International Conference on QiR (Quality in Research), QiR 2015 - In conjunction with 4th Asian Symposium on Material Processing, ASMP 2015 and International Conference in Saving Energy in Refrigeration and Air Conditioning, ICSERA 2015* (Institute of Electrical and Electronics Engineers Inc.) pp 175–9

[3] Prasetya R B, Fariza A, Hasim J A N and Basuki A 2016 River flood spreading prediction system using cellular automata (Case study Bengawan Solo River) *Proceedings of 2015 International Conference on Data and Software Engineering, ICODSE 2015* (Institute of Electrical and Electronics Engineers Inc.) pp 201–6

[4] Sukaridhoto S, Pramadihanto D, Taufiqurrahman, Alif M, Yuwono A and Funabiki N 2015 A design of radio-controlled submarine modification for river water quality monitoring *2015 International Seminar on Intelligent Technology and Its Applications, ISITIA 2015 - Proceeding* (Institute of Electrical and Electronics Engineers Inc.) pp 75–80

[5] Halim K and Rakun E 2019 Sign language system for Bahasa Indonesia (Known as SIBI) recognizer using TensorFlow and long short-term memory *2018 International Conference on Advanced Computer Science and Information Systems, ICACSIS 2018* (Institute of Electrical and

Electronics Engineers Inc.) pp 403–7

[6] Boyko N, Basystiuk O and Shakhovska N 2018 Performance Evaluation and Comparison of Software for Face Recognition, Based on Dlib and Opencv Library *Proceedings of the 2018 IEEE 2nd International Conference on Data Stream Mining and Processing, DSMP 2018* (Institute of Electrical and Electronics Engineers Inc.) pp 478–82

# Simple Registration Software for Health Clinic

**W Kartika[1], F N Fauziah[2], N H Wijaya[1]**

[1]Universitas Muhammadiyah Yogyakarta, Indonesia
[2]Universitas Respati Yogyakarta, Indonesia

*wisnu2007@umy.ac.id

**Abstract**. In this era, there are many people who lives on Earth. Many among them get sickness, for example influenza, cold, etc. Recovering the sickness they must go to a hospital or mini clinic. The focus of this research is how to solve to reduce the use of paper on a small clinic. Therefore, the small clinic has much outcome to cover the daily clinic operation. Usually small clinic the patients must fill a paper form to meet the doctor. Our research is to back up the paper-based fill form. We propose a new software using Visual Basic 2010. Our method use Visual Basic .NET 2010 programming language. Our research is to focus on Visual Basic 2010 to entry the patient's data to Excel automatically. The advantages of this software is that we can fill the data form directly to the computer and then save them to the computer. The saving format from the computer is Microsoft Excel Format.

## 1. Introduction

In this era, many people in the world would love to be healthy. Unfortunately, some of them have severe sicknesses. It will become a problem. Usually people solve those problem by using a medical treatment. In the form of herbal or medicine. Medical treatment using medicine is administered when they go to hospital. But for the first medical treatment, they can choose by themselves. They can go to small health clinic near the home. The registration for the medical treatment [1] usually require an ID card (Kartu Tanda Penduduk/KTP) and BPJS card. The registration takes four to six minutes to fill [2] the patient's data. Inserting the patient's data can be a photo, image fingerprint [3], etc [4]. To save time we propose a new software registration to fill the patient's data to Excel on real time [5] by using Visual Basic .NET 2010 [6]. Visual Basic .NET is used on the another application and field. The registration is used for new student [7] for University [8] and new member of Aptisi [9].
This paper is consisted by six parts. They are Introduction, System Design, Methodology and The Result, Conclusion, Acknowledgement and References.

## 2. System Design

Our proposed scheme is by using the Visual Studio 2010 on Visual Basic .NET 2010 [10] as previewed on Figure 1 to make the software. This software is compatible to many platform. It is easy to use and has interesting GUI (Graphic User Interface).

**Figure 1.** Visual Basic .NET 2010 Visual Studio

## 3. Methodology and Result

After our proposed research had been designed, we made several experiment. This were the experiments. First of all, we opened the software (Visual Basic . NET 2010 and Microsoft Excel). Then we run them sequentially.

This is our flowchart process as seen on Figure 2. It describes the flow of working principle about the software. First, we open the software. After that we can see on the screen that there are columns for Nama, Alamat, Status, and Faskes Tingkat. The patient fills in his data to be collected by administrative staff on front office. The administrative staff then inserts the patient's data to this software. After fulfilling all data, the staff click on "Tampilkan Saat Ini" button. It will show the present date and time. The date and time will be recorded manually by staff to the board note.

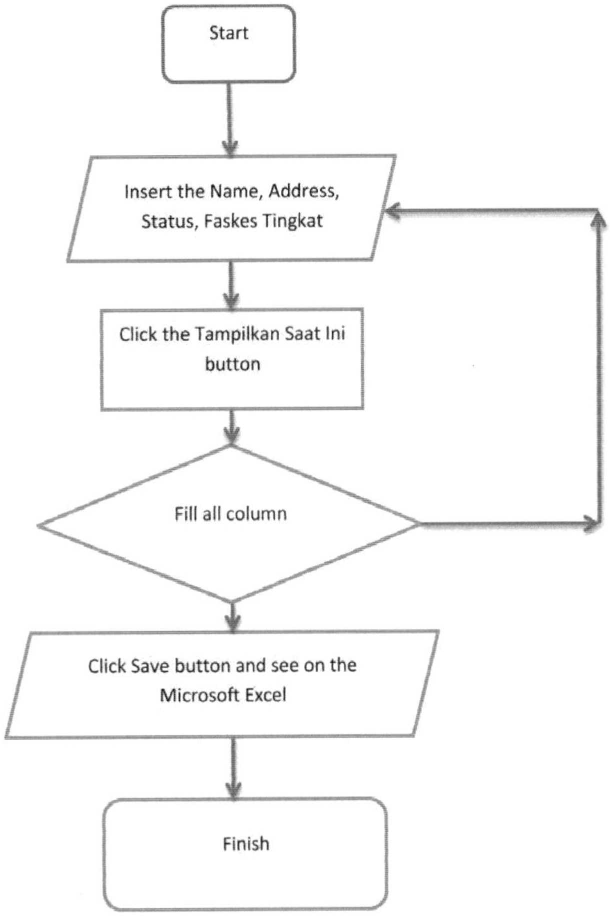


**Figure 2.** Process Flowchart

**Figure 3.** Software Registration Form

The software is consisted of ID, Nama, Alamat, Status, and Faskes Tingkat as shown at the Figure 3. The Figure 3 shows the based original form of our software. This is the authority of clinic staff to fill this form. There is the explain of the blank box of the software. The ID column is identifed as the patient entry. The Nama column is name of the patient. The Alamat column is the address of the patient. The Status identities whether he is BPJS's patient or general patient. At the last, Faskes Tingkat identify the health facility of the BPJS's patient. The BPJS facility is consisted of several grade. This is shown on Figure 4.

**Figure 4.** Software after filled out by data

After that, we should click the button Tampilkan Saat Ini. This is indicate and show the time and date now below the "Tampilkan saat ini" button. This is shown on Figure 5. The function of this button is to record the patient registration and apply to the doctor. From this record, we as staff can know the patient's time record. For the example, we know the patient named Wisnu always go to the clinic every week on certain days to his suffer of sickness.

**Figure 5.** Tampilkan Saat ini on Software

After that, we click the Save button to save the patient's data to Microsoft Excel 2010, and when it is successful "Successfully" mark will appear on the software as seen in Figure 6. This is also indicate that the patient data has succeed to be recorded to the system. The system uses the Microsoft Excel 2010. The column is splited into several column. The first column is the ID of patient, then patient's name, then patient's address, the patient's status and the last is grade of BPJS facilities (for BPJS card holder only). If the patient don't uses the BPJS it can be filled by strip symbol.

**Figure 6.** Entry Succesfully

after that, we shift to Microsoft Excel to see the result as shown on

Figure 7.

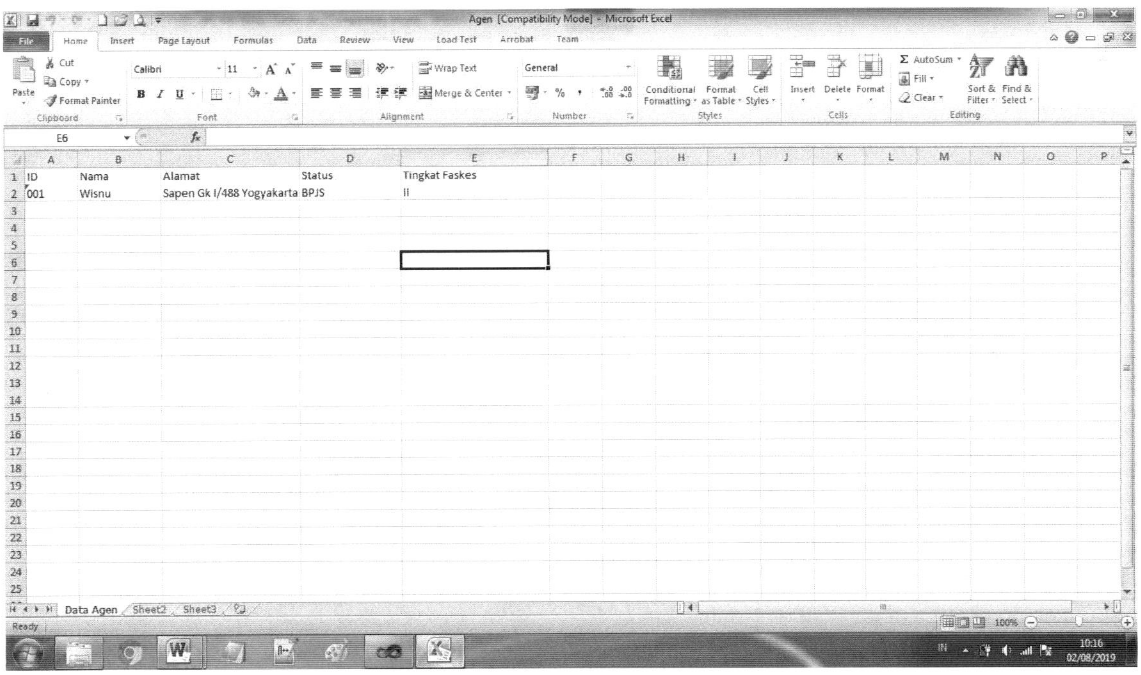


**Figure 7.** The Record result on Microsoft Excel 2010

After above step, we can save it with another name. It can be one file for a whole day. The result on Microsoft Excel can be seen on

Figure 7. It shows the collected data to the Microsoft Excel. This worksheet can be save every two hours, or every day. To keep this file safe, we need to make the back up data on the computer.

After that, back again to the Visual Basic .NET software again and press the Reset button, it will make all of the text blank again. So we can fill it with another patient, and just do like before.

## 4. Conclusion

From our research, we can make a conclusion that

a. The proposed scheme can replace the paper form on the clinic and the system design work effectively.
b. It will reduce the use of paper. It will reduce the cost of the clinic operation gradually. The paper can be used for another use.
c. It will make the patient data inserting faster than before.
d. The save data will be sequentially and up to date. It can be updated everyday. We can also distribute the patient and classify them.
e. It will be easy, to search for the history of a patient on that clinic.

## 5. Acknowledgement

We appreciate our study program which has helped a lot on this research and also another colleague on D3 Teknik Elektromedik Universitas Muhammadiyah Yogyakarta.

## References

[1] Du Y, Si X, Fang X and Ma Z 2018 The Online Medical Physics Curriculum-Group-Construction Under the Background of Modernization of Traditional Chinese Medicine and Emerging Engineering Education *2018 9th International Conference on Information Technology in Medicine and Education (ITME)* (IEEE) pp 661–3

[2] Ruiqing D, Sanjian S, Longbin O and Ruiqing D 2012 Applications of Excel VBA in Data Processing of Magnetic Survey *2012 International Conference on Computer Science and Service System* (IEEE) pp 2157–60

[3] Ma J, Jiang J, Zhou H, Zhao J and Guo X 2018 Guided Locality Preserving Feature Matching for Remote Sensing Image Registration *IEEE Trans. Geosci. Remote Sens.* **56** 4435–47

[4] Pavan N L and dos Santos D R 2017 A Global Closed-Form Refinement for Consistent TLS Data Registration *IEEE Geosci. Remote Sens. Lett.* **14** 1131–5

[5] Vijayalakshmi C 2013 An integrated development environment to establish web services in grid computing Using .NET framework 4.5 and Visual Basic 2012 *2013 World Congress on Computer and Information Technology (WCCIT)* (IEEE) pp 1–8

[6] Othman K A, Isa M A H M, Baharuddin M A, Ghazali M A, Khan Z I and Zakaria N A 2018 Forest Monitoring System Implementation using Visual Basic and Android Application *2018 18th International Symposium on Communications and Information Technologies (ISCIT)* (IEEE) pp 447–51

[7] Chandra D, Babar Z, Ng S X and Hanzo L 2019 Near-Hashing-Bound Multiple-Rate Quantum Turbo Short-Block Codes *IEEE Access* **7** 52712–30

[8] Nopriandi H 2018 Perancangan sistem informasi registrasi mahasiswa *J. Teknol. dan Open Source* **1** 73–9

[9] Rahardja U, Lutfiani N and Alpansuri M S 2018 Pemanfaatan Google Formulir Sebagai Sistem Pendaftaran Anggota Pada Website Aptisi.or.id *SISFOTENIKA* **8** 128

[10] Hamilton J P 2012 Data Selection Query with Visual Basic.NET *Practical Database Programming with Visual Basic.NET* (Chichester, UK: John Wiley & Sons, Ltd) pp 241–416

# Body temperature monitoring based on telemedicine

**H R Fajrin[1*], M R Ilahi[1], B S Handoko[2], I P Sari[3]**

[1]Department of Medical Electronics Technology, Vocational Program, Universitas Muhammadiyah Yogyakarta, Bantul, Yogyakarta, Indonesia
[2]Panembahan Senopati Bantul Public Hospital (RSUD), Yogyakarta, Indonesia
[3]Department of Informatics and Computer Engineering, Jakarta State Polytechnic, Depok, Indonesia

*hanifah.fajrin@vokasi.umy.ac.id

Abstract. In general, medical equipment at the Hospital to monitor body temperature such as a thermometer is less effective for nurses and doctors because it takes time to come to the patient's room to retrieve body temperature data. Then, when the patient's condition suddenly worsens, it still takes time for the treatment process. Therefore, the authors intend to make a tool that can help nurses monitor the patient's body temperature through an Android smartphone in real time and equipped with telemedicine that can send notifications in the form of SMS when the patient's body temperature is outside the normal limit. The working principle of this tool is that the body temperature data received by the DS18B20 sensor that will be processed by the Atmega328 microcontroller, then displayed on the OLED LCD and sent to an Android smartphone via Bluetooth HC-05. If body temperature data is outside the normal range, the Android application will send an SMS to another recipient's smartphone. Based on the results of the tool testing, the highest data error results is 0.829% and the lowest is 0%. For the process of sending data using Bluetooth HC-05 by taking measurements every 1 meter, the maximum distance obtained is 13 meters with a hitch. In the application of telemedicine in the form of automatic SMS sending successfully sends body temperature to other recipient's cellphone.

## 1. Introduction

Examination of vital signs is a measurement of the most basic functions of the body to find out clinical signs and is useful to strengthen the diagnosis of a disease and in determining appropriate medical action planning [1]. Vital signs of human health can be identified from body temperature, heart rate, respiratory, and blood pressure. From these vital signs, body temperature is a sign that needs most attention. Body temperature is the difference between the amount of heat produced by body processes and the amount of heat lost to the outside environment. Normal temperatures in adults range from 36.5 °C - 37.5 °C [2]. If the body temperature is below 36 °C, a hypothermia is indicated, but if the body temperature is above 37 °C, it is indicated that one suffers from hyperthermia.

Body temperature is an equally important parameter in checking body condition. Unhealthy physical conditions are reflected in poor body temperature. Many people are not aware of the importance of keeping body temperature normal. Any changes in environmental temperature will cause changes in one's body temperature[3].

The importance of checking body temperature causes everyone to be aware of their body condition. However, medical equipment used in hospitals to monitor or check body temperature is a thermometer that is less effective for nurses and doctors because it takes time to reach the patient's room to collect the body temperature data. Then, when the patient's condition suddenly deteriorates, it also still needs time to process it, so that a device that can immediately display patient's data without having to go to the patient's roomis needed [4].

Based on the data from Asymco analyst - Horace Dediu, Indonesia is the fifth ranked country with the most smartphone users in the world and the world most Android users with 1 billion users [5]. Utilization of Android technology today can be used as a means to support health information technology (Telemedicine), as well as monitoring or examination technology from Android to determine heart health and body temperature.

From the problems stated hereinabove, the author intends to design a device that can be used to help nurses or doctors when they monitor or retrieve patient's data who are being treated by utilizing Android as the displayer making it easier for nurses for data retrieval process. And this device also uses a Telemedicine system that is sending short messages or SMS about the information of the patient's condition data that beyond the normal body temperature to the doctor or nurse.

## 2. Related Work

Mansor et.al. conducted a project to design and develop body temperature measurement device that can be observed by the doctor in real time as well as history data via internet with an alarm/indication in case of abnormalities. That projectwas only focus on body temperature wireless monitoring system. The temperature sensors will send the readings to a microcontroller using Xbee wireless communication. To send the real-time data to health monitoring database, wireless local area network (WLAN) has been used. Arduino with Ethernet shield based on IEEE 802.11 standard has been used for this purpose [6]. Test results from a group of voluntary shows the real-time temperature reading successfully monitored locally (at home) and remotely (at doctor's computer) and the readings are comparable to commercial thermometer. Yet, this system needs to be developed using portable basis. For example: using smartphone as an interface. Therefore, it is easy to carry everywhere.

A wireless wearable device for monitoring internal temperature a few centimeters deep in the body is presented by Will Haines et.al [7]. A radiometer operating in the 1.4-1.427 GHz quiet band is used with a circular patch probe to measure the thermal radiation emitted by the body. The output is digitized and transmitted over Bluetooth by a TI CC2541. The wearable device is powered by a 3.7V Li-Ion battery, through three buck-conversion circuits. The sensor design trades performance (continuous calibration) for simplicity to reduce size and power consumption. Validated measurement data of water temperature inside the cheek demonstrates the feasibility of radiometric internal temperature measurement in a wearable platform. However this tool does not provide Telemedicine system that can send patient's body temperature to paramedic.

Previous research has been developed by [8] who designed "Heart Rate Measurement and Body Temperature with Data Storage". From the results of measurements of data taken from 20 people with 5 measurements and compared with a comparison device. The error value obtained still meets the standard because it is still below the maximum error of 5%. The disadvantage of this tool is that it still used LCD Character displays to display temperature results.

Muhlis et.al. developed a tool entitled "Implementation of a System for Monitoring the Heartbeat and Temperature of the Human Body Wirelessly". This tool takes data from 10 people and compared with a comparison tool, and the results of the errors obtained still meet the standard because it is still below the maximum error of 5%. The device has also used the NRF24L01 wireless system as the sender of data to a personal computer [1]. However, this tool is still not practical because it only displays data on personal computers only and has not been equipped with Telemedicine.

So the author designed and made a body temperature monitoring device using the DS18B20 waterproof sensor so that when the measurements of human body temperature that are usually affixed to the armpit make the sensor's performance uninterrupted. The tool that is made can also display data on the Android display with the help of Bluetooth and also equipped with telemedicine in the form of SMS with a system of sending information on patient condition data if at any time the patient experiences an increase or decrease in body temperature.

## 3. Research Method

### *3.1 Hardware Design*

Hardware design uses a number of circuit modules including the controller system circuit and DS18B20 temperature sensor circuit.

*3.1.1 Controller Circuit.* It can be seen from Figure 1 that working principle of controller circuit is utilizing the ATmega328 IC to make orders to other circuits such as the temperature sensor and OLED LCD. This circuit requires an input voltage of +5V and ground to work. The Program Input Connector is useful for downloading Arduino program to be installed into microcontroller. OLED LCD connectors are useful for connecting LCD with microcontrollers to display body temperature data output. Bluetooth connector is used to connect the HC-05 Bluetooth module with TX and RX pins on the microcontroller for the process of sending body temperature data to android. DS18B20 sensor connector is used for connecting body temperature sensors with microcontroller which output from the sensor will enter the PB5 pin. The reset circuit in which a reset button exists is useful to repeat the program work. LED is useful as indicator that voltage has entered the minimum system circuit and the system works.

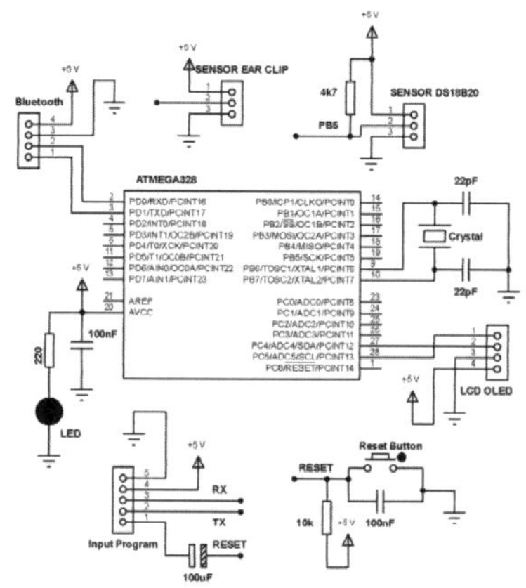


Figure 1. *Minimum system circuit*

*3.1.2 DS18B20 Sensor Circuit.* Figure 2 shows the working principle of DS18B20 sensor, there are 3 pins such as Vcc pin, ground, and data output. The voltage needed by this sensor to work is +5 V. Data output from the sensor will enter the digital microcontroller pin, namely PB5 pin.

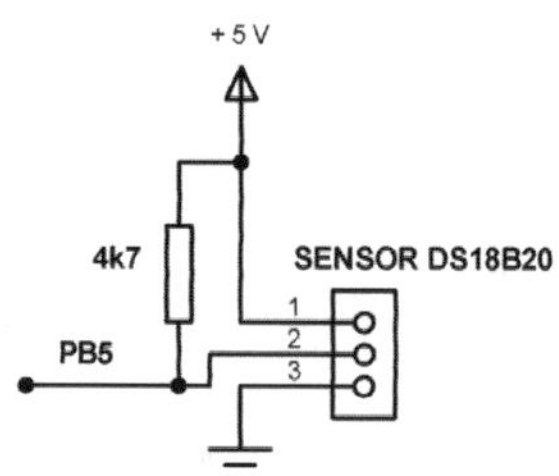


Figure 2. DS18B20 sensor circuit

### *3.2 Software design*

The software used is Arduino programming software as a tool for data processing and MIT Inventor app as a data processor in the Android application.

*3.2.1 Block diagram.*The patient will be fitted with a DS18B20 temperature sensor placed in the patient's armpit. Afterwards, the sensor will measure the patient's body temperature and the results will be sent directly to the microcontroller. After entering the microcontroller, the results will be processed and displayed on the LCD. The results that have entered the microcontroller will also be sent to the Android application via Bluetooth and will appear on the Android application interface. If the results of the body temperature show results below 35°C or above 37.5°C, the Android application will send an SMS to other cellphone users (doctors, nurses, or patient families) to determine the patient's condition and to take medical action quickly. Block diagram is as follows in Figure 3.

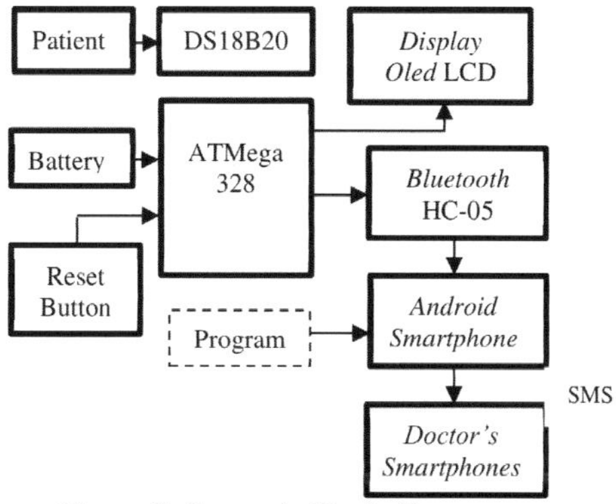


Figure 3. System's block diagram

*3.2.2 Flowchart.* In Figure 4, the working process of the device starting from taking body temperature data by the sensor and displayed on the OLED LCD, then sending temperature data to the android application for display and sending SMS.

When the device is switched on firsthand, it will initialize the data. Then, the temperature sensor will work to read the body temperature of the patient. Afterwards, the microcontroller will perform data calculations to calculate data retrieved from the patient and will be displayed on the OLED LCD display on the tool. Data will also be inputted into Bluetooth circuit to be sent to the Android application and will be displayed on the application.

When the Android application is activated, the application will perform a data initialization process. When the application has initialized and is connected to the device by Bluetooth, the application automatically receives data and displays the data. The process of connecting between the sender and receiver of data is by using and pairing the sender's Bluetooth ID. If the Bluetooth ID of the sender has been found on Android, then the recipient can choose the Bluetooth ID and once connected, the data can be approved. If the temperature is below 35°C or above 37.5°C, the application will immediately send an SMS to the destination number. For the process of sending abnormal data by SMS, you must request the contact number of the receiver beforehand in the column on the android application.

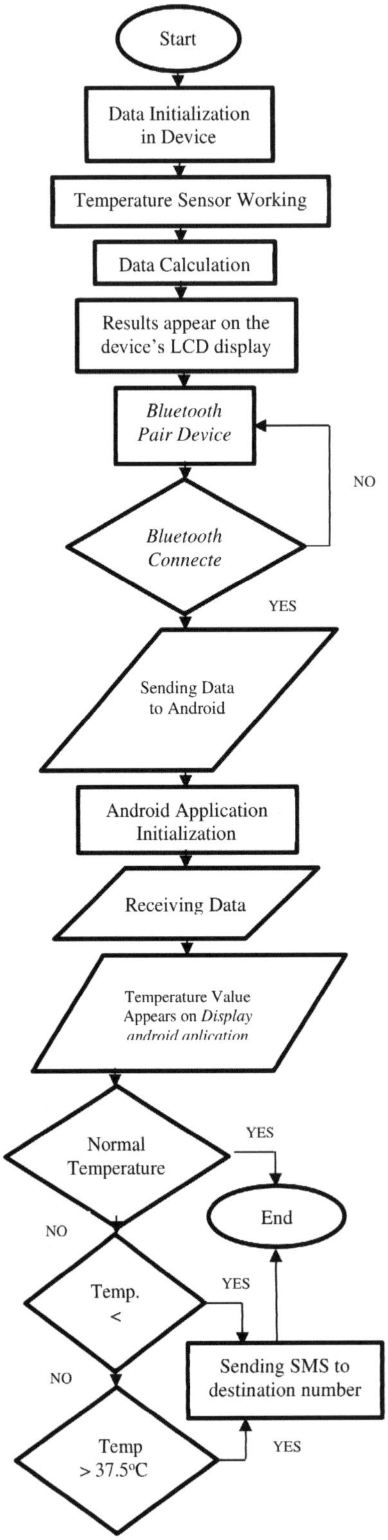


Figure 4. Tool's Flowchart

## 4. Research Results And Discussion

### *4.1 Body Temperature Measurement*

The following data in Table 1 are measured data taken from 10 respondents by measuring body temperature in the armpit area for 1 minute and comparing it with a digital thermometer.

Table 1. Body Temperature Measurement

| No. | Equipment | Measurement Result (°C) | | | Average | Error (%) |
|---|---|---|---|---|---|---|
| | | 1 | 2 | 3 | | |
| 1 | Thermo | 36.3 | 36.3 | 36.3 | 36.3 | 0.367 |
| | Module | 36.1 | 36.2 | 36.2 | 36.17 | |
| 2 | Thermo | 35.5 | 35.8 | 35.9 | 35.73 | 0.653 |
| | Module | 35.3 | 35.6 | 35.6 | 35.5 | |
| 3 | Thermo | 36.1 | 36.1 | 36 | 36.07 | 0.462 |
| | Module | 35.9 | 35.9 | 35.9 | 35.9 | |
| 4 | Thermo | 36.2 | 36.2 | 36.2 | 36.2 | 0.829 |
| | Module | 35.9 | 35.9 | 35.9 | 35.9 | |
| 5 | Thermo | 34.8 | 34.8 | 34.8 | 34.8 | 0.67 |
| | Module | 35 | 35.1 | 35 | 35.03 | |
| 6 | Thermo | 36.5 | 36.5 | 36.5 | 36.5 | 0.548 |
| | Module | 36.2 | 36.3 | 36.4 | 36.3 | |
| 7 | Thermo | 36.9 | 36.9 | 36.9 | 36.9 | 0 |
| | Module | 36.9 | 36.9 | 36.9 | 36.9 | |
| 8 | Thermo | 36.7 | 36.7 | 36.7 | 36.7 | 0 |
| | Module | 36.7 | 36.7 | 36.7 | 36.7 | |
| 9 | Thermo | 36 | 36 | 36 | 36 | 0.556 |
| | Module | 36.2 | 36.2 | 36.2 | 36.2 | |
| 10 | Thermo | 36.2 | 36.2 | 36.2 | 36.2 | 0.276 |
| | Module | 36.1 | 36.1 | 36.1 | 36.1 | |

From the results of the tool measurement with a comparison, it shows that the change in temperature value on the tool is close to the temperature value displayed by the comparison device. Based on the measurement results between the module and thermometer (comparator), the data results of the device error is 0.829% as the largest and 0% as the smallest. The amount of uncertainty obtained is due to several factors such as less precise sensor placement and ambient temperature influence.

### *4.2 The Measurement of Bluetooth Performance*

The process of sending data from the device module to Android via Bluetooth is by setting the position of the Android smartphone that has been connected to the Bluetooth connection and moving it every 1 meter.The specific testing results are as follows in Table 2.

Table 2. The Measurement Result of Data Transmission via Bluetooth

| Distance (m) | Data Result | Description |
|---|---|---|
| 1 | Sent | Succeed |
| 2 | Sent | Succeed |
| 3 | Sent | Succeed |
| 4 | Sent | Succeed |
| 5 | Sent | Succeed |
| 6 | Sent | Succeed |
| 7 | Sent | Succeed |
| 8 | Sent | Succeed |
| 9 | Sent | Succeed |
| 10 | Sent | Succeed |
| 11 | Sent | Succeed |
| 12 | Sent | Succeed |

| | | |
|---|---|---|
| 13 | Sent | Succeed |
| 14 | Sent | Unstable |
| 15 | Sent | Unstable |
| 16 | Not Sent | Disconnected |
| 17 | Not Sent | Disconnected |
| 18 | Not Sent | Disconnected |
| 19 | Not Sent | Disconnected |
| 20 | Not Sent | Disconnected |

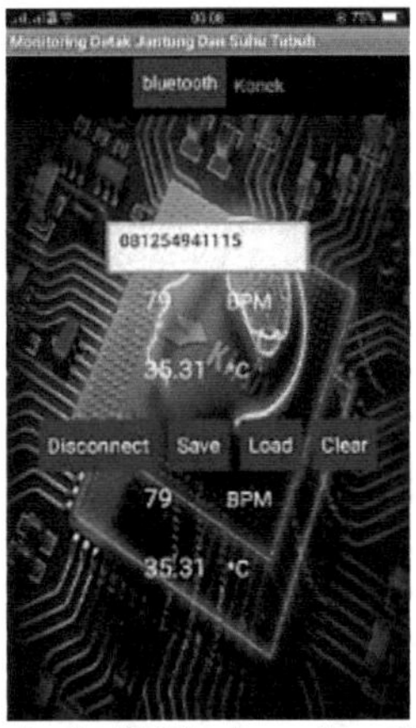


Figure 5. Android Display when Connected

Based on the measurement results of the above data transmission it is known that HC-05 Bluetooth can only send data with a distance of 1 m to 13 m stably, whereas at a distance of 14 m to 15 m the connection will be disrupted or unstable and at a distance of 16 m to 20 m, it is disconnected. This can be due to the distance limit of the Bluetooth module and the presence of signal blockers such as walls. Figure 5 is a Telemedicine application which content some features and can display the data when the system is active.

*4.3 The Measurement of SMS Delivery Performance*

Based on the measurement results of the Telemedicine data in Table 3, it is known that the data will be successfully sent if the temperature is below 35 °C and above 37.5 °C. This happens because it has been set in the Telemedicine application program to send messages in the form of SMS when the temperature is beyond normal. Figure 6 shows the SMS sent to a phone which the phone number has been saved.

Table 3. The Measurement Results of SMS Delivery Performance or Telemedicine

| Temperature (°C) | Data Process | Description |
|---|---|---|
| 43.69 | Sent | Succeed |
| 47.44 | Sent | Succeed |
| 46.50 | Sent | Succeed |

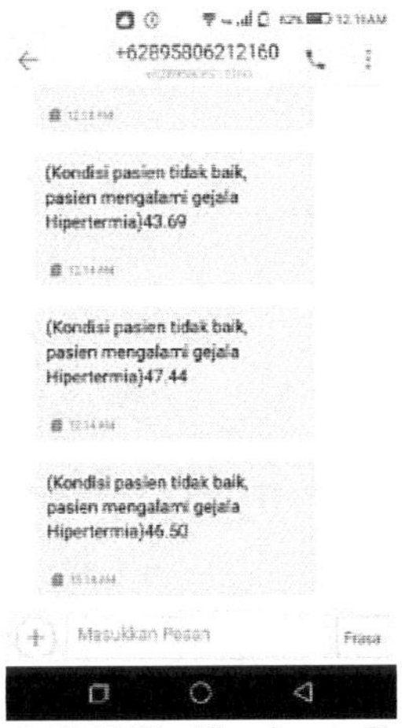


Figure 6. SMS Notification to Receiver

## 5. Conclusion

Based on the device testing and data collection on 10 respondents then compares it to digital thermometer, the highest error value obtained is 0.829%. The device module uses Bluetooth that is able to send data from module to android application. The process of data retrieval and transmission is by putting the android smartphone that has been connected to the device and moving it every 1 meter. The data transmission distance has maximum limit of 13 meters with wall as obstacle. Android application can display measurement data from the module as long as Bluetooth stays connected and saves the temporary recent data. And there is also column for cell phone number as Telemedicine operation that is useful for sending SMS to the smartphone user automatically when the patient's body temperature is below or above the normal range. The application of Telemedicine can send notification in form of SMS to the receiver's number or other smartphone regarding the illness suffered by the patient and when the temperature data is below 35 °C or above 37.5 °C. Overall, the tool that author design can work well based on the minimum error value obtained from tool testing.

## References

[1] Saputro, Agung Muhlis , Widasari Edita Rosana and F H 2017 Implementasi sistem monitoring detak jantung dan suhu tubuh manusia secara wireless *J. Pengemb. Teknol. Inf. dan Ilmu Komput*

[2] Damayanti, Nezwa Nadia , Rahmawati T and R M 2015 *Wireles Monitoring BPM dan Suhu Dilengkapi Nurse Call Berbasis PC* (Poltekes Kemenkes Surabaya)

[3] Jones W D 2006 Taking body temperature, inside out *IEEE Spectr.* **43** 13–5

[4] Kim H, Min Y and Choi B 2019 Real-time temperature monitoring for the early detection of mastitis in dairy cattle: Methods and case researches *Comput. Electron. Agric.* **162** 119–25

[5] Horace Dediu 2013 Global smartphone penetration nearing 10%

[6] Mansor H, Shukor M H A, Meskam S S, Rusli N Q A M and Zamery N S 2013 Body temperature measurement for remote health monitoring system *2013 IEEE Int. Conf. Smart Instrumentation, Meas. Appl. ICSIMA 2013* 1–5

[7] Haines W, Momenroodaki P, Berry E, Fromandi M and Popovic Z 2017 Wireless system for continuous monitoring of core body temperature *IEEE MTT-S Int. Microw. Symp. Dig.* 541–3

[8] Alvian A G 2016 *Alat Ukur Detak Jantung dan Suhu Tubuh dilengkapi Penyimpanan Data* (Universitas Muhammadiyah Yogyakarta)

# Analysis of business intelligence system design for student performance monitoring

**S Anardani[1*], L Sofyana STT[1], A Maghfur[1]**

Informatics Engineering University of PGRI Madiun

*anardani@unipma.ac.id

**Abstract**. The Informatics Engineering Study Program at the University of PGRI Madiun is one of the institutions that provide education services. The increasing number of students each year requires supervision and measurement of student performance in order to create an optimal process. The optimal process will produce output that is in accordance with BAN-PT standards related to student performance. Looking at these problems, it is necessary to have an information system that can present student performance data for the purposes of accreditation and evaluation of student performance. The purpose of this research is to analyze and design student performance monitoring information systems by applying the Business Intelligence concept. The systematics of this study refers to the Bussiness Intelligence Roadmap Method of Larissa T Moss and Shaku Atre. The results of this study are a design of data sources and system architecture that will be used in the system development process in the future.

## 1. Introduction

BAN-PT accreditation is one form of evaluation to determine the feasibility of a university or study program determined by the National Accreditation Agency. Student data processing will experience difficulties if the Study Program does not provide a means of access to all data needed to meet the standards of BAN-PT.

The Informatics Engineering Study Program at the University of PGRI Madiun is one of the institutions that provide education services. Currently the Study Program has students with a range of 450 students. With the increasing number of students each year, the study program needs to monitor and measure student performance in order to create an optimal process. The optimal monitoring and measurement process will produce output that is in accordance with BAN-PT standards related to student performance.

Looking at these problems, it is necessary to have an information system that can present student performance data for the purposes of accreditation and evaluation of student performance. The results of monitoring student performance will be the basis for making decisions quickly and precisely. Asstated in research[1] . Business Intelligence-based information system can collect, store, organize and provide the information needed to input student data according to BAN-PT standards

Business Intelligence System is one of the information technologies that is used as a tool to increase the value of business organizations or companies[2]. Business Intelligence is a concept to improve the quality of business decision making based on data-based systems that are commonly applied by business actors or companies[3]. This monitoring system with the Business Intelligence concept can process extensive data so that the values contained in the data can be represented in several concise pages and display information needed by decision makers[4].

Therefore, the implementation of this study aimed to analyze and design student performance monitoring information systems by applying the Business Intelligence concept. This student performance monitoring information system can process and present data for standard needs of BAN-PT forms. The development of this software system used the Bussiness Intelligence Roadmap method of Larissa T Moss and Shaku Atre.

## 2. Study ofLiterature

### *2.1. Business Intelligence*

Universities are good candidates for the application of the BI system, this is because competition in the education sector is very competitive in recruiting students both domestically and abroad. The BI system is able to carry out the process of collecting information correctly and precisely so that it can achieve its objectives because the decision-making process runs well [5].

Business Intelligence systems must have the ability to allow leaders to view data in different perspectives. Data generated by the BI system must be able to navigate business processes so that leaders can anticipate and estimate organizational changes[6].

Business Intelligence System is a technology that functions as a tool to convert data into information. This information will be used by the leadership to guide organizational leaders to make better business decisions[7].

BI is an important component of Higher Education which has a function to measure, monitor and manage the performance of Higher Education management more effectively based on its resources and business processes[8].

### 2.2. *LarissaAtreBusiness IntelligenceRoadmap*

*Larissa AtreBussiness IntelligenceRoadmap*is a visual guide in developing effective Bussiness Intelligence systems through 6 stages, namely Justification, Planning, Business Analysis, Design, Construction, Deployment[1].

The integrated BI decision support system environment cannot be built in an instant, data and system functionality must be updated repeatedly. The BI Roadmap does not discuss organizational change, the BI roadmap aims to define BI project activities needed to support organizational system infrastructure both technical infrastructure and non-technical infrastructure[9].

## 3. Research Metodology

The systematics of this study refers to the Bussiness Intelligence Roadmap Method of Larissa T Moss and Shaku Atre. This method explains that the business intelligence information system development process consists of 6 phases, namely justification, planning, business analysis, design, construction and deployment. The stages of the phases in the Business Intelligence Roadmap can be illustrated in Figure 1 below:

Journal of Physics: Conference Series **1381** (2019) 012015 doi:10.1088/1742-6596/1381/1/012015

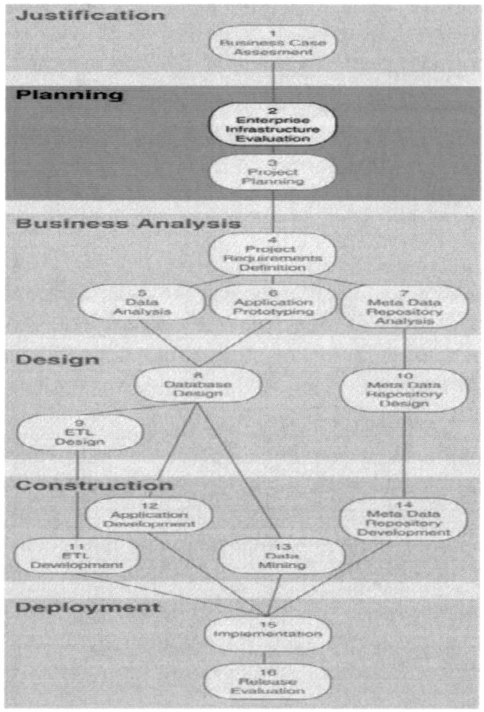


**Figure 1.** *Larissa Atre Bussiness Intelligence Roadmap*Method

The activities carried out at stagesofLarissa Are roadmap are as follows:

1. Justification
   Defining business problems and opportunities as a basis for determining problem solving with Business Intelligent application solutions.
2. Planning
   At this stage it is divided into two activities, namely Enterprise Infrastructure Evaluation and Project Planning. The Enterprise Infrastructure Evaluation activity is analyzing infrastructure requirements to support the development of Business Intelligent systems. The infrastructure analyzed consists of two parts, namely technical infrastructure and non-technical infrastructure. The second activity is Project Planning, which is to estimate the activity schedule.
3. Business Analysis
   The activity in this phase is to make documentation of needs that have been previously identified. Documents contain business data analysis, system specification analysis both functional and non-functional, prototype design in the form of software and hardware.
4. Design
   At this stage the activity carried out is to make a database scheme design that is able to integrate data and facilitate decision making. This stage also designs data sources and system architecture.
5. Construction
   It is the stage of developing a Business Intelligence system with planned software.
6. Deployment
   The final stage of the Larissa Atre roadmap is implementing a system that has been built and tested for the system evaluation process.

In this study based on the research objectives, the stages of the Larissa Atre roadmap adopted only reached the fourth stage, namely the design stage. The framework in this study is as shown in Figure 2 below:

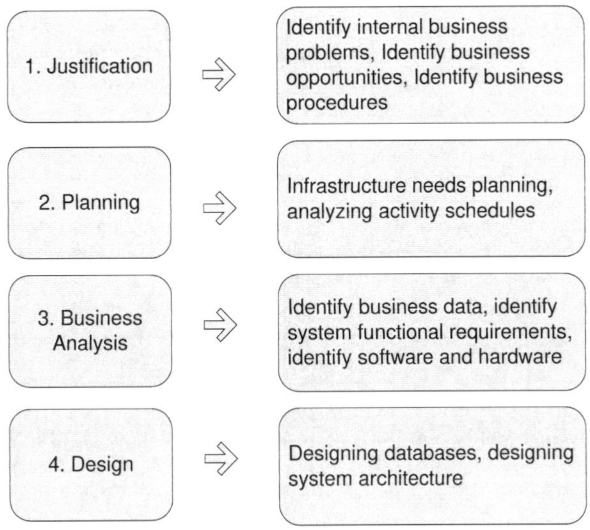


**Figure 2.** ResearchFramework

## 4. Analysis and Discussion

After determining the research framework, the next stage is to create a design analysis document for the Business Intelligence System application based on the Larissa Atre roadmap in Figure 2 above. The results of the design analysis can be explained as follows:

1. Justificationstage
   At this stage, identification of the problems faced by the Informatics Engineering Study Program at the University of PGRI Madiun was identified. Problems were analyzed with the SWOT approach method. The results of the SWOT analysis are as follows:

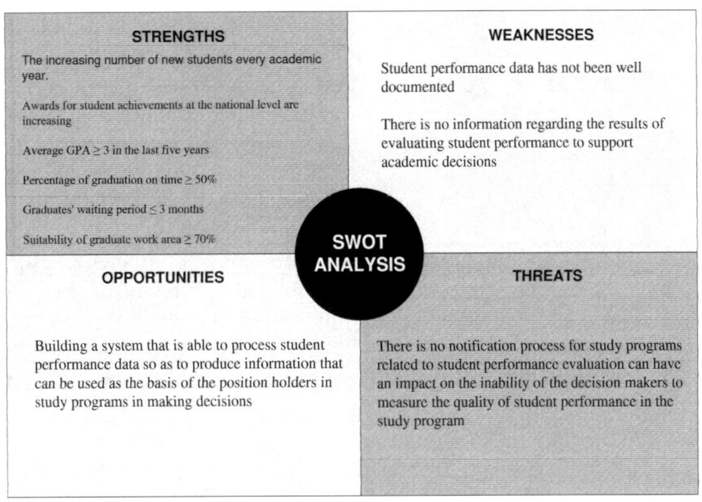


**Figure 3.** SWOT Analysis

From the results of the mapping in Figure 3 in the form of a SWOT analysis that there are problems that are student performance data not well documented in the study program so that there is no information related to the results of student performance evaluations that can support the stakeholders to make improvements in the study program. Data processing must be well organized through software development that is able to provide student performance data so that it can support input of accreditation forms quickly and accurately.

2. Planning Stage
The first activity at this stage is to evaluate the condition of the current technical infrastructure and then analyzing the infrastructure requirements that will be needed for the software development process. There are several problems that will be analyzed, namely the analysis of hardware requirements, analysis of HR requirements, analysis of network conditions, analysis of Database Management Systems and other non-functional requirements needed for system development. At this stage also planned the need for non-technical infrastructure through the ETL process (Extract, Transform, Load) which aims to extract data from data sources and then transfer it to data warehouse. Before the data was inputted into data warehouse, a data cleaning process was carried out. The evaluation steps taken are first assessing the current technical infrastructure platform, then evaluating and selecting the new products needed. Assessment results are written in technical infrastructure assessment reports. Based on the results of these assessments, ordering, installing and testing can be performed.

3. Business AnalysisStage
In the business analysis stage, the first activity carried out was to analyze the functional requirements of the system in managing student performance. The results of the analysis of the functional requirements of the system can be seen in Table 1 below:

**Table 1**. Analysis of system functional requirements

| No | Business Process | Data needed by business people |
|---|---|---|
| 1 | Manage the ratio of prospective students to capacity | prospective student data<br><br>class capacity data |
| 2 | Manage average GPA data | data on average student GPA<br><br>data on the average GPA of graduates |
| 3 | Manage appreciation for student achievement | data on scientific competition achievements, sports competitions, art competitions at the national or international level |
| 4 | Manage graduation percentage data on time | Graduation data |
| 5 | Manage waiting period for graduate work | Data on graduate work waiting period |
| 6 | Manage data on the suitability profile of graduates in accordance with the study program | data on the suitability profile of graduates in accordance with the study program |
| 7 | Manage data on students dropping out or resigning | data on students dropping out or resigning |

Based on the mapping in Table 1 above, it can be concluded that there are 7 business processes conducted by study programs to measure student performance according to the

Accreditation Form. Based on the business process, data needs can be identified related to student performance management.

4. Design stage
   At the database design stage, there are several things that must be considered, namely consideration of performance, selection of DBMS and HR. Data sources to be used in the BI system are student performance adjusted to the needs of accreditation forms, while the identification of types of data can be seen as follows:

**Table 2.**Identification of Data Types

| No | Information Needed | Data Type | Data Sources |
|---|---|---|---|
| 1 | Student Data Information | New student data | DbSIA |
| | | Old student data | |
| | | graduate data | |
| | | student data drop out | |
| | | Resignedstudent data | |
| 2 | Student GPA Information | Student GPA data | |
| 3 | Alumni Data Information | Alumni Data | DbTracerStudy |
| | | WaitingPeriod Data | |
| | | Alumni Job Data | |
| 3 | Student Achievement DataInformation | StudentAchievement Data | DbPPBK |

It is known that the type of data needed for managing student performance is student data consisting of new student data, old student data, graduate data and data on students who resign. In addition to inputting accreditation forms, the data also requires student GPA data, alumni data and alumni waiting period, alumni job data and student achievement data. All of the data is sourced from the SIA database, the tracerstudy database, and the PPBK database.
The entire operational data will be managed into the student performance BI system for ETL processes (Extract, Transform, Load) and the results will be saved to data warehouse. Here is the ETL Process Flow Diagram of the BI system student performance can be seen in Figure 5 below:

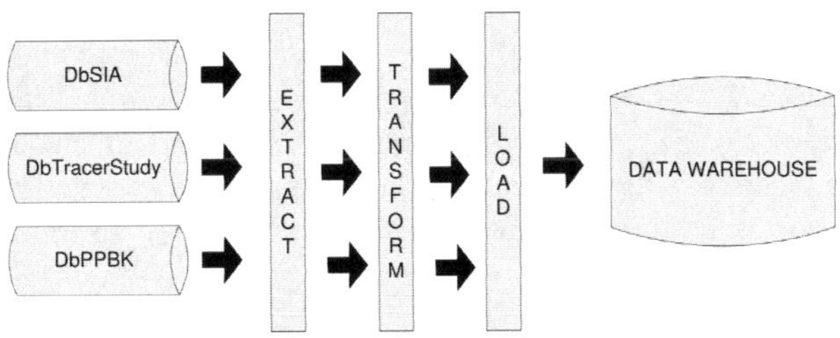


**Figure 4.** ETL Process Flow Diagram ofBI System Student Performance Monitoring

The ETL process begins with an ETL machine performing the process of extracts, cleanses and transforms on operational data. Then the middleware will help bring data to the BI database. Finally, when the data requested by the user will be mapped by the application by running queries and printing reports. After the ETL process is implemented, the next step is to design the BI system architecture that will be developed. The system architecture that will be built is as follows:

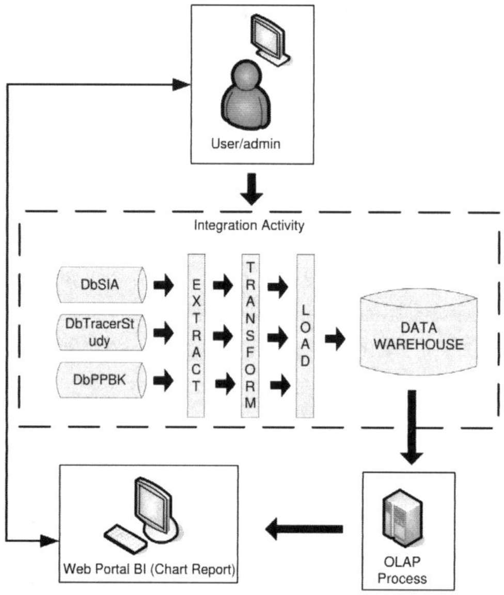


**Figure 5.** BI System Architecture

In Figure 6, the admin or user processes the ETL data source from the database that has been built. In the integration process, the admin will update the data that will be stored in the data warehouse according to the analysis needs. The next step after the integration activity is logging into the BI portal web system to select the desired data in the form of graphical reports or statistics. The data entered will be processed by the system in the process of On-Line Analytical Processing (OLAP).
In the BI system that will be built after the process of integration and data analysis, the system will print output in the form of reports in the form of graphs and tables in the form of student performance data. With the display of this report, the decision makers in the study program will be easier to evaluate the quality of student performance and then be reported in the preparation of accreditation forms.

## 5. Conclusion

Based on the results of the analysis of system requirements and design that has been carried out, this study succeeded in making the design of the Business Intelligence information system architecture design of student performance monitoring according to the needs of accreditation. The limitations of the results of this study are still limited to the design concept as a form of recommendation for the future system development process.

## References

[1] Agustiono W and Nasrullah W A 2017 Desain Academic Business Intelligence UntukAkreditasi: StudiKasusUniversitasTrunojoyo*Seminar NasionalMatematika dan Aplikasinya* (Surabaya) p 527 - 435
[2] Fatimah F N N, Murahartawati and Widjajarto A 2014 Penerapan Business Intelligence Pada Aplikasi Dashboard Monitoring PerformansiMahasiswa Dan LulusanBerdasarkanStandar 3 BAN-PT Program StudiSarjanaMenggunakanMetode Scrum *J. RekayasaSistem&Industri*(2014)1 1
[3] Harijanto B and BudiprasetyoG 2013PerancanganAplikasi Business Intelligence Hasil Proses BelajarMengajar (StudiKasus Program StudiManajemenInformatika) *J.ELTEK* (2013)1101
[4] Ariani T R, Tania K D and Indah D R 2016 Penerapan Business Intelligence Pada

SistemInformasiPenjualanBarang PT. Winsa (StudiKasus Di PT. Winsa Palembang) KonferensiNasionalTeknologiInformasi dan Aplikasinya (Palembang) p D103 – D110

[5] Yusof A F, Miskon S, Ahmad N, Alias R A, Hashim H, Abdullah N S, Ali N M AndMaarof M A 2015 Implementation Issues Affecting The Business Intelligence Adoption In Public University *J. of Engineering and Applied Sciences* (2015)1023

[6] Lupu A R, Bologa R, Sabau G and Muntean M 2007 Influence Factors of Business Intelligence in the Context of ERP Projects *International Journal Of Education And Information Technologies* (2007)21

[7] Ramanigopal C S, Pallaniapan G AndHemalatha N 2012 Business Intelligence For Infrastructure And Construction Industry *International Journal of Business Economics & Management Research* (2012)26

[8] Mustafid 2013 Sistem Business Intelligence UntukMendukungPerguruan Tinggi yang Kompetitif*J.SistemInformasiBisnis* (2013)01

[9] Larissa T M And Atre S 2003 *Business Intelligence Roadmap* (Boston:Addison-Wesley)

# The Designing of Website-Based Learning Media as an Alternative for Online Learning for Student Practicing at SMKN 1 Sawoo Ponorogo

**Y P Yuda*,M N L Azis, I D Kurniawati**

Universitas PGRI Madiun, Jl. Setiabudi NO. 85, Indonesia
*yogaprisma@unipma.ac.id

**Abstract.** SMKN 1 Sawoo is a vocational high school that requires each student at a certain class level to do an internship. These activities require students to carry out these activities outside of school and of course the learning process does not stop completely. Therefore, we need a learning media that can help teachers in online learning. So that the learning process can continue. One of them uses website-based learning media. The website-based learning media at SMKN 1 Sawoo aims to help teachers and students in carrying out teaching and learning activities remotely. The research method used is the waterfall method. The media needed is a learning media website that can be used to deliver material and conduct learning evaluations. The material that can be delivered is of course multimedia based, which in addition to text can be in the form of images, videos and audio. In addition to being used for delivering material, this online learning media website can be used as a medium to do assignments given by each subject teacher.

## 1. Introduction

Developing technology covers various fields, starting from business activities, information, entertainment and also education. The development of a technology has a purpose as a media or tool to facilitate users in doing their jobs. Among them a job is possible to be done in a different place or done remotely (Online). One of the things that can be done is the scope of education.

SMKN 1 Sawoo Ponorogo is a vocational high school that allows for the distance learning process. This needs to be done because the policy of each Vocational School requires each student at a certain class level to carry out business activities or PKL (Field Work Practices). The activity requires students to run outside the school and of course the process of learning is not stopped completely. Thus the students will need more time to take part in learning and of course it will be difficult for students who run Pratice activities outside the City.

Based on the above exposure, a medium that is capable of assisting students to obtain learning materials Online during the process of Prakerin activities takes place. The media is a learning media website that can be used to convey material. The material that can be submitted must be various in addition to the text, the material that is sonicated can be images, video and audio. In addition to being used for material delivery, this online learning media website can be used to work on and accomplish the tasks provided by teachers of each subject.

Website is chosen as an alternative media learning because according to some studies, the media website has various advantages including in the form of e-learning so that the learning process is more efficient and effective because it can be done anywhere and anytime.[1]

Therefore, the author plans to design a website "The Designing of Website-Based Learning Media as an Alternative forOnline Learning for Student Practicing at SMKN 1 Sawoo Ponorogo ", which is made especially for students of SMKN 1 SawooPonorogo. With hope later can assist in the learning process Online and develop the advancement of learning media technology in the future.

## 2. Learning Media

Media can also be called audio-visual equipment, meaning tools that can bring up images and sound, which can be used in the learning process with the intent to make the way to communicate more effectively and efficiently. With the use of a media teacher and students can communicate more accurately and the interactions are many directions or long distance.[2]

According to Daryanto (1993:1), that media is a tool that can help teach learning activities that serve to clarify the meaning of the message conveyed so that the teaching purpose can be delivered better. Media in the educational technology area is a source of learning that is a combination of materials and equipment.[3]

## 3. Website

According to Hidayat, R. (2010), the website or website is a collection of pages that serves to display text, image, animation, sound, video, and other information both static and dynamic, each of which is connected to the Various page networks. The relationship between one Web page and the other Web page is called Hyperlink, while the text being used as a link media is called Hypertext. [4]

## 4. Research Methodology

The research methodology used in the app-based learning media design process in SMKN 1 Sawoo Ponorogo consists of several phases. Starting from the library study to collect various literature, followed by observation methods and interviews. Observations were conducted at the scope of teachers and students regarding teaching and learning process during the practice work, and then the interviews were conducted on teachers and students. Next is the design stage, in the process of designing conducted research using the Waterfall method and using various tools, among others, Context Diagaram, DFD, ERD and so on.

### *4.1. Literature Review*

Literature riview is done by searching literature related to the issues studied, namely the literature on learning media, alternative media, website designing. Some sources of literature on this study were obtained from books, journals and some online media.

### *4.2. Observation*

Observation is done on teacher and student learning activities during practice work. To get the curriculum information and learning system in SMKN 1 Sawoo and to obtain information and classification of problems in SMKN 1 Sawoo Ponorogo, this obtaining information on system needs tailored to the design System for Learning Media at SMKN 1 Sawoo.

### *4.3. Interview*

The interview was conducted to several teachers and students of SMKN 1 Sawoo Ponorogo, to obtain information of teachers and students needs tailored to the needs of the system design.

## 5. Analisys

### *5.1. Analysis* of Interview Results

From the interview results of several teachers at SMKN 1 Sawoo. That the teacher needs an online medium that can be used to efficiently study the learning process for students during an off-school practice activity process. So students can get learning during a practical practice.

The results of the analysis carried out to several students, that students need learning media that can help them get learning materials during the process of practical activities, without having to come to school. So that practice activities outside the school can run optimally and still get learning material.

### *5.2. SystemRequirements Analysis*

The results of analysis on the system that will be made that there is an administrator in charge of managing the learning media system, then each teacher can private access to be able to manage the subjects and classes according to the task chasing him. It is able to add text, image and video materials to the subjects and classes they teach.

Students get private access based on their parent number to be able to view material that has been shared by the teacher of the respective subjects according to the set schedule. Each student is required to do online learning activities during the practice.

### *5.3. Hardware Analysis*

To support the running of the learning media that is designed it takes a hardware device, namely a complete computer. At PonorogoSawoo 1 Vocational School already have a computer device that is quite complete so that it can help the system run. Based on the results of the analysis the user needs a computer that is used to manage and run a learning media website with minimum specifications as written in the table below:

**Table 1.**Hardware Analysis

| No | Component | Spesification |
|---|---|---|
| 1 | Processor Intel | Core i3 |
| 2 | Memory (RAM) | 4 GB |
| 3 | Harddisk | 500 GB |
| 4 | Mouse | Usb Mouse |
| 5 | Audio | Speaker |
| 6 | Lan | Internet Connection |

From the results of the analysis of hardware requirements above, that SMK 1 Sawoo already has computer peripheral components that meet the system requirements.

### *5.4. Software Analisys*

Software requirements used to support the running of the system with minimum specifications using the Windows 7 Ultimate 64 bit operating system. Windows 7 Ultimate was chosen because of the fairly lightweight specifications and can be used for the latest generation of software and the older generation of software.

## 6. System Design

### *6.1. Conteks Diagram*

The context diagram is the highest level part that describes all inputs to the system or output and the system that describes the whole system. Context diagram is the highest level in the data flow diagram and only contains one process. [5]

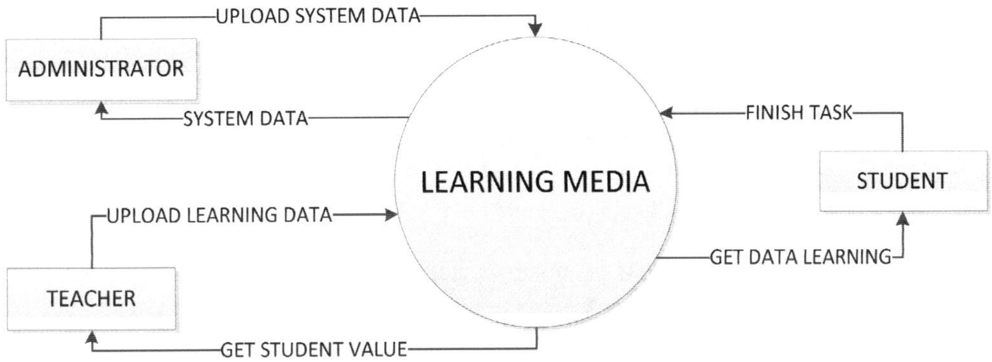


**Figure 1.** Context Diagram

The context diagram above explains that in the system designed there are several users, namely:

- Administrator
  The task is to manage the online learning media system, including the input of teacher data and student data at SMK 1 SawooPonorogo.
- Teacher
  Duty to input material and assignments according to the subjects taught based on a predetermined schedule. Give an assessment of each student who does the task through online learning media.
- Students
  Receive material from subject teachers of each class and do assignments given by subject teachers through online learning media.

## 7. Interface Design

### *7.1. Login Design*

In designing the login page there are two columns that function to input the id number and password of the users starting from administrators, teachers and students. All users can enter based on the id number that has been adjusted by the administrator. Following is the design of the login page in explained in Figure 2.

**Figure 2.**Login page design

### *7.2. Administrator Page Design*

The admin page is used for users as administrators, administrators have the right to full access to the learning media system. Administrators are in charge of input and manage all data requirements needed to run a learning media system, starting from teacher data, student data, learning schedules and much more. The administrator page design as in Figure 3.

**Figure 3**. Administrator page design

*7.3. Teacher Page Design*

Teacher pages are used by teaching teachers to input learning material according to the subjects being taught, assign assignments and provide assessment to students according to the results of doing the assignments. Next is the teacher page design as in Figure 4.

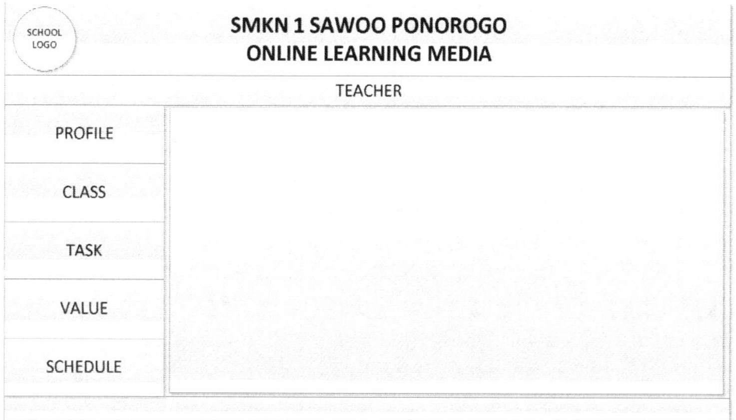


**Figure 4.** Teacher page design

*7.4. Student page design*

Students can only access student pages provided by online learning media systems. Each login is in accordance with the specified learning schedule, automatically students also attend attendance. On this page students will get learning material that has been shared by subject teachers. Students can work on assignments and can see the schedule of lessons that have been determined by the school. The following is the design of the student page as in Figure 5.

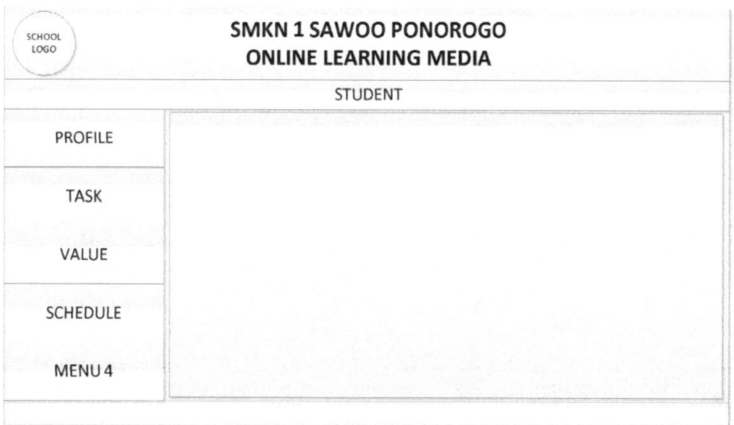


**Figure 5.** Student page design

## 8. Conclusion

1. Conventional learning is less effective if done when students are running practical work outside of school.
2. System design can help facilitate programmers in making online learning media. So that later this online learning media website can be used in the distance learning process at SMK 1 SawooPonorogo.

## 9. Suggestion

1. Better learning media are needed to improve the process and quality of learning at SMK 1 SawooPonorogo.
2. Development of an online learning media system is needed at SMKN 1 SawooPonorogo Vocational High School from the design section, complete system and detailed system requirements. So that it becomes an optimal online learning media system.
3. With the design of online learning media at SMK 1 SawooPonorogo, it is expected to be developed into a learning media application that can run on various devices.

## References

[1] Lovy Herayanti, Muhammad Fuaddunnazmi, H. 2017. *Pengembangan Perangkat Pembelajaran Fisika Berbasis Moodle*. Jurnal Pendidikan Fisika Dan Teknologi, *3*(2), 107–206.

[2] W. Wibisono and L. Yulianto 2010, *Perancangan game edukasi untuk media pembelajaranp pada sekolah menengah pertama persatuan guru Republik Indonesia Gondang Kecamatan Nawangan Kabupaten Pacitan*,J. Speed Sentra Penelit. Eng. dan Edukasi, vol. 2, no. 2, pp. 37–42.

[3] I. Trisna Santi 2014, *Pembuatan Game Pembelajaran Pengenalan Huruf Hijaiyah Di Taman Kanak-Kanak (TK) Az-Zalfa Sidoharjo Pacitan,*J. Speed-Sentra Penelit. Eng. dan Edukasi, vol. 6, no. 2, p. 7.

[4] Hidayat, R. 2010. *Cara praktis membangun website gratis*. Elex Media Komputindo.

[5] Muslihudin, M. 2016 *Analisis Dan Perancangan Sistem Informasi Menggunakan Model Terstruktur Dan UML*. Penerbit Andi.

# Android Radio Streaming Apps for Songgolangit FM Ponorogo

**D Muriyatmoko[1*], N A S Asy'ari[2], M S Arif[1]**

[1]Department of Informatics Engineering, University of Darussalam Gontor, Siman Road Km.6, Siman, Ponorogo, East Java, Indonesia 63471
[2] Department of Communication Science, University of Darussalam Gontor, Siman Road Km.6, Siman, Ponorogo, East Java, Indonesia 63471

*dihin@unida.gontor.id

**Abstract**. Songgolangit FM Radio is one of the information media in Ponorogo society. The broadcast signal has limited covered areas because of the geographical condition of Ponorogo. Then, the other media is needed to solve a problem. This research aims to expand the broadcast area using Android radio streaming apps. This research used the waterfall steps. The test using black box shows that the apps running well and have no error. The feedback from Songgolangit FM peoples expected that these apps hope could expand new user with especially to Android users. These apps can be download by all Android phone through the Google Play Store (https://play.google.com/store/apps/details?id=com.amoled.radiosonggolangit&hl=en). On the next research can measure the effectivity of radio streaming Android apps. Besides, future research also can develop its apps in other operating systems such as iOS, Tizen, Hongmeng and Blackberry mobile apps.

## 1. Introduction

The Songgolangit Radio is a radio located on Jalan Halim Perdana Kusuma Number 12 Ponorogo. Radio Songgolangit FM broadcasts at Frequency 99.2 FM. This radio which has the motto "Information and Solution, Pride of the Ponorogo Radio" hopes to become the proud radio station of the City of Ponorogo and the people of Ponorogo. With a varied program to entertain the entire Ponorogo community. Radio Songgolangit FM also gives a warm feel to listeners by providing services through SMS and social media [1].

In the era of technology 4.0, Songgolangit FM radio still has some problem. One of them is the limited broadcast area. It can happen because Ponorogo has many various highlands and some mountains. Its geological condition can block the broadcast signal. Songgolangit FM already has web-based streaming radio, but it's still felt lacking because most Ponorogo people use Android phone. This research offers a solution using the Android mobile streaming application. Android was chosen because in 2017 have 73% of Global mobile devices use Android [2].

Many previous types of research to develop radio streaming. The first from Hatta, Gusfiannur, and Fahrul is building Android-based radio streaming application "TDA Samaradio" is expected to be a media to share information that can help members to find out various knowledge about entrepreneurship [3]. The secondary research from Said using streaming application for mobile devices using Java 2 Micro Edition (J2ME) has been successfully simulated, but if an experiment is conducted on mobile phones that have Java and GPRS facilities, often the connection to the server fails even if Java is successfully downloaded [4]. The third research from Simatupang is web-based radio streaming application using the website is show buffering will occur if the internet connection and or shout cast server conditions are unstable [5]. The fourth research from Maulina and Kurniawan was building radio streaming application using the website will be used by Suara Bangkalan FM radio connoisseurs so as not to be obstructed by geographical and weather factors [6].

This research aims to expand the broadcast area using radio streaming Android apps. Its apps expected able to increase the number of listeners. Its apps hope to help listener, especially users in an area with many various geological problems. Besides, this application is expected to increase the number of listeners with an Android phone.

## 2. Methods

The Android radio streaming apps used for the system development Waterfall System Development Life Cycle. The waterfall model where this illustrates a systematic and sequential approach to software development, starting with the specifications of user requirements [7]. The Waterfall model phases consist of 5 stages, as shown in Fig. 1.

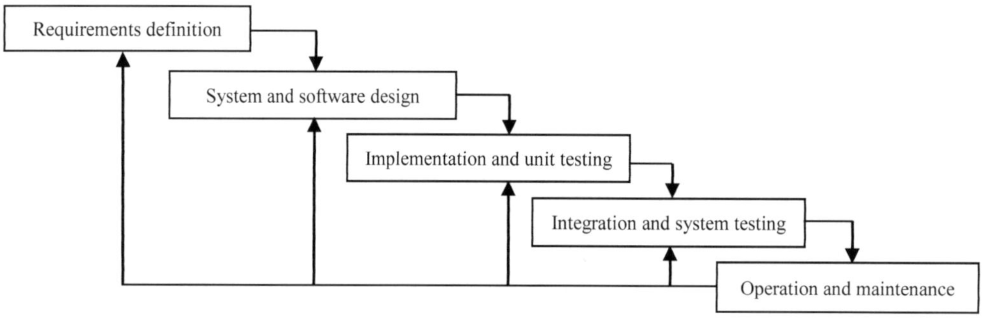


Fig. 1. Waterfall System Development [7]

### *2.1 Requirements definition*

The first stage is requirements definition is performed the service, limitation, and objective. It also identifies a problem related to data of how many the radio in Ponorogo used Android streaming. The requirement in journal source results shows that radios in Ponorogo, on average, have converged media to support radio broadcasting activities. New media used are websites, radio streaming, android applications, YouTube and social media (Facebook, Instagram, BBM and WhatsApp's) [8].

But in fact, Songgolangit FM radio does not have a radio streaming mobile application. At the interview with them, they wanted the Android radio application. According to that case, Ponorogo people have a smartphone, especially Android. In another case, Songgolangit FM have many problems such as the basic lack of knowledge on all employee, lack of room facilities, have no IT division, etc.

### *2.2 System and Software design*

The locations studied were Radio Songgolangit FM in Jalan Halim Perdana Kusuma Number 12, Siman Subdistrict, Ponorogo Regency. Data analysis and processing to build Android apps at the University of Darussalam Gontor Ponorogo. Tools in this research are using computer hardware and software such as Java language programming, Android Studio, and XML language.

Fig. 2. A use case diagram for Songgolangit FM

Based on fig. 2, in the Songgolangit FM application, users can play or stop streaming. Besides, user can view about Songgolangit FM, website Songgolangit FM, share apps, and about apps. While edcast can play playlists by specifying the settings beforehand which will be sent to the Ice cast service which will then be broadcast and can be received through the application. In this research only explain about making applications from radio listeners

## *2.3 Implementation and Unit Testing*

The results have appeared in Fig. 3 (c) is the main page that contains four navigation menus. It is a home, about radio, website radio, share apps, and about apps. Since the server is on, the user can listen to the broadcast program through the streaming. It will display with play and stop command. So, the user can stop and continue the broadcast streaming as needed.

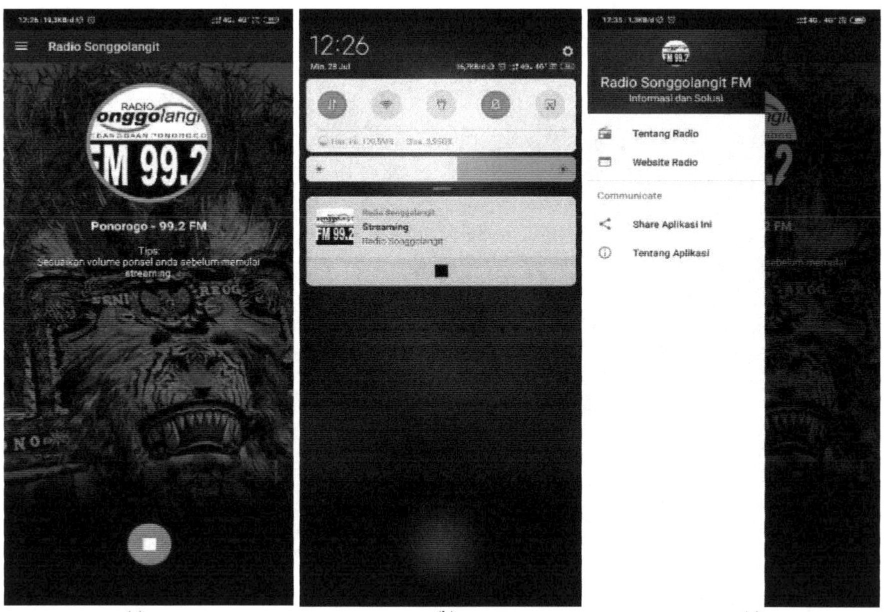


(a) (b) (c)

Fig. 3. Songgolangit FM Android radio streaming apps interface view[7][7][7]

Based on Fig 3. (a) shows that the application is online. Fig 3. (b) While the apps are online also can see in the Android interface on the setting menu. It also shows that the apps are running in the background. The user can stop the broadcast streaming through the Android interface on running background menu. Fig 3. (c) the share menu button can be used by users who want to share Songgolangit FM applications with other users through various social media, such as email, Facebook, Twitter, Instagram, etc.

*2.4 Integration and system testing*

If all parts and requirements for system specifications are met, the integration stage can be continued. Data that has been displayed by programmer page will be visible directly in Google Playstore. In this part of the system, Android radio streaming apps are can be tested directly by the user.

*2.5 Operation and Maintenance*

In the last part of this Android radio streaming apps system are operation and maintenance. All data starts from the information about the company until social media pages have shown directly. Furthermore, Songgolangit FM management can try and see instantly so that the data displayed is valid. So that the information on these apps can be useful for the community, especially user in Ponorogo area.

## 3. Result and Discussion

This research projection of Android streaming mobile apps for Songgolangit FM divides into two main page functions. The first menu is the main menu of these apps, namely the play button and stops streaming. The second menu is general information consisting of four buttons. The button is about Songgolangit FM radio, share applications, about applications and the Songgolangit FM radio website.

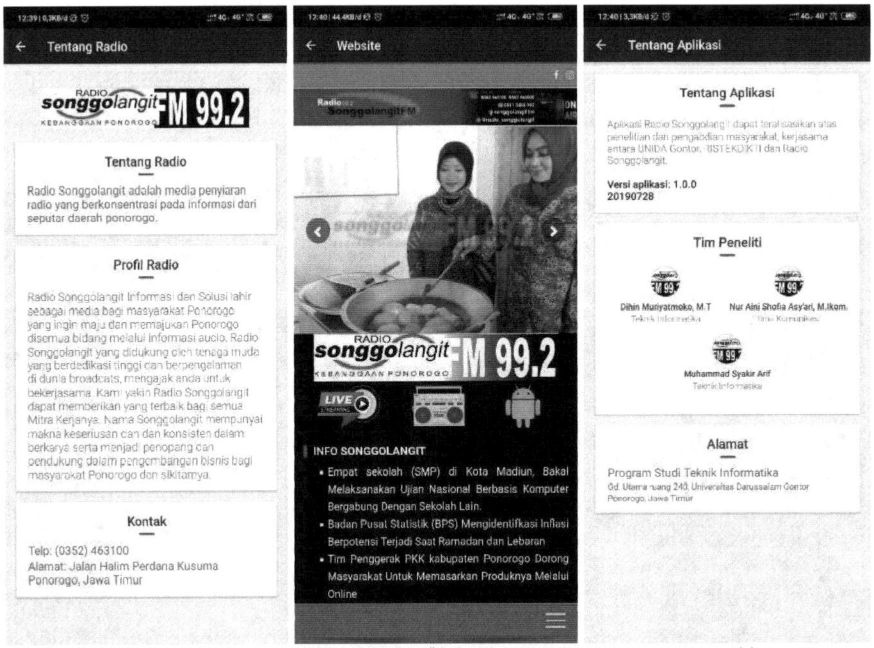


(b) (b) (c)

Fig. 4. Songgolangit FM Android radio streaming apps interface view by the listener[7][7][7]

In Fig. 4 (a) is a menu about radio contains profile information, addresses, contacts, and a glimpse of Songgolangit radio. In Fig. 4 (b) is website menu, if the user selected this button, it will go directly to the Songgolangit FM radio website address, https://www.songgolangitfm.com. In Fig. 4 (c) is about apps contains version and date of this apps, developer team and developer address.

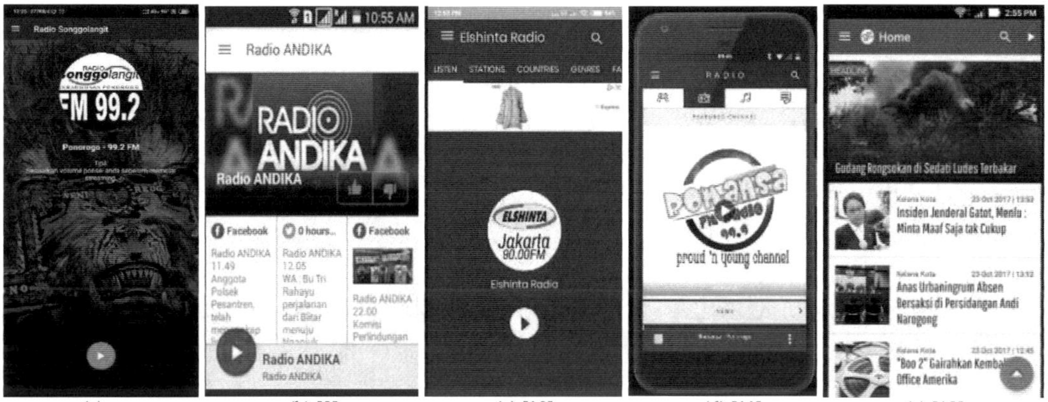


(a) (b) [9] (c) [10] (d) [11] (e) [12]

Fig. 5. Songgolangit FM Android radio streaming apps interface view by listener

The functional test results with black box show that this application runs smoothly and without errors. The results of the response from Songgolangit FM give the impression that hopefully, this application can increase the number of listeners. When compared with other radio streaming this application has features that are not much different. The advantages of this application compared to others, is almost non-existent because all the applications have the same function, namely as an Android-based streaming media.

In Fig. 5 (a) is an Android application belonging to Songgolangit radio Ponorogo 99.2 FM led by Setyo Budi. In Fig. 5 (b) radio Andika Kediri 105.7 FM led by Rofik Huda [9]. In Fig. 5 (c) Elshinta radio Jakarta 90.0 FM led by Iwan Haryono [10]. In Fig. 5 (d) Romansa radio Ponorogo 98.8 FM led by Zahtul Azyhary [11]. In Fig. 5 (e) Radio Suara Surabaya 100 FM, led by Edy Prastyo [12]. It four radios, Andika FM and Suara Surabaya FM radio, were examples and comparative studies in the development of Songgolangit radio management [13].

## 4. Conclusion

The research produces an Android radio streaming apps for Songgolangit FM with have no error and running well. The apps can be download on Google Playstore with URL Address https://play.google.com/store/apps/details?id=com.amoled.radiosonggolangit&hl=en. Its apps expected to help Songgolangit FM increase the listener that have an Android phone. In addition, its apps expected to help the user in various highland and mountains area. Next step, the researcher can try to develop the similar apps in a different operating system, such as: iOS, Tizen, Hongmeng and Blackberry mobile phone. Next research also can continue to measure how far this apps can be effectively used by Ponorogo society.

## Acknowledgements

The research reported in this publication is supported by the Research Grant Program of the Republic of Indonesia Ministry of Technology and Higher Education in 2018. This content is entirely the responsibility of the author and does not always represent the official view of the Ministry of Republic of Indonesia Technology and Higher Education.

## References

[1] Suharmanto H 2017 *Strategi Penyiaran Radio Songgolangit FM Ponorogo dalam Memberikan Informasi Seputar Ponorogo pada Program Acara Graha Warta* (Muhammadiyah University of Ponorogo)

Journal of Physics: Conference Series **1381** (2019) 012017 doi:10.1088/1742-6596/1381/1/012017

[2] Anon 73% Perangkat Mobile Global Menggunakan Android | Databoks
[3] Heliza Rahmania Hatta, Gusfiannur G and Agus F 2015 Rancang Bangun Radio Streaming Berbasis Android *J INTAKE* **6** 66–81
[4] Said I M 2005 Aplikasi Untuk Perangkat Bergerak Menggunakan Java 2 Micro Edition ( J2ME ) *Semin Nas Apl Teknol Inf* **2005** 117–21
[5] Simatupang R M 2014 Penerapan Metode User Centered Design Untuk Perancangan Aplikasi Radio Streaming Berbasis Web *Inf dan Teknol Ilm* **3** 1–5
[6] Maulina Sinta F and Kurniawan A 2019 Pengembangan Aplikasi Radio Streaming dan On-Demand Berbasis Website di Radio Suara Bangkalan FM *J Manaj Inform* **9** 8–17
[7] Sommerville I 2010 *Software Engineering* (USA: Addison-Wesley Publishing Company)
[8] Asy'ari N A S 2018 Broadcasting Radio Ponorogo dalam Kajian Konvergensi Media *Petanda J Ilmu Komun dan Hum* **1** 37–43
[9] Huda R 2017 Radio ANDIKA - Apps on Google Play
[10] Haryono I 2018 Elshinta Radio Jakarta - Apps on Google Play
[11] Azyhary Z 2019 Romansa FM - Ponorogo - Apps on Google Play
[12] Prastyo E 2018 Suara Surabaya Mobile - Apps on Google Play
[13] Songgolangit R T 2019 Radio Songgolangit FM Kebanggaan Ponorogo, Belajar Bareng Mr Errol Jonathans (RadioMan) CEO Radio Suara Suara Surabaya - Radio Songgolangit

# The Concept and Implementation of Smart Room using Internet of things (IoT) for Cost Efficiency and Room Security

**R Y Endra[1*], A Cucus[1], F N Affandi[2]**

[1]Informatics Study Program-Faculty of Computer Science Universitas Bandar Lampung (UBL) Jl. Zainal Abidin Pagar Alam no 26 Bandar Lampung
[2]Informatics Study Program- School of information and computer management Bandar Lampung

robby.yuliendra@ubl.ac.id

**Abstract**. The Internet of Things (IoT) is a concept where internet connectivity can exchange information with each other with objects around it. The essence of IoT is interconnected devices that produce and exchange observation data, facts, and other data, so that it is available to anyone. In this paper we present how the smart room model is designed using sensors and micro-controllers to automate the use of electronic devices and the security of a room using the concept of the Internet of Things. Implementation of the smart room concept from the results of this study, we hope that the concept in this smart room can be implemented and the automation process in this smart room can have a major impact on the efficiency of operational costs, especially electricity payments and improve home security because there is automatic control.

## 1.Introduction

Changes in the dynamic times, especially in the field of communication and information technology, can not be avoided by anyone. At present we enter the era of the industrial revolution 4.0 which requires every human being to make dynamic changes in carrying out activities in everyday life. This certainly changes the way of human thinking for the present as well as the future.

Internet of things (IoT) has become one of the most popular technologies. The idea behind this is to have objects from everyday life connected to the internet. This allows objects to continue sending data to the web and can be accessed from anywhere. It will also be possible to interact directly with one another, this is called Machine to Machine communication or M2M. [1]

Nowadays, efficient and sustainable energy usage is an urgent necessity that influences technology development in numerous areas (e.g. building, transport and energy production and conservation systems). In this context, emerging Smart Grid technologies have been developed to reduce the energy consumption of electric devices installed at home, to seek out the lowest rates and contribute to the smooth and efficient functioning of the electric grid. Although many Smart Grid devices today are commercially available, the limited service scalability, the complexity of configuration and the low usability prevent their mass adoption.[2]

A smart home is one that has a highly sophisticated automated system to control multi-media equipment for monitoring and enable the security forces (alarm and alert) associated with the windows and doors, lighting and temperature and many other functions. A smart home appears "intelligent" because of a computer system that can monitoring many aspects of daily life.[3] The conclusion is that a smart home exists because of a computer system that can monitor many aspects of everyday human life.

Why does that matter? The main reason we make smart rooms for resource efficiency is that many electronic devices that forget to turn off will be on for 24 hours. This is a waste of electricity. Many electronic equipment is still on when there are no lectures. This causes high electricity bills to be paid. Then the background of the importance of smart rooms for home security in particular is the leakage of gas in the house. The house becomes one of the important parts in human life because the house becomes a shelter and also a place to spend time on their daily activities. However, based on statistical data in Figure 1.1 at this time the house is only used as a place to sleep at night and left by the owner in the morning until the evening, making the house empty or uninhabited. Such circumstances make the house vulnerable to danger, both unexpected disasters and security for the items contained in the house. There are many cases of theft and fire in an empty house because the owner left it. It can be seen from Figure 1.1 that the statistics of the number of fires based on their causes from 2014-2019. The cause of the fire in the statistics in Figure 1.1 comes from electricity, cigarettes, stoves and others. This makes us as homeowners must be vigilant at all times.

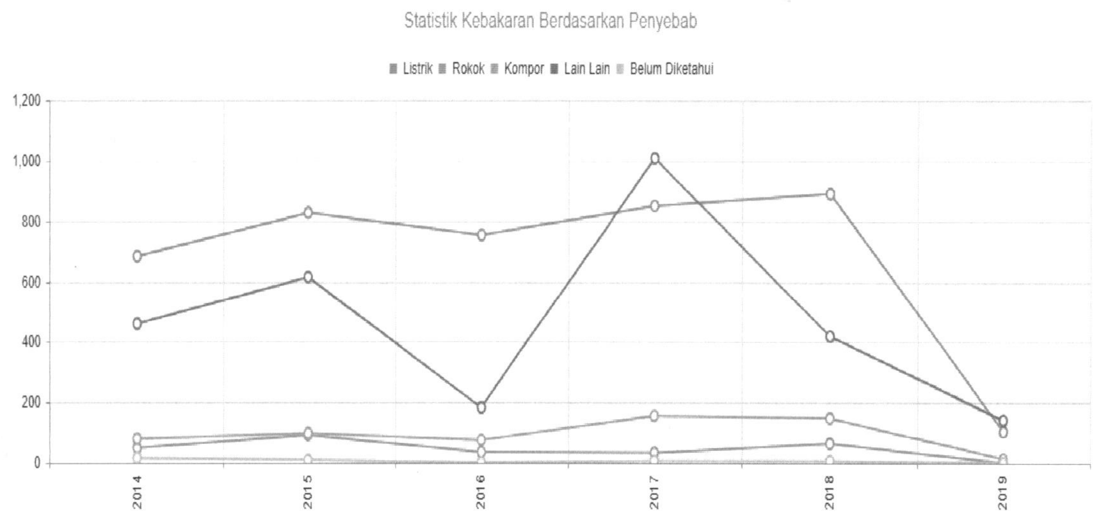


**Figure 1.** Fire Statistics Based on Cause [4]

The solution for solving these problems is a smart room system model using Arduino micro-controller and several sensors such as PIR sensor, temperature sensor, and ultrasonic sensor.

The Arduino is basically open-source electronics prototyping platform that is made up of two essential parts: the hardware, which is the Arduino board, and the software, the Integrated Development Environment (IDE). At the heart of every Arduino board is a Micro-controller Unit (MCU), a type of Integrated Circuit (IC) with a processor, embedded memory, and programmable I/O peripheral devices. [5]

## 2.Related Work

Research entitled "Smart Home: Architecture, Technologies and Systems" explains that smart home is an important part in the concept of an automatic network. The automation process between humans and electricity can be done in real time. The application of smart home can increase the efficiency of electricity usage due to the increased use of electricity networks and the pattern of users or humans in the use of this electricity network. Broadly speaking, this study describes the architectural design in making the concept of smart home or smart room. Technologies that can be used to implement the concept of smart home include the Internet of Things (IoT) platform, computer technology, control technology, image display technology and communication technology.[6]

Another study entitled "**Smart homes and the control of indoor air quality**" is motivated by global climate change, demographic change, and advanced mechanization of daily life that will go hand in hand with new ways of living. Extreme temperatures, an aging society and higher demands for a comfortable life will lead to the implementation of sensor-based networks to create acceptable and improved living conditions. The idea of smart home mainly serves the efficient use of energy and optimization of ventilation technology that is connected with new ways of building a structure (low energy and passive houses respectively).[7]

Another study entitled "**IoT Big Data Analytics for Smart Homes with Fog and Cloud Computing**" explains that the analytics of an Internet of Things system is an important means of gaining knowledge and supporting applications for making smart home / smart rooms. The equipment connected to the smart home system generates big data about users and their daily activities. IoT analytics can help personalize applications that benefit both homeowners and industry that can benefit from consumer profiles. This research proposes the use of fog nodes and cloud systems to enable data-based services and overcome the challenges of complexity and resource demands for the processing, storage, and classification of online data and analysis offices.

## 3.Methodology

### *3.1.Research Method*

In this study we used applied research where we focus on the application of the results of research conducted in the community to solve problems, not to focus on development and the impact that can be felt directly. Applied research is done with the aim of implementing, testing, and evaluating the ability of a theory to be applied in solving practical problems.[8]

Applied research is done a lot to solve practical problems of human life, and applied research usually refers to theories produced by basic research. Researchers who develop certain products have a purpose to solve existing and useful problems in people's lives.[9]

### *3.2.Groove system*

The following is a prototype system flow chart of the smart-home application for home lighting automation, garden lighting, and gas leak detection using sensors and Internet of Things (IoT) technology. The first thing to do is that the user must log in to enter the main menu. After logging in, the user will be able to access the main menu of automation.

In the menu there are several choices of automation buttons namely the on and off button for home lighting, the on and off button for garden lights, and the on and off button for the blower (fan) and also there is a place for notifications. If the homeowner wants to turn on or turn off the lights from outside the house, he can use the Smart-home application, and if the homeowner is at home then he can use an automatic system using sensors.

Furthermore, if a gas leak occurs, the MQ-2 sensor will detect and automatically the buzzer along with the blower (fan) will turn on. After that the system will automatically send notifications to homeowners.

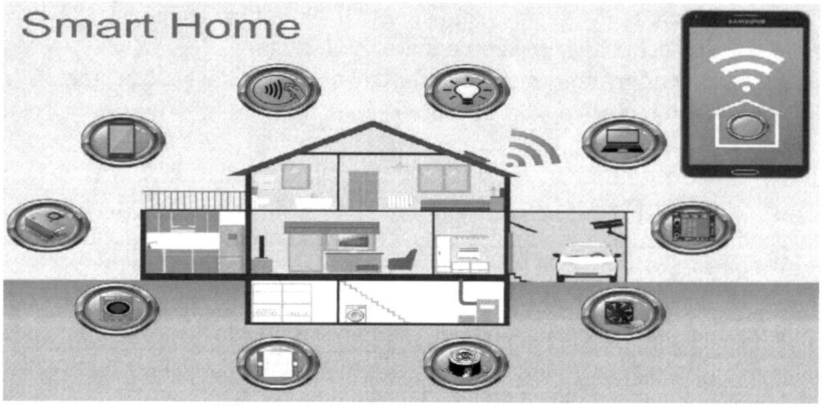


**Figure 2.** Smart Home Master Plan

*3.3. General system design*

In the general design of a smart room system, it can be explained as follows: sensors currently in use such as the MQ-2 Sensor, PIR Sensor, LDR Sensor and fire sensor are inputs that will send data through the node MCU. NodeMCU receives a command and then checks the command in the program. If the command exists, nodeMCU will send commands to Relay and send commands to firebase. The relay will automatically turn on or turn off tools such as home lights, garden lights, fans, and buzzers. Meanwhile firebase will send notifications to the android application. The general system design scheme can be seen in the image below:

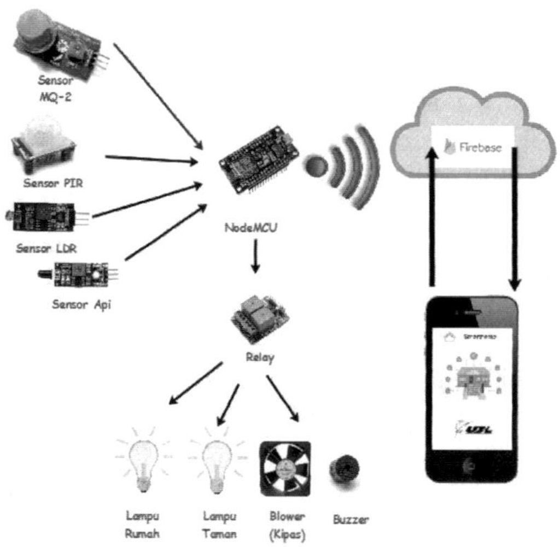


**Figure 3.** General system design

## 4. Results And Discussion

*4.1. The design of Smart Home Prototype uses the MCU Node*

The prototype design in this study uses the NodeMCU micro-controller as the centre of Controlling. The output processors are motion sensors, light sensors, gas sensors (MQ-2) and fire sensors. The output generated from NodeMCU is in the form of LED (Light Emitting Diode) lights that light up, representing room lights and garden lights. The gas sensor will produce output in the form of a live blower. The following is a design drawing of a prototype smart home using NodeMCU:

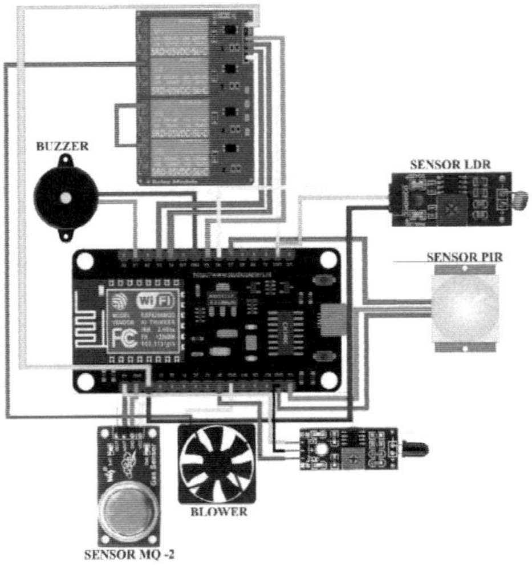


**Figure 4.** NodeMCU Schematic Series

The picture above declares the pins connected to the MCU node from existing devices such as PIR sensor, LDR sensor, MQ-2 sensor, buzzer, blower, male-male and male-female jumper cables, and 1 relay module with a total of 4 channel.

*4.2.Coding of Arduino IDE (Integrated Development Environment)*

The Arduino IDE application is used to run programs from the results of input devices in the form of sensors to be processed and these tools produce an output. After all the hardware circuits are finished, the Arduino program is run by uploading it to NodeMCU.

In this study Arduino IDE Software (Integrated Development Environment) uses version 1.8.4. By using the Arduino IDE (Integrated Development Environment) we can embed commands into the NodeMCU micro controller. The following is an Arduino 1.8.4 interface:

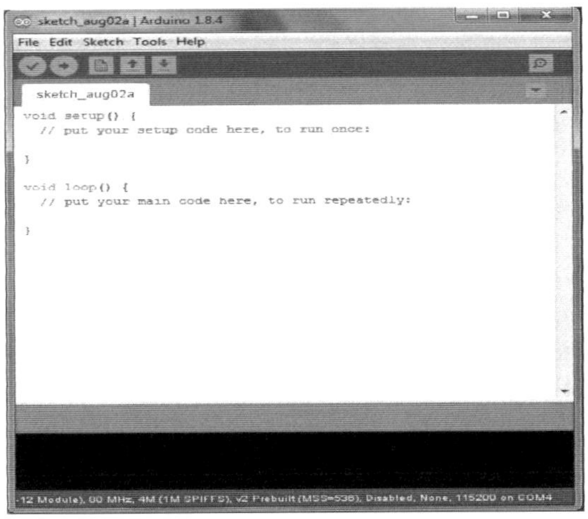


**Figure 5.** Arduino IDE Interface Display

To embed commands into Arduino, we use the syntax in Arduino called *Sketch.*

```
void setup() {
Serial.begin(9600);
```

Journal of Physics: Conference Series **1381** (2019) 012018 doi:10.1088/1742-6596/1381/1/012018

```
    pinMode(buzzer, OUTPUT);
    pinMode(smokeA0, INPUT);
    pinMode(inputPin, INPUT);
    pinMode(lamp1, OUTPUT);
    pinMode(lamp2, OUTPUT);
    pinMode(kipas, OUTPUT);
    pinMode(ldr, INPUT);
}
```

*4.3.Pre Test on resource efficiency*

Electronic devices in each class in the Faculty of Computer Science are 6 lamps, 1 Air Conditioner (AC), and 1 LCD projector.

**Estimated power usage :**

a. 1 Light : 20 Watt
   1 kWh : Rp 1.400
   Power of 1 day : 20 Watt x 8 hours = 320 Wh = 0,32 kWh
   Power of 1 month : 0,32 kWh x 20 days = 6,4 kWh
   Cost of 1 month : 6,4 kWh x Rp. 1.400 = Rp 8.960/light
   Rp 8.960 x 6 lights = Rp 53.760
b. Air Conditioner (AC)
   1 PK : 750 Watt
   2 PK : 750 Watt x 2= 1500 Watt
   Power of 1 day : 1500 Watt x 8 hours =12000 Wh = 12 kWh
   Power of 1 month : 12 kWh x 20 days = 240 kWh
   Cost of 1 month : 240 kWh x Rp 1.400 = Rp 336.000
c. 1 LCD projektor : 250 Watt
   Power of 1 day : 250 Watt x 8 hours =2000 Wh = 2 kWh
   Power of 1 month : 2 kWh x 20 days = 40 kWh
   Fee of 1 month : 40 kWh x Rp 1.400 = Rp 56.000

total costs incurred : Rp 53.760 + Rp 336.000 + Rp 56.000
= Rp 445.760

*4.4 Post Test on resource efficiency*

Electronic devices in each class in the Faculty of Computer Science are 6 lamps, 1 Air Conditioner (AC), and 1 LCD projector :

Estimated power usage :

a. 1 light : 20 Watt
   1 kWh : Rp 1.400
   Power of 1 day : 20 Watt x 6 hours = 120 Wh = 0,12 kWh
   Power of 1 month : 0,12 kWh x 20 days = 2,4 kWh
   Cost of 1 month : 2,4 kWh x Rp. 1.400 = Rp 3.360/light
   Rp 3.360 x 6 lights = Rp 20.160
b. Air Conditioner (AC)
   1 PK : 750 Watt
   2 PK : 750 watt x 2= 1500 Watt
   Power of 1 day : 1500 Watt x 6 hours =9000 Wh = 9 kWh
   Power of 1 month : 9 kWh x 20 days = 180 kWh
   Cost of 1 month : 180 kWh x Rp 1.400 = Rp 252.000

c. 1 LCD projektor : 250 Watt
   Power of 1 day : 250 Watt x 6 hours =1500 Wh = 1,5 kWh

Power of 1 month : 1,5 kWh x 20 days = 30 kWh
Cost of 1 month : 30 kWh x Rp 1.400 = Rp 42.000
total costs incurred : Rp 20.160 + Rp 252.000 + Rp 42.000
= Rp 314.160

From the calculation of estimated costs incurred during the month, a comparison can be found in the graph below:

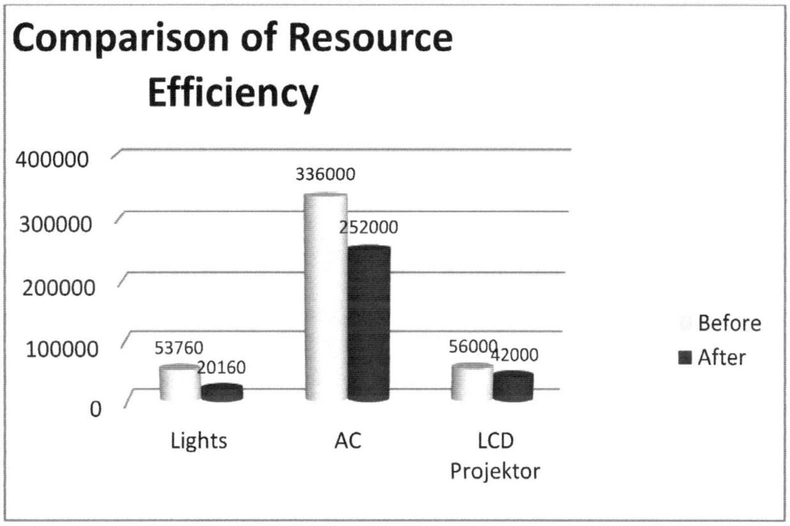


**Figure 6.** Percentage of Cost Efficiency

## 5. Conclusion

In the research conducted on automatic control of electronic devices, we draw the conclusion that:

a. Control of electronic devices can automatically manage resources more efficiently.
b. A lamp can be activated automatically when an object (Human) is detected by the camera and can be automatically deactivated when no object (Human) is detected in the room, such as the design in Figure 3.6 Modelling of smart room control. When the camera manages to detect the presence of an object (Human) it will automatically activate the DHT11 sensor as a temperature and humidity sensor which serves as an indicator to activate the B lamp. Then when there are more than 2 objects detected by the camera it can activate the C light.
c. The air conditioner can activate automatically when the room temperature exceeds 250 Celsius, and it will automatically deactivate when the room temperature is below 170 Celsius.
d. The design of the control system on the smartphone can run well based on the results of black box and white box testing..
e. Control of home lights, garden lights, and blowers can be done in 2 ways, namely by automatic systems through sensors and manual systems through smartphone applications.
f. The condition of the gas level can be monitored via smartphone; if it has exceeded the normal limit, then the application will send a notification that a gas leak has occurred.
g. Control of this lamp can make the occupant's working time become efficient and the use of electric power can also be suppressed because residents can monitor the condition of the lights from a smartphone.

**Acknowledgements**

We would like to thank the Chancellor of University of Bandar Lampung and his staff who have fully supported our research process as well as many parties who have supported this research both from a technical or technical aspect, which we cannot mention one by one. We hope our research becomes one of the scientific references in the field of computer science and is beneficial to society.

**References**

[1] Schwartz M-O 2014 *Internet of Things with Arduino* (England)
[2] Ceccacci S and Mengoni M 2017 Designing Smart Home Interfaces: Traditional vs Virtual Prototyping *Proc. 10th Int. Conf. PErvasive Technol. Relat. to Assist. Environ.* 67–74
[3] Bregman D 2010 Smart home intelligence - The eHome that learns *Int. J. Smart Home* **4** 35–46
[4] Jakarta Fire 2019 Statistik kebakaran berdasarkan penyebab *Jakartafire.net*
[5] Meruje M, Samaila M G, Franqueira V N L, Freire M M, Ricardo P and Inácio M 2018 A Tutorial Introduction to IoT Design and Prototyping with Examples 153–89
[6] Li M, Gu W, Chen W, He Y, Wu Y and Zhang Y 2018 Smart home : architecture, technologies and systems *Procedia Comput. Sci.* **131** 393–400
[7] Schieweck A, Uhde E, Salthammer T, Salthammer L C, Morawska L, Mazaheri M and Kumar P 2018 Smart homes and the control of indoor air quality *Renew. Sustain. Energy Rev.* **94** 705–18
[8] Sugiyono 2014 *Metode Penelitian Kuantitatif Kualitatif dan R&D* (Alfabeta)
[9] Azuar Juliandi, Irfan S M 2014 *Metodologi Penelitian Bisnis: Konsep dan Aplikasi* (Medan: UMSU Press)

Journal of Physics: Conference Series **1381** (2019) 012019 doi:10.1088/1742-6596/1381/1/012019

# Compilation Technique Learning Design Using Automatic Lessimic Analysis Method

**S Nita*, E R N Sari, O Prismatura**

Teknik Informatika Fakultas Teknik Universitas PGRI Madiun

*nita@unipma.ac.id

**Abstract**. The compilation process is the process of translating programs written in high-level programming languages (source programs) into machine language (object programs). Source programs such as, pascal, basic, c++ while the program objects such as compilation files have an extension file such as*.exeor*.com.The file can run on its own without he help ofthe source program (the file has property executable). Translators are divided in to two types of interpreter sand compilers.Both functions are the same, the difference lies in the translation process. In this research, a compilation machine called Automatic Lessimic Analyzer which can be used as an analysis machine. There are two job descriptions of the analysis machine, in the front as an analysis and behind as a synthesis function. For example, a simple program is included (c++) can be analyzed successfully and it can detect errors as a source program. While at the end, it will generate code generation and program object optimization.

## 1. Introduction

In learning compilation techniques a lot of references are needed can support to understand the work process of compilation machines. The compilation engine starts working when a program was run by the user. The program was made by a programmer who has mastered programming techniques. Many programs were made with different programming languages (source languages), one of them with c ++ programming languages. Programming is required to be able to analyze program algorithms are in accordance with the rules of the specified algorithm. According to Donald Ervin Knuth (1968) in his book states that Algorithms are a sequence of steps in problem solving in the form of sentences with a limited number of words and arranged logicallya and systematically [1].

When the program starts (running) it will be seen how the quality of the algorithm has been translated into programming languages. The faster the results of a program the better than this program and it is clearly free from errors in supported programs. An important factor which the programmer must know is a successful compilation technique. FirarrUtdirartatmo (2005) said that, compilation technique is a technique in reading a program written in the source language, then translated into another language called the target language. The translation process carried out by the compiler is called the compilation process (compiling)[2].

For this reason, the author asks the program to complete and manage the process to make the program inside, which contains errors and must be able to resolve the error.

Journal of Physics: Conference Series **1381** (2019) 012019 doi:10.1088/1742-6596/1381/1/012019

The problem of the question writer is quickly and precisely the results of the compilation process with the application help. The author found a reference of the help application to test the support of compilation programs, namely by Automatic Lessimic Analyzer [3].

2. Theoretical Review

2.1. Compilation Process.

The compilation process was called the translation process. So,compilation programs were a translation process from programs written in high-level languages to low-level languages (program objects)[4]. A good compilation (Compiler) can work well on computer machines and it does not require a long process. There were 3 ways to make compilation, and the language used : [5]

a. Machine language, its characteristics are very difficult to understand by humans, because it is very dependent on the engine
b. Assembly language, the features of the facility are limited, the order is brief.
c. High Level Language, character is easily understood by humans, more facility

The main task of the compiler is divided into 2, namely [6]:

a. The front end function, the task of decomposing the source program into its basic parts.
b. Synthesis (back end) function, the task is to generate and optimize the object program.

The compiler task is illustrated in the form of a compiler model as below:

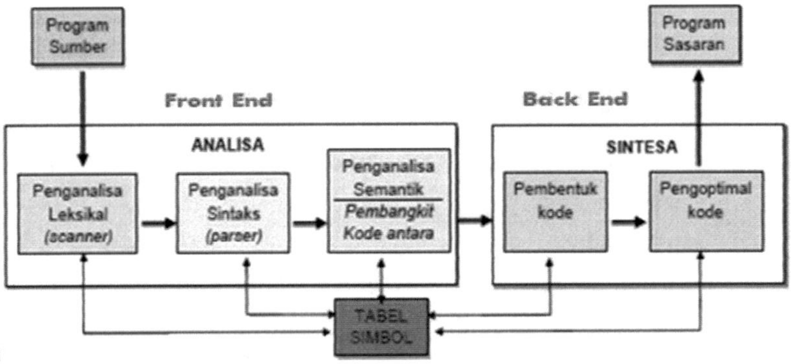


Figure 1. Compiler Model

First Function Analysis was divided into 3 stages:

a. Lexical Analyzer (Scanner), decomposes the source program into small parts.
b. Syntactic Analyzer (Parser). Syntax is the arrangement of sentences and the rules in forming sentences are called grammar. The syntax analyzer in the compilation field was often called a Parser. To analyze sentences, parse-tree help was usually used.
c. Semantic Analyzer Intermediate Code Generator, usually in realization it will be combined with an intermediate code generator (the part that functions to generate the intermediate code) [7]

The Second Function of Synthesis was divided into 2 stages:

a. Code Generator, generates object code.
b. Code Optimizer, minimizes results and speeds up the process.

The following figure was the phases of a compilation process[8]:

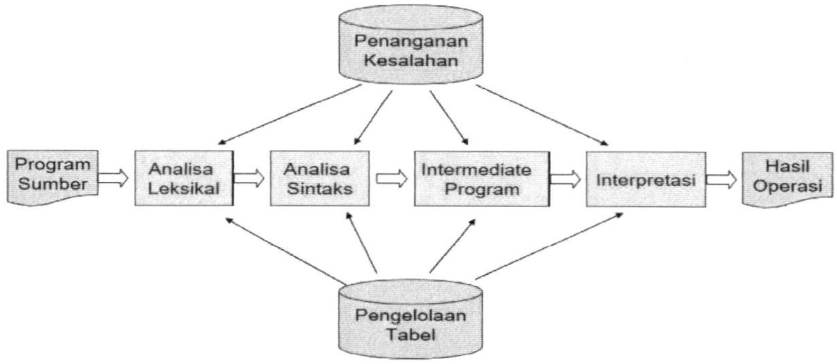


Figure2.Compilation Process Phase

Quality Compiler depends on 3 factors a.l [9]:
a. Speed & compilation time (quality depends on writing compiler algorithm)
b. Program Quality Objects, determined by the size (produced) and the speed of execution of the object program.
c. Integrated Environment has facilities integrated in the compiler.

The compiler as a compilation machine was divided into 2 types [10]:
a. Compiler
b. Interpreter

Both have the same function, it is as translators, the difference is in the compilation process. If the compiler works when the data and source code were processed at different times, while the source code interpreters and data was processed at the same time.

The following is an illustration of the compilation process and the working interpreter.

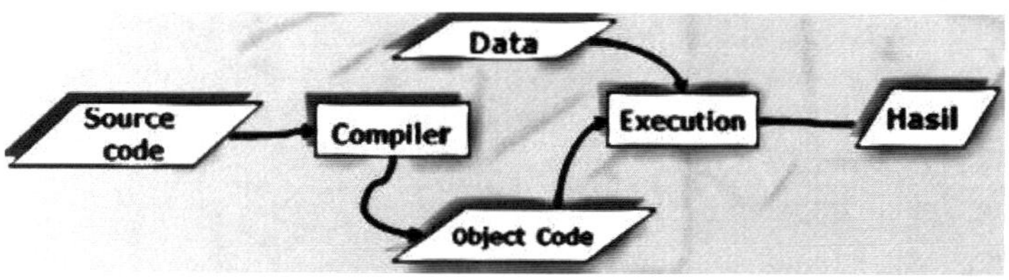


Figure3. Compiler process

Another difference from the interpreter does not produce a form of object code, but the result of the translation is only in an internal form, where the parent program must always exist, different from the compiler.

Figure 4. Interpreter process

2.2. Black Box Testing.

Black box testing was a test which done only observing the results of execution through test data and functional checking of software. It can be analogous to seeing the shape of a black box only; can only see its outer appearance, without knowing what is behind the black box. So, it can be interpreted that black box testing only evaluates on its external appearance (interface), without knowing what actually happens in the detailed process (in short, only knowing inputs and outputs).

So, this Black Box testing method performs software testing on application functionality, does not involve internal structures. Special knowledge of the application code/ internal structure and programming knowledge are generally not needed. Test cases are built based on specifications and requirements, namely the application what should be done. Use an external description of the software, including specifications, requirements, and designs to reduce test cases. This test can be functional or non-functional, although usually functional. The designer tests by selecting valid and invalid inputs and determining the correct output. The picture below shows the scheme of black box testing:

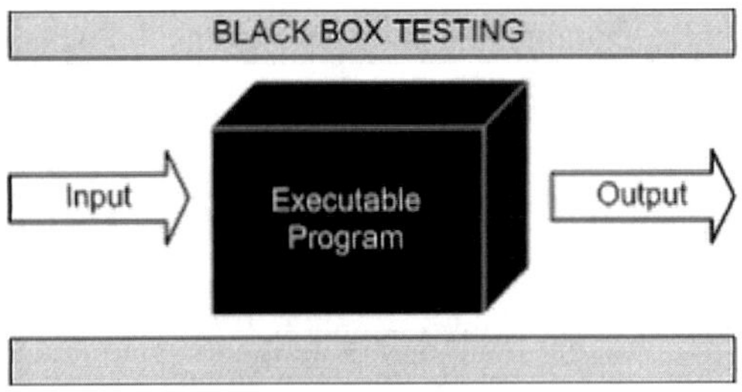


Figure 5. Black Box Testing

Testing on the Black Box will find errors such as:
a. Incorrect or missing functions
b. Interface error
c. Errors in data structures or external database access
d. Performance error
e. Initialization and termination errors

Score limits for testing, including several values, namely
a. Minimum value of input variable
b. Score above minimum score
c. Normal value
d. Score below the maximum score
e. Maximum score

The advantages of black box testing:
a. Program specifications can be identified at the beginning
b. It can be used to assess program consistency
c. Testing is carried out based on specifications
d. There is no need to look at the detailed program code

The disadvantages of black box testing: if the program specifications are made less clear and concise, it will be difficult to make documentation properly [11].

3. Methodology

There are 2 methods used in this study, namely:

a. Descriptive method, where in descriptive writing the results of the study will be written clearly and describe the essence of writing
b. Software Development Method, the method used is SDLC quoted from Sukamto and Shalahuddin's book, in the year of 2013 inYokiFirmansyah[12]. The figure below shows the parts of the SDLC method.

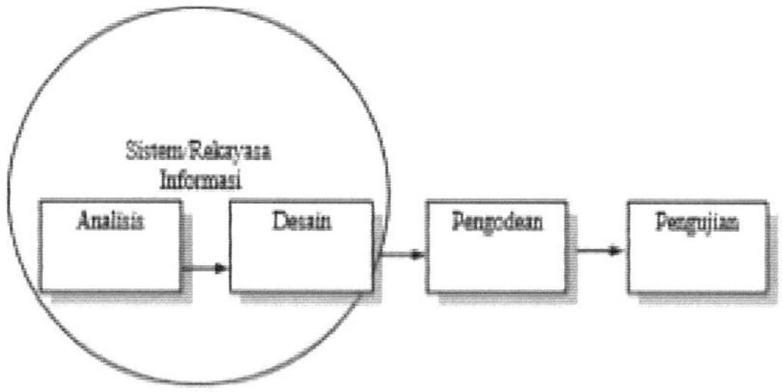


Figure 6. SDLC Method
Source:SukamtodanShalahuddin (2013:29)

While the explanation of the SDLC method in figure 6 above is:

a. Analysis, the analysis phase of system requirements, both hardware and software needed in making applications include:
   1)Hardware: Minimum laptop 4 GB RAM, 100MB ROM space and quad-core 2.2 processorGHz.
   2)Software: Visual Studio 2010 for making the Automatic Lessimic Analyzer interface
b. Design, stages include the interface design for Automatic Lessimic Analyzer using
c. Visual Studio 2010 and some tools consist of labels, textbox, group box and button.
d. Encoding, in the Automatic Lessimic Analyzer application using a programming languageC ++.
e. Testing, the final stage in the SDLC method uses black box testing and evaluates if there are errors in the application that has been made to be repaired.

4. Results and Discussion

4.1.Lexical Analysis (scanner).

Lexical Analysis is identifying all the magnitudes that build a language. In this case the character flow that forms the source program is read from left to right and grouped called tokens, which are character sequences in a single unit that has its own meaning. This analysis of translating inputs into forms is more useful for the next compilation stages. Lexical analysis is the interface between source code and parser analysis. The scanner checks characters per character in the input text, breaking the source of the program into parts called tokens. Lexical Analysis works on grouping character sequences into the main components: identifier, delimiter, operator symbols, numbers, keyword, word noise, blanks, and comments, and so on to produce a Lexical Token that will be used in Syntactic Analysis. Scanners must also be able to identify tokens in full and distinguish keywords and identifiers. For this reason the scanner requires a symbol table. The scanner enters the identifier into the symbol table, enters literal and numeric constants into the symbol table itself after conversion into an internal form.

The main tasks of this lexical analysis include:

a. Read the source code by tracing character by character.
b. Identify lexic quantities (identifiers, keywords, and constants).
c. Transform into a token and determine the type of token.
d. Send tokens.
e. Remove or ignore white-space and comments in the program.
f. Deal with errors.
g. Handle the symbol table.

Example : case studies of source programs from c ++ programming languages
Known: statements in a source program that support arithmetic expressions

Fahrenheit: = 35 + celcius * 1.8 ;

Translation results into tokens in lexical analysis as in table-1.

Table 1: Lexic Magnitude

| No | Lexic Magnitude | Source Program | Sum |
|---|---|---|---|
| 1 | Identifier : | | |
| | - Keyword | - | 0 |
| | - Name | Fahrenheit, celcius | 2 |
| 2 | Konstanta | 35 , 1.8 | 2 |
| 3 | Operator | := , + , * | 3 |
| 4 | Delimeter | ; | 1 |
| 5 | White-space | - | 0 |
| | | **Total** | **8** |

4.2. Syntax Analyst (parser)

Tree syntax (Tree) is a non-circular connected graph that has one node / node / vertex called Root (root), from which it will have a path to each node. Tree syntax / tree decline / parser tree is useful for describing how to obtain strings by lowering terminal symbols, where each symbol variable will be lowered into a terminal, until nothing has been replaced [13].

A context-free grammar with production rules (symbol initial symbolized S):

$S \rightarrow AB$ , read S produces AB orSreduce AB

Requirement : $A \rightarrow aA \mid a$

$B \rightarrow bB \mid b$

*The Final Result :**a a b bb***

The parser formation is in the form of a tree as shown in Figure 7 below.

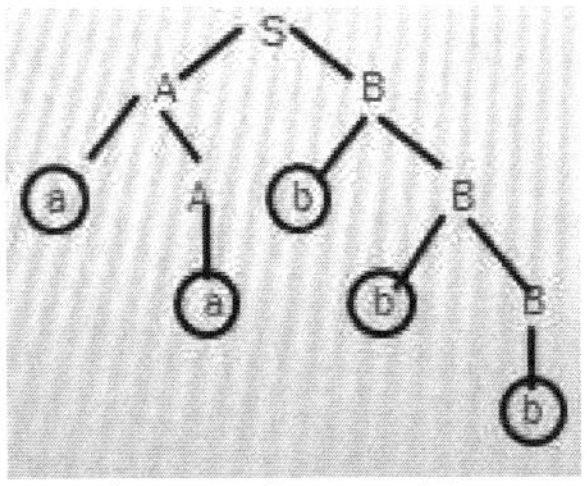


Figure7.Syntax Tree

4.3. Semantic analysis produces the syntax tree generated in the parsing process. The general analysis of semantics is to determine the meaning of the agreement provided in the source program. So in essence, this semantic analysis uses the syntax tree produced by the Parsing process. The syntax of syntactic analysis and semantic analysis is two processes that are very easy to understand and difficult to approve. Semantic analysis, often also combined in thegeneration of intermediate code that produces Intermediate Code output, which will be used in the next compilation process.

There are two kinds of Intermediate Codes, namely: Postfix Notation and N-Tuple :

a, Notasi Postfix.

Syntaxin the form of Arithmetic expressions:

<operan><operan><operator>

Example :

(a+b)*(c+d) → *Postfix* :ab + cd + *

F:=(A+B)*(C+D) → *Postfix* : *AB*+ CD+* F:=

In the form of logic:

Logic Expression ( syntax ) : IF <exp> THEN <stmtT1> ELSE <stmt2>

PostfixForm :<esp><label1> BZ <stmtT1><label2> BR <stmt2>

b. Notasi N-Tuple

If Postfix each instruction line consists of only one Tuple, while in N-Tuple line consists of several Tuples:

Syntax Notasi N-Tuple is: operator ................ N-1 operand

c. Triple Notation (prefex)

Syntax : <operator ><operand><operand >

Example :

| A:=D*C+B/E | Hirarkie Operator : |
|---|---|
| 1. * , D , C | * |
| 2. / , B , E | / |
| 3. + , (1) , (2) | + |
| 4. := , A , (3) | - |

d. Qudrapules Notation

Syntax : <operator><operand><operand><result>

The result is a temporary variable that can be placed in the memory or register. Existing problems how to manage temporary variables (results) to a minimum.

Example Instructions: A := D * C + B / E

When the between Code is made:

1. * , D , C , T1 → T1 := D*C

2. / , B , E , T2 → T2:= B/E
3. + , T1 , T2 , A → A := T1+T2

e. Code Generation

The results of the analysis phase will be received by the code generation section. Here the Between Code program is usually translated into assembly language or machine language.

Example: (A + B) * (C + D)

Where is the intermediate code in the form of quadruples:

1. + , A , B , T1 → (A+B) = T1
2. + , C , D , T2 →(C+D) = T2
3. * , T1 , T2 , T3 → (T1*T2) = T3

Can be translated into assembly language with a single accumulator:

LDA A LDA T1
ADD BA+B=T1 MUL T2 T1*T2=T3
STO T1STO T3
LDA C →The LDA command loads the contents of the register/memory.
ADD D C+D=T2 →To the accumulator *(load to accumulator)*
STO T2 →The STO command stores the contents of the accumulator to memory/register (store from accumulator)

f. Error Handling

The form of an error that occurs in the source code in the form of an error message will be displayed in the Automatic Lessimic Analyzer application. The compiler will handle errors from the lexical, syntactic and semantic analysis processes.

g. The results of Design

The user interface of the Automatic Lessimic Analyzer application as in Picture 8 below consist of 8 columns, they are program code, error messages, intermediate code, table identifier, keyword table, data type table, operator table, delimeter table.

The following is the initial display image of the Automatic Lessimic Analyzer application :

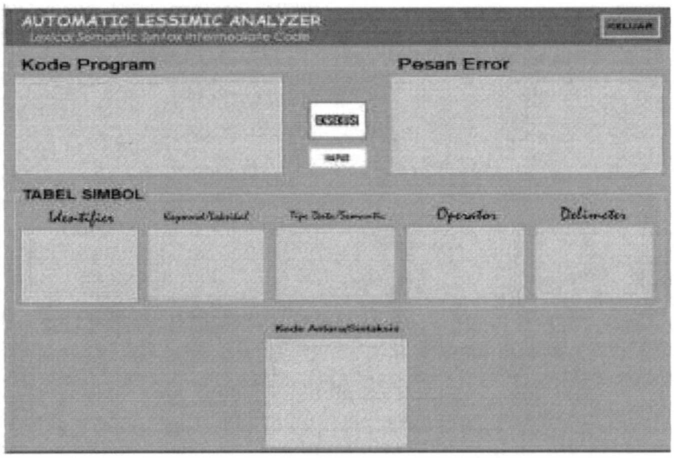


Figure 8. Initial appearance of the AtomaticLessimic Analyzer machine

If the source code is included and performs compiling the program, if there are no errors in writing the program then look like in picture 9 below:

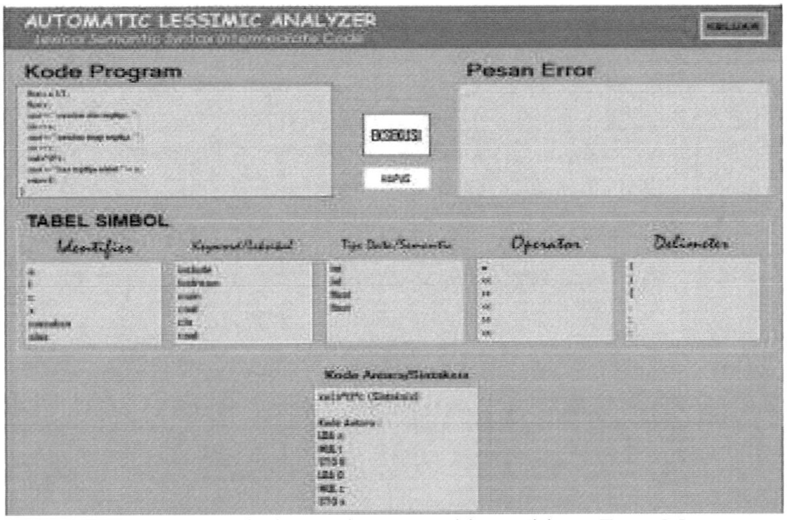


Figure 9.Automatic Lessimic Analyzer Machine Without Error Messages

In Picture 9 it can be seen that the Error Message textbox does not have any contents (blank), meaning that the program code entered by the user has no error. But if the program code entered is not correct, then the textbox in the Error Message will be filled in according to the errors that occur, for example can be seen in Picture 10 below.

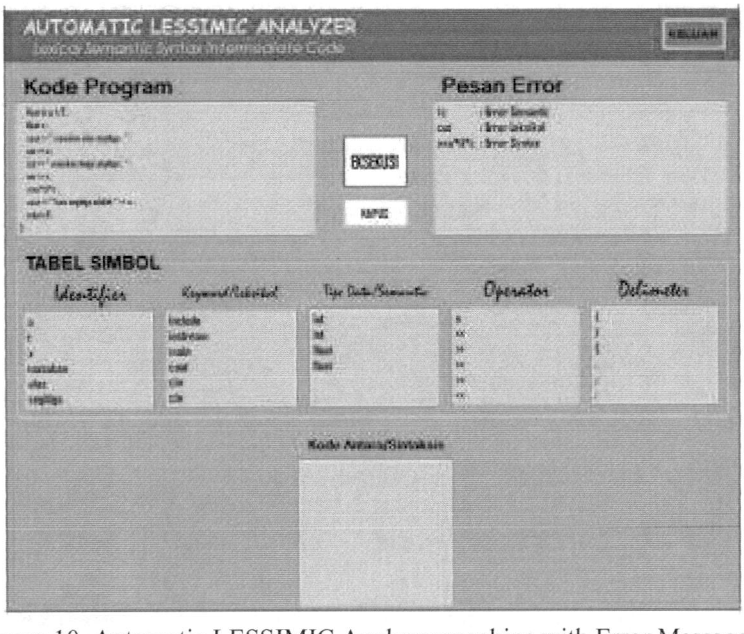


Figure 10. Automatic LESSIMIC Analyzer machine with Error Message

While the results obtained from the application research are explained in Table 2 below.

Journal of Physics: Conference Series **1381** (2019) 012019 doi:10.1088/1742-6596/1381/1/012019

Table 2. Results of Discussion of Applications with Black Box Testing.

| Fase | Source Code | Aplikasi *Automatic LESSIMIC Analyzer* | Keberhasilan Fungsi Aplikasi |
|---|---|---|---|
| Leksikal (menganalisa bahasa pemrograman) | if (temp2 != "") kumpulankata.Add(temp2);<br>for (int i = 0; i < kumpulankata.Count; i++)<br>{<br>if (_keyword(kumpulankata[i]))<br>{ cekkumpulankata.Add(kumpulankata[i] + "" + System.Environment.NewLine);<br>}<br>}<br>if (error2) { }<br>else<br>{<br>for (int i = 0; i < cekkumpulankata.Count; i++)<br>{<br>result2 += cekkumpulankata[i] + "";<br>}<br>} tabelkey.Text = result2; | 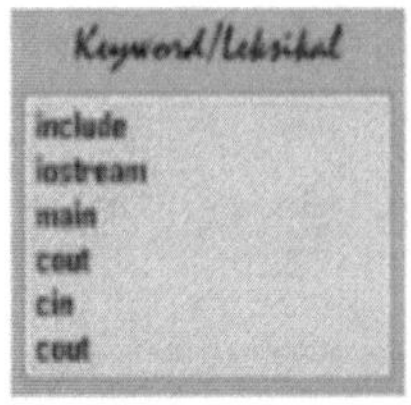<br><br>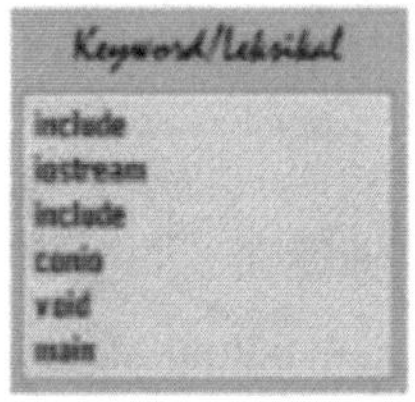<br> | Berhasil |
| Sintaksis (menganalisa aritmatika) & Kode Antara (mentranslasi bahasa sumber, menjadi bahasa target) | if (temp7 != "") kumpulankata.Add(temp7);<br>for (int i = 0; i < kumpulankata.Count; i++)<br>{<br>if (_kodeantara(kumpulankata[i]))<br>{ cekkumpulankata.Add(kumpulankata[i] + " " + Environment.NewLine");<br>}<br>}<br>if (error7) { }<br>else<br>{<br>for (int i = 0; i < cekkumpulankata.Count; i++)<br>{<br>result7 += cekkumpulankata[i] + "";<br>}<br>} tabel_antara.Text = result7; | 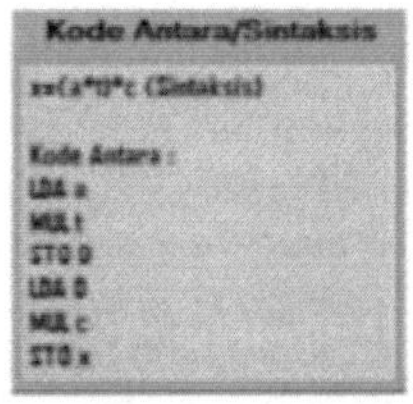<br><br>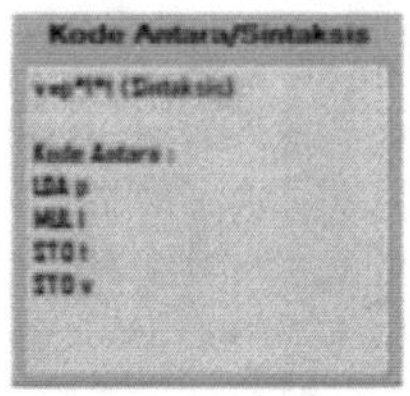<br> | Berhasil |
| Semantik (menganalisa variabel) | if (temp6 != "") kumpulankata.Add(temp6);<br>for (int i = 0; i < kumpulankata.Count; i++)<br>{<br>if (_int(kumpulankata[i]))<br>{ cekkumpulankata.Add(kumpulankata[i] + "" + System.Environment.NewLine);<br>}<br>}<br>if (error6) { }<br>else<br>{<br>for (int i = 0; i < cekkumpulankata.Count; i++)<br>{<br>result6 += cekkumpulankata[i] + "";<br>}<br>} tabeltipe.Text = result6; | 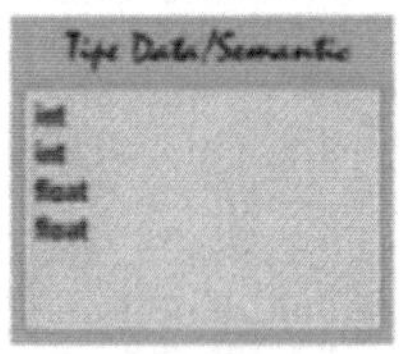<br><br><br> | Berhasil |

Table 2 explains the discussion of testing (trials) using Automatic Lessimic Analyzer with the Black Box testing process. Black Box testing in the application of Automatic Lessimic Analyzer is a test that is carried out to determine the functions that exist in the application has been running accordingly or not.

## 5. Conclusion

Based on the final results of the test stated that:

a. With the application of Automatic Lessimic Analyzer it has succeeded in carrying out three analyzes they are lexical, syntax, semantic and intermediate code and error handling with source code using the C++ programming language.
b. It has been successful in using Black Box testing for lexical, syntactic, semantic analysis and intermediate codes and error handling according to the input given.

So, by using the Automatic Lessimic Analyzer application we can know and understand the performance of compiling programs effectively and efficiently.

## REFERENCES

[1] Knuth D, 2003,*The Art of Computer Programming*, Fundamental Algoritms Third Edition, Volume 1, Addison-Wesley, United States, 0-201-03801-3
[2] FirrarUtdirartatmo, 2005, *TeknikKompilasi*, GrahaIlmu, Yogyakarta,Isbn 979-756-058-1, hal xiv+188
[3] AndezApriansyah. A, dll., 2019, *DesainMesin Compiler untukPenganalisaLeksikal, Sintaksis, Semantik, KodeAntaradan Error Handling PadaBahasaPemrogramanSederhana*, Journal of Applied Informatics and Computing (JAIC) Vol.3, No.1, pp. 01~07 e-ISSN: 2548-6861.
[4] Utdirartatmo, F., 2011, *TeknikKompilasi*, J & J Learning Jogjakarta
[5] RinaldiMunir, 2015, *TeoriKomputasi IF5110*,Bandung : ITB,.
[6]M. SidiMustaqbal, dll, *PengujianAplikasiMenggunakan Black Box Testing*,Boundary Value Analysis, ISSN : 2407 – 3911
[7] YokiFirmansyah&Udi,2018, *PenerapanMetode SDLC Waterfall DalamPembuatanSistemInformasiAkademikBerbasis Web StudiKasusPondokPesantren Al-HabiSholehKabupatenKubu Raya, Kalimantan Barat*, JurnalTeknologi&ManajemenInformatika, Vol. 4 No.1
[8] J. Suciadi, 2001, *StudiAnalisisMetode-Metode Parsing danInterpretasiSemantikPada Natural Language Processing*, J. Inform., vol. 2, pp. 13–22,.

Journal of Physics: Conference Series **1381** (2019) 012020 doi:10.1088/1742-6596/1381/1/012020

# Indonesian High-Speed Railway Optimization Planning for Better Decentralized Supply Chain Implementation to Support e-Logistic Last Miles Distribution

**E B Setyawan*, N Novitasari**

Department of Industrial Engineering, Telkom University, Bandung, Indonesia

*erlangga.setyawan@gmail.com

**Abstract**. E-logistics must have better strategy because they must satisfy unpredictable demands which came from customer. Other consumer behaviours which affect e-logistics must work more efficiently are the requirement of last miles fast delivery and more competitive pricing. Decentralized supply chain model can handle all these issues and complex logistics activity although it only has one echelon model in the supply chain structure. Decentralized supply chain model doesn't need a distribution centre but requires inventory sharing policy. Key to success inventory sharing policy is low lead time delivery. In Indonesia, only two transportation type which can handle product in low lead time, i.e., train cargo and aircraft cargo. Train cargo still needs high lead time but has low operation cost. Aircraft cargo can handle product in low lead time but need high operation cost. The trade-off in this condition is e-logistics need low lead time transportation tenant to satisfy demand until last miles delivery but has low operation cost not as expensive as aircraft cargo. Indonesia is developing high-speed train which connects Jakarta to Surabaya. In this research, we will review the level of effectiveness high-speed train in Indonesia to realized decentralized supply chain distribution network for handling e-logistics last miles shipping using dynamic system simulation. The output of this research is efficiency review of Indonesian high-speed railway cargo to support decentralized e-logistics supply chain distribution network, compared by aircraft cargo. After we perform simulation using dynamic system simulation, delivery of products between regions by using fast train is declared feasible to deliver products with fast delivery type if in one day there are six departure time (delivery batch). If less than six departures, the railway is declared unfeasible to deliver products with fast delivery type. However, the train can save about 37% batch delivery when compared to cargo aircraft.

## 1. Introduction

Supply chain management plays an important role in fulfilling the demand. There are two types of demand that must be met, namely, demand growth in the upstream supply chain, which fulfils the demand for the production process, and the fulfilment of demand in the downstream supply chain, where the demand for the distribution process to the consumer.

In the development of the modern era, downstream supply chain requires significant improvement and innovation. This is because in the modern era there is also a shift in the culture of public spending, from conventional shopping culture, where consumers come to a shopping canter, now turn to the culture of online shopping or called e-commerce. The change in shopping culture is evidenced by the research that has been done by e-commerce foundation which is the world's e-commerce survey institute [1]. From the results of this study, there is an increasing trend in e-commerce across the country from 2014 to 2016, where countries in the Asia Pacific and Latin America region occupy the largest e-commerce increase, i.e., 28% [2]. Figure 1a is a graph showing the growth of transactions on global e-commerce from 2013 to 2016. While Figure 1b shows the comparison of growth of e-commerce transactions in each country by 2015.

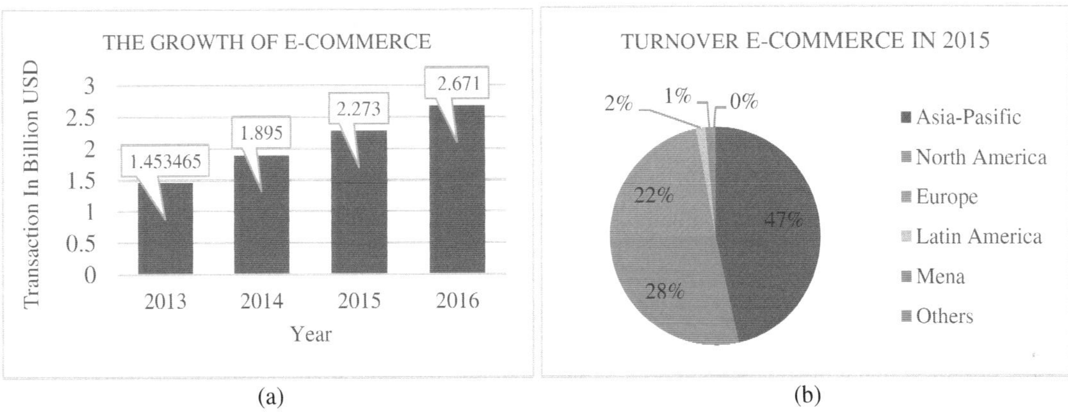


**Figure 1.** (a) The Growth of Global E-Commerce; (b) Turnover Rate e-Commerce in 2015

The rapid development of e-commerce will indirectly trigger the development of marketplace that acts as a container of transactions, that is, where sellers meet with buyers. Marketplace requires supply chain role to handle consumer demand, so the term e-logistics is developed. E-logistics integrates logistics activities with multiple electronic activities, such as, managing demand in e-commerce and logistics activity logging in ERP systems aimed at increasing fulfillment of incoming demand from consumers [3].

E-logistics must work hard to meet the unexpected demand of its consumers. Fulfilling demand for such inter-region, e-logistics must be balanced with fast delivery of goods and have a more competitive price [4]. The purpose of this study is to show the level of efficiency and minimum needs of high-speed railway departures every day. The level of efficiency and the minimum requirement of high-speed railway departures every day have an effect on to the level and cost of fulfilment request by e-logistic, the more the number of high-speed railway departures in a day, the more desirable to substitute the aircraft cargo that has been used to deliver inter-region on the island of java. If the high-speed railway performance is below the prescribed minimum limit, then the high-speed railway is not feasible to substitute aircraft cargo because it cannot fulfil the consumer demand maximally.

## 2. Research Methodology

Things needed to understand and make improvements to the system are system dynamics, systems thinking, and soft operations research (soft OR). System dynamic is a description that leads to model equations, simulations to understand dynamic behaviour, alternative policy evaluation, education and

policy options, and better implementation [5]. System dynamics is a simulation method used to solve systemic and complex problems by considering both quantitative and qualitative methods. System dynamics involves interpreting real life systems into computer simulation models that allow one to see how the structure and decision-making policies in a system create its behaviour. The method used at this time can assess dynamically changing systems by simulating to learn and plan future events. In addition, this method can also be used for decision making according to the system [6].
Tools that are often used to simulate is Powersim. Powersim is a tool used to do business modelling or simulation. These tools are designed to handle to cope with real - world business situation, addressing complexity as well as dynamic and risk related issue [7].

## 3. Discussion and Result

The aim of this paper is to study the efficiency level of the Jakarta - Surabaya fast train that will be built by the Indonesian government in performing logistics cargo services. We will compare the rate and speed of cargo delivery service for inter-region within Java Island between the use of aircraft cargo and fast train. Since the rapid train has not been realized, the data used in this discussion is not real data, but the operational data shown in the Jakarta - Surabaya rapid train planning.
To perform the efficiency test described earlier, the authors use case studies of an e-logistic in handling the product ordered by the customer of an e-commerce ABC. In the case study, the authors will examine the reliability level of an e-logistics in sending the product with the type of fast delivery service (delivered on the next day) from Jakarta to Surabaya, which is the farthest delivery point on the island of Java (long delivery miles). The use of aircraft cargo that is normally used to handle shipments of this type of service, in this study will be substituted by the use of fast train cargo.

Some of the assumptions used in this study are:

- Operational cost and fast travel time Jakarta - Surabaya uses fast train planning data that has been prepared by the government.
- E-logistics XYZ handles e-commerce ABC in delivering goods.
- Business processes for handling product shipments using business processes from e-logistics XYZ.
- The type of shipping service used in the case study is a fast delivery service (delivered up to last miles delivery on the next day).
- If the product does not reach the consumer the next day, it will get a penalty in the form of decreased consumer loyalty to e-commerce ABC. A further consequence is that e-commerce ABC penalizes penalty charges to e-logistics XYZ.
- The delivery time is assumed in regular time based on KPI of e-logistics XYZ, not considered extra time caused by any disruption.

Figure 3 shows the number of ordered items by the consumer by day of the month. Based on historical data, most customers choose to make purchases on e-commerce ABC on Friday and it's patterned in the following weeks of the month (see Figure 3). This condition becomes a challenge for e-logistics XYZ where if there are buyers who choose shipping with fast delivery service type, they must ensure the goods can arrive at the destination location on the next day and when late, ordered items will arrive to the costumer on Monday. It causes disappointment experienced by consumers so that will have an impact on decreasing consumer loyalty shopping on e-commerce ABC.

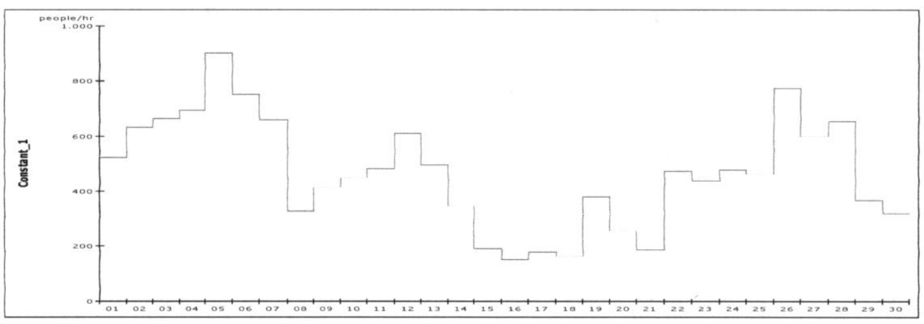


**Figure 2.** Number of Ordered Item in e-Commerce ABC in One Month

To perform a dynamic system simulation, the author uses the business process of cooperation between e-commerce ABC with e-logistics XYZ which can be seen in Figure 4.

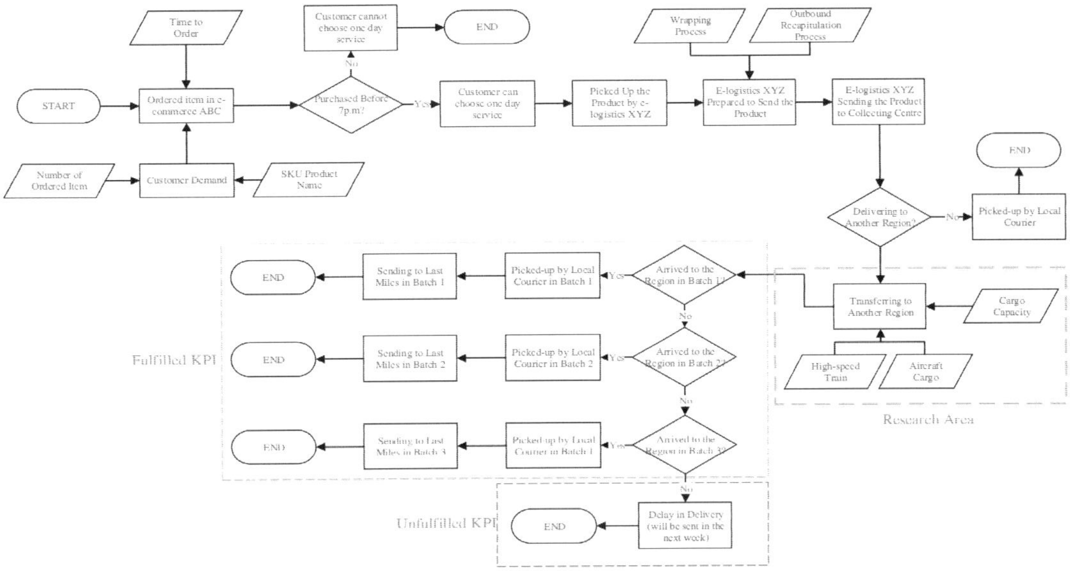


**Figure 3.** Business Process of Order Handling from e-Commerce ABC to e-Logistics XYZ

Due to the high interest in e-commerce ABC's customer spending on Friday, the authors will assess the reliability level of demand fulfillment by sending the fast delivery type (delivered in the next day) to the inter-region if substituted with fast train mode. Customers can only choose the delivery mode with fast delivery type when finished shopping and make payment by 19:00 on Friday. After customer e-commerce ABC verifies the payment, e-logistics will start processing the product by wrapping and recapitulating the outbound process. The product will be sent to the collecting center in each region. When the product is delivered to the region, the product will be transferred to another region.

This study focused on the mode of transportation used to deliver products to the inter-region. Aircraft cargo modes that are normally used to deliver products between regions within an island, will be substituted with cargo from fast trains due to minimizing shipping costs. After arriving at the destination area, the product will be forwarded to last miles delivery using local courier service. On Saturday, there are three batches of delivery. Delivery is "fulfilled" if the product reaches the customer in the three batches. If the product is not shipped in all three batches, the shipment is declared "late" and the product will be delivered to the consumer the following week on Monday. The delay is greatly avoided by XYZ e-logistics because it will get a penalty from e-commerce ABC caused by buyers of e-commerce ABC feel disappointed the goods cannot be sent on the next day. The disappointment will decrease interest in purchasing costumer e-commerce ABC.

To start the simulation, besides the company's existing business processes, we map out the weaknesses and advantages of each mode of transportation (aircraft cargo and fast train cargo). The advantages of aircraft cargo are lower travel times and more flexible schedules (many departure schedules). However, some of these advantages must be paid at an expensive price and a relatively small aircraft hull capacity. While fast train cargo, longer travel time and departure schedules between regions are not available in many schedules (not too flexible). However, fast trains have a much larger capacity and with relatively cheaper logistics cost.

We will simulate the demand between 500 until 1500 ordered items in the Friday. In each item, based on regulation from e-commerce ABC, they categorized into large item and small item. Small item is the item which have average size no larger than 30cm x 30cm x 30cm (l x w x h). Large item is the item which have average size no larger than 50cm x 50cm x 50cm (l x w x h). Parameter used in this paper can be seen in Table 1.

**Table 1.** Parameter Setting for Dynamic System Simulation

| **Number of Ordered Items** | **500** | **500** | **750** | **750** | **1000** | **1000** | **1250** | **1250** | **1500** | **1500** |
|---|---|---|---|---|---|---|---|---|---|---|
| Number of Small Item | 72% | 28% | 72% | 28% | 72% | 28% | 72% | 28% | 72% | 28% |
| Number of Large Item | 28% | 72% | 28% | 72% | 28% | 72% | 28% | 72% | 28% | 72% |
| Number of Small Item Demand | 360 | 140 | 540 | 210 | 720 | 280 | 900 | 350 | 1080 | 420 |
| Number of Large Item Demand | 140 | 360 | 210 | 540 | 280 | 720 | 350 | 900 | 420 | 1080 |
| Pallet Size L (cm) | 120 | 120 | 120 | 120 | 120 | 120 | 120 | 120 | 120 | 120 |
| Pallet Size W (cm) | 150 | 150 | 150 | 150 | 150 | 150 | 150 | 150 | 150 | 150 |
| Pallet Size H (cm) | 100 | 100 | 100 | 100 | 100 | 100 | 100 | 100 | 100 | 100 |
| Pallet Capacity Small Item | 66 | 66 | 66 | 66 | 66 | 66 | 66 | 66 | 66 | 66 |
| Pallet Capacity Large Item | 14 | 14 | 14 | 14 | 14 | 14 | 14 | 14 | 14 | 14 |
| Pallet Requirements For Small Items | 6 | 3 | 9 | 4 | 11 | 5 | 14 | 6 | 17 | 7 |
| Pallet Requirements For Large Items | 10 | 26 | 15 | 39 | 20 | 52 | 25 | 65 | 30 | 78 |
| Total Pallet Requirement for Ordered Items | 16 | 29 | 24 | 43 | 31 | 57 | 39 | 71 | 47 | 85 |
| Capacity Requirement Aircraft Cargo Boeing 747-xxx (pallet) | 8 | 8 | 8 | 8 | 8 | 8 | 8 | 8 | 8 | 8 |
| Capacity Requirement Fast Train in Planning Concept (pallet) | 15 | 15 | 15 | 15 | 15 | 15 | 15 | 15 | 15 | 15 |

To start the simulation, besides basic parameter setting and business processes between e-Commerce ABC with e-logistics XYZ, the authors mapped the number of products ordered by customers on Friday using a fast delivery service. The product must be delivered the next day, Saturday. The number of ordered products using this delivery service has a normal distribution (see Figure 5). Some additional parameters, such as capacity fulfillment center and processing time at the collecting center are also considered. The inventory level of ordered products is assumed always available so there is no backorder. From the simulation results using input data varies between 500 to 1500 product orders with fast delivery type, the largest batch delivery required for high speed train to deliver inter-regional products is 6 batches (if there are 1500 products using fast delivery service). In the existing conditions, delivery of all products between regions can be handled by aircraft cargo, although it requires a larger amount of batch delivery. For more details the simulation results can be seen in Table 2.

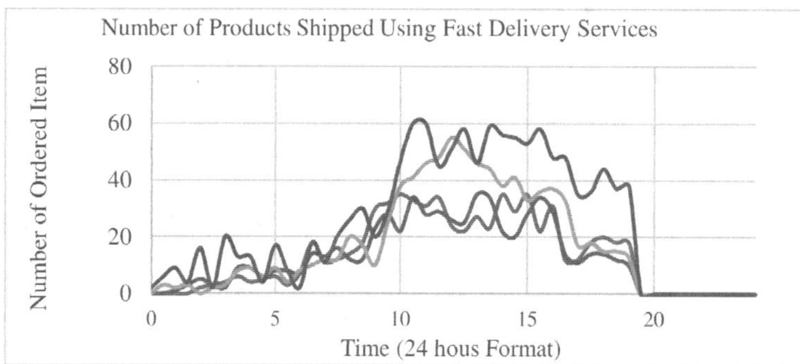


**Figure 4.** Number of Products Shipped Using Fast Delivery Services

**Table 2.** Simulation Result

| Cargo Type | Aircraft | Train | Aircraft | Train | Aircraft | Train | Aircraft | Train | Aircraft | Train |
|---|---|---|---|---|---|---|---|---|---|---|
| Number of Ordered Items | 500 | | 750 | | 1000 | | 1250 | | 1500 | |
| Ritation Needed For Aircraft | 2 | 4 | 3 | 6 | 4 | 8 | 5 | 9 | 6 | 11 |
| Ritation Needed For Train Cargo | 2 | 2 | 2 | 3 | 3 | 4 | 3 | 5 | 4 | 6 |
| Efficiency | 100% | 50% | 67% | 50% | 75% | 50% | 60% | 56% | 67% | 55% |
| Average Efficiency | 37% | | | | | | | | | |

## 4. Conclusion

E-logistics, which is a third party handling the distribution of goods from an e-commerce, is required to work faster and more efficiently in establishing the product up to last miles delivery. The choice of transportation modes to make the delivery must be done properly so that the product reaches the hands of the consumers with the time in accordance with the company KPI. In this paper, we study the efficiency level of the Jakarta - Surabaya fast train that will be built by the government when it is used to deliver products between regions in Java. To do this condition, we use dynamic system simulation. Dynamic system simulation can simulate unpredictable demand time by time. The result is high-speed train can save about 37% batch delivery when compared to cargo aircraft. For the next research, we suggest to review the effectiveness of high-speed train cargo for same day delivery services.

## References

[1] N. Singh, "Global E-Commerce: a Portal Bridging the World Markets," *J. Electron. Commer. Res.*, vol. 11, no. 1, p. 1, 2010.

[2] K. S. Nair, "Impact of E-Commerce on Global Business and Opportunities - A Conceptual Study International Journal of Advanced Engineering and Management Research IMPACT OF E-COMMERCE ON GLOBAL BUSINESS AND OPPORTUNITIES - A CONCEPTUAL STUDY," no. April, 2017.

[3] Y. Yu, X. Wang, R. Y. Zhong, and G. Q. Huang, "E-commerce Logistics in Supply Chain Management: Practice Perspective," *Procedia CIRP*, vol. 52, pp. 179–185, 2016.

[4] A. Ghezzi, R. Mangiaracina, and A. Perego, "Shaping the E-Commerce logistics strategy: A decision framework," *Int. J. Eng. Bus. Manag.*, vol. 4, no. 1, pp. 1–13, 2012.

[5] J. W. Forrester, "soft OR," *Syst. Dyn. Rev.*, vol. 10, no. January, pp. 245–256, 1994.

[6] Y. Feng, "System Dynamics Modeling for Supply Chain Information Sharing," *Phys. Procedia*, vol. 25, pp. 1463–1469, 2012.

[7] M. Myrtveit and M. Bean, "Business modelling and simulation," *Wirtschaftsinformatik*, vol. 42, no. 2, pp. 156–160, 2000.

Journal of Physics: Conference Series **1381** (2019) 012021 doi:10.1088/1742-6596/1381/1/012021

# Decision Making in Inventory Policy Determination for Each Echelon to Stabilize Capsicum Frutescens Price and Increase Farmers Share Value Using Discrete Event Simulation

**Nia Novitasari***, **E B Setyawan**

Department of Industrial Engineering, Telkom University, Bandung, Indonesia

*nia.novitasari.93@gmail.com

**Abstract**. Capsicum frutescens is one of the vegetable commodities that can affect the dynamics of the national economy. The dynamics of the national economy is due to significant fluctuations in chili prices in Indonesia at all times. Price fluctuations occur due to lack of information either in the form of demand information or Chilean price information about consumers received by farmers for marketing red pepper. Besides, the length of the distribution of red pepper (farmer-the middleman- wholesaler- retailer- consumer) results in significant price changes from farmers to consumers at each echelon or distribution level. Such price changes result in a decline in the farmer's share. Farmer's share has a negative relationship with marketing margins. the negative relationship is the higher the marketing margin, the lower the share received by the farmer. from the explanation, the concept shows that the victims of the length of the distribution channel are farmers and consumers. Where farmers will bear a high business risk, and consumers will get high chili prices. The research will discuss how to stabilize the price of red pepper and improve farmer's share by making the best decision using discrete simulation using Monte Carlo, so it can determine the supply policy by the distribution at each level aimed at reducing the operational cost. Besides, this can also reduce the hoarding activities undertaken by the relevant distribution. The hoarding can disrupt the stability of the existing chili price.

## 1. Introduction

Chili red pepper is a vegetable commodity that is often used as a flavouring food. Therefore, the needs of chili in large cities are about 800,000 tons / month or 66 tons / month. From these data can be known, that chili is one of vegetable commodities that strategic enough to affect the dynamics of the national economy. Chili prices in Indonesia fluctuate every year. Such price fluctuations occur quite significantly, especially when it comes to the celebration of the big day. This is due to the demand for red chilies that increased by about 10% to 20% of normal needs. [1]

The fluctuation in the price of red pepper causes uncertainty of income earned by the farmers. The uncertainty of the revenue is caused by the lack of cooperation between the farmers and other distributors. Lack of cooperation resulted in a lack of information received by farmers on the marketing of red pepper. So, farmers cannot determine the bargaining position in price formation. From this problem indicates that the occurrence of dependence between farmers with other distribution parties. The inadequacy and transparency of communication and information received by these farmers resulted in significant price changes from farmers to consumers. Not only that, if farmers directly distribute chili to consumers, farmers do not have a transportation infrastructure, pricing information system, and adequate market facilities. This is where third parties arise. The appearance of these third parties is what causes the length of the distribution flow of chili pepper. These third parties are often referred to as middleman (perpetrators of distribution). The number of perpetrators of this distribution is uncertain, depending on the region of each. In the present study, the perpetrator of this distribution consists of three parties (the middleman - wholesalers - retailers). So, if the written flow distribution is as follows.

Fig 1 Perpetrators in Red Rice Chilli Distribution

If the ineffectiveness and transparency of an information, the price changes will result in a decrease in the farmer's share. has a negative relationship with the marketing margin means that the higher the marketing margin, the share received by the farmers is lower. The concept shows that the victims of the distribution channel are farmers and consumers, where farmers will bear a high business risk, and consumers will get a high price of chili. In Figure 2 below shows a comparison of farmer's selling price and buying price by consumers in 1983 - 2014. The graph shows a very significant difference.

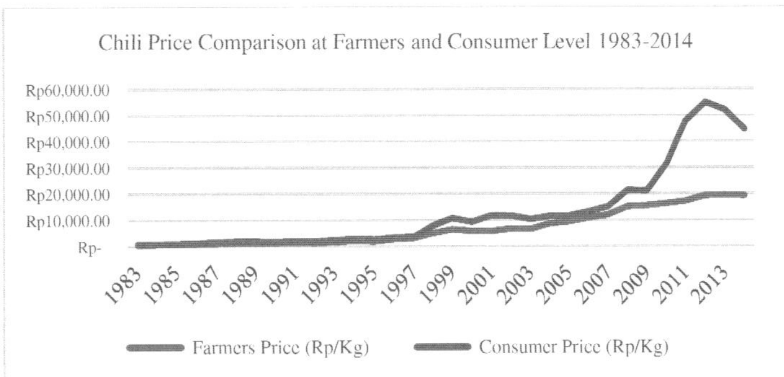


Fig 2 Chili Price Comparison at Farmers and Consumer Level 1983-2014

In the previous study, entitled "Marketing System of Red Rice (Capsicum Frutescence) in Cigedug Village, Cigedug Sub-district, Garut Regency" [2] discusses the analysis of marketing channels, marketing function, market structure, and market behaviour of red chilli in Cigedug Village District Cigedug Garut regency and marketing margin analysis, farmer's share, profit and cost ratio, and cohesiveness of vertical market of red pepper between market at farmer level in Cigedug Village with Kramat Jati Parent Market.

The differences in the research made now are very significant, the differences that occur include differences in different viewpoints in analysing the chili distribution channel chain. In this research will discuss how to stabilize the price of red pepper and improve farmer's share by making the best decision using Monte Carlo simulation so that it can determine the supply policy by the distribution at each level aimed at reducing the operational cost. The operational costs will be computed with the profit margin on each echelon, so it will obtain the stability of the selling price of chilli on each echelon. In addition, the research undertaken can reduce the stockpiling activities performed by the related distribution. This is done because the government is vigorous-incessant to reduce this activity. The hoarding can disrupt the stability of existing chili prices, because the chili that has been stored will be sold when the price begins to rise. As a result of the hoarding, chili will have a poor quality caused by the duration of hoarding.

## 2. Research Methodology

### *2.1 Analysis of Performers and Distribution Patterns of Marketing*

According to [3] the marketing channel is a series of commercial institutions that pass the goods in their distribution from producers to consumers in which they engage several marketing institutions. In this study, the actors involved include: Farmers (suppliers), medicines (the middleman - wholesalers - retailers), Consumers. Each of these actors has their own marketing distribution pattern. The pattern of red pepper distribution in Figure 3.

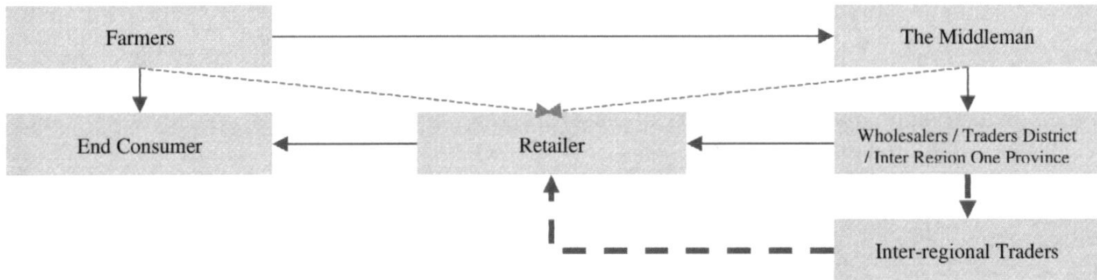


Fig 3 Distribution Channel Level of Red Chili

Information:

| | | |
|---|---|---|
| ⟶ The marketing process tends to be continuous | - - -→ The marketing process tends to be relatively (sometimes) | - - -→ Beyond the scope of the study |

The picture shows the distribution of the chili can actually be done from farmers directly to consumers, but it is rare. What happens is that the distribution of this chili has been trapped in the circle mafioso, the syndicates who seek maximum profit at the expense of the farmers and final cosmos. The trading mafia was deliberately created with various tendencies surrounding it, so the high margin of red pepper is only enjoyed by the traders involved, but the price at the farm level is very cheap. Then the next victim is the end consumer who usually consumes this chili in sufficient quantities, but now becomes a very limited quantity of consumption with weakening purchasing power. Another victim is a good retailer trader in the market, peddler which originally per day can sell at least 20 kg, currently only maximum 5 kg / day. This condition indicates that the morals and ethics of traders experience the process of degradation [3]

### *2.2 Operational Cost Analysis*

The main purpose of this study was to stabilize the price of chili and increase farmer's share by reducing operational costs. The variables included in the operational costs have been mentioned previously. These variables influence the perpetrators of the distribution of chili. For example, the cost of harvesting is an operational cost that is only found in farmers. For more details can be seen in the table below.

Table 1 Variable of Operational Costs

| Variable Cost | Cost of Harvest | Freight Charges | Packing Cost | Cost of Middleman | Levy Fees | Cost of depreciation | Cost of Loading & Unloading | Sorting Fee | Rental Rental Cost | Cost Disadvantages | Margin Cost |
|---|---|---|---|---|---|---|---|---|---|---|---|
| Farmers | v | v | v | | | v | | | | v | v |
| Middleman | | v | v | | v | v | v | v | | v | v |
| Wholesaler | | v | v | v | v | v | v | v | v | v | v |
| Retailers | | v | v | v | v | v | | | | v | v |

### *2.3 Farmer' Share Analysis*

Farmer's share has a negative relationship with marketing margin, which means that the higher the marketing margin, the share received by the lower farmers. *Fs* is percentage received by farmers, *Pf* is prices at farm level, and *Pr* is prices at the consumer level, the mathematically farmer's share can be formulated as follows:

$$Fs = \frac{Pf}{Pr} \times 100\%$$

### *2.4 Inventory Management*

Inventory management is used to perform inventory levels at each red chili supply chain level. The continous review inventory system is used at the retailer level. This is because the demand that comes at this level comes from the end customer. While we can not regulate people's habits in determining when to buy red chili. Continuous review allows the fulfillment of requests at any time, because it does not pay attention to the demand fulfillment period. While at the level of wholesalers and middlemen, we use simulated inventory system periodic review. Retailers may only purchase to wholesalers within a period of seven days and wholesalers may only buy into a middleman within a seven day period as well.

### *2.5 Monte Carlo Simulation*

Simulation is a technique in making a model of a real system or proposed system so that the behaviour of the system can be learned [4]. The simulation used is Monte Carlo simulation using Crystal Ball module. Monte Carlo simulation is one method that can be used to analyse business planning dynamically by building random scenarios according to the business risk analysis [5]. The risk analysis used can be from historical data and expert estimation data in the form of probability distribution. One example of a Monte Carlo simulation application that uses data estimation was developed by [6] using Monte Carlo simulations to predict the level of human resource performance within a company. Given these characteristics, Monte Carlo simulations are quite appropriate to be used in quantitative analysis of organizational change risk, where available data are based on expert estimates.

## 3. Discussion and Problem Solution

In the discussion and problem-solving section, the first step is to analyses the inventory of red chili. The inventory data of red chili is divided into four inventory scenarios. These four scenarios are obtained through trial and error method. The purpose of finding this inventory scenario to find out the minimum and maximum supply of red chili per day. So, in Table 2 it explains the minimum inventory and maximum inventory, for example retailers with a minimum inventory of 190,000 kg and a maximum supply of 220,000 kg, as well as for wholesalers and middlemen. From the inventory data combined with the cost of red chili each level. After that both data are solving using the simulation as much as 10,000 times (trial), resulting in the data in Table 3.

After knowing inventory levels and red chili cost/kg of each scenario, then the next step is to arrange alternatives based on the cheapest to highest cost scenario. Alternatives taken as many as 4 alternatives are shown in Table 4. The results of alternative calculations 1, 2, 3, 4 are shown in Table 5 and Table 6. The calculation of these alternatives compares the selling price of red chili and red chili each level. Then will be obtained margin, that is percentage comparison of pepper cost per kg divided by selling price. For example, in alternative 1, the selling price of chili at retailers is Rp 374.256.01, - while the cost is Rp 271.300.53,00 so that the margin of 34.38% is obtained. the next step is to calculate the farmer share's by comparing the purchase price of the consumer level with the price of the farmer. In the same alternative that is alternative 1, can know the value of farmer share's equal to 29,47%. And so, on alternatives 2 to 4.

Table 2 Red Chili Supplies/Stock

| | Chili stock (kg/days) | | | |
|---|---|---|---|---|
| Retailers | 190000 | 200000 | 210000 | 220000 |
| Wholesaler | 1505000 | 1510000 | 1515000 | 1520000 |
| Middleman | 1506000 | 1507000 | 1509000 | 1510000 |

Table 3 Result of Chili Simulation (Cost/Kg Red) (10.000 Trials)

| | Results of Chili Simulation (Cost / Kg) (10,000 Trials) | | | | | | | |
|---|---|---|---|---|---|---|---|---|
| | Scenario 1 | | Scenario 2 | | Scenario 3 | | Scenario 4 | |
| | Minimum Inv. Level | Cost /Kg | Minimum Inv. Level | Cost /Kg | Minimum Inv. Level | Cost /Kg | Minimum Inv. Level | Cost ;/Kg |
| Retailers | 190,000 | Rp 341,168.211 | 200,000 | Rp 4,156.547 | 210,000 | Rp 22,682.616 | 220,000 | Rp 27,169.952 |
| Wholesaler | 1,505,000 | Rp 147.700 | 1,510,000 | Rp 170.292 | 1,515,000 | Rp 232.826 | 1,520,000 | Rp 324.542 |
| Middleman | 1,506,000 | Rp 33.363 | 1,507,000 | Rp 23.461 | 1,509,000 | Rp 23.653 | 1,510,000 | Rp 33.802 |

Table 4 Alternatives Arrangement based on Cheapest Cost – Most Expensive Cost Scenario

| | Compilation of Alternatives Based on the Cheapest Charge Scenario sd. Highest | | | |
|---|---|---|---|---|
| | **Alternative 1** | **Alternative 2** | **Alternative 3** | **Alternative 4** |
| **Retailers** | Skenario 2 | Skenario 3 | Skenario 4 | Skenario 1 |
| **Wholesaler** | Skenario 1 | Skenario 2 | Skenario 3 | Skenario 4 |
| **Middleman** | Skenario 2 | Skenario 3 | Skenario 1 | Skenario 4 |
| **Farmers** | - | - | - | - |

Table 5 The Calculation of Alternatives 1 And 2

| | Alternative 1 | | | | Alternative 2 | | | |
|---|---|---|---|---|---|---|---|---|
| | Inventory Level | Cost / Kg | Margin | Selling Price | Inventory Level | Cost / Kg | Margin | Selling Price |
| Retailers | **200,000** | Rp 4,157 | 34% | Rp 15,272 | 210,000 | Rp 22,683 | 34.38% | Rp40,203 |
| Wholesaler | **1,505,000** | Rp 148 | 18% | Rp 7,208 | 1,510,000 | Rp 170 | 18.00% | Rp 7,235 |
| Middleman | **1,507,000** | Rp 23 | 32% | Rp 5,961 | 1,509,000 | Rp 24 | 31.78% | Rp 5,961 |
| Farmers | | Rp 1,406 | 69% | Rp 4,500 | | Rp 1,406 | 68.77% | Rp 4,500 |
| Farmer's Share Index | | 29.47% | | | | 11.19% | | |

Table 6 The Calculation of Alternatives 3 And 4

| Alternative 3 | | | | Alternative 4 | | | |
|---|---|---|---|---|---|---|---|
| Inventory Level | Cost / Kg | Margin | Selling Price | Inventory Level | Cost / Kg | Margin | Selling Price |
| 220,000 | Rp 27,170 | 34.38% | Rp 46,352 | 190,000 | Rp341,168 | 34.38% | Rp468,443 |
| 1,515,000 | Rp 233 | 18.00% | Rp 7,324 | 1,520,000 | Rp 325 | 18.00% | Rp 7,433 |
| 1,506,000 | Rp 33 | 31.78% | Rp 5,974 | 1,510,000 | Rp 34 | 31.78% | Rp 5,974 |
| | Rp 1,406 | 68.77% | Rp 4,500 | | Rp 1,406 | 68.77% | Rp 4,500 |
| | 9.71% | | | | 0.96% | | |

After performing the calculation of the margin, the share farmer 's on each alternative, it is known that the highest value of the farmer's share in alternative 1 is equal to 29.47%. While alternative 2 is 11,19%, alternative 3 that is 9,71%, and alternative 4 that is 0,96%. So, the proposed selling price on alternative 1, with a comparison of farmer 'share of 4.97%. Can be seen in Table 7.

Table 7. Comparison of Existing and Proposed on the chosen alternative

| | Comparison of Existing and Proposed | |
|---|---|---|
| | Existing Selling Price | Proposing Selling Price |
| Retailer | Rp 18,000 | Rp 15,272 |
| Wholesalers | Rp 10,000 | Rp 7,208 |
| Middleman | Rp 7,500 | Rp 5,961 |
| Farmers | Rp 4500 | Rp 4,500 |
| Farmer's Share Index | 25.00% | 29.47% |

## 4. Conclusion

This study was conducted, because there is a gap of red chili prices at the level of the farmer to the consumer level each period. These price fluctuations caused by price increases that occur at each level of the distribution supply chain. Therefore, to reduce the significant price increase then the authors do

Journal of Physics: Conference Series **1381** (2019) 012021 doi:10.1088/1742-6596/1381/1/012021

research on the distribution of red chili. This study will discuss how to stabilize the price of red chili and improve farmer's share by making the best decision using Monte Carlo simulation in order to determine the inventory policy by the distribution at each level aimed at reducing the operational cost. The operational costs will be computed with the profit margin on each echelon, so it will obtain the stability of the selling price of red chili on each echelon. From the simulation, the best result is to use alternative 1, using scenario 2 at retailer level, scenario 1 at the merchant level, and scenario 2 at the middleman level. Thus, obtained value of the farmer's share of 29.47%. These results are the highest results of other alternatives. By using these alternatives, the proposed price at retailers amounts to Rp 15,272, -, at large traders of Rp 7.208, -, at the middleman level of Rp 5.961, -, and at the farmer level of Rp 4,500, -. So, when compared to the previous farmer's share amounted to 4.97%.

## References

[1] Suwandi 2015 Outlook Cabai 2015 *Pus. Data Dan Sist. Inf. Pertan.* 79
[2] Asmayanti 2012 Sistem Pemasaran Cabai Rawit Merah (Capsicum Frutescens) Di Desa Cigedug Kecamatan Cigedug Kabupaten Garut 1–127
[3] Hadi S And Susetyo A B 2012 Analisis Margin Pemasaran Cabe Merah ( Capsicum Annum L ) Di Kabupaten Jember Analysis Of Marketing Margin Of Red Pepper ( Capsicum Annum L .) In Jember Regency 55–62
[4] Chu Y And You F 2014 Simulation-Based Optimization For Multi-Echelon Inventory Systems Under Uncertainty Yunfei *2014 Winter Simul. Conf.* 1539–48
[5] Mun J 2006 *Modeling Risk: Applying Monte Carlo Simulation, Real Options Analysis, Forecasting, And Optimization Techniques*
[6] Aderibigbe A 2018 Monte Carlo Simulation

# Development of the E-Government Evaluation Model on City Level Through Pattern System Approach

**A Sukma*, T D Susanto**

Institut Teknologi Sepuluh Nopember, Surabaya, Indonesia

*alifian.sukma16@mhs.is.its.ac.id

**Abstract**. The evaluation process of e-government continues to be investigated in various parts of the world. It becomes an important thing to do an evaluation as a benchmark for the government as a developer of the e-government to find out how the current conditions of e-government with the objectives to be achieved, namely to get a more efficient and effective government process. Many studies tried to develop an evaluation model but only focus on a number of aspects from e-government. Some of them tried to examine the efficiency of e-government through outcomes in the form of services and the impact from using e-government from the perspective of users and providers of e-government services. In this study, evaluation of e-government is based on a systems pattern approach through input - process - output - impact from the development of e-government. In this study will use a case study from the city of Surabaya as one of the cities with a very good predicate by the Indonesian e-Government Ranking (PeGI). The final results of this study are in the form of a comprehensive model for evaluating e-government at the city level.

## 1. Introduction

E-Government or Digital Government is a form of governance that utilizes ICT resources to improve the governance of a region or country so that citizens can participate in electronic / digital governance [1]. E-government itself continues to evolve in terms of service and governance. There are several stages that occur during an evolving e-government, starting from the initial stage, namely as a platform to get information online, input a two-way interaction or in the form, a transaction or service done online, to a one-stop system -Government which all administrative processes and services are carried out online and are mutually integrated between governmental units / units [2].

Evaluation using the ranking of Indonesian e-government (PeGI) aims to map the conditions and utilization of existing ICT resources in each region. Evaluation using PeGI is not necessarily done by all provinces in Indonesia. In 2011 PeGI ranked 26 provinces, 24 provinces in 2012, 21 provinces in 2013, 22 provinces in 2014, and 20 provinces in 2015 [3]. Measurements using PeGI itself only focus on the readiness of an area in implementing e-government systems in their area. The uneven development of e-government in the regions is certainly the reason why there is a need for a tool or benchmark to find out how the sustainability or success rate of an e-government area with other regions

[4]. Research conducted to developed a framework to measure the maturity of an e-government system by combining PeGI, Framework COBIT 5, and Janssen indicators to determine the level of readiness of both resources (input), efficiency of productivity (processes), benefits obtained (output) [5]. Until finally the government through Ministerial Regulation Number 5 of 2018 issued a guideline for evaluating the Electronic-Based Government System (SPBE). The SPBE evaluation itself is expected to be able to produce an index value that is able to describe the maturity level of e-government in this case called SPBE both in central agencies and regional agencies [6].

## 2. Literature Review

Evaluations of e-government systems need to be carried out with the aim of being able to assess the development of e-government implementation. There are several methods that already exist regarding e-government evaluation methods that have been applied both in the world and in Indonesia itself. The ranking of Indonesian e-government (PeGI) evaluates by measuring the dimensions of policy, institutions, infrastructure, applications, and planning [7], and currently using an Electronic-Based Government System (SPBE) evaluation tool issued by the Government through the Ministry of Administrative Reform Country. The SPBE evaluation tool measures through 3 aspects of government, namely, aspects of policy, governance, and SPBE services from each government organization [6].

In a previous research study, the main focus of the development of an e-government framework was to provide a value for current e-government position. The purpose of getting this value is nothing but to find out the position and know the steps and strategies for developing so that it can become a government system that is transparent, effective, and efficient [8]. Using a system approach used in the evaluation process of the Colombian government on e-government, Cordoba (2009) explains that there are aspects of the input-process-output-impact in the e-government implementation process [8,9]. Previous research focuses on each aspect starting from readiness, process, output, and impact with the aim of being able to improve the efficiency of the use of e-government [9]. Based on the system approach, this study measures the maturity level of implementing e-government at the city level comprehensively through the aspects of input, process, output and impact. Table 1 explains how the mapping of several studies on aspects of the system approach.

**Table 1** Mapping Prior Research Based on Aspects in the System Approach

| No. | Author | e-Government's Aspects | | | |
|---|---|---|---|---|---|
| 1 | Cordoba, 2014 | | | Output | Impact |
| 2 | Berntzen, 2014 | | Usage | Output | |
| 3 | Puron-cid, 2014 | | | | Impact |
| 4 | Munoz, et al, 2014 | | | | Impact |
| 5 | Inkinen, et. Al, 2014 | | | Output | Impact |
| 6 | Janssen, 2003 | Input | Usage | Output | Impact |
| 7 | Susanto, 2015 | Input | Usage | Output | Environment |
| 8 | Kim & Grant, 2010 | Input | Proses | Output | |
| 9 | Osman, et al, 2014 | | | | Impact |
| 10 | Evi, 2017 | Input | Proses | Output | |
| 11 | Joshi & Islam, 2018 | Input | Proses | Output | Impact |

## 3. E-Government Evaluation Models

The model proposed in this study will be explained as a conceptual framework. The conceptual framework in this study was obtained from various previous studies, specifically related to the development of e-government evaluation models. This section examines the evaluation process of e-government and how the development of evaluation models is based on several related studies. The government through e-government will need some input as a basis and support for the implementation of e-government, which can then carry out the implementation process, to produce outputs that can be in the form of public services, so that they can ultimately have an impact on stakeholders from e-government. Based on the previous literature mapping, it is able to explain that in the e-government evaluation process it is necessary to cover from several aspects ranging from input, process, output to the impact of the existence of an e-government. In accordance with Figure 1 this system approach is used to cover all aspects of e-government development starting from the input consisting of readiness in implementing e-government, IT governance processes in e-government, e-government outputs, to the impact of running e-government .

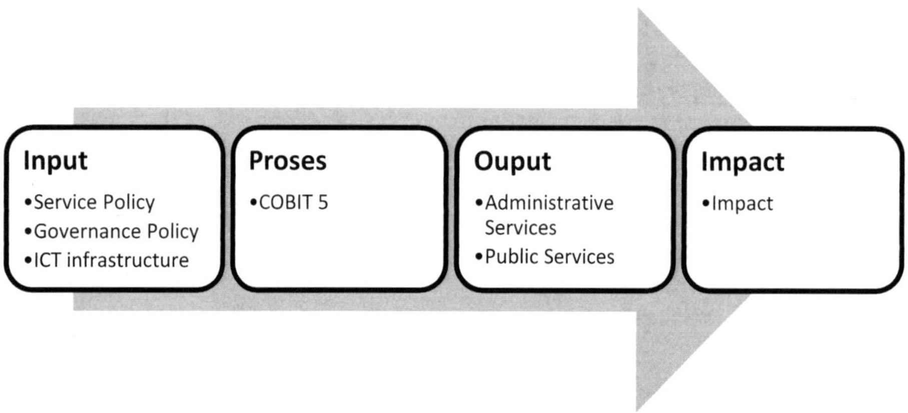


**Figure 1:** Proposed E-Government Evaluation Model.

The approach of this system is also used in research conducted by Wahyuni (2017) using input - process - and output aspects which are the result of the development of performance measurement. Based on the system approach, some of the research that has been collected is then mapped based on the aspects discussed or researched for further use as a basis in developing the conceptual framework or conceptual model in this study.

## 4. Domains and Attributes

### *4.1. Input*

Some research that has been collected explains that in measuring the performance or maturity of an e-government implementation can be seen in several aspects. The input aspects examined by Janssen (2004) explain that with information related to how much readiness or investment of the resources needed in the development of e-government implementation will affect the level of success and influence how decision makers in determining the next steps in e-government development [10]. Furthermore, through several perspectives explained that there is a need for commitment from the government to be able to support the running of e-government in order to achieve the desired goals [11]. This became the basis for Permenpan No. 5 of 2018 in developing the SPBE Internal Policy domain as a form of commitment of the Indonesian government in developing and implementing SPBE [6]. As a form of support, the existence of adequate infrastructure will be very useful in improving the performance or the performance of e-government services provided [12].

**Table 2** Mapping Prior Research Based on Aspects in the System Approach.

| Variable | Indicator | Reference |
|---|---|---|
| SPBE Governance Internal Policy | • Internal Policy of the SPBE Steering Team of Government Agencies,<br>• Internal Policy Integrated Business Process Innovations,<br>• Internal Policy of the Government Agency SPBE Master Plan,<br>• ICT Internal Budgeting and Expenditure Policy,<br>• Internal Data Operations Policy,<br>• Internal Application System Integration Policy,<br>• Internal Policy on Use of General Applications for Sharing | [6] |
| SPBE Service Policy | • Internal Policy of Official Script Services<br>• Internal Staffing Management Services Policy<br>• Internal Planning and Budgeting Management Services Policy<br>• Internal Financial Management Services Policy<br>• Internal Policy on Performance Management Services<br>• Internal Procurement Services Policy<br>• Internal Public Complaints Service Policy<br>• Internal Documentation Services and Legal Information Services<br>• Internal Policy for the Whistle Blowing System<br>• Internal Government Service Public Policy | [6] |
| Information & Communication Technology Infrastructure | • Interoperability<br>• Interconnectivity | [12] |

### *4.2. Process*

Aspects of the e-government process in several previous studies explained at the time of the process the measure is how services are provided [4,10,13]. This was also explained by Kim & Grant (2010) who from the perspective of capability or capability of e-government implementation, which later became the basis of Permenpan No. 5 of 2018 through the Governance domain. Meanwhile another study conducted by Evi (2017) tried to measure how the level of e-government maturity by using COBIT 5 which is the best practice in measuring the performance of IT governance in an organization. COBIT 5 as a framework that can help organizations or companies in ensuring the creation of value and the distribution of these values in order to achieve the goals of the company. Through the capacity model COBIT measures various processes in 37 process domains so that the organization or company can know the level of capability of the organization or company in managing information systems.

### *4.3. Output*

The output aspect of e-government in accordance with the approach taken by Cordoba (2009) explains that the output (output) of an e-government system are services available to stakeholders to be used [9]. Various studies also mention the importance of measuring the outcome of e-government systems as a measure of the success of e-government systems [4,10,12,13]. Various indicators of service outcomes are proposed by several studies with the aim of knowing how the services are used, and the level of efficiency of services compared to services before the use of e-government systems. Janssen (2003) explains that service availability is one of the factors that measure the success of e-government [10]. Susanto (2015) also explains how outputs are measured through the ease of use and availability online to be a factor that needs attention in the development of e-government [4].

**Table 3** Details of the output domain of the e-government evaluation process.

| Variable | Indicator | Reference |
|---|---|---|
| Electronic-based Government Administration Services | • Service Script Services<br>• Staffing Management Services<br>• Planning Management Services<br>• Budgeting Management Services<br>• Financial Management Services<br>• Performance Management Services<br>• Procurement Services | [6] |
| Electronic-Based Public Services | • Public Complaints Services<br>• Legal Documentation and Information Services<br>• Whistle Blowing System service<br>• Public Service Government Agencies | [6] |

### *4.4. Impact*

Impact aspect is one of the aspects that is difficult and complex in measuring the level of efficiency of e-government. Several studies that focus on examining how the impact of implementation on various stakeholder perspectives to find out how the level of success and efficiency of e-government. Berntzen (2014) explains in knowing how the level of e-government implementation needs to be seen through the efficiency achieved [13]. The intended efficiency is obtained from the measurement of several indicators ranging from improving service performance from the point of view of the user, and the government. Indicators that become important points in the assessment of efficiency are seen based on how time, cost, and effort savings from using e-government services [12,13].

**Table 4** Details of the impact domain of the e-government evaluation process

| Variable | Indicator | Reference |
|---|---|---|
| Impact | • Improved service efficiency in terms of time<br>• Increased service efficiency in terms of cost<br>• Improved Service Efficiency in terms of business | [8,10,12,14] |

| Variable | Indicator | Reference |
|---|---|---|
| | • Is there information from the SPBE used for decision making? | |

## 5. Conclusion and Future Work

Evaluation is carried out starting from how the readiness or input that is in the agency to the implementation of e-government, then how the process of governance of the resources in the agency, the output or output of the e-government system, to how the impact or impact of e-government government against stakeholders of e-government itself. With the evaluation through a system approach it is expected to provide optimal and comprehensive measurement and evaluation information so that it can be the basis of government information in determining strategies or next steps in e-government development.

In the system approach, measurement of e-government is seen based on aspects of input, process, output, and impact. More and more studies are developing measurements on each aspect. In the input aspect, measurements were made to focus more on the readiness of e-government implementation, starting from supporting government policies, information technology infrastructure, and data and application readiness in the e-government system that will be used [15]. For aspects in the process that are measured based on the COBIT 5 framework which is the best practices in assessing and measuring the maturity of processes in governance and management of the use of information technology [5]. Output aspect by using how services are provided by using guidelines made by the Indonesian government in Ministerial regulation number 5 of 2018 SPBE, then the impact is to look at the benefits of each service used both from the user's perspective as well as from the perspective of service provider view [9].

The model developed is expected to provide value or level of maturity of e-government at the city level by looking at several aspects ranging from readiness, governance processes, services, and the impact of e-government. In addition, the results of this study can provide advice and recommendations in increasing the level of maturity of e-government so that it can provide optimal and efficient services to each stakeholder through a model or methodology in measuring or evaluating e-government. So it is necessary to do an empirical trial to get results that can represent the value of the maturity level of a city's e-government.

## References

[1] Veit D and Huntgeburth J 2014 Introduction to Digital Government pp 1–17
[2] Anon Gartner's Four Phases of E-Government Model
[3] Novi Prisma Yunita 15917221 2018 Kondisi Terkini Perkembangan e-Government di Indonesia: Praktik Pemerintah dan Persepsi Publik
[4] Susanto T D and Ph D 2015 Measuring E-Government Implementation Program : A Case Study of Surabaya City , Indonesia *Int. J. Inf. Syst. Eng.* **1** 1–8
[5] Wahyuni E V I D W I and Susanto T D W I 2017 *Pengembangan Model Pengukuran Kematangan E-Government Level Kota*
[6] Birokrasi Reformasi D A N K P 2018 Pedoman Evaluasi Sistem Pemerintahan Berbasis Elektronik Instansi Pusat Dan Pemerintah Daerah 2017
[7] Fitriansyah A, Budiarto H and Santoso J 2013 Metode Pemeringkatan e-Government Indonesia (PeGI) Untuk Audit Tata Kelola Teknologi Informasi *Semin. Nas. Sist. Inf. Indones. 2013* 10–20
[8] Alcaide-Muñoz L, Hernández A M L and Caba-Pérez C 2014 Public Managers' Perceptions of e-Government Efficiency: A Case Study of Andalusian Municipalities pp 135–56
[9] Córdoba J-R 2014 Systemic Patterns of Practice to Improve e-Government Evaluation pp 9–24
[10] Yasin M, Vorhies D W, Morgan N a., Tölösi P, Lajtha G, Thompson K, Schofield P, Public

Policy Polling, Pinkney S, Marsden G, Marshall B, Magd, Link M W, Oldendick R W, Kozak M, Kouzmin A, Löffler E, Klages H, Korac-Kakabadse N, Janssen D, Rotthier S, Snijkers K, Its P W, Pinkney S, Holzer M, Yang K, Gogozan A and February R 2015 National Survey Results *Benchmarking An Int. J.* **44** 121–30

[11] Kim D and Grant G 2010 E-government maturity model using the capability maturity model integration *J. Syst. Inf. Technol.* **12** 230–44

[12] Joshi P R and Islam S 2018 E-government maturity model for sustainable E-government services from the perspective of developing countries *Sustain.* **10**

[13] Berntzen L 2014 Electronic Government Service Efficiency: How to Measure Efficiency of Electronic Services pp 75–92

[14] Ahuja A and Ahuja N 2008 Why Enterprise Architecture is must for One-Stop e-Government? *Emerg. Technol. E-Government* 235–41

[15] Iribarren M, Concha G, Valdes G, Solar M, Villarroel M T, Gutiérrez P and Vásquez Á 2008 Capability maturity framework for eGovernment: A multi-dimensional model and assessing tool *Lect. Notes Comput. Sci. (including Subser. Lect. Notes Artif. Intell. Lect. Notes Bioinformatics)* **5184 LNCS** 136–47

[16] Purón-Cid G 2014 Measuring e-Government Efficiency from the Perspective of Implementers: The Case of e-Budgeting in Mexico pp 111–34

# Docking application for food tenacity protection

**B N Kholila[1], D N Martono[2] and T E B Soesilo[2]**

[1]Magister Programe of Environmental Science, School of Environmental Science, University of Indonesia, Salemba, Indonesia.
[2] School of Environmental Science, University of Indonesia, Salemba, Indonesia
*Corresponding author e-mail: nowo2003@yahoo.com

**Abstract**. Sustainable agriculture is based on the concept of organic agriculture. However, the sustainability of agriculture system depends on the system and the people. Therefore, the human knowledge on concerning the sustainable environment is the main contribution towards achieving food security. One of the systems that could be utilized for the preservation of food security is through '*in silico*' approach with docking application. The production under organic agriculture is more resilient against threat, it is also more stable thus the food security and the environmental condition could be preserved. This research aims to recommend computerized technology (docking) as a supporting device with a function to view the effectivity of the natural basic materials on the creation of biopesticide. Therefore, the utilization of docking application for the creation of biopesticide is able to contribute to the alleviation of environment quality degradation in order to preserve food security as well as to enhance the knowledge on agricultural behavior. The result found in this research is the computerized analysis outcome for the past four years using docking application that has obtained the value Ki converted as $IC_{50}$ and the value of $LC_{50}$ so that it could find its effectivity and toxicity.

## 1. Introduction

The increasing population becomes one of the causes of land use change such as agricultural lands. Empirically, paddy field area is an area that is most vulnerable towards land use change from agricultural land [1]. Records in BPS in 2017 stated that the reduced paddy fields has caused a decrease in production or harvest from 29,245 tons to 4,054 tons [2]. The increasing population is line with the number of food production required to be supplied, but the fact is that the land use change still continues. This has force the agricultural actors to be able to intensively farm thus initiating the efforts to minimize the yield loss due to plant pest organisms by utilizing pesticides [3].

Pesticide is chemical mixture that is bioactive, but produce many contamination and residual effects on soil, land ecosystem as well as waters. [4]. Therefore, the indecent use of pesticides will cause serious threat. A country's food security target is achieved when the food security is at a regional or national level, especially in household or individual level is achieved [5]. The fulfillment of food domestic food security must be encouraged with new innovation to obtain sustainable quality and quantity standards. The continuity of agricultural system depends on the current system and the people. Therefore, the basic knowledge affecting the behavior to choose and the right implementation of pesticides or a switch to an environmental-friendly pesticide/ biopesticide to create sustainable agriculture.

Sustainable agriculture is based on the concept of organic agriculture with the principle of ecosystem sustainability, production result as well as safe products [6]. Organic agriculture is defined as an agriculture that does not include synthesis elements [7]. One of them is by utilizing biopesticide that is effective for agriculture pest control without causing any damage to the ecological chain or worsen the environmental pollution [8], thus the ecosystem cycle could continue to be sustainable in order to fulfill the food security with superior products.

*In silico* approach is one of the alternatives to actualize organic agriculture capable to achieve sustainability that is efficient in terms energy, time and cost in terms of the creation of biopesticide. *In silico* performed through molecular docking is one of the modelling approaches to predict ligand affinity towards protein targets [9]. The function of molecular docking consists of the conformation from the ligand and the protein (position/pose) that might occur to the protein binding pockets. Generally, molecular docking performs a position search with a pattern of flexible ligand and rigid protein that furthermore on every position is assessed based on the form and characteristics thus finding the most appropriate position.

Molecular docking has been widely used for drug discovery as well as its development by utilizing the research algorithm components and algorithm that creates the complex probability between protein and ligands thus the ligand binding affinity could be measured [10]. Studies about insect hormones have been considered important due to the ability to be used in the insecticides and pesticides, these hormones become the base of the molecular docking as the macromolecule or protein. In the context of biopesticides, ligand is based from compounds found in plants. Therefore, a series of phyto—pesticides formula preparation with the cost and time spent as well as the remainder of the formulation which are wasted could be controlled.

This research aims to show that docking application could contribute to support food security and sustainable agriculture with the knowledge of Ki value converted into $IC_{50}$ and $LC_{50}$ values obtained through computerized interaction of molecular docking shown in the last four years of research. Not only the food security, but also in terms of environment, contamination could be maintained due to the accurate result of the molecular docking containing compounds from an environmental-friendly material. Therefore, agriculture based on sustainable agriculture with the implementation of ecological principles in the design and management of agricultural ecosystem [11] would reach a high level of environmental performance and productivity [12].

## 2. Materials and Methods

This research uses methods of literature study about docking application utilized to show the interaction occurring between the ligands and macromolecules. The chosen result is a research conducted for the last four years determined as the limitation. The chosen interaction is a specific interaction for the utilization of docking application supporting the agricultural activities that show an interaction and inhibitor value produced through molecular docking.

### *2.1. Article Analysis*

Article analysis is performed based on these following categories: docking application, Ki value or $IC_{50}/LC_{50}$ value, country of study, and agricultural supports. Those categories could contribute to comprehend the advancement of docking application in the agricultural sector from year by year. These data will give an overview that molecular docking approach could contribute in supporting food security with the knowledge of interaction obtained from ligands and macromolecules so as to increasing knowledge for the agriculture actors about environmental-friendly materials. Therefore, the system and the people occurring could support each other to support the creation of food security.

## 3. Result and Discussion

Journal of Physics: Conference Series **1381** (2019) 012023 doi:10.1088/1742-6596/1381/1/012023

### *3.1. Result*

Based on our method analysis and variable could create basic categories for the study on the utilization of the molecular docking as the agricultural activity support for the past four years. The study on molecular docking is related to the agricultural utilizing ligands and macromolecular that is in line with the agricultural problems. Ligands illustrate compound contained in the plants utilized as biopesticide and macromolecule materials or protein indicating plant-disturbing organisms or pests that are often detrimental to agricultural activities.

Table 1.1 Study of molecular docking in the sector of agriculture for the past four years

| Year | Reference | Country | Docking Application | Result |
|---|---|---|---|---|
| 2019 | Abdelgaleil *et al.* [13] | Egypt | *Discovery Studio 3* software | Interaction between several monoterpenes ((1,8-cineole,(-) citronellal, limonene, α-pinene ,pulegone and 4-terpineol) indicating the high acaricidal activities by fumigant and the direct contact action on *Tetranychus urticae* Koch. Pulegone has the highest toxicity level ($LC_{50}$=3,81 mg/L of air). Citronellal has the lowest fumigant toxicity with the value of $LC_{50}$=15,5 mg/L of air. Limonene has the highest value of inhibitors, namely $IC_{50}$ = 11,37 mg/L. |
| 2019 | Raghav, D., Susobhan M., & Krishnan [14] | India | *Autodock 4.2* | The interaction occurring between Bovine Serum Albumin (BSA) and Karanjin has shown that Karanjin is bound closer. The interaction indicating that Karanjin has the lowest toxicity towards the human cervix cancer cells and L929 normal mouse fibroblast cells or in other words could hamper the cell growth of *B. subtilis* and *E. coli*. The pesticide works by way of hampering the proliferation of HeLa cells depending on the concentration of $IC_{50}$ value. $IC_{50}$ value that was obtained amounting to 0.22 mM. With the concentration of 5 mM could hamper 100% proliferation of HeLa cells producing acute cytotoxicity. |
| 2018 | Warsito *et al.* [15] | Indonesia | *Autodock Tools* | The optimization results conducted on the citronellal ligands from citronella oil, eugenol ligands from clove oil respectively are 2,18 mM and 1,14 mM. The interaction received from each ligands and macromolecules of 2DJC produced Ki amounting to 1.75 mM. From such Ki value, an insecticide formulation ratio of 4:1 was obtained |

| | | | | |
|---|---|---|---|---|
| | | | | (citronella oil : clove oil) that contains $LC_{50}$ value with the concentration of 0.05% (toxic contact) and 0.31% (leaf contact) |
| 2017 | Kholila *et al.* [16] | Indonesia | *Autodock Tools* | The interaction occurring on geraniol ligands, citronellal, citronellol, anethole and cineol through multiple docking that obtained value of Ki Geraniol 3.29 mM, Citronellal 2.18 mM, Citronellol 5.13 mM, Anethole 1.18 mM, and Cineol 1.75 mM. Ki from each of those compounds through multiple docking of Citronellal-anethole with 2DJC macromolecule produced $IC_{50}$ value of 3.2 mM. |
| 2016 | Gowrilakshmi, G. & M.S. Nalina S. [17] | India | *ArgusLab* | The ligand interaction of Lipase, Amylase, Protease and Chitinase with the protein of Cytochrome b (H6TG75) indicating the minimum binding value (kcal/mol) amounting to -9.14 kcal/mol for Cytochrome b-Lipase; Cytochrome b-Amylase -8.12 kcal/mol: Cytochrome b-Protease -13.63 kcal/mol; Cytochrome b-Chitinase -12.27 kcal/mol. The binding value produced indicated that the interaction between the protein and ligands caused the cell degradation as well as killing the insect pests. |

### *3.2. Discussion*

The food security of a country is achieved when the food security at a regional or national level especially in the household and individual level are met [5]. The food security has a strong correlation with the environmental condition and quality, meaning that the environment becomes one of the factors supporting food security. The environmental quality that could support food security is influenced by the system and people managing it. Hence, the human knowledge to preserve the environmental quality has a big role, in other words food security is closely related to knowledge. In addition to knowledge, in order to maintain food security requires new innovations that could immediately support food security. *In silico* approach becomes one of the alternatives that could be chosen to support food security. *In silico* approach uses the application docking that is used as a means for the knowledge of interaction occurring between the ligands and macromolecules [9]. This could aid determining binding value occurred between the ligands and macromolecules thus knowing the its potential toxicity. Literature study for the past four years was conducted as the limitation of study selection.

In the research for the past four years has shown a role of docking application to indicate the interaction occurring desired compounds. Despite of utilizing different software, but the results obtained is relatively showing results of binding energy as well as toxicity value of Ki value (Inhibition Constant), $IC_{50}$ (Inhibitory Concentration 50) as well as $LC_{50}$ (Lethal Concentration 50). Ki value is a value that shows the strength of the interaction between the ligands and macromolecules. This could be seen from the Ki value obtained from the docking result, where if the Ki value is smaller means the interaction occurring between the ligands and macromolecules is strong and only has a small inhibitor [16]. Ki value becomes the base to determine the $IC_{50}$ value which is a concentration that could hamper the interaction occurring amounting to 50%. In contrast to $LC_{50}$ that shows concentration causing death of 50% of the target organisms obtained from the calculation to determine the liveliness from a compound or extract in less than 24 hours [18]. This means by knowing the Ki value through molecular docking, the toxicity value could be discovered. The toxicity results through molecular docking proves the compound implementation as the potential for biopesticide [17].

A research in 2019 by Abdelgaleil *et al.* of the docking application utilizing the software of Discovery Studio 3 obtained $LC_{50}$ value from Pulegone amounting to 3,81 mg/L and $LC_{50}$ value from Citronellal, namely 15.5 mg/L on the interaction between several monoterpenes (1.8-cineole, citronellal, limonene, α-pinene, pulegone and 4-terpineol) showing the high acaricidal activity by fumigants and direct contact on *Tetranychus urticae* Koch. From the result of molecular docking conducted has shown that monoterpene Pulegone has a higher level of toxicity on *Tetranychus urticae* Koch than Citronellal[13]. This means to create biopesticide to eradicate *Tetranychus urticae* Koch (Red mites) requires natural raw materials containing Pulegone. In the same year as the research by Raghav, D., Susobhan M., & Krishnan, the docking application with the software of *Autodock 4.2* is utilized to determine the interaction between Bovine Serum Albumin (BSA) and Karanjin and $IC_{50}$ value was obtained amounting to 0.22 mM that could hamper cell growth of *B. subtilis* and *E. coli* [14]. This shows that the inhibitor value produced was small thus the interaction between BSA and Karanjin is strong to hamper cell growth of *B. subtilis* and *E. coli*.

The utilization of docking application in 2018 by Warsito *et al.* and Kholila *et al.* in 2017 a result of molecular docking was obtained using the application *Autodock Tools* comprising Ki value which then is converted into $LC_{50}$ and $IC_{50}$ values. In the research of Warsito *et al.*, Ki value was processed thus receiving ratio of biopesticide formulation from citronella oil and clove oil (4 : 1) thus receiving $LC_{50}$ value with a concentration of 0.05% (toxic contact) and 0.31% (leaf contact) [15]. This shows that the formulation produced has mortality value amounting to 50% only with the formula concentration of 0.05% and 0.31%. Whereas in the research of Kholila *et al.*, the interaction on geraniol ligand, citronellal, citronellol (citronella oil), anethole and cineol (basil oil) through multiple docking based on the lowest Ki value, namely 2DJC macromolecule citronellal-anethole producing $IC_{50}$ value of 3.2 mM [16]. The lowest Ki value chosen for multiple docking was performed as the smaller the Ki value, the inhibitor value is also smaller, this means the interaction on ligands and macromolecule simulated has a strong interaction so that the toxicity value is also significant. In 2016 Gowrilakshmi, G. & M.S. Nalina S. Utilization the docking application of ArgusLab to observe the interaction of ligands *lipase, amylase, protease and chitinase with protein Cytochrome b (H6TG75)*. The result obtained comprises of the minimum binding value (kcal/mol) amounting to -9.14 kcal/mol for *Cytochrome b-Lipase; Cytochrome b-Amylase* -8.12 kcal/mol: *Cytochrome b-Protease* -13.63 kcal/mol; *Cytochrome b-Chitinase* -12.27 kcal/mol [17]. The binding value produced shows that the interaction between the protein and ligands have caused cell degradation as well as bringing death to the insect pests.

Based on the research conducted for the past four years, utilizing docking application through *in silico* approach could determine the interaction between the ligands (compounds contained in the natural raw materials) and macromolecule (plant disturbing organisms) created in Ki or $LC_{50}$ value. The utilization of docking application proved to be effective to be used to support the manufacture of biopesticide formulation. This application will be more efficient as it does not need the preparation of a series of formulas. Therefore, the utilization of docking application also functions to reduce experimental waste to obtain the appropriate composition in the making of biopesticide.

Molecular docking application which is widely used for the past for years is *Autodock Tools* in several versions. *Autodock Tools* is a docking program with the technique of flexible docking and grid-based method for energy evaluation [19]. The process of application usage *Autodock Tools* has many steps but rendering with high quality results [16]. This application calculated intermolecular energy, internal energy as well as the output which later forms docking energy as well as binding energy [20]. This is to make the application *Autodock Tools* be used by many researchers to perform molecular docking process. The use of molecular docking has been developed in several pharmaceutical field in development of new drug discovery and planning. The principle used in molecular docking, namely the placement of ligand interaction to the active side of the receptor which then is simulated based on the structural conformation and the nature of electrostatic [21].

*In silico* approach conducted to utilize the docking application could support the existence of food security through the accurate biopesticide modelling. Through this approach, the potential for the creation of organic agricultural is also higher due to the use of natural raw materials as biopesticide was proven in real terms with Ki value produced from the molecular docking process. On the other hand, with the knowledge of the appropriate toxicity value, this could be a trigger agricultural actors to be more interested or possibly switching to use environmental-friendly materials for agricultural activities. Therefore, the occurrence of environmental quality degradation will be smaller and the agricultural production could be maintained resulting in food security.

## 4. Conclusion

The food security threatened due to the agricultural land use change, especially the paddy fields being the matter that needs to be solved. *In silico* becomes one of the innovations to preserve food security. Docking application used for the past four years to know the conformity between the ligands and target protein (pests) in the agricultural activities is proved to have produced Ki, $IC_{50}$ as well as $LC_{50}$ values/ 50% toxicity value for agricultural problems. Molecular docking data results could be the knowledge base for the agricultural actors that support food security. Not only does this potentially could preserve food security, but also reduce environmental quality degradation so as to the creation of sustainability.

**Acknowledgments**

This paper is supported and funded by the International Indexed Publication Grant for Student Final Project (PITMA) of the University of Indonesia 2019.

## References

[1] Iqbal M and Sumaryanto 2007 Strategi Pengendalian Alih Fungsi Lahan Pertanian Bertumpu Pada Partisipasi Masyarakat *Anal. Kebijak. Pertan.* **5** 167–82

[2] Central Bureau of Statistics of Bekasi Regency 2018 *Bekasi Regency in Figures*

[3] Kardinan A and Suriati S 2012 Effectivity of Botanical Pesticides Against Tea (Camellia cinensis L.) Pest Attack *Balai Penelit. Tanam. Rempah dan Obat* 148–52

[4] Carvalho F P 2017 Pesticides, Environment, and Food safety *Food Energy Secur.* **6** 48–60

[5] Rachman H P S, Purwoto A and Hardono G S 2005 Kebijakan Pengelolaan Cadangan Pangan pada Era Otonomi Daerah dan Perum Bulog *Forum Penelit. Agro Ekon.* **23** 73–83

[6] I Wayan Supartha N L K 2016 Konsep dan Strategi Pengembangan Pertanian Oganik di Bali *Seminar Regional Pertanian Organik dan Prospek Pengembangannya di Bali* (Denpasar)

[7] Kremen C and Miles A 2012 Ecosystem Services in Biologically Diversified Versus Conventional Farming Systems: Benefits, Externalities, and Trade-Offs *Ecol. Soc.* **17**

[8] Shilpi Sharma P M 2012 Biopestcides : Types and Applications *Iunternational J. Adv. Pharmacy, Biol. Chem.* **1** 508–15

[9] Mahobia A and Jha Z 2018 Docking Studies of Andrographolide and Its In Vitro Validation *Int. J. Chem. Stud.* **6** 1164–6

[10] R J M and S N J 2014 Molecular Docking and Binding Energy Studies on Nuraminidase of h1n1 Reveal Possible Answer to its Resistance for Oseltamivir *Int. J. Drug Discov.* **1** 34–9

[11] Dalgaard T, Hutchings N J and Porter J R 2003 Agroecology, Scaling and Interdisciplinarity *Agric. Ecosyst. Environ.* **100** 39–51

[12] Schutter O De 2014 *The Transformative Potential of the Right to Food*

[13] Abdelgaleil S A M, Badawy M E I, Mahmoud N F and Marei A E S M 2019 *Acaricidal activity, biochemical effects and molecular docking of some monoterpenes against two-spotted spider mite (Tetranychus urticae Koch)* vol 156 (Elsevier Inc)
[14] Raghav D, Mahanty S and Rathinasamy K 2019 Biochemical and toxicological investigation of karanjin, a bio-pesticide isolated from Pongamia seed oil *Pestic. Biochem. Physiol.* **157** 108–21
[15] Warsito, Utomo E P, Ulfa S M, Kholila B N and Nindyasiwi P 2018 Green Chemistry: Strategy in Essential Oils Sustainability by Development of Insecticide Using Docking Method *IOP Conf. Ser. Mater. Sci. Eng.* **299**
[16] Kholila B N, Warsito and Ulfa S M 2017 In Silico Method on Natural Insecticide Formulation Based on Citronella Oil and Basil Oil and Activity Test on Insects Ulat Grayak ( Spodoptera litura ) *International Conference of Essential Oil (ICEO)* pp 126–32
[17] Gowrilakshmi S and Nalina Sundari M S 2016 In Silico Docking Studies of Four Enzymes of Beauveria bassiana on the Subcellular Protein of Stored Product Pest, Corcyra cephalonica *Int. J. Curr. Res. Biosci. Plant Biol.* **3** 91–105
[18] Sulmartiwi L, Umami M R, Alamsjah M A and Subekti S 2014 Bioaktivitas minyak Atsiri Daun Bandotan (Ageratum conyzoides) terhadap respon fisiologis (Glukosa darah dan tachiventilasi) Ikan Koi (Cyprinus carpio) *J. Ilm. Perikan. dan Kelaut.* **6** 81–6
[19] Luo W 2014 Design, Synthesis and Evaluation of Genistein-polyamine Conjugates as Multi-functional Anti-Alzheimer Agents *Med. Chem* **4**
[20] Sharma P and Awasthi P 2015 Molecular Modeling Study of Various Synthetic Jhas Containing Varied Functionality 1687–93
[21] Ikhlas N 2013 Uji Aktivitas Antioksidan Ekstrak Herba Kemangi (Ocimum americanum Linn) dengan Metode DPPH (2,2-Difenil-1-Pikrilhidrazil) 78

# Maturity Level Enterprise Academic Information System Based on Cobit

**T Lestariningsih[1*], Y Afandi[2], L D Setia[1], Q Qimyatuss'aadah[1], M E Echsony[1], T Prihatinta[1], M Taali[1], S Srimiatun[1]**

[1]The State Polytechnic of Madiun, Jl. Serayu No. 84 Madiun East Java Indonesia
[2]The State Polytechnic of Malang, Jl. Soekarno Hatta No. 9 Malang East Java Indonesia

*trilestariningsih@pnm.ac.id

**Abstract.** The success rate of the implementation of information systems applied to the public and private universities are necessary to determine the extent of the maturity level of the college information System management, allowing Management of the College knows what deficiencies are and in what direction the information system should be developed and managed. To find out what action to do for higher education advances need to evaluate maturity level, one of the frameworks used for evaluation is COBIT.This research uses the COBIT framework with the domain deliver and Support (DS) and Monitor and Evaluation (ME) to evaluate the maturity level of the college information system. The result of this research is the level of maturity of the college Informasai system is on good but there are some parts that need improvement, the improvement focused on the documentation of each service process on the academic information system and Oversight or control of system security as well as intense training and socialization.

## 1. Introduction

Implement information technology today has spread to all areas including in public and private colleges because information technology combines all processes from procurement, processing up to the dissemination of Information.[1] Implementation of the information system needed framework as a reference from planning to evaluation so that the implementation of the informations system can be in accordance with good governance objectives. Governance is an executive and board responsibility that involves leadership, organizational structure, and processes in ensuring that IT becomes a supporter and part of the realization of the strategy and achievement of organizational objectives[2].

(Benaroch, Chernobai&Goldstein, 2012), (Ward and Peppard, 2002), (Pollack, 2009), (Wardani, S and Puspitasari, M. 2014), Start if the information technology is said to be good and aligned between businesses with technology when managed well.[3][4][5][6].

Universities in Indonesia have already implemented an academic information system. The academic system is used to facilitate administrative and academic activities in universities, examples of universities ' administrative and academic activities include new student Admissions (PMB), Study plan cards (KRS), Processing, the management of data Lecturers & students. So that the academic system in the college can be beneficial to support the analysis of data used determines or decision making universities.

An important Factor that supports the utilization of information technology in universities is IT Governance.. To realize an IT Governance in a good university need to be done control or audit information systems. Information systems on the college authors are audited by using the COBIT 4.1 framework with Delivery and Support (DS) and Monitoring and Evaluate (ME) domains. The focus taken by the authors for the Delivery-Support (DS) domain is on the service process and the operation of the college academic system so that the service and operation process runs effectively and efficiently, being for the domain Monitoring-Evaluation (ME) focuses on the supervision and evaluation of the academic system of the college to measure the maturity level of the college academic System management of appropriate or not with the aim of the school of Focus taken by Author for Delivery-Support domain (DS) is in the process of service and operation of the college academic system so that the service and operation process runs effectively and efficiently, is for domain Monitoring-Evaluation (ME). Measuring the College's academic information system is conducted to evaluate the maturity level of the appropriate college academic system management process or not with the purpose of the college. [7].

## 2. Research Method

Research conducted is a descriptive research, where the results of research conducted by the authors are presented in the form of descriptions, both qualitative and quantitative descriptions. Research conducted using college case studies in Indonesia. Measuring the maturity level of the academic information system in Indonesia aims to make the service process in the College academic Information system in accordance with the framework of COBIT version 4.1, using primary and secondary data. Use of primary data acquired authors by conducting interviews with the management team, IT Manager and the system operator either directly or with a questionnaire, the author also conducts surveys and observations on the implemented academic system of information. For seconds the data authors earn from various sources such as reports and publications that are relevant to the research the authors do. Measurements in this study used 2 domains namely Delivery-Support (DS), and Monitoring-Evaluation (ME) domains. The measurement of each domain uses the COBIT Framework version 4.1 statement as below:

Table 1. Sub-domain on COBIT 4.1

| Domain | Sub-domain | Jumlah Pertanyaan |
|---|---|---|
| *Delivery-Support* (DS) | DS1 | 31 |
| | DS2 | 33 |
| | DS3 | 33 |
| | DS4 | 40 |
| | DS5 | 49 |
| | DS6 | 31 |
| | DS7 | 34 |
| | DS8 | 32 |
| | DS9 | 25 |
| | DS10 | 28 |
| | DS11 | 32 |
| | DS12 | 33 |
| | DS13 | 38 |
| *Monitoring-Evaluation* (ME) | ME1 | 33 |
| | ME2 | 32 |
| | ME3 | 26 |
| | ME4 | 44 |
| Total | | |

Then each sub-domain will be the assessment of IT Maturity Model, this assessment determines the level of maturity (maturity level) of each selected process is implemented through the questionnaire then grouped at maturity level such as Below table:

Table 2. Index Maturity[8]

| *Index Maturity* | Description |
|---|---|
| 0.00 – 0.50 | Non-existent |
| 0.51 – 1.50 | Initial / Ad Hoc |
| 1.51 – 2.50 | Repeatable but Intuitive |
| 2.51 – 3.50 | Define Process |
| 3.51 – 4.50 | Manage and Measureable |
| 4.51 – 5.00 | Optimized |

Once known the level of maturity then determine the position of the company. The position that occurs in the current company (As-is) and the desired position (To-be) that will be the reference to the IT governance model to be developed.Analysis gaps. This process is for determining the IT processes with COBIT 4.1 what is required to improve IT management.Make an action proposal improvements to the of IT governance. This process to make management guidelines refers to COBIT 4. 1

## 3. Result and Analysis

The analysis of the data in this study is divided into 2 (two), namely analysis of the maturity level of the current, the expected maturity analysis.

1. The Current level of maturity analysis (as is)

Based on data questionnaire results are conducted analysis to assess the current level of maturity (as-is) for the DS and ME processes.At the current level of maturity analysis (as-is), assessment of each activity. As for the result of the maturity questionnaire, there will be 6 options for the answer with a value of 0 – 5. The maturity attribute level is gained from the calculation of the total answer option of the questionnaire with the formula and the weighted answer options as follows:

$$\text{Attribute Maturity Index} = \frac{\sum(\text{ Total Answers} + \text{weights})}{\text{Amount of respondents}}$$

Determination of the level of maturity gained from the dissemination of the questionnaire with a the reponden of 125 people. Recapitulation of the questionnaire as follows:

Table 3. Questionnaire Recapitulation

| Domain | Process | Total |
|---|---|---|
| DS1 | Define and Manage Service Level | 120 |
| DS2 | Manage Third-Party Services | 448 |
| DS3 | Manage Performance and Capacity | 601 |
| DS4 | Ensure Continuous Service | 254 |
| DS5 | Ensure System Security | 128 |
| DS6 | Identify and Allocate Costs | 321 |
| DS7 | Educate and Train Users | 388 |
| DS8 | Assist and Advice Customers | 446 |
| DS9 | Manage the Configuration | 134 |
| DS10 | Manage Problems and Incidents | 207 |
| DS11 | Manage Data | 198 |
| DS12 | Manage Facilities | 182 |
| DS13 | Manage Operations | 240 |
| ME1 | Monitoring and Evaluate IT Performance | 161 |
| ME2 | Manage Third-Party Services | 367 |
| ME3 | Ensure regulatory Compliance | 298 |
| ME4 | Provide IT Governance | 373 |

On the delivery and support the domain there are 13 processes, with 3 processes at the level of maturity 4 (managed and measurable), 6 processes are at the level of maturity 3 (defined process) and 4 other processes with Maturity Level 2 (Repeatable boots Biofermentation. In the Monitoring and Evaluate domain There are 4 processes with 3 processes being at the Maturity Level 3 (defined process) and 1 other process has a maturity level of 2 (Repeatable but Intuitive), as shown in table 4 and Figure 1.

Table 4. *Current maturity*

| Domain | Index | Level |
|---|---|---|
| DS1 | 3.52 | 4 |
| DS2 | 2.58 | 3 |
| DS3 | 2.67 | 3 |
| DS4 | 3.19 | 3 |
| DS5 | 2.57 | 3 |
| DS6 | 3.53 | 4 |
| DS7 | 1.98 | 2 |
| DS8 | 2.22 | 2 |
| DS9 | 3.55 | 4 |
| DS10 | 2.33 | 2 |
| DS11 | 3.25 | 3 |
| DS12 | 2.15 | 2 |
| DS13 | 2.91 | 3 |
| ME1 | 3.02 | 3 |
| ME2 | 3.18 | 3 |
| ME3 | 2.45 | 2 |
| ME4 | 3.14 | 3 |

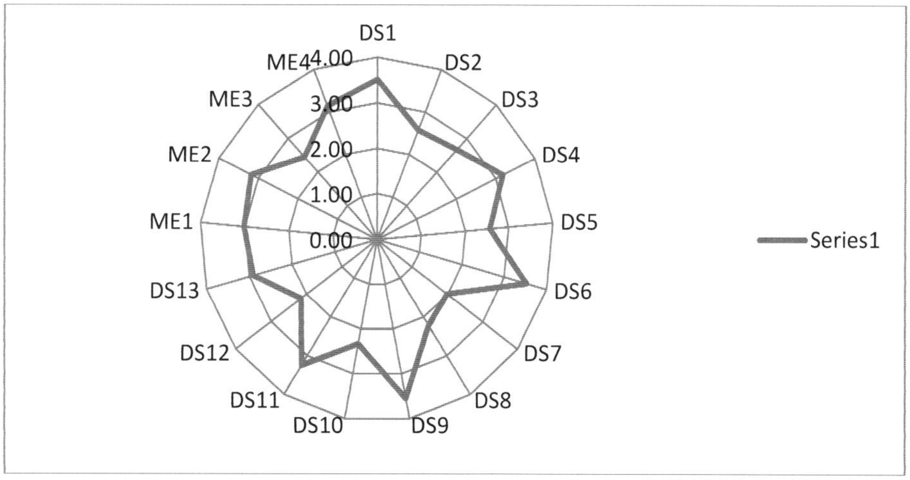


Figure 1. *Current maturity*

Based on the results of table 4 and picture 1 in the general the implementation of academic information system in universities both in the domain deliver and support and monitoring evaluate is at level 3 (defined process).This means that activities or standards related to the implementation of academic information systems in universities are in accordance with the standards, documented and have been communicated with the training, but for the implementation of documentation is still limited by formality and the procedure is not clear and perfect, so it is still possible to the occurrence of malpractice.

2. Level of hope assessment (to-be)

The expected level of assessment (to-be) provides a reference to the development of IT governance in college. Development of college academic information system is expected to refer to the College statute of Vision and mission, college objectives. The level of maturity expected by the college can be seen in table 5 and Figure 2.

Table 5. GAP *Maturity Level*

| Domain | Maturity level | | |
|---|---|---|---|
| | As is | To be | Gap |
| DS1 | 3.52 | 5 | 1.48 |
| DS2 | 2.58 | 3 | 0.42 |
| DS3 | 2.67 | 3 | 0.33 |
| DS4 | 3.19 | 4 | 0.81 |
| DS5 | 2.57 | 3 | 0.43 |
| DS6 | 3.53 | 5 | 1.47 |
| DS7 | 1.98 | 3 | 1.02 |
| DS8 | 2.22 | 3 | 0.78 |
| DS9 | 3.55 | 5 | 1.45 |
| DS10 | 2.33 | 3 | 0.67 |
| DS11 | 3.25 | 4 | 0.75 |
| DS12 | 2.15 | 3 | 0.85 |
| DS13 | 2.91 | 4 | 1.09 |
| ME1 | 3.02 | 5 | 1.98 |
| ME2 | 3.18 | 4 | 0.82 |
| ME3 | 2.45 | 3 | 0.55 |
| ME4 | 3.14 | 4 | 0.86 |

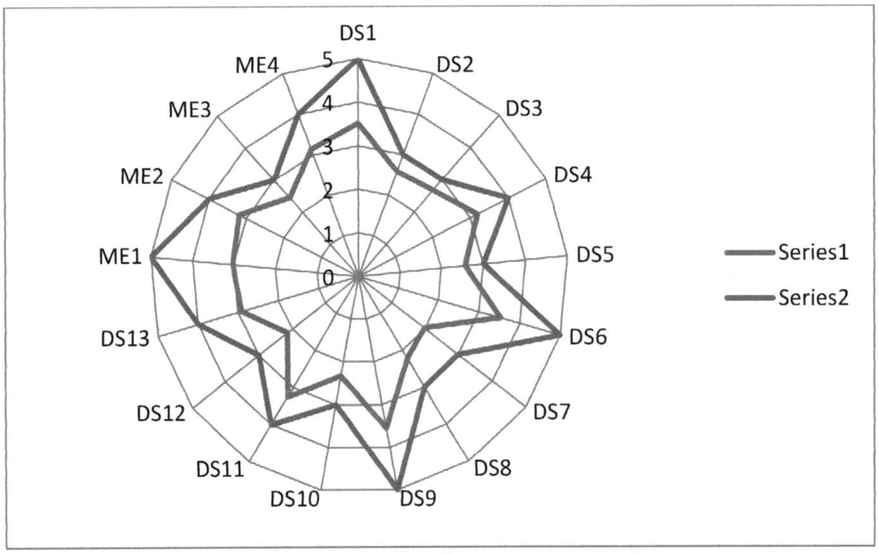


Figure 2. GAP *Maturity Level*

Based on table 5 and Figure 2 then the recommendation for the College's academic information system is intensive IT management to the use of college academic information system.Related management must have a commitment to the level of security by documenting each service activity in accordance with the COBIT standards to allow the information system on the college to implement the academic information system of IT management Good governance.

## 4. Conclusion

The level maturity level for academic information systems in colleges uses the COBIT Framework 4.1, using domains DS (Delivery and Support ) and ME (Monitoring and Evaluation) with the following conclusions:

1. The level of maturity of the college's academic information system is at level 3 (defined process), thereby means that the activities or standards relating to the implementation of the academic information system in college have fulfilled Standards, documentation has been conducted but still not in accordance with the procedure and is not formal, while for the improvement of competence has been conducted with the training, so the possibility of misuse or The irregularities still exist.
2. The improvement that needs to be done is in the formal documentation process and made the procedure of documenting its documentation for each service process on the academic information system. And need to be done supervision or control over each service process by increasing the level of system security

## References

[1] Lucy A 1993 An Introduction to Computer- Based Library System *3rd ed.; Chichester: John Wiley & Sons* p 3

[2] ITGI 2007 ver.4.1: Framework, Control Objectives, Management Guidelines, Maturity Models. Rolling meadow

[3] Benaroch, M., Chernobai & Goldstein J 2012 An Internal Control Perspective on the market value consequences of IT operational risk events *Int. J. Account. Inf. Syst.* **13(4), 357**

[4] Ward J and Peppard J 2002 *Success Factors in Strategic Information Systems*

[5] Pollack. 2009 Linking the Hierarchical Service Quality Model to Customer Satisfaction and Loyalty *J. Serv. Mark.* **23(1):** 42–50

[6] Wardani, S and Puspitasari M 2014 Audit Tata Kelola Teknologi Informasi MengunakanFramework Cobit Dengan Model Maturity Level (Studi Kasus Fakultas ABC) *Jurnal.Teknologi* **07** 38–46

[7] Gustiarni, I., and Putra, S. A. M Pengembangan Aplikasi e-university : Sistem Infromasi Pengelolaan Audit Teknologi Informasi Berbasis Risiko Menggunakan Framework COBIT Versi 4.1

[8] Krisanthi G.A.T.,Sukarsa I.M., Bayupati I.P.A. 2014 Governance Audit Of Application Procurement Using COBIT Framework *J. Theor. Appl. Inf. Technol.*

Journal of Physics: Conference Series **1381** (2019) 012025 doi:10.1088/1742-6596/1381/1/012025

# Elayah: Mobile Based Media For Al-Qur'an Memorization Using Takrar Method

**F R Pradhana[*], A Musthafa, T Harmini, M Dedy Setiawan**

Universitas Darussalam Gontor, Main Street Siman-Ponorogo KM.06, Ponorogo

*faisalrezapradhana@unida.gontor.ac.id

**Abstract**. As a pesantren based campus,University Of Darussalam Gontor requires their student to memorize Al-Qur'an as part of learning activities. However, the facts shows the level of memorization of students has not been able to achieve the target. This media aims to help students improve the quality of memorizing Al-Qur'an with practical features in a mobile application. This media uses the takrar method which has been used since the time of the Prophet Muhammad S.A.W. The takrar method is used by repeating the reading 40 times each sentences. This media based on mobile android application and created using waterfall model. Testing is done in two ways, the suitability of the application with an Android devices that shows the suitability of a device measuring at least 5 inches, and the distribution of questionnaires to the examiner or Muhafidz resulted in the suitability of the functions and materials as much as 93%, and the average value of students as much as 97.3% in terms of ease of use. The results of the questionnaire indicate the suitability of the media to the user needs.

## 1. Introduction

Al-Qur'an is the holy book of Muslims throughout the world. Al-Qur'an was given to humans not only as teaching material, but also as a miracle for the prophet Muhammad S.AW and proof of his prophethood.[1] As a Muslim, it is a must to study Al-Qur'an so that it is not wrong in understanding the meaning contained therein. One of the virtues that can be obtained from the Koran is to become a memorizer.[2] As mentioned in one of the verses *"But it is clear revelations in the hearts of those who have been given knowledge, and none deny Our revelations save wrong-doers."* Al-Ankabut:49.

Indonesia is a unique country, one of its uniqueness is the existence of pesantren, pesantren is an educational institutions that are more focused on religious education and the morals of their students.[3] one of the pesantren models in Indonesia is the pesantren *Tahfidz Al-Qur'an* which specifically teaches memorization of Al-Qur'an, each pesantren has its own methods of memorizing Al-Quran. some methods are *juz'i, sima'i, tasmi ', muraja'ah, jama', takrar and kitabah.*[4] This research uses the *takrar* method as the Al-Qur'an memorization method and the waterfall method in its design because it is considered to be in accordance with the memorization system carried out at the university, and there are previous studies that have verified their strengths and weaknesses so as to provide convenience in future development.

University Of Darussalam Gontor as one of the pesantren model university requires all students to memorize Al-Qur'an, in fact memorization of Al-Qur'an is used as one of the requirements for graduation.This is in accordance with the principle of University of Darussalam Gontor which applies the practice of dzikir, thought, sports and feeling as a principle in living everyday life.[5] The memorization system of al-quran which has been implemented at the University of Darussalam Gontor is by visiting examiners or muhafidz regularly. The weakness that arises is that there is no memorization method of Al-Qur'an specifically applied at the university, so the expected results have not reached the target, even the data in 2016 and 2017 showed only 48% of students who managed to achieve the memorization target.

The target of this media is Darussalam gontor university students and every muslims who have difficulty in memorizing Al-Qur'an, because they do not have the correct method.

This media is designed to be used on Android-based smartphones, because the device is the most popular among students, the Android device is also the most widely used device in the world today, more than 190 countries are recorded using the Android. In addition to practicality and ease of use, Android provides convenience in terms of developers because all the code in Android has an open source license from Google so that it gives freedom for anyone to develop it.[6]

in other research about the development of mobile based assistive applications in memorizing Al-Qur'an has produced applications that can be accepted by its users, also the usefulness of the features contained in the application make it easier for users to memorize.[7] Secondly, research on the use of the *takrar* method in memorizing the Koran has given positive results, the *takrar* method is proven to have advantages in strengthening one's memory of memorization, with the recitation method which is done repeatedly, making it easy for the memorizer to memorize verses so that he can know verses by paragraph in more detail.[8]

The novelty of this research is to focus on the use of features provided in Android that are tailored to the needs of the Takrar method, so that the results obtained can be appropriate and better than the use of the Takrar method directly

## 2. The Methods and Materials

### *2.1. Materials*

This research uses java programming language for back end and XML for front end design, Corel Draw X7 software is used to design images and logos. Open source software from Google Android Studio and SQL Lite to store application databases.

### *2.2. Methods*

The Takrar method is derived from the Arabic word *takrar* which means to repeat something, to do it repeatedly. Based on the above understanding, the takrar method is a systematic practice that is carried out repeatedly regularly, orderly and think well to obtain the expected results. Takrar method in memorizing the Qur'an based on the verse in the Surah Al-Furqan verse 32 states: *"And those who disbelieve say: Why is the Qur'an not revealed unto him all at once? (It is revealed) thus that We may strengthen thy heart therewith; and We have arranged it in right order".*

In the implementation of the takrar method, the memorizer reads repeatedly 40 times, after which the memorizer starts the memorization in accordance with what has been repeated. in the takrar method there are several conditions for easy memorizing Al-Qur'an, namely thought, sincerity, patience, istiqomah, good character, parental consent. In addition to the memorization model, there are several strategies that support the successful implementation of this method, namely:

- Double looping strategy
- Never move on to the next verse before the verse being memorized is completely memorized
- Memorizing the sequences that are memorized in a single unit after the number of memorized verses.
- Using just one mushaf
- Understand the verses that are memorized
- Pay attention to similar verses

- Deposit to someone who is able to memorize the Qur'an.
- Manners to memorize Al-Qur'an.[8]

The design uses the waterfall model with the stages of requirement, design, implementation, verification, and maintenance[9] as shown in Figure 1, all the results in each of these stages will be explained in the implementation chapter.

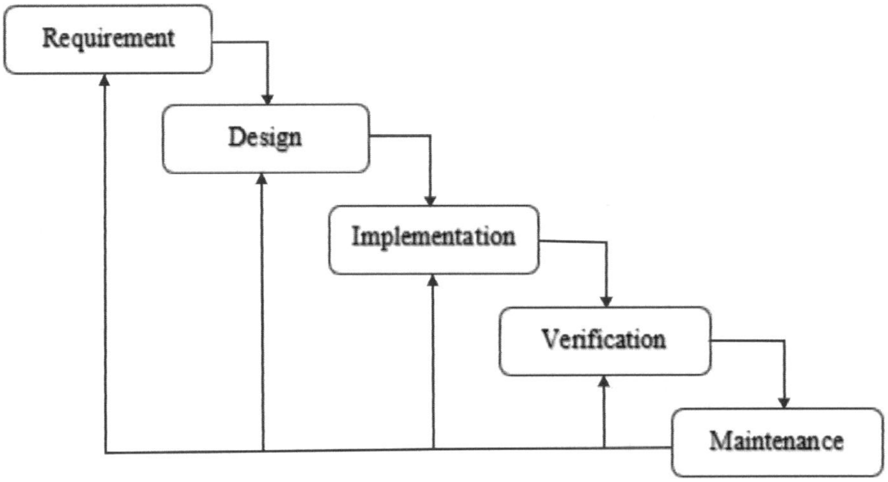


**Figure 1.** Waterfall Model

## 3. Implementation and Result

### *3.1. Requirement*

At this stage the initial analysis phase begins, such as the formulation of the problem, until the technological solution that will be developed to complete the development of mobile-based quran memorization media. Needs based on the needs of students in obtaining a memorization method that can help improve the quality and quality of their memorization of the Koran, the solution is combined with the latest Android technology that is easy to use anywhere and anytime.

**Table 1:** Application development needs

| No | Activities | Result |
|---|---|---|
| 1 | identification of problems | Many students have difficulty in completing the target of memorizing the Koran because they do not have the correct memorization method |
| 2 | Application content | The procedures for memorizing Al-Qur'an using the takrar method |
| 3 | The memorization method | Takrar Method |
| 4 | Offered solution | Elayah mobile based media for Al-Qur'an memorization using takrar method |
| 5 | The used technologies | Android based mobile application |

### *3.2. Design*

The next stage of the content that has been obtained previously, the development of the design flow of the use of mobile applications to be made. Included are juz inputs, audio inputs and clock inputs etc, as explained in Figure 2 below:

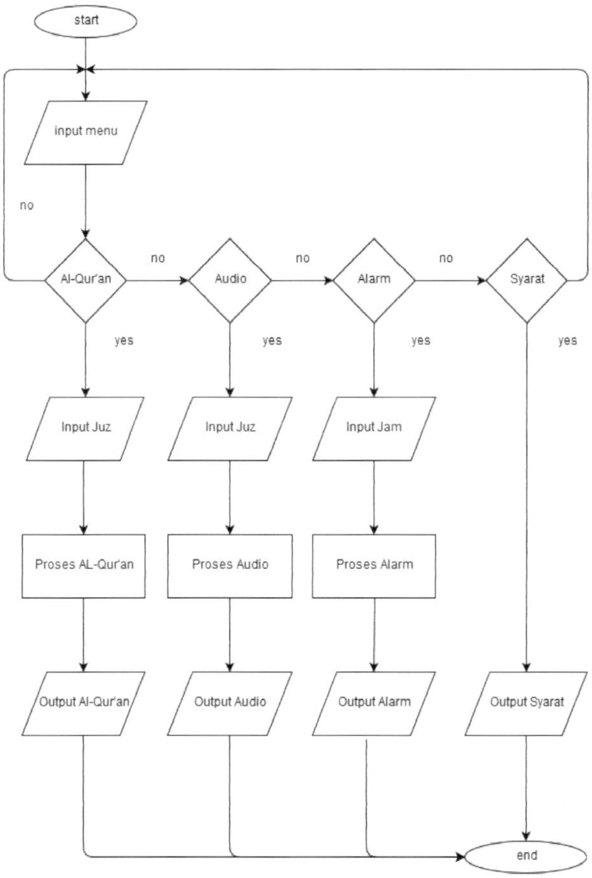


**Figure 2.** Flowchart steps for using Al-Quran memorization media

### *3.3. Implementation*

The next stage is application development based on the flowchart that has been made before, some of the results of application development can be seen in Figures below.

**Figure 3.** Elayah welcome screen

In figure 3, welcome screen page is intended for new users who install the Elayah application, will show some of the application available services, such as the Al-qur'an, alarm, audio and terms

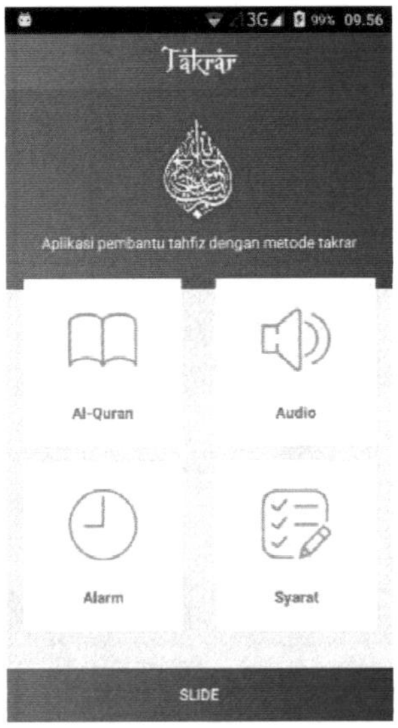


**Figure 4.** Elayah main Menu

The main menu displays the main features of the application area that can be used by each user. Users can choose each menu according to their desired needs.

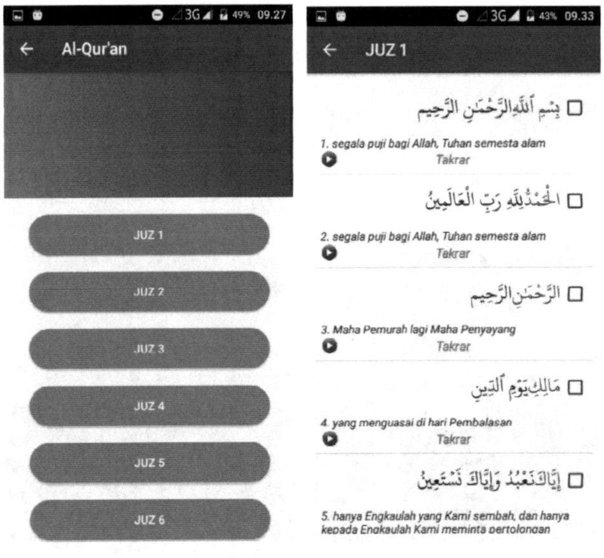


**Figure 5.** Al-Qur'an And Juz pages section

In figure 5, explained when the user selects the Al-Qur'an menu on the main page, it will be directed to the Al-Qur'an page, there is a button for each juz in the Al-Qur'an. When user selects one of the juz of Al-Qur'an, then the entire verse list will be displayed as much as one juz, and in each verse list there is audio, translation, checklist buttons, and different colors in the verse letters. the core of this media usage is on the juz page, when a check mark on a verse is turned on, only the red paragraph appears. This

Journal of Physics: Conference Series **1381** (2019) 012025 doi:10.1088/1742-6596/1381/1/012025

applies until the user squeezes the verse up to 40 times in accordance with the takrar method. after that, then the next verse will appear, and so on until the whole verse is raised.

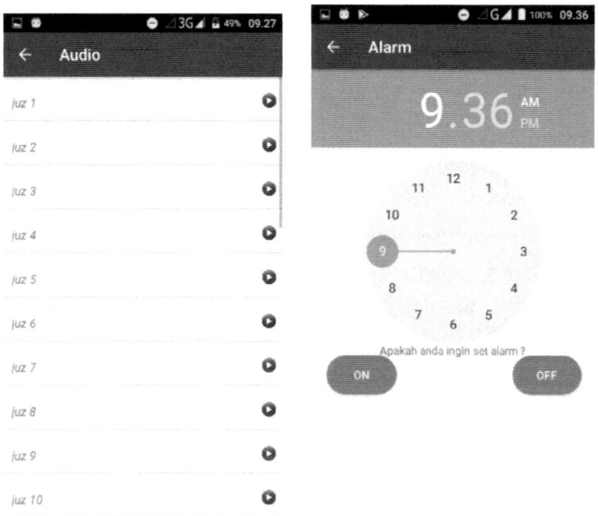


**Figure 6.** Audio and alarm pages section

In figure 6, display the audio and alarm pages. On the audio page, there is audio divided into chapters with a play and stop buttons. and on the alarm page there is a clock that has an on and off button. If the user sets an alarm the system will display a notification and a ringtone and the alarm will stop when the off button is pressed.

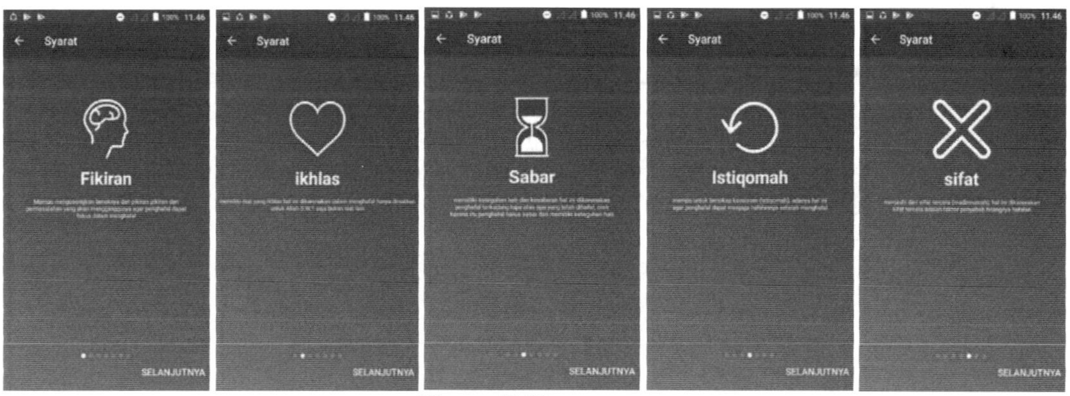


**Figure 7.** Terms page

On the Terms page, there is a slide there is a picture and description of the memorization requirements. On this page there are 7 slide conditions for memorization, namely mind, sincerity, patience, istiqomah, good character, parental consent, can read the Qur'an.

*3.4. Verification*

Testing is done in two ways, namely the suitability of the application with an android device and the distribution of questionnaires to prospective users, with the following results

*3.4.1. Software Functionality Test.* At this stage, testing the suitability of the software when used on a variety of different Android devices. the results obtained by the software can work well on all android devices with a minimum size of 5 inches. Details of the test are in table 2 below.

**Table 2.** The result of software functionality test

| No | Brand and type of devices | Screen size | Suitability of function |
|---|---|---|---|
| 1 | Samsung A20 | 6,4 inc | Yes |
| 2 | Xiaomi Redmi Note 7 | 6,3 inc | Yes |
| 3 | Samsung J5 | 5,2 inc | Yes |
| 4 | Vivo V15 | 6,3 inc | Yes |
| 5 | Samsung Galaxy Tab A6 | 7 inc | Yes |

*3.4.2. Application Material Test.* This test aims to check the results material / content in the application as a material consideration for users. This Testing is given to the examiners or muhafidz to be able to assess the suitability of the contents to the needs of users. All muhafidz are expert and work at Al-Qur'an and language department. The results of the test can be seen in table 3 below.

**Table 3.** The result of application material test

| No | The elements | Result |
|---|---|---|
| 1 | Clear content presentation | 9 |
| 2 | Material compatibility with the needs of memorizing Al-Qur'an | 9 |
| 3 | Suitability and ease of media flow | 8 |
| 4 | Interactive and communicative presentation | 9 |
| 5 | The media can encourage users to increase the quantity of memorization | 10 |
| 6 | The media does not cause ambiguity and confusion | 10 |
| 7 | Clarity of function of each feature contained in the media | 9 |
| 8 | Compatibility of the contents of the features with the purpose of making media | 10 |
| 9 | The media can increase user understanding the importance of the topic | 9 |
| 10 | The media can help achieve the target of memorization | 10 |
| | Average | 93% |

In addition, testing is also carried out on students, as potential application users and get 97% of a satisfaction rate. The results of the test can be seen in figures 8 below.

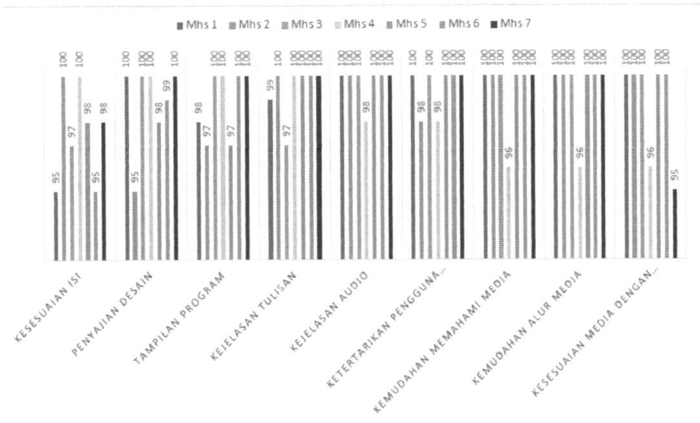


**Figure 8.** The result of user satisfaction level

### *3.5. Maintenance*

the final stage was repairs and maintenance of application based on feedback from the Muhafidz and students from the results of testing application and questionnaire distribution. one of which is the addition of the guide feature, to help users who are confused in running the media.

## 4. Conclusion

Mobile based media for Al-quran memorization elayah is proven to be compatible with android devices with a screen resolution of at least 5 inches, and the contents and features in it can be accepted by students and recognized by Muhafidz in helping to achieve memorization targets. but this media can not automatically increase the number of memorization, unless it is balanced with the intention and sincerity of users in carrying out memorization of Al-Qur'an in an orderly and disciplined manner.

## References

[1] Abdurrahman 2016 Mukjizat al-qur'an dalam berbagai aspeknya *J. Pusaka* **8** 68–85
[2] Anon 2016 Mengungkap Kemukjizatan Ilmiah Dalam Al-Qur'an *J. An-Nur* **4** 167–86
[3] Muhajir 2014 Pesantren Sebagai Instiusi *J. Saintifika Islam.* **Volume 1** 1–18
[4] Muthoifin M, Anshori A and Suryono S 2016 Metode Pembelajaran Tahfiz Al-Quran di Madrasah Aliyah Tahfiz Nurul Iman Karanganar dan Madrasah Aliyah Al-Kahfi Surakarta *Profetika, J. Stud. Islam* **17** 29–35
[5] Penyusun T 2017 *Panduan Universitas Darussalam Gontor* vol 1
[6] Developer A Developer Guides | Android Developers *Android Dev. Guid.*
[7] Herdiansyah M Y and Afrianto I 2013 Pembangunan Aplikasi Bantu Dalam Menghafal Al-Qur ' an Berbasis Mobile *J. Ilm. Komput. dan Inform.* **2**
[8] Gade F 2014 Implementasi Metode Takrar Dalam Pembelajaran Menghafal Al-Qur'an *J. Ilm. Didakt.* **14** 413–25
[9] Bassil Y 2012 A Simulation Model for the Waterfall Software Development Life Cycle **2**

# Improving The Ease of Digital Marketing Learning Using Mobile Learning

**A Kurnianti, M Erfiana*, and T W Wijaya**

Department of Information Technology, Universitas Muhammadiyah Yogyakarta, Jl. Brawijaya, Geblagan, Tamantirto, Kec. Kasihan, Bantul, Daerah Istimewa Yogyakarta 55183

email:erfiana19@gmail.com

**Abstract.** Digital marketing is a new way of advertisement that has been in high demand by the public lately. Slowly, people are beginning to abandon conventional marketing methods and switch to digital marketing. But in general, they are self-taught. People who sell online really need digital marketing concepts to increase their understanding. If knowledge increases, marketing success increases. The main obstacle for people to get digital marketing courses is limited time and cost. Many digital marketing courses are offered but at high prices. In this study, we develop the digital marketing learning model that uses mobile technology and feedback learning method. By using this technology, people are expected to be able to increase their knowledge of digital marketing principles at any time at a very affordable cost. In this module, there are four main themes presented interactively, namely think, create, engage, and optimize. We design the mobile content using revised Bloom's concept that divided into four knowledge domains: factual, conceptual, procedure, and metacognitive. Along the learning process, the students are under monitoring, evaluating, and supervising their teacher. The results of the study show that mobile learning shows very attractive and exciting. The interesting finding in this research that the feedback part of the teaching is the most significant influence on student performance.

## 1. Introduction

One of the most common course materials today is digital marketing. But unfortunately, in Indonesia, the cost of this course is still high. The community requires knowledge about digital marketing because it has become a necessity to carry out business pursuits. In Indonesia, there are so many mobile devices for many backgrounds of users. It's means that, if we build an online learning system with an approach of smartphone devices is considering very suitable. If there does a digital marketing learning system that is practical, easy, inexpensive, this will be very beneficial for the community. Mobile learning offers the convenience of carrying out learning activities anytime and anywhere without the need for classrooms [1][2]. The Android-based mobile learning system is now a tool that is widely used for teaching and learning in Indonesia [3].

This paper discusses the responses of users or participants in digital marketing lectures who use mobile devices or better known as mobile learning. In this study, mobile applications are used to facilitate online learning for people who need to learn digital marketing. The main reason for using an Android application-based system is because this tool has high flexibility and the only tool that is mostly used every day by people today. The understanding of mobile learning in this paper is mobile technology as a learning tool [4]. There are many varieties of smartphone technology for mobile learning applications. In this study, we developed based on the Android operating system. We prefer using Android rather than others, because of this already well known. Android is the most popular smartphone in the world; there are 73% Android users [5][6].

Android applications have been used successfully for various purposes, such as transportation and commerce. But for learning it is still at an early stage, so it still requires a lot of research. Smartphones have a high degree of flexibility, but if related to shopping, there are some disadvantages, including memory limitations, computational limitations, and interface convenience. This study aims to examine whether the use of mobile learning can increase the ease of participants in understanding digital marketing material. This research very important because everyone has different preferences in terms of how to process information [7]. On the other hand, each lecture material has different delivery characteristics so that the way to present it on a smartphone is also different. There are learning materials that do not require teaching media, but there are those that need learning media [8].

The results of this investigation are beneficial for the improvement of learning models, especially mobile learning. People favor to use a mobile device because of its simplicity, but in education, other factors require attention, namely the element of comfort and effectiveness. If we can extract these two elements, people will find it easier to improve their knowledge and eventually business in the digital age is growing.

## 2. Method

This experiment aims to measure the ease and effectiveness of using mobile learning applications. We tried the system with 85 participants in digital marketing lectures. The convenience of display demonstrated the easy of the module's operation, the effectiveness of the teacher's mentoring, and the effectiveness of the content. We collect the research data using questionary and testing. The students' accomplishments show the effectiveness of the entire system. We process data using Smartpls. Data obtained from purposive sampling to determine the characteristics of the ease of learning using applications. This trial uses purposive sampling because, in sampling based on a specific consideration [9][10], we use consideration of the similarity of learning interests and habits of using mobile devices. We chose students who take digital marketing courses. Because the number of samples is limited, we chose to analyze the sample using SmartPLS. SmartPLS is well known as Statistical Software for the Partial Least Square Structural Equation Modeling method [11][12][13][14]. Structural equation modeling with partial least square has three components namely, structural model (inner model), measurement model (outer model), and weighting scheme (weight relation) [15][16]. With modeling like this, it is expected to know the factors that influence the ease and effectiveness of using mobile learning.

Each student is allowed to use this application for one to two months with the assistance of a lecturer. For the first time, students are required to download the material first and then proceed with the registration process. Course material for one or several chapters stored on a smartphone and students can use it to study anytime and anywhere. Tn students are online. Then, the lecturer can check what students already did and how far the students can understand the lecture. The activities were always monitored and evaluated by the lecturer. Lecturers are directed to always look for the students who are low motivated and lack of understanding. Generally, less motivation is close to low activity. We can found low student motivation from the frequency clicks during the use of the application. We can find out less understanding by the low task scores. Each learning module has a quiz so that the lecturer can know the level of effectiveness of learning.

## 3. System Development

Most people today always take the time to use a smartphone wherever and whenever they have time. Easy to find, in general, they use smartphones for entertainment, so it is natural to feel at the screen for hours. What should we do if this applied for learning? Of course, it requires concepts, methodologies, and technologies which a specific designed so that the learning objectives achieved. The concept is to answer the challenges of the learning culture of Indonesian people who are generally comfortable in a face-to-face manner. The methodology is a way to run a mobile device-based learning process. Technology is a characteristic of digital marketing content and mobile systems that are built to support concepts and methodologies.

*3.1. Concept*

This study uses the concept to provide practice facilities for students so that they are interested in using it. Practicality, we can obtain the aim by utilizing the advantages of Android smartphone applications. We can design the most important of learning elements from learning material arrangement on the smartphone screen. In this experiment, we arranged the learning material based on the knowledge domain of the revised bloom theory. The screen presents conceptual knowledge in text or image form as the central display. For factual knowledge, we presented in text or picture form in the sub-menus. Finally, the procedural and metacognitive knowledge presented in an interactive form such as video or animation. We conclude that the use of text, images, audio, video, animation, or interactive screens depends on the cognitive domain. Students will learn subject matter starting from concepts, but they can jump to factual or procedural. The screen layout can follow what the student learning sequence wants.

*3.2. Methodology*

The methodology is an important part. Students willing to study independently depend on how learning methods can influence them. The methodology was chosen to overcome student learning habits. Every student in Indonesia is comfortable with a face-to-face model. The methodology used here to direct the learning process. Students can study on their own but remain under the supervision and guidance of lecturers. There are two main learning factors which using for feedback learning mechanisms, namely motivation, and understanding. These two factors we must control so that each student achieves mastery. During the student learning to use the application, all the activity data automatically stored on the smartphone device. The system will send the data in an LMS (Learning Management System) server through a synchronization process. This system provides a synchronization process to anticipate data loss due to damage or replacement of smartphone devices. All activity and score data are used by lecturers to conduct the evaluation, monitoring, and treatment (mentoring). By using the monitoring module, lecturers can contact students directly through the WhatsApp application.

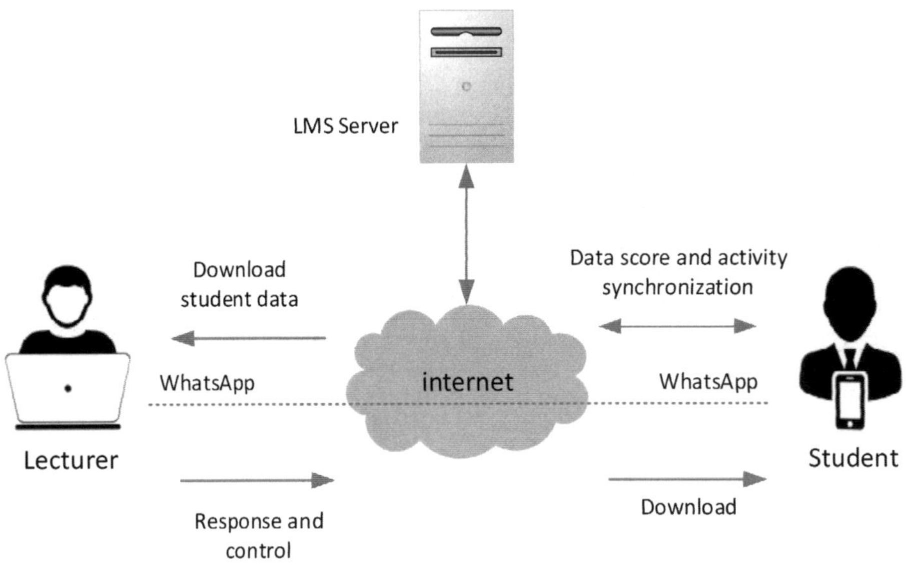

**Figure 1. Monitoring, Evaluation, and Treatment**

### *3.3. Technology*

The technology was chosen and developed to overcome the cost control and internet connection. In most parts of Indonesia, the internet quality is low. In the other situation, most students still consider the price of the internet still expensive. We design technology to solve those problems and also this is easy in development. In this study, selected mobile applications based on HTML5, XML, and JavaScript. This selection considers the time needed to develop an application module. Each module is created using CourseLab authoring tools version 2.7 and PhoneGap. We choose Moodle as a server to collect data and control each module, both the lecturer and student modules. The system was developed using architecture, as shown in **Figure 2**.

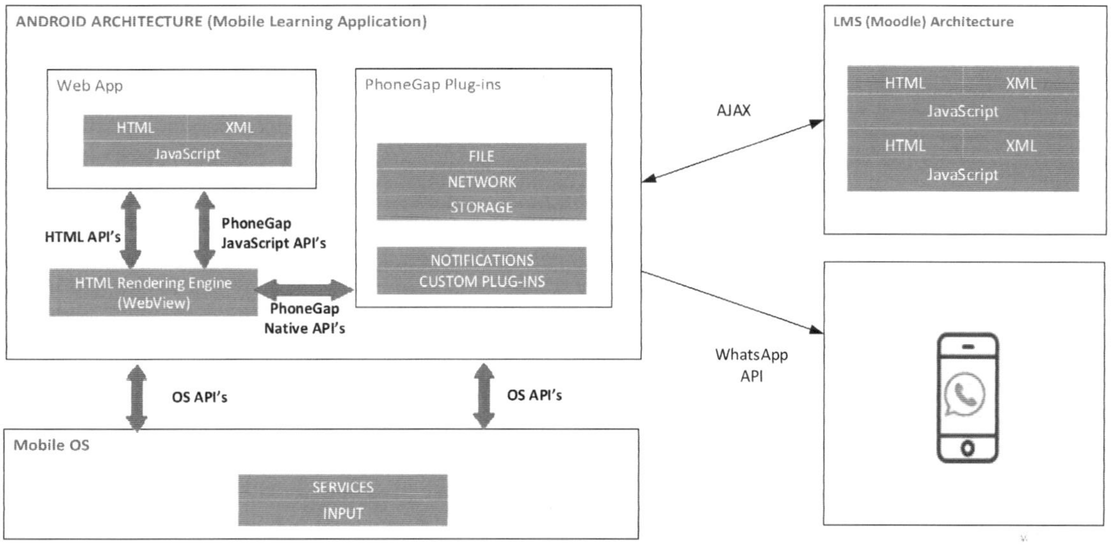


**Figure 2. Mobile learning architecture for digital marketing**

**Figure 2** shows an architecture consisting of three blocks, Android application, LMS Moodle, and WhatsApp API. We choose the Android application module architecture that meets with the Moodle architecture so that they can communicate with each other using AJAX. We also use WhatsApp API so the communication can initiate from the monitoring module's link. This architectural design makes communication easy by combining mobile applications with Moodle without involving a web browser. The advantage of this architectural design without involving a web browser. The ease of design of this architecture to support the low-cost system learning process. Students only need one login, students can save material files into smartphone memory, and using it anytime with or without the internet. The synchronization and notifications make it easier for lecturers to monitor and control the students learning activity. The use of PhoneGap is a useful solution for building mobile applications using modern web programming languages, such as HTML, HTML5, CSS, and JavaScript rather than using other known languages such as C# or Java [17]. This design gives many benefits, especially to the development process, which an essential part of overall system development and operational [18].

## 4. Result and Discussion

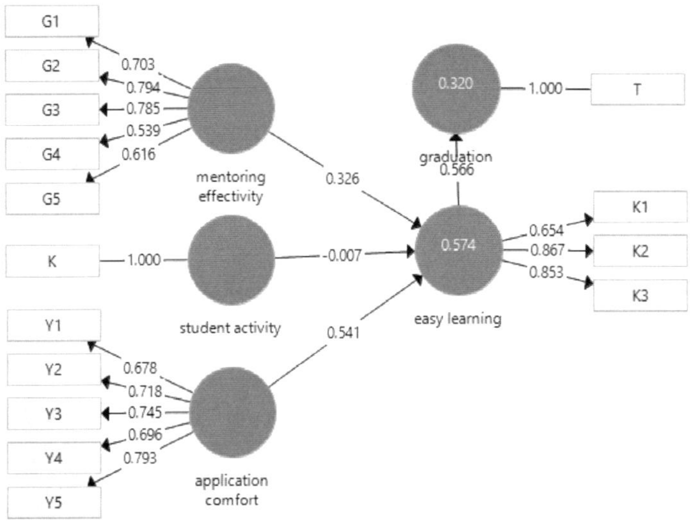


**Figure 3. SmartPLS model calculation**

From the results in **Figure 3** explain the results of calculations using the SmartPLS application. G1, G2, G3, G4, and G5 show the results of lecturer supervision, both in terms of criticism that also motivates lecturers who can pay attention, so it is beneficial in understanding the material. K is the number of clicks on the application module to measure student activity. Y1, Y2, Y3, Y4, and Y5 are the results of testing the convenience of application use; both in terms of appearance, an easily memorized menu, and a sequence of material that is easy to understand. K1, K2, and K3 are testing the ease of use of application modules, such as applications that formally can run anytime and anywhere, applications that can run on various types of smartphones and also the ease of understanding material without a lecturer explanation whereas is an indicator of the passing grade after using the digital marketing application module. Calculation results show that the model meets convergent validity because all loading factors are above 0.50 [19]. The data states that the use of this application can facilitate the learning of digital marketing material. The results that can show that the digital marketing application module can facilitate students in learning the material. The lecturer is only as a companion in charge of supervising. The final value looks quite increased after the use of digital marketing application modules.

As mentioned in the guide, the purpose of this research is to facilitate students in learning digital marketing material. The increasing of the number of smartphone users can provide widely learning manner possibility. It is meant that the concept of mobile learning has no limitations of time, space, and place [20][21].

We get the ease of learning from the results of the data obtained through a questionnaire. K1, K2, and K3 show the simplicity of the module, namely the ease of use of applications. It is mean that we can operate the module anytime and anywhere. Also, applications can run on various types of smartphones. Finally, the student feels comfortable to understand learning materials. We can say it is better than the lecturer's explanation. The results obtained by the average student show that the use of this application can facilitate digital marketing learning. The results also supported by other data, namely G1, G2, G3, G4, and G5 as a form of mentoring for lecturers, Y1, Y2, Y3, Y4, and Y5 are testing the comfort of application use, and also K as a form of test results on the number of clicks on the application module, while T represents the passing value obtained after passing the final test contained in the module. The data obtained states that 51% of the results of this application test can facilitate students in learning the material, and 49% influenced by other variables that are not yet known. We can say that the students not yet familiar with using mobile learning.

From the above paragraph, we summarized that mobile learning has the potential to be developed as a digital marketing learning tool so that many people can use it. From the experiment, we found that learning can run well even though lecturers must proactively monitor and communicate with students. The existence of this application module greatly helps lecturers who carry out supervision activities. Lecturers do not need to collect and process data manually, so data that is synchronized automatically makes it very easy for lecturers to make assessments. The lecturer observes the data as a basis for evaluating and motivating according to student learning problems. According to the results of testing, this application can improve student grades. The application can also run well on any Android-based device, be it a smartphone or tablet. Lecturers as supervisors can also check the extent to which students understand the material. The lecturer assistance system is still very much needed to carry out the learning process using mobile learning, although theoretically the use of car learning is aimed at making it easier for students to be able to learn anytime, anywhere, and from any device. This application still needs further testing.

## 5. Conclussion

The use of mobile learning for digital marketing has the potential to facilitate learning communities. The use of mobile devices is also beneficial for lecturers to accompany students. Learning activities become easy because this system utilizes internet and smartphone facilities. This model very useful because we use Android smartphones that most popular in Indonesia. The digital marketing mobile learning application does not limit students to study; the students can study online or offline. The application proved to be able to run various android devices. With the use of digital marketing application modules that can run both online and offline, it helps students in easy learning anytime, anywhere, also using any Android device. Easy use anywhere and anytime also addresses the problem of students who are not comforted or not easy to catch the lecturer's explanation. So that the use of this application makes students comfortable in its use and can give a more in-depth understanding, students can improve grades with this application.

On the lecturer side, the use of this system makes it easy to communicate and provide assistance without being bound by place and time. Lecturers can see the results of student activities and performance. Lecturers can evaluate and find learning problems. The interesting thing found in this research is the use of mobile learning still requires adaptation, first to eliminate face-to-face learning habits, overcome the habits of smartphone students as an entertainment tool, and get the lecturers to monitor and assist learning.

## 6. Reference

[1] Nurhadryani Y, Sianturi S K and Hermadi I 2013 Pengujian Usability untuk Meningkatkan Antarmuka Aplikasi Mobile Usability Testing to Enhance Mobile Application User Interface *J. Ilmu Komput. Agri-Informatika* **2** 83–93

[2] Aripin I 2018 Konsep Dan Aplikasi Mobile Learning Dalam Pembelajaran Biologi *Bio Educ.* **3** 1–9

[3] Abdullah M R T L and Azelin M N 2010 M-learning: Changing Roles of Instructors and Learners *Int. J. Arts Sci.* **3** 83–95

[4] Osman M, Cronje J C, El-hussein M O M and Cronje J C 2016 International Forum of Educational Technology & Society Defining Mobile Learning in the Higher Education Landscape Published by : International Forum of Educational Technology & Society Defining Mobile Learning in the Higher Education Landscape *J. Educ. Technol. Soc.* **13** 12–21

[5] Rahadi D R 2014 Pengukuran Usability Sistem Menggunakan Use Questionnaire Pada Aplikasi Android Interface pengguna Android didasarkan pada manipulasi langsung

menggunakan masukan sentuh yang serupa dengan tindakan di dunia nyata , seperti menggesek ( swiping ), mengetuk **6** 661–71
[6] Katadata.co.id 2017 73% Perangkat Mobile Global Menggunakan Android 1
[7] Widayanti F D 2013 Pentingnya Mengetahui Gaya Belajar Siswa Dalam Kegiatan Pembelajaran Di Kelas *Erud. J. Educ. Innov.* **2**
[8] Asyhari A and Silvia H 2016 Pengembangan Media Pembelajaran Berupa Buletin dalam Bentuk Buku Saku untuk Pembelajran IPA Terpadu *J. Ilm. Pendidik. Fis. Al-Biruni* **5** 1
[9] Miftahul Ulum, IMade Tirta D A 2014 ANALISIS STRUCTURAL EQUATIONMODELING ( SEM ) UNTUK SAMPEL KECIL DENGAN PENDEKATAN PARTIAL LEAST SQUARE ( PLS ) ( STRUCTURALEQUATION MODELING ( SEM ) ANALYSIS FOR SMALL SAMPLE WITH PARTIAL LEAST SQUARE ) 1 Pendahuluan *Pros. Semin. Nas. Mat. Univ. Jember* 1–15
[10] Sugiyono 2010 Metode Penelitian Bisnis. Pendekatan Kuantitatif, kualitatif dan R & D *Bandung Alf.* **15** 90
[11] Kwong-Kay Wong K 2013 Partial Least Squares Structural Equation Modeling (PLS-SEM) Techniques Using SmartPLS *Mark. Bull.* **24** 1–32
[12] Leguina A 2015 *A primer on partial least squares structural equation modeling (PLS-SEM)* vol 38
[13] Hair, J. F., Hult, G. T. M., Ringle, C. M., & Sarstedt M 2014 *Partial least squares structural equation modeling (PLS-SEM)*
[14] Wong K K 2019 *Mastering Partial Least Squares Structural Equation Modeling (PLS-SEM) with SmartPLS in 38 Hours*
[15] Monecke A and Leisch F 2012 The Journal of Statistical Software *J. Stat. Softw.* **1** 2012
[16] Nurimawati U 2014 *Membuat Skripsi, Tesis, dan Desertasi dengan Partial Least Square (PLS-SEM)*
[17] Wang Y and Ma J 2014 Mobile social networking and computing: A multidisciplinary integrated perspective *Mob. Soc. Netw. Comput. A Multidiscip. Integr. Perspect.* 1–386
[18] Myer T 2012 *Begining Phonegap*
[19] Ghozali and I 2006 *Structural Equation Modeling Metode Alternatif dengan PLS*
[20] Iskandar D 2010 Pengembangan Aplikasi Berbasis Teknologi Mobile untuk Pembelajaran
[21] Hermawan B and Leonardo O P 2017 Penggunaan Keefektivitasan Media Mobile Learning Dalam Meningkatkan Pelafalan Hanyu Pinyin Bahasa Mandarin *Paramasastra* **4**

## Acknowledgment

This research is part of the research Growing Self Directed Learning (SDL) Using Gamma Feedback Learning Model (GFLM). We want to say many thanks to the Ministry of Research, Technology and Higher Education of the Republic of Indonesia based on the Decree of the Director-General of Research and Development Strengthening number 7/E/KPT/2019. We are also grateful to the University of Muhammadiyah Yogyakarta for fully supporting this research by providing additional funding and research facilities.

# Work posture analysis of gamelan craft center workers using quick methods of ergonomic risk assessment

**R A A Rahma*and I Faiz**

Occupational Safety and Health Program, Universitas Darussalam Gontor, Indonesia

*Corresponding author: ratihandhika@unida.gontor.ac.id

**Abstract**. Gamelan craft center in Ponorogo Regency has many hazard factors and risks, one of which is ergonomic risk factors. Workers do work with bent postures, awkward postures, and frequent repetitive movements for a long time. This research aims to analyze the work posture of gamelan craft center workers using quick methods of ergonomic risk assessment. This research used the observational analytical study. Analysis of the work posture by using three methods are Rapid Entire Body Assessment (REBA), Rapid Upper Limb Assessment (RULA) and Quick Exposure Checklist (QEC). Eight jobs were selected are divided into two task elements: the dominant task and risky task. Analysis of the research data using the Friedman non-parametric test. The result of work posture analysis shows that the final score in the three methods gets a high-risk level that is 56.25% (REBA), 43.75% (RULA), and 62.50% (QEC). Friedman test results showed the value of Chi-Square = 13.630, and the significant score is 0.003 ($p < 0.05$). So it can be concluded that the three ergonomic risk assessment methods have different effectiveness. Quick Exposure Check (QEC) is the best method among all three methods for the ergonomic risk assessment of gamelan center workers.

## 1. Introduction

Small and micro enterprises (SMEs) have difficulties managing Occupational Safety and Health (OSH) and hence, may experience poor OSH conditions and often lack systematic OSH management [1]. According to Sutjana, the challenge of applying OHS in the workplace is the OHS aspect does not provide benefits to industrial owners. OHS management priorities are still low and require a promotive program about OHS in the workplace [2]. Lack of the application of OSH to SMEs can result in increased workplace accidents and occupational diseases [3].

One of the SMEs in Ponorogo is a gamelan craft center that used zinc, brass, and bronze plate resources in its production process [4]. Most workers work in a seated position such as the process of cutting material (iron plates), forming the material into gamelan and the process of perfecting the gamelan (finishing). Workers often do their work with bent work postures, awkward postures, and doing repetitive movements for a long time. Un-ergonomic work posture is the ergonomic risk factors that have the potential to cause interference with skeletal muscles characterized by fatigue and pain in certain body parts [5].

The complaints caused by ergonomic risk factors can be prevented by conducting an ergonomic risk assessment. However, in the process, ergonomic risk cannot be assessed directly. Because it can disrupt the activities of workers so that most small and medium business owners, including gamelan craft, have never evaluated the work posture of their workers. Therefore, one way to resolve this is by using quick ergonomic risk assessment methods. Popular methods used are Rapid Entire Body Assessments

(REBA), Rapid Upper Limb Assessment (RULA) and Quick Exposure Check (QEC). These three methods have been widely used in identifying the level of ergonomic risk in various occupations with the primary factor being assessed is work posture [2,5,6]. This research aims to analyze the work posture of gamelan craft center workers using quick methods of ergonomic risk assessment.

## 2. Methods

This research used the observational analytical study. Analysis of the work posture by using three methods are REBA, RULA and QEC. The jobs selected to assess work posture are material preparation, cutting, moving, forming the iron plate, welding, combining gamelan, finishing, and rolling the gamelan. Dominant task and risky task are the basic task among those jobs. The dominant task is commonly done on the type of work. The risky task means an action tends to lead injury on the job and requires a lot of energy during the work [7]. The data of the research through the observation, documentation, interviews, and measurement of the components of:

### *2.1. Rapid Entire Body Assessment (REBA)*

The work posture analysis using the REBA Assessment Excel Worksheet. The body part score that assessed was Group A (trunk, neck, legs) and Group B (upper arms, lower arms, wrists). An activity score represented type of muscle activity. Moreover the score is added to give the final REBA score. The observer converts the final REBA scores to the action levels (Table 1). These are bands of counts corresponding to increase urgency for the need to make changes [8].

**Table 1.** REBA Action Levels

| REBA Score | Risk Level | Action Level | Action (including further assessment) |
|---|---|---|---|
| 1 | Negligible | 0 | None necessary |
| 2-3 | Low | 1 | Change may be needed |
| 4-7 | Medium | 2 | Further investigation, change soon |
| 8-10 | High | 3 | Investigate and implement change |
| 11-15 | Very high | 4 | Implement change |

### *2.2. Rapid Upper Limb Assessment (RULA)*

The measurement of work posture using the RULA Assessment Excel Worksheet. The body part score that assessed was Group A (arm and wrist) and Group B (neck, trunk, and legs). The postures are measured utilizing scoring sheet, body-part diagrams, and tables. Furthermore, the scores were transformed to be one of four action level [6].

**Table 2.** RULA Action Levels

| RULA Score | Action Level | Action (including further assessment) |
|---|---|---|
| 1-2 | 0 | Acceptable if it is not maintained or repeated for long periods |
| 3-4 | 1 | Further investigation is needed, and changes may be required |
| 5-6 | 2 | Investigation and changes are required soon |
| 7 | 3 | Investigation and changes are required immediately |

### *2.3. Quick Exposure Check (QEC)*

The measurement of work posture using the QEC Calculation Tool Template. Assessment of the body part at the QEC methods includes back posture, back movement, shoulder/arm posture, shoulder/arm movement, wrist/hand posture, wrist/hand movement, neck, and worker's assessment. QEC assesses whether an ergonomic intervention can effectively reduce these exposure levels [6]. Assessment of QEC has important role where exposure level calculated. The exposure level E is provided in percentage rate that determined the actual exposure score X and max possible total Xmax depicted in table 3.

$$E\ (\%) = \frac{X}{Xmax} + 100\%$$

**Table 3.** Preliminary Action Levels for the QEC

| QEC Score (E) (percentage total) | Action | Equivalent RULA Score |
|---|---|---|
| ≤ 40 % | Acceptable | 1-2 |
| 41-50 % | Investigate further | 3-4 |
| 51-70 % | Investigate further and change soon | 5-6 |
| >70% | Investigate and change immediately | 7+ |

The collected data of the work postures with three ergonomic risk assessment methods, including the risk levels for each posture, were entered into SPSS version 20 for the next analysis [2]. At first, descriptive analysis was performed. The data were analyzed with the Shapiro-Wilk test (sample < 50) [9] for checking the normality distribution of data. Then Friedman non-parametric test for investigating the significant differences between the values.

## 3. Result and Discussion

### *3.1. Work Posture Assessment*

The results of work posture assessment using the REBA Assessment Excel Worksheet, RULA Assessment Excel Worksheet, and QEC Calculation Tool Template. Point 1 means dominant tasks, and points 2 means a risky task is explained in table 4.

**Table 4.** Final Score REBA, RULA, QEC

| No | Task | $REBA_1$ | $REBA_2$ | $RULA_1$ | $RULA_2$ | $QEC_1$ | $QEC_2$ |
|---|---|---|---|---|---|---|---|
| 1 | Material preparation | 3 | 3 | 2 | 2 | 2 | 2 |
| 2 | Cutting | 4 | 4 | 3 | 3 | 2 | 3 |
| 3 | Moving the iron plate | 3 | 3 | 2 | 3 | 2 | 2 |
| 4 | Forming the iron plate | 2 | 3 | 1 | 1 | 1 | 2 |
| 5 | Welding | 3 | 3 | 2 | 2 | 2 | 2 |
| 6 | Combining the gamelan | 2 | 2 | 1 | 1 | 1 | 1 |
| 7 | Finishing | 3 | 3 | 2 | 2 | 2 | 2 |
| 8 | Moving the gamelan | 4 | 4 | 3 | 3 | 3 | 3 |

Based on table 4, the highest final scores are the task of cutting the iron plate and moving the gamelan. The work postures that are often done at the tasks are awkward and extreme posture, repeated positions for a long time and heavy load of the gamelan. The final score distribution in each method, as shown in table 5, 6, and 7.

*3.1.1. REBA*

**Table 5.** Final Score Distribution of REBA Assessment Excel Worksheet

| Final REBA Score | N | Percentage | Action Level | Risk Level | Action |
|---|---|---|---|---|---|
| 4-7 | 3 | 18.75 % | 2 | Medium | Further investigation, change soon |
| 8-10 | 9 | 56.25 % | 3 | High | Investigate and implement change |
| 11-15 | 4 | 25 % | 4 | Very High | Implement change |
| Total | 16 | 100 % | | | |

Table 5 shows that nine workers (56.25%) received a final REBA score of 8-10 with a high-risk level, so an investigation and implement change was needed.

*3.1.2. RULA*

**Table 6.** Final Score Distribution of RULA Assessment Excel Worksheet

| Final RULA Score | N | Percentage | Action Level | Risk Level | Action |
|---|---|---|---|---|---|
| 3-4 | 4 | 25 % | 1 | Medium | Further investigation is needed, and changes maybe required |
| 5-6 | 7 | 43.75 % | 2 | High | Investigation and changes are required soon |
| 7 | 5 | 31.25 % | 3 | Very High | Investigation and changes are required Immediately |
| Total | 16 | 100 % | | | |

Table 6 shows that twelve people (75%) received a high and very high-risk level that needed investigation and changes immediately.

*3.1.3. QEC*

**Table 7.** Distribution of QEC Calculation Tool Template

| QEC Score (E) (percentage total) | N | Percentage | Action Level | Risk Level | Action |
|---|---|---|---|---|---|
| 41-50 % | 3 | 18.75 % | 1 | Medium | Investigate further |
| 51-70% | 10 | 62.5 % | 2 | High | Investigate further and change soon |
| >70% | 3 | 18.75 % | 3 | Very High | Investigate and change immediately |
| Total | 16 | 100 % | | | |

Table 7 shows that ten workers (62.5%) received a QEC Score of 51-71% with a high-risk level that needed to investigate further and change action soon.

*3.2. Analysis*

The data were obtained from the three methods REBA, RULA, and QEC, as shown in Figures 1. In these figures, the colors indicate the action levels: the blue, red, and green colors show action levels 1, 2, and 3, respectively, in the three methods.

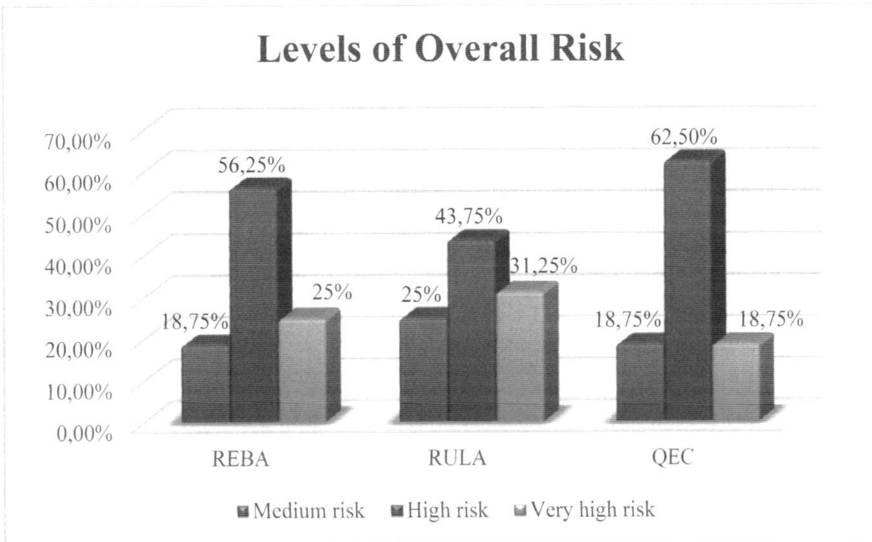


**Figure 1.** Levels of Overall Risk

Based on figure 1, ergonomic risk assessment using REBA, RULA, and QEC shows the final score in the three methods is at a high-risk level of 56.25% (REBA), 43.75% (RULA) and 62.50% (QEC). The action that needs to be done is to investigate further and change soon.

The results of the Shapiro-Wilk test showed that the distribution of data REBA, RULA, and QEC final score was not normal with Sig score is 0,001 ($p < 0,05$). Therefore, to find out the differences in the three methods used by the Friedman non-parametric test.

**Tabel 8.** Result of Friedman Non-Parametric Test

| Methods | N | Std. Deviation | Chi-Square | Asymp. Sig |
|---|---|---|---|---|
| REBA | 20 | 2.28 | | |
| RULA | 20 | 2.83 | 13.630 | 0.003 |
| QEC | 20 | 2.88 | | |

The result of the Friedman test showed the value of Chi-Square = 13.630, and the significant score is 0.003 ($p < 0.05$). So it can be concluded that the three ergonomic risk assessment methods have different effectiveness. The highest assessment of work posture effectiveness is by using the QEC method (std. deviation = 2.88).

Based on sensitivity and usability, QEC has a high level position. This makes QEC mostly acceptable inter and intra-observer reliability [6]. It leads that QEC has widely used of tasks based on field studies [6]. The statement supports the results of the work posture assessment in which the QEC method is most useful for assessing the work posture of gamelan craft center workers in Ponorogo Regency..

The tasks that get a high-risk level is cutting the iron plate and moving the gamelan. The work postures that are often done at the tasks are awkward and extreme postures, repeated position for a long time and heavy load of the gamelan. Besides, most of the work is done in a sitting position. Possible actions to reduce the high-risk level are using an auxiliary machine to lifting the gamelan and suggesting frequently breaks allowed to reduce risk level of repeated tasks.

A study conducted by Ferdiansyah claimed suggested that Strain Index should be used to measure work. This fact had been shown by validity, sensitivity, and reliability testing on ergonomic evaluation methods. This indicated the primarily task using dominant hand movements [10]. Moreover, the REBA is also more sensitive compare to other methods which measure risk elements [10]. Other research that was also evaluating the work posture risk level has shown that there is no significant difference between the REBA, OWAS, and QEC methods. Also, there is a positive correlation between the three methods. The research is about comparing the sensitivity of the REBA, OWAS, and QEC methods in analyzing postures for workers in WL Aluminum [11]. Based on these research studies shows that work posture can be assessed adequately according to task characteristics using quick ergonomic risk assessment methods such as REBA, RULA, QEC, etc.

## 4. Conclusions

The result of work posture analysis shows that the final score in the three methods (REBA, RULA, QEC) gets a high-risk level. The action that needs to be done is to investigate further and change soon. Quick Exposure Check (QEC) is the best method for the ergonomic risk assessment of gamelan center workers among all three methods (std. deviation = 2.88).

## References

[1] Hasle P, Refslund B, Antonsson A-B, Ramioul M and Walters D 2017 Safety and Health in Micro and Small Enterprises in The EU : from Policy to Practice *EU-OSHA* 0–282 p

[2] Yazdanirad S, Khoshakhlagh AH, Habibi E, Zare A, Zeinodini M and Dehghani F. Comparing 2018 the Effectiveness of Three Ergonomic Risk Assessment Methods—RULA, LUBA, and NERPA—to Predict the Upper Extremity Musculoskeletal Disorders *Indian J Occup Environ Med* 22(3):8–13.

[3] Rahma RAA, Afif D and Bintoro E 2018 Analisys of Dust Levels and Masks Usage Behavior On Respiratory Disorder Complaints Of Rice Milling Workers *Elseveir* 01:97–101

[4] Mersida 2016 Studi Kelayakan Tentang Pengembangan Industri Kecil Menengah (IKM) Kerajinan Gamelan Tresno Laras Di Kelurahan Paju, Kabupaten Ponorogo (Ponorogo: University of Muhammadiyah Ponorogo)

[5] Yassierli, Oktoviona D and Ulin Na I 2016 Hubungan Antara Indikator Pengukuran Kelelahan Kerja Dan Metode Cepat Penilaian Risiko Ergonomi *J Ergon dan K3* 1(1):1–5.

[6] Stanton N, Hedge A, Brookhuis K, Salas E and Hendrick H 2018 Handbook of Human Factors and Ergonomics Methods (London: CRC Press) 91–120 p.

[7] Bao S, Spielholz P, Howard N and Silverstein B 2009 Application of the Strain Index in multiple task jobs *Appl Ergon* 40(1):56–68.

[8] Kong YK, Lee S yong, Lee KS and Kim DM 2018 Comparisons of ergonomic evaluation tools (ALLA, RULA, REBA and OWAS) for farm work *Int J Occup Saf Ergon* 24(2):218–23.

[9] Dahlan MS 2011 Statistik Untuk Kedokteran Dan Kesehatan Seri 3 (Indonesia: Penerbit Salemba)

[10] Syahril F and Sonjaya E 2015 Validity, Sensitivity, and Reliability Testing by Ergonomic Evaluation Methods for Geothermal Task *World Geotherm Congr* (April):4.

[11] Ramadhan and Baroroh DK 2017 Perbandingan Sensitivitas Metode REBA, OWAS dan QEC dalam Evaluasi Tingkat Risiko Postur Kerja. (Studi Kasus di WL Aluminium Giwangan) *Semin Nas Tek Ind Univ Gadjah Mada* (November):ER93–102.

# The Use of ISO and COBIT for IT Governance Audit

**Y Aprilinda[1*], A K Puspa[2], F N Affandy[3]**

[1]Informatics department, Universitas Bandar Lampung, Bandar Lampung, Indonesia
[2]Information system departement, Universitas Bandar Lampung, Bandar Lampung, Indonesia
[3] Technical Information department, STMIK Tunas Bangsa, Bandar Lampung, Indonesia

*ayukartikapuspa@ubl.ac.id

**Abstract**. along with the growing number of practices in the use of technology for business processes, there is a need for IT governance for the private and government sectors. Company leaders implement IT governance in order to ensure that the investment spent by the organization for IT can provide benefits or optimal value for the organization and reduce the lowest risk possible. In practice IT governance requires standards. There are several standards that are often used the most popular are COBIT and ISO, both of which are known as IT governance best practices, but can any company apply the two standards? In this paper we will discuss the use of COBIT and ISO in corporate governance, the strengths and weaknesses of each standard and the merging of the two standards for more comprehensive governance practices.

## 1. Introduction

The most significant change that occurred in the technological era was the change in the role of technology that occurred in organizations, if previously information technology was widely used for automation of manual processes, along with the times and complexity of processes in organizations making IT applications more widely used. The IT function extends not only at the operational level but to the strategic level, from analyzing data, forecasting trends to decision making by leaders.

In some information technology industries can be competitive resources to differentiate and provide competitive advantage, in some other industries information technology can also be a solution to minimize risk. This causes the use of IT to spread in all lines of business, with a lot of investment, IT governance is certainly necessary in accordance with the organization's business objectives without leaving applicable international standards.

Fox and Zonneveld [1] concluded that in good governance the role of IT Governance is very important. The IT Governance process begins by setting targets for IT, providing initial guidance. After that, an iteration is continually formed, performance is measured and compared to the initial goal, resulting in a directional return from the required activity and a corresponding change in objectives. When goals are the primary responsibility and measure of management performance, it must clearly be developed with good planning so that the targets are affordable and the size describes the goals precisely.

And according to Fox and Zonneyeld [1] , the final reason for IT Governance is important because of the gap between expectations and reality. The leaders always expects management to providing IT solutions with good quality, on time, and efficiently.Increase theutilization of IT provides business value returns and increase the utilization of IT to improve efficiency and productivity when managing risk.The ineffectiveness of IT Governance allows the causes of negative experiences in the use of IT, including:

1. Business loss, damage to reputation or weak competitive position.

2. The deadline is not reached, the cost is higher than the desired expectations
3. Efficiency and processes have a negative impact on the low quality of IT use.
4. Failure of IT initiatives can bring about promised innovations and benefits.

In practice , the most widely used international standards for IT governance are COBIT and ISO, but not a few also combine these two standards, this is due to the limitations possessed by both. In the next section of this article the literature study will be presented from several articles that use both standards to assess the maturity of IT governance, the limitations of the two standards, the strengths they have to conclusions on the needs of using each standard.

## 2. Literature Review

The following describes some of the writings that contain the use of COBIT or ISO and articles that contain both standards

### *2.1. Service Performance Measurement in Semarang City 1 Samsat Using COBIT 5 and Quality Perception Method by Dina [2]*

This research is about measuring the performance of internal parties using COBIT 5. The research is entitled "Performance Measurement at Semarang City Samsat 1 using COBIT 5 and the Quality Perception Method". The study was conducted to determine the capability level of service satisfaction in Semarang City Samsat 1. This study aims to help improve service productivity and align the vision and mission of the City of Semarang 1 by optimizing resources. The methodology used is the COBIT 5 framework for measuring capability levels, which focuses on the DSS02, DSS03, and EDM04 processes. measurement results of the capability level achieved indicate that the capability level of the DSS02 and DSS02 processes = 4.00 (predictable); the EDM04 process = 3.00 (established). And the level of customer satisfaction = 3.91 (satisfied) with the satisfaction indicator is disconfirmation.

### *2.2. Information Technology Governance Audit Using the Cobit 5 Framework (Case Study: Central Lampung Sea Aquaculture Fisheries) by Suryono [3]*

In the journal Management of Information Technology Audit Using the Cobit 5 Framework (Case Study: Lampung Maritime Aquaculture Center) it can be seen that the Lampung Maritime Cultivation Center has used the e-SKP (electronic Employee Performance Target) system. At present information security governance activities have not been carried out to the full. To anticipate the occurrence of obstacles such as human resources who do not understand the e-SKP application so that the potential for errors in the application, then the information technology management method is implemented using the COBIT framework. The processes used in this study are EDM03, APO13, APO12, BAI06, DSS01, DSS02, DSS03, DSS05, MEA01, MEA02. Data analysis uses Maturity Level and Gap Analysis to determine the level of maturity.

### *2.3. Security Audit on Loan Debit Network Corporation System Using Cobit 5 and ISO 27001: 2013 by Fathoni [4]*

Stakeholders ina a company have right knowing about optimizing information security management. It can affect a company's performances and reputation. Information is the biggest business driver in an organization or company. This research aims to measure the capability of the company which is implemented information security governance that impacts on enterprise risk. The Loan Debit Network Corporation System is the main system that supports the company's business process for corporate lending transactions management. The capability level measurement is based on COBIT 5.0 for Information Security and ISO 27001: 2013 guidelines as a value against the Information Security Governance Component Rating. It starts with aligning organizational goals from COBIT 5.0

perspectives to obtain five COBIT 5.0 IT processes. The Current capability is at level 2.5. Improvement recommendation from level 2.8 to level 3 refers to best practice recommended by COBIT 5.0 for information Security.

## 3. COBIT and ISO

### *3.1.* COBIT

COBIT (Control Objectives for Information and Related Technology) is an IT governance framework and a set of devices that support and enable managers to bridge the gap existing between control requirements, technical issues and business risk [5 ] . COBIT provides clear guidelines for controlling IT management in companies with a measure of maturity at each level, this helps organizations to increase the value to be achieved by the use of IT .

COBIT first appeared in 1996, COBIT version 1 which emphasized on audit, COBIT version 2 in 1998, which emphasized the control phase, COBIT version 3 in 2000 which was management-oriented, COBIT version 4 which was more directed towards IT Governance, and The last released was COBIT version 5 in 2012 which led to governance and management for IT company assets. [5]

Although COBIT is known as best practices in IT governance, in fact COBIT has weaknesses, namely:

1. General
   COBIT has 5 domains and many control objectives in it, but COBIT is not intended for certain types of companies. This is a good thing because COBIT can be used in all types of companies, on the other hand this is also a disadvantage because there are several companies with special characteristics such as SOEs or financial service companies so that additional standards are needed that are more detailed.
2. More often used for audits
   COBIT is very popular to be used for audits, where the audit is usually placed at the end of an activity as a reflection and evaluation of activities that are already running. The complexity of COBIT also causes audits not to be done often, usually held every 6 months or once a year. So it can be concluded that COBIT is rarely used as a daily operational standard

### *3.2.* ISO

ISO published by the International Organization for Standardizations (ISO) and the International Electrotechnical Commission (IEC), Founded on February 23, 1947, ISO sets world industrial and commercial standards. ISO, which is an international non-profit institution, was originally formed to create and introduce international standardization for anything. Standards that we are familiar with include standard types of photographic films, phone card sizes, Bank ATM cards, paper sizes and thickness and others. In setting a standard they invited representatives from 130 countries to sit on the Technical Committee (TC), Sub-Committee (SC) and Working Group (WG).

Although ISO is a non-governmental organization, its ability to set standards that often become law through approval or national standards makes it more influential than most other non-governmental organizations, and in practice ISO becomes a consortium with strong relationships with government parties. ISO participants include one national standard body from each country and large companies.

ISO itself is a code of practice for providing a framework as a standard. This means providing a high level, general description of the area that should be considered when starting to implement, implement or maintain an activity in management. ISO now does not cover all areas but is still undergoing careful revisions

The use of ISO related with public trust, a company that already uses the ISO certification is often included in the promotion and profiles. ISO certified companies indicate policies and procedures that are carried out. An organization that applies ISO will have a tool to measure, manage and

control assets that are important for the operation of their systems. This in turn can lead to customer trust, efficiency and effectiveness.
Even though it is expensive, ISO is a popular standard because the application of ISO in a company is useful for:

1. Improve company's image
2. Improve the company's environmental performance
3. Increase the efficiency of activities
4. Improve organizational management by implementing planning, implementation, measurement and corrective actions (plan, do, check, act)
5. Improve the arrangement of the provisions of the legislation in terms of environmental management
6. Reducing business risk
7. Increase competitiveness
8. Improve internal communication and good relations with various interested parties
9. Get the trust of consumers / work partners / financiers

## 4. Result

As previously explained, COBIT is a framework within the scope of IT that is the main reference in IT Governance. COBIT is a tool to support and connect the gap between control equipment, technical issues, and business risk. Simply put, COBIT can explain how we solving the problem of securing company assets, technical issues, and risks that can still be measured in a comprehensive manner Basically, COBIT is a framework (can be interpreted as a guide) is generic (general, with a broad scope) that provides direction on "how should companies / organizations manage IT" (from infrastructure, the people in it, especially related to responsibilities and roles in IT, to the process of acquisition, development, maintenance and documentation to the security of Information Technology in a corporate environment). However, it should be noted. COBIT just shows how to achieve good IT governance criteria. COBIT does not provide details of the implementation of good IT governance. Details of the implementation of good IT governance are left to the user of COBIT. Compared to COBIT , ISO is a more specific, specific, narrow, and detailed framework . ISO provides direction along with guidance and examples in its implementation. Therefore, implementing ISO tends to be easier than COBIT 5, which has a slightly "abstract" language. For example, ISO 17799 talks about IT Security . ISO 17799 is an international standard that provides instructions and controls for regulating information security (Information Security Management Standard / ISMS). Meanwhile, COBIT 5 does not discuss things as detailed as at ISO 17799. However, this does not mean that the two frameworks (COBIT 5 and ISO) are incompatible. In certain cases, COBIT 5 and ISO have the same function. Nevertheless, COBIT 5 and ISO do not clash with each other, instead complementing each other. This can be seen in the Figure below.

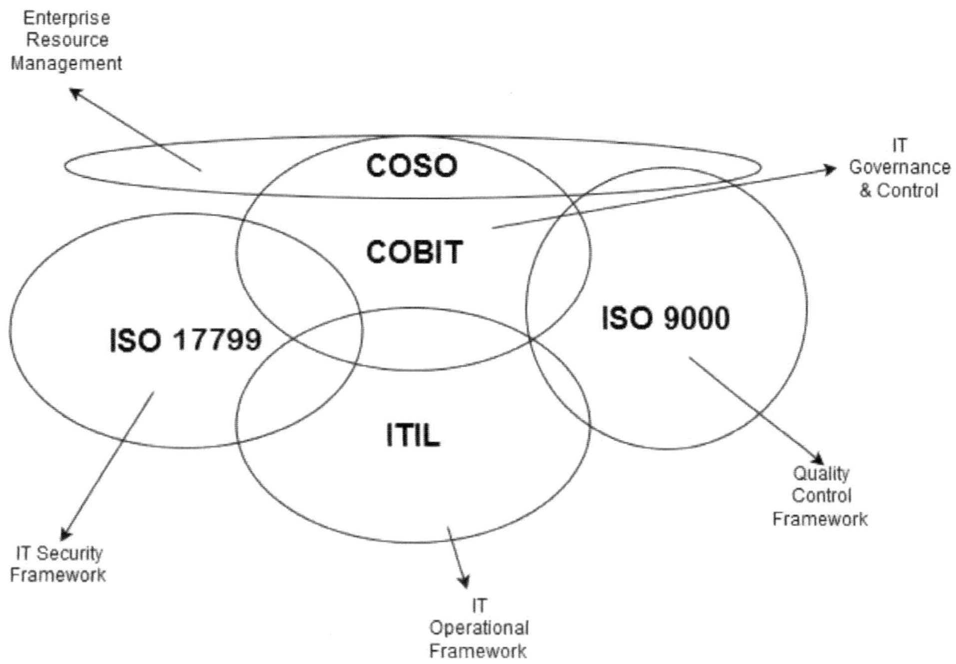


**Figure 1.** Compatibility in IT Governance Framework

COBIT discusses business and IT functional areas in a company and considers IT related to the interests of internal & external stakeholders. The 5 COBIT principles are based on five key principles for corporate governance and IT management as follows :

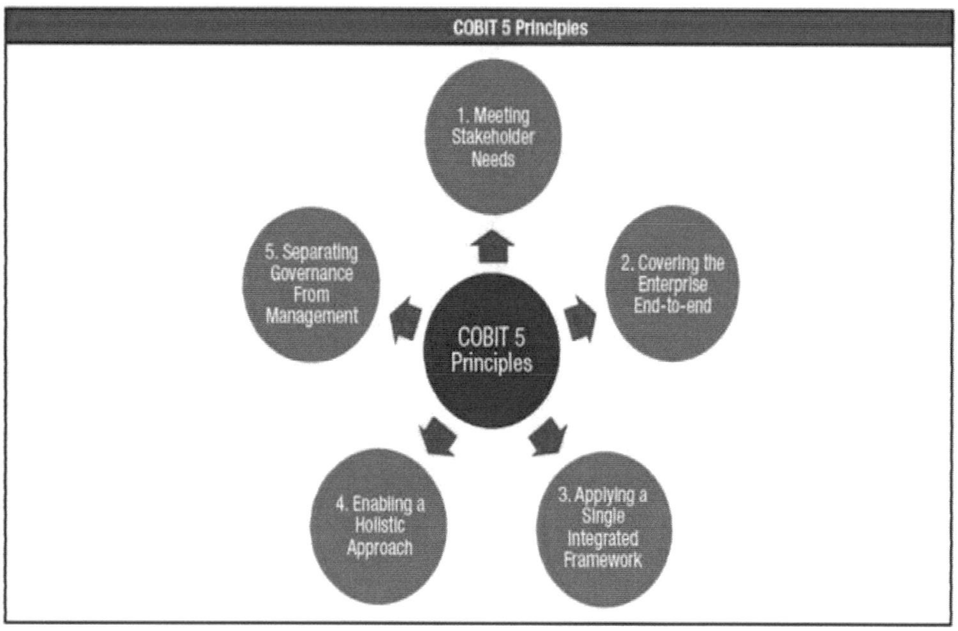


**Figure 2.** COBIT 5 Principles

In the third principle, Applying a single, COBIT 5 integrated network can be adapted to other standards and frameworks, and allows companies to use standards and other frameworks as the scope of management frameworks for enterprise IT. COBIT 5 for Information Security brings knowledge from previous ISACA versions such as COBIT, BMIS, Risk IT, Val IT with guidance from the ISO / IEC 27000 standard which is the ISF standard for information security and the US National Institute of Standards and Technology (NIST) SP800-53A .

## 5. Conclusion

COBIT provides parameters for assessing how high and how good IT management is at an organization by presenting enterprise stakeholder value, and good governance and management of information assets and technology is needed, including information security arrangements . One of which is included in the COBIT additional standards is ISO , COBIT and ISO in practice can be used together and complement each other because of the characteristics of both. COBIT is a general framework with broad scope so it is more often used at the strategic level, while ISO is a more specific, specific, narrow and detailed framework so that it can be used at the operational level.

The second combination can produce a sustainable framework where ISO which regulates more detailed matters is used in daily operations and guidelines, and COBIT as a global framework that regulates decision making and other things on the strategic side.

## References

[1] Fox 2003 Rolling Meadows IL: IT Governance Institute.
[2] Dina A 2016 *J.of Information System.*
[3] Suryono 2018 Jurnal TEKNOINFO
[4] Fathoni 2019 J. Phys.: Conf. Ser. 1196 012033
[5] ISACA 2019 COBIT

# Development of Qur'an Memorization Learning Model Based on Mobile Learning

**D Purbohadi, B RN Rahmawati*, and H Setiyawan**

Department of Information Technology, Universitas Muhammadiyah Yogyakarta, Jl. Brawijaya, Geblagan, Tamantirto, Kec. Kasihan, Bantul, Daerah Istimewa Yogyakarta 55183

*aik.ridwana@gmail.com

**Abstract.** Memorizing the Qur'an is the Sunnah of Prophet Muhammad, may peace be upon him. The most crucial part of remembering the Quran is the process of learning. In general, this activity is conducted privately memorizing, one teacher for one student. One obstacle of memorizing the Quran is the limited number of teachers and many students. This private model undoubtedly ineffective. The teacher is difficult to handle the activities. Also, the teacher cannot control the student to memorize more verses because teachers must hear one by one to the student's reading voice. Information technology has the potential benefit to innovate the learning models of memorizing the Quran. In this study, we develop a learning model using mobile technology. We design mobile learning that students can memorize independently; however teachers can still listen to reviews their reading sample, evaluate, and give guidance. The student module comes with a paragraph of verse view, including the translations. The student module also features a reading reference from international Qori. Besides that, the student module comes with a voice recording facility so students can compare with the reference reader. Teachers can assess every student until students good in reading and more verses to memorize. The primary findings of this study indicate that the teacher can handle more with better student performance. Students are highly motivated to memorize and self-evaluate their reading because they always feel under supervising by the teacher.

## 1. Introduction

Islamic religious education is mandatory for all levels of formal education of Indonesian Islam. The Qur'an is one of the ingredients that compulsory theme in Islamic studies because the Qur'an is the holy book of Muslims [1] [2]; currently, one of the subject matter of Islam being developing is a lesson "memorize." This material is an essential part of the learning activities of the Qur'an [3], Memorizing teaching is not memorization process only, but also the aspects of the method or the art of memorizing the verse by verse [4],

There are several methods of memorizing the Quran, but the most widely used method is "murojaah" or more commonly known as a method of "memorize demonstration." Many teachers use this method because it is proven to help the memorization process. Murojaah method directs students for memorizing gradually by the division of chapters and increase repeatability. Murojaah is a traditional method, and this is a private teaching model. Students must be facing the teacher. Firstly, every student must open the Holy Qur'an than the teacher will divide the chapters and give a number target of verses that student must remember. We call it a student assignment. The student continues to memorize by repeating verses to memorize. When a student feels successful in remembering and aiming the target, students will face the teacher to demonstrate the verses reading and remembering. Meet often students feel lazy and bored using this method because it must queue for demonstrating how many verses has memorized.

The main drawback is that the teacher murojaah method can only monitor one student at a time. This condition causes the intensity of the meeting of teachers and students logically low. This method is challenging to implement in schools that have only one or two teachers, with hundred of the student. In order to memorize culture can implement widely in school, mostly rote learning and innovation is needed, one of them by utilizing smartphone and internet technology. In this study developed a mobile learning technology as a new way to memorize the Qur'an to the school with a limited number of religious teachers.

Mobile Learning (M-Learning) is a technique that uses mobile technology and the Internet for learning and education. M-Learning allows students to build their learning experience in a group learning environment [5]. Learning innovation is indispensable because information technology plays a beneficial role to work and learn, information technology must be optimized to bring progress in the education system [6]. The education must be able to absorb new technologies to improve the quality of learning [7]. Further, M-Learning is one manifestation of an irreversible change due to the influence of smartphone technology and the Internet [8], Several studies have shown that many learners who tend to use their mobile devices for music, social media/networking, and gaming but not for education [9]; therefore, we need a system that is designed to be efficient to be able to resolve the problems of education problems [10]. People adopt technology to blend in or connect with others [11], so we need a unique design for rote learning to be active and efficient. Teachers can handle many students to optimize the timing of holidays or when the student at home. Expectations of effort confirm that when technology is easy to use and requires less effort is one reason why users adopt a system [12].

## 2. Design and Architecture

Research into design has been approached both from the general perspective of design guidelines and frameworks (Parsons, Ryu & Cranshaw, 2007) [13].We use architecture as the fundament of the mobile learning environment at the beginning of the study. Another goal is to attract the interest of students in memorizing the Quran and improve its capabilities. In fact, when smart phones and tablets land in the hands of teachers and students and a number of sites offer on-line education at no cost, there is an assumption that adopting for it will take a leap. However when EDX was launched even Har-vard's faculty argued about its impact on the current system [14]. In addition, because technological gadgets can influence students' social emotions, balancing and controlling
that use is necessary [15]. Teachers can handle many students by optimizing the free time at home with formed mobile devices, the internet, and social media. In this study, the system is designed to create mobile applications that are connected and controlled to the server (Figure 1). There are two modules, the first so-called mobile application or modules. There are two modules, namely students module that used for memorizing and the other module called the teacher module that used for monitoring.

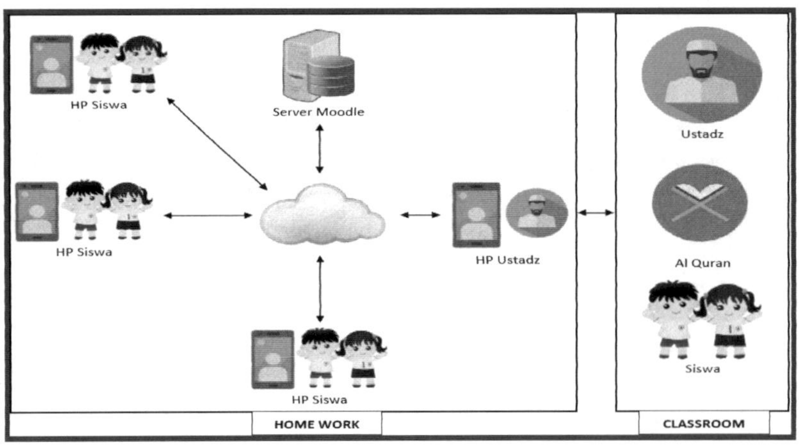


**Figure 1.** M-Learning Architecture

We design the student module, as shown in Figure 2.A. Social influence technology is no longer significant [16].The designed is unique which student can be used to memorize repeatedly. While the student is practicing, the system randomly sampling the student's voice. The sampling voice automatically sends to the server as the MP3 file. Student module have a recording system and international qory sound. Student can replay the recording voice for self-evaluating by comparing with the reading voice of international qory; it is mean that student can do the self-voice correction. The module also records the mouse "click" as student activity detection.

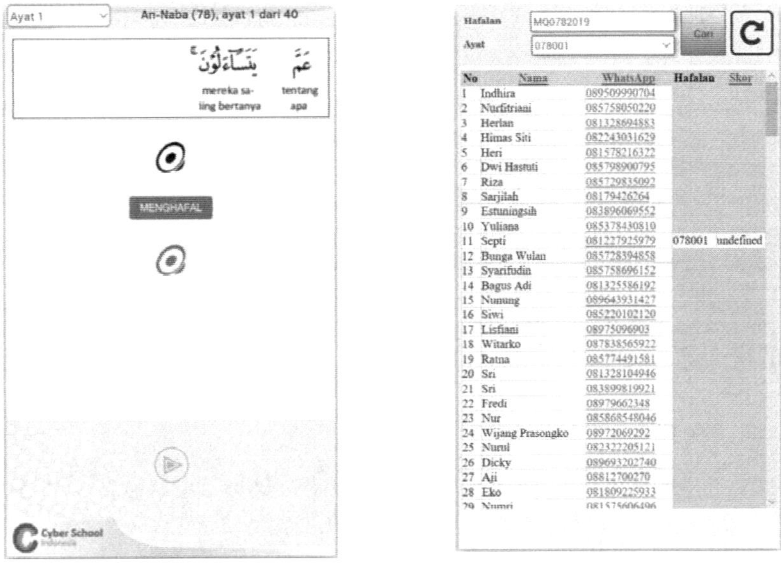


a b

**Figure 2.** Module Memorize (a) And The Monitoring Module (b)

Teacher module (Figure 2b) have the facility to view a listing of students who are monitored and mentored teachers. We call this module as a monitoring module. In recent years, projects that have been carried out for the purpose of both instructional support and improving functionality have increased interactions between teachers and students [17].This module, equipped with a button for verse selection a verse and assignment link for detail checking. In the list of students, there is a link to activate WhatsApp API as a means to communicate privately. Teachers can monitor and check the details of student activities, including sampling sound rote. The teacher can see the time, the number of clicks, and which verse already read by the students. We design the interface as simple as possible but handy to carry out monitoring and mentoring activities. Teachers are very free to provide feedback, confirming readings, motivating individually on each student by using WhatsApp.

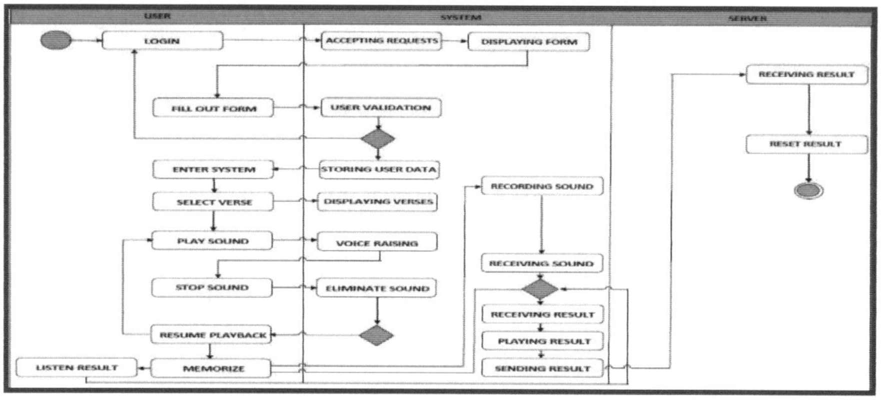


**Figure 3.** Activity Diagram

Figure 3 shows the activity diagram of the modules. There is three essential part of the diagram activity, including user, system, and server. Those three components are inter-related; the user can select multiple menus in the system. The system implements the method using some features; the first feature is images showing, how to display an image is using the HTML code. The "<imgsrc=">" is very useful here for connecting the image and an action. There is so many "action," one of its is to hide or show the image. We can see in the mobile display the red and black loudspeaker. The black image is for playing the cory's sound while the red one for playing the recording. In this module, we cannot hear the qory sound and recording sound together. One image disables other images and vice versa. This feature is very use full when the user will make memorization. When we press the button to record the user voice automatically picture paragraph hide. The steps to hide the image of the verse the image hide and show mechanism. We also use a try-catch. This mechanism removes error state, which can stop the application. Another feature is the sound randomly sampling system. This mechanism will initiate if the phone detects a specific minimal sound level. The system saves the sound as an MP3 file and sends it to the server using AJAX mechanism. The system using Javascript code for all mechanism handling, including multimedia controlling, file saving, file reading, synchronization, and mouse action-response mechanism.

## 3. Result

Research carried out by tests on 32 students in the school as a user. Before students use the application, the teacher asks her students to use Manuscripts memorize then deposited rote. Here are the results of a comparison test using the Manuscripts and applications.

| USING MUSHAF | Speed | Comprehension | Correction | Attraction | Frequency |
|---|---|---|---|---|---|
| Average Score | 2 | 3 | 1 | 2 | 2 |
| USING APPLICATION | Speed | Comprehension | Correction | Attraction | Frequency |
| Average Score | 2,5 | 3 | 2 | 3 | 2,5 |

**Figure 4.** Standard Measurement

### Experiment Result

We see in the control group. Figure 4 shows a standard measurement from points 0 to 4. The average student memorizing using the Qur'an book gets point 2 for reading speed. It was seen that there was change from point 3,5 to 2, meaning that students had not fully memorized using the Mushaf.
Meanwhile for reading comprehension using Manuscripts students get point 3; there seems to be a decrease from point 3.5 to 3. This means that the level of reading comprehension of students is good

though not perfect. If they can master reading and reading correctly, it is very likely that students' memorization will be correct. For self-memorization, the average student gets point 1. There is a decrease from point 3.5 to 1, it means that the level of correction of students is practically low. For the average interest students get point 2; seen a decrease from point 3.5 to 2 means that their level of interest is still lacking. The average memorizing frequency students get point 2, from point 3,5 to 2 means the memorization frequency is less.

Interviews with teachers to know their response after the use of the application. We obtained information from the teacher after running the system. They expressed interest to use because it is an alternative to conducting recitation of the Qur'an. Students will use current technology to improve their efficiency **[18]**. However, it is still not fluent because the first time using a system like this. There is no obligation or additional work of teachers to supervise and inspect the activities and results of student records. For teachers who try this application are happy to use because it allows present, but they still thinking about how to start its program, mainly related to the school's academic rules.

## 4. Discussion

The purpose of this study was to gather information about the effectiveness of the application for students that includes speed memorization, reading comprehension, error correction, interest memorize, and frequency (activity). Students taught in traditional methods are less enthusiastic and motivated during the learning process.Tests conducted for a middle school learning group in two days activity. We divided the activity into two different class activities, those act as a control group and an experiment group. The control group teaches using mushaf (book) and murojaah method, while the experiment group teaches using mobile learning system (application). After all the experiments done, we summarize the performance of students verses memorizing. Figure 4 shows the experiment result. The average student from control group using the "murojaah" method is slower than the experiment group.

Next we look at the group experiments to see memorization performance using the application. Figure 5 shows the average student gained point 2.5 for speed; seen changes in data from point 3.5 to 2.5. This shows the speed at which students memorize better than the control group at position 2. For reading comprehension the average student gets point 3, there appears to be a change from point 3.5 to 3. This means that students have a pretty good reading comprehension; from the results of the trial using the Manuscripts and the application there is no difference because getting the same point is 3. For the point of reading errors using the application the average student gets 2, while in the control group students get point 1. This shows the qori reading facility is useful for correcting memorization mistakes. For students' interest in getting point 3, it means that students are more interested in using the application compared to books because of some features that facilitate the memorization process. Finally, for the frequency of memorization activities using the application the average student gets point 2.5; this is higher than using books where student points are 2; it means students memorize using the application more often than using books.

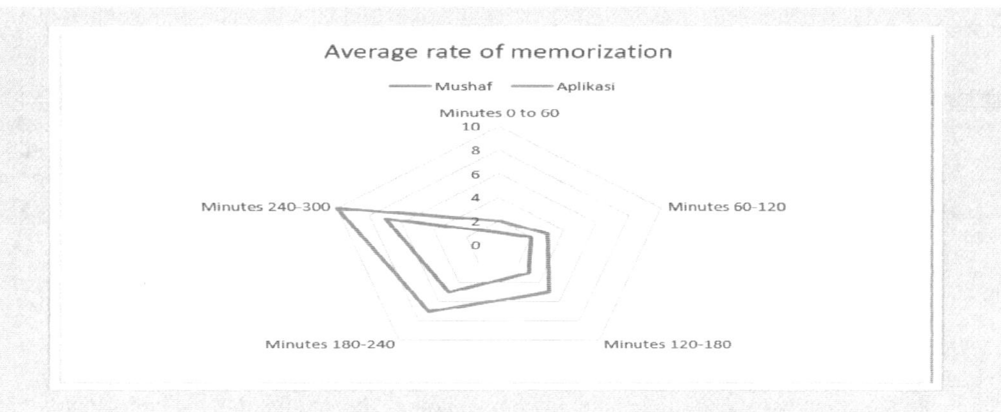


**Figure 5. Result Rate Memorization Students**

In Figure 5 described the average deposit rote students during the experiment, from experiments conducted over four days, we specify the hour intensifies About 5 hours, start from 0 to 300 minutes, the experiment results using the Manuscripts minute average 0-60 students can collect as much as 1 verse memorization, at 60-120 minutes Stratas average students can deposit as much as 2 verse memorization, in minutes to 120-180 students can deposit as much as 3 verse memorization, at minute average 180-240 students can deposit as much as 5 verse memorization, at minute average 240-300 students can deposit rote 7 paragraph. Of the two pictures data image number 4 and 5 can be concluded that the effectiveness of the application does not need to doubt, in addition to the two above data, the results of our review of the student feels supervised by teachers, because there are features that send the rote last into the server, then connect to the teacher module.

From the experimental results, we conclude that the student module in the form of mobile learning to memorize the Koran has the potential to increase the speed of memorizing, increasing the frequency of memorizing, making students more actively to try the application, make it easier for teacher performance, and increase students' understanding of how to read Qur'an is excellent and accurate. In general, students memorize not monitored by anyone; it is most likely with the development of mobile learning to memorize the Koran can help students to be more frequent. Mobile devices have various distinctive features such as individualized interfaces, real-time access to information, context sensitivity, instant communication, and feedback **[19]**. The experimental results are consistent with the user's comments that the student after trying, comment students when compared to using the old methods, modules, students will be more promising. After seeing the test results of students feel motivated, they will use this application as well as possible.

Features of the module students are still in the development stage, there are still many in need of testing and analysis of data, so far there has been no shortage of significant applications, but we need to add the feature to display the score or similarity value with a qory, so students do not need to be corrected by means of manual. With features that are currently available, the student measure his ability to listen to the recording, while hearing students also sometimes no harm, for example, the reading "wamaa" requires two beats long, but by the students only read one beats long, and most students do not realize their mistakes, so that if one reads it will happen when the correct on memorization,

In the development of mobile applications that are used to memorize the Holy Qur'an. Studies report that resistance to change plays an important role in accepting technology in education **[20]**. the ultimate goal is to increase students' interest in memorizing, increase the frequency of memorization, and increase the speed of memorizing. So with this study, students are expected to be motivated in the process of memorizing the Quran. Coupled with the advantages of using mobile learning as described previously, this application quiet intended for students, so that can be used anywhere and anytime, especially outside of school hours. Making these applications, we use Courselab software to integrate the learning material in a learning module. Besides that, the framework to convert from the desktop to mobile use software framework PhoneGap and HTML5 + JavaScript as the programming language.

## References

[1] Mawar Sari, D. N. A., Nailil Muna, A., & Susilayati, M. (2019). Integration of Islamic Religious Education and Natural Science on Photosynthesis Experiments. *International Journal of Students' Research in Technology & Management*, *7*(1), 17–22. https://doi.org/10.18510/ijsrtm.2019.714

[2] Jahangir, S. F., & Nawaz, N. (2014). Effects of memorizing Quran by heart (hifz) on later academic achievement. *Journal of Muslim Mental Health*, *8*(2), 58–64. https://doi.org/10.3998/jmmh.10381607.0008.208

[3] Sedek, A., Mustaffa, A., & Khadher, A. (2015). Method on Memorization the Quran in Malaysia : A Study in Darul Tuba Institute , Malaysia. *Proceedings of the International Conference on Global Business, Economics, Finance and Social Sciences (GB15_Thai Conference)*, (February), 9. https://doi.org/10.1007/s13398-014-0173-7.2

[4] Jamaluddin, I. Z., Othman, M. K., & Yusoff, M. Z. M. (2017). The Art Of Memorizing Qurân: Applications Of Lahjat Al-Huruf And Lahjat Al-Tarannum. *International Journal of Academic Research in Business and Social Sciences*, *7*(8), 495–504. https://doi.org/10.6007/ijarbss/v7-i8/3256

[5] Sarrab, M., & Elgamel, L. (2013). M Obile L Earning ( M-L Earning ) and. *International Journal of Distributed and Parallel System*, *3*(4), 31–39. https://doi.org/10.3991/ijim.v3i4.986

[6] Kumar, S. (2013). M-Learning: A New Learning Paradigm. *International Journal on New Trends in Education and Their Implications*, *4*(2), 24–34.

[7] Camulleri, M.A., Camilleri, A.C. (2017). The technology acceptance of mobile application in educaation. *Proceedings of the 13th International Conference on Mobile Learning 2017, ML 2017*, 41–48.

[8] Delgado Cepeda, F.J. (2017). Small private online research: a proposal for a numerical methods course based on technology use and blended learning. *Proceedings of the 13th International Conference on Mobile Learning 2017, ML 2017*, 41–48.

[9] Li, Z. (2017). Mobile learning in the theater arts classroom. *Proceedings of the 13th International Conference on Mobile Learning 2017, ML 2017*, 95–99.

[10] Hajri, I. (2019). *Sepenuhnya Automated Kelas Sistem Absensi*. *13*(8), 95–106.

[11] Pedersen, P. E., & Ling, R. (2003). Modifying adoption research for mobile Internet service adoption: Cross-disciplinary interactions. *Proceedings of the 36th Annual Hawaii International Conference on System Sciences, HICSS 2003*. https://doi.org/10.1109/HICSS.2003.1174217

[12] Sripalawat, J., Thongmak, M., & Ngramyarn, A. (2011). M-banking in metropolitan bangkok and a comparison with other countries. *Journal of Computer Information Systems*, *51*(3), 67–76.

[13] Hopkins, G., & Fitzgerald, E. (2014). *of Mobile and Blended Learning A Mobile Learning Overview by Timeline and Mind Map*. *6*(4). https://doi.org/10.4018/ijmbl.2014100101

[14] W. RJ , "The Harvard Crimson," The Harvard Crimson, 12 12 2012. [Online]. Available: http://www.thecrimson.com/article/2012/12/12/faculty-edx-courses- debat /? page = 1 . [Accessed 27 6 2016].

[15] Ashari, Z. M., Ngadiman, A. A., Zainudin, N. F., & Jumaat, N. F. (2018). The relationship between knowledge and attitude towards technology gadget usage with students' socio-emotions development. *International Journal of Interactive Mobile Technologies*, *12*(7), 152–163. https://doi.org/10.3991/ijim.v12i7.9711

[16] Wolfgang, A., & State, P. (2010). *Hierarchical Bayesian Modeling*. 336–358. https://doi.org/10.1002/9780470317105.ch14

[17] Göksu, İ., & Atici, B. (2013). Need for Mobile Learning: Technologies and Opportunities. *Procedia - Social and Behavioral Sciences*, *103*, 685–694. https://doi.org/10.1016/j.sbspro.2013.10.388

[18] Tarhini, A., El-Masri, M., Ali, M., & Serrano, A. (2016). Extending the UTAUT model to understand the customers' acceptance and use of internet banking in Lebanon. *Information Technology & People*, *29*(4), 830–849. https://doi.org/10.1108/itp-02-2014-0034

[19] Sung, Y. T., Chang, K. E., & Liu, T. C. (2016). The effects of integrating mobile devices with teaching and learning on students' learning performance: A meta-analysis and research synthesis. *Computers and Education*, *94*, 252–275. https://doi.org/10.1016/j.compedu.2015.11.008

[20] Al-Hunaiyyan, A., Alhajri, R. A., & Al-Sharhan, S. (2018). Perceptions and challenges of mobile learning in Kuwait. *Journal of King Saud University - Computer and Information Sciences*, *30*(2), 279–289. https://doi.org/10.1016/j.jksuci.2016.12.001

**Acknowledgement**

This research is part of the research Growing Self Directed Learning (SDL) Using Gamma Feedback Learning Model (GFLM). We want to say many thank to the Ministry of Research, Technology and Higher Education of the Republic of Indonesia based on the Decree of the Director-General of Research and Development Strengthening number 7/E/KPT/2019. We are also grateful to the University of Muhammadiyah Yogyakarta for fully supporting this research by providing additional funding and research facilities.

Journal of Physics: Conference Series **1381** (2019) 012030 doi:10.1088/1742-6596/1381/1/012030

# A Hybrid Approach on Single Image Dehazing using Adaptive Gamma Correction

**O V Putra**[*], **A Musthafa, and F R Pradhana**

Universitas Darussalam Gontor, Ponorogo, Indonesia

[*]**oddy@unida.gontor.ac.id**

**Abstract.** Since the last eruption of Mt. Kelud, a surveillance has been conducted using camera to observe the crater lake and its surroundings. The key point of this observation is based on the captured haze-free images. However, these images suffer from hazy condition because of degassing. Degassing obscures the camera vision so that the observer barely maintain to keep track of lake phenomenon. Furthermore, even many dehazing techniques have been proposed to tackle this such problems, it still leaves some other problems such as oversaturation, color distortion, and halo effects. In this paper, a hybrid method is proposed. This work incorporates two the state-of-the-art visibility restoration and contrast enhancement methods which are color attenuation prior (CAP) and adaptive gamma correction (AGC). This method is separated into three different modules. They are disparity estimation (DispE) module, transmission map enhancement (TME) module, and hazy image restoration module (ImRec). The DispE module adapts the color attenuation model for its powerful depth estimation for handling outdoor image. Subsequently, the TME module is enhanced using AGC. And the last module is calculated from the modified hazy model. The experimental results are quite impressive. It is able to maintain edge and minimize image distortion qualitatively. And the fog density measured with Fog-Aware Density Estimation (FADE) estimated at 1.448.

## 1. Introduction

Human vision can see through any object in a clear scene condition. An object is visible to human eye because it reflects or scatters the light that come from the sky. The ray propagates from the object through the particle in the air and reaches the eye [1,2]. On the other side, an image which is taken under hazy condition loses its contrast and visibility. Haze often obscure the vision.

Generally, many concept of image dehazing were proposed to overcome the vision problem depend on the portion of the skylight [2–4]. They use outdoor image in a proper composition. The image may consist of three main parts, which are the land, the object, and the sky region. Ideally, these main parts are divided evenly. Thus, the proportion of scene within image is adequate. However, at certain outdoor photograph condition, the proportion may lack something, such as the existence of sky region with. For example, our prior research uses the crater lake images as subject for visibility restoration [5]. The image is captured using CCTV. Prior to the application of CCTV, the Ministry of Energy and Mineral Resources of Indonesia conducted the program for volcano disaster mitigation [5]. This observation is in purpose of monitoring the lake activity. At certain time, the sulphur dioxide starts to emerge and diffuse with particle in the air [6,7]. This process is mainly caused by the volcanic activity. With the degassing moving upward, the camera visibility is obscured. Obviously, one of the solution is to restore the camera visibility.

By the camera is directly facing toward the lake, it does not cover the sky region. Another problem is that the restored scene radiance color is affected by global color luminance. Which results the restored color tends to be bluish. This condition compared to other outdoor images makes the dehazing difficult to handle. Since most of the dehazing is relied upon the light source. Restoring the visibility is not only haze removal but also color adjustment.

In this paper, an adaptive dehazing method using gamma correction is proposed to restore the visibility and adjust the color of hazy image. By incorporating the concept of dark channel prior (DCP) for atmospheric light estimation, color attenuation prior (CAP) [8] for the depth estimation, and adaptive gamma correction [9] for color adjustment, we can restore the hazy scene images.

The main contribution of this paper is the adaptive gamma correction incorporated with DCP and CAP to improve the previous works [8,10,11] including the conference version [5]. Our new method is able to preserve the color of restored images. Thus, the image is no longer affected by global scene radiance.

## 2. Related Works

A number of dehazing researches have been conducted for about 20 years. Many approaches and assumptions are built during these development. Fattal *et al*, have been proposed a dehazing technique based on Independent Component Analysis (ICA) [3]. He assumed that the transmission medium and surface shading are locally uncorrelated. Fattal also refined the hazy image model which used in [1]. The vector of constant albedo R and scalar l are added to the hazy model. These variables redefine the non-haze image. By using Gauss-Markov random field, Fattal's algorithm can be used for transmission map estimation. The white light proportion which reflected by the surface is estimated using this algorithm. Finally, the transmission map can be refined along with estimation of albedo scene. However, the shortcomings of this method are the long process of refinement and the weakness for handling dense haze.

Not only the global contrast is higher on clear scene compared to hazy scene but also the local contrast. Tan *et al* has proposed a dehazing method based on Markov Random Field (MRF) [12]. The local contrast of a scene is maximized using his method. Restoring the visibility of hazy scene is quite a problem. This is due to the non-existence of atmospheric light. The overcast caused the light dimmer. On the other side, the light globally remains constant. Instead of estimation the atmospheric light, the remaining task is focused on light chromaticity estimation. There are several conditions that must be met in order to dehaze a hazy scene. First, the haze-free image must have higher contrast than the hazy one. Second, the depth image affects the atmospheric light variance. Therefore, at some patches, the atmospheric light and the depth are likely to be similar. Third, the outdoor hazy images usually disrupted by bad weather. Therefore, the recovered image must satisfy the clear image features. The restored image yields good result but susceptible to oversaturation. This is caused by high distinct contrast in adjacent block window patches.

At some colors in most patches of natural outdoor scene, there is one color channel which has low intensity. Some colorful outdoor object such as tree leaves, yellow flowers, and green bushes have low intensity from one channel of their RGB color space [2,13] When the color channels are observed from the green leaves, the green color obviously has higher intensity than the remaining two channels. This observation also valid for the other colorful objects. The low intensity channel within colorful object is called the dark channel. This assumption was introduced by He *et al*. He called this as the dark channel prior (DCP). One of the concept of dark channel is that its capability for estimating the light source. If the colorful object at least has one channel with low intensity then the light source has high intensity on each channel. This claim can be proven by using the dark channel assumption. The DCP takes a certain patch size. From each patches, it takes the minimum intensity color channel within a patch to make a block-like window sliding throughout the image. The side effect of this process will create some block artifacts for the restored image. To handle this problem, He utilized the soft matting algorithm to refine the restored image. However, the soft matting take long time to process. In the next two years, He

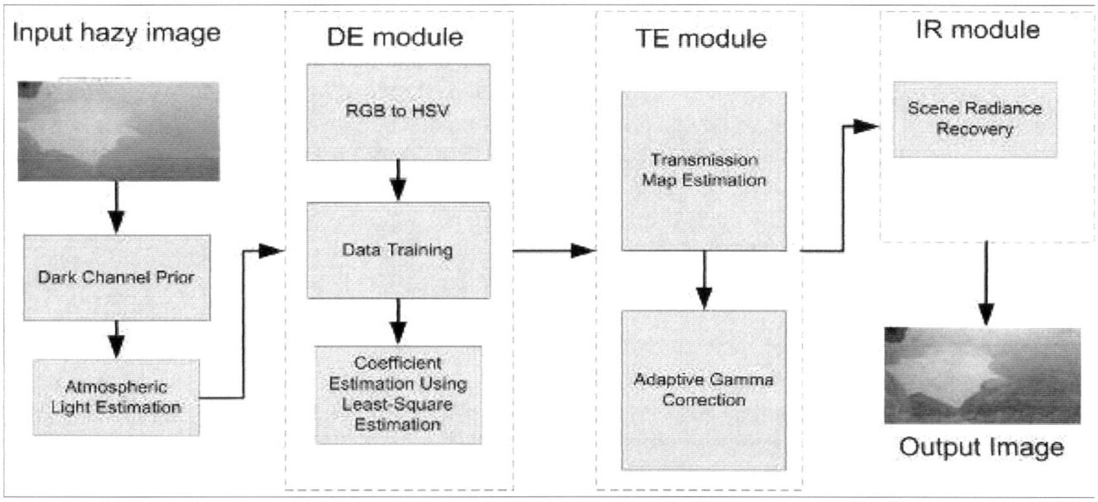


Figure 1. Proposed method using AGC

proposed an image filter method called guided image filter [14]. This method works faster than soft matting. The DCP gives impressive result on dehazing of hazy images. It can handle denser haze. However, it may results oversaturation when there is exist an object whose color intensity similar or equal to the light source.

At the same year when the guided filter was proposed, a new adaptive method was also proposed by Meng *et al [15]*. He modified the hazy image model geometry. Two new components was added prior to the model. They are the boundary constrain C0 and C1. This method is used for transmission map estimation and based on L1-norm for weighting the contextual regularization. The concept of contextual regularization is originated from observation that neighboring pixels share same depth at local patch. Meng's method refines the DCP. However, it may fail on recovering the halo effect.
The new prior was proposed by Zhu *et al[8]* was called color attenuation prior (CAP). Zhu has found the depth model for outdoor hazy image. He found a correlation between saturation and brightness. The correlation is based on observation of outdoor image. He took samples of different distance from an outdoor image. The samples was separated into three patches. Each patches represents the image depth. The first to the third patch represents near, mid, and far distance respectively. They are observed in the HSV color space. From this observation, the saturation gradually decreased along with the distance. And vise versa, the brightness statistically increased. In near distance, we can clearly distinguish objects with different color. The colorfulness of object represent high saturation. When we see objects in far away, it is hard to see them because of the obscuring haze. Therefore, the depth of image, the concentration of haze, and the brightness are positively correlated. The depth model was trained using maximum likelihood estimation (MLE). He gathered data from Google dan Flickr. The CAP algorithm is the fastest compared to the previous state-of-the-art algorithms and able to to preserve edge information. However, there is a shortage of this algorithm. When an object close to the camera, the restored object is darker.

In the next year, Chen *et al [11]* proposed a gain intervention refinement (GIR) filter for transmission refinement. This filter idea came from the shortage of DCP which are the halo effect. As mentioned earlier, the main problem of dehazing is the halo effects and block-like artifacts [11,13]. The use of filters such as softmatting and guided filter avoids the generation of halos and artifacts. However, the computational intensifies because of the overlapping window in local linear model. Compared to the previous two filter of DCP, soft matting and guided filter, GIR utilizes the gain intervention of the dark channel images. This filter surpassed the previous filter in term of transmission map refinement due to its fast computation. However, oversaturation exists when dealing with an object that as bright as the light source.

In our previous work [5], an incorporated haze removal was implemented using DCP and CAP. The result is quite impressive. However, the restored images tend to be bluish.

## 3. Proposed Works

According to previous research, dehazing problem mainly focused on restoring the visibility. This issue works well when there is no influence of global color radiance. Since the global color mostly affects the restored image, the remaining problem is adjusting the restored color. We will discuss this issue more details later.

Mostly, the previous state-of-the-art methods [2,8,12] recover the obscured image with proper proportion of the background. This proper proportion is composed by objects and sky. Since the camera Field of View (FOV) covers only the lake and its surroundings, our data does not contain the sky region. With this condition, it is quite problematic for estimation of the atmospheric light.

The output dehazed images come in different issue according to the method. Almost every dehazing algorithms can bring up the vivid color of hazy images. However, a problem may come when dehazing problem only depends on haze removal and is affected by the global luminance. The restored image does not show its original color.

Occationally, the surrounding lake color varies. The green color appears when it is cloudy. The yellow color also appears when the weather is clear. Sometimes, it becomes dark brown when it is overcast. In order to overcome this problem, in this paper, a hybrid single image haze removal using CAP dan Adaptive Gamma Correction (AGC) is proposed. This method is applied to the CCTV images of the lake crater of the mountain Kelud. The proposed method shows impressive results.

The proposed method consist of three modules as seen in **Error! Reference source not found.**. The first module is the disparity estimation module (DispE). The second module is the transmission enhancement (TME) module. And the last module is the image recovery module (ImRec). These three modules will be elaborated later.

### *3.1. DispE Module*

The hazy image degradation model is proposed by McCartney [16]. Later, this model is further improved by [1]. The image degradation formula can be written as:

$$I(j) = \Phi(j)\big(t(j)\big) + A\big(1 - t(j)\big) \tag{1}$$

where I(x) is the output hazy image which is formed by the right term in equation **Error! Reference source not found.**. The $\Phi(x)$, A, $t_0(x)$, and x in right term are the haze-free image, the atmospheric light, the transmission medium, and the position within images respectively. The main objective of dehazing is to extract $\Phi(x)$ from equation 1. The haze-free image can be achieved once the remaining variables I(x), A, and \\$t_0(x)$ are satisfied. The physical model of hazy image can be seen in Figure 2.

The transmission map can be regarded as image transparency level [2]. There is another definition of transmission map according to [3,10,17]. It is defined as unscattered portion of light that reaches the camera. Mathematically, the tranmission map is formulated as:

$$t_0(j) = exp\left(-\int_0^m \beta\,(z)dz\right) \tag{2}$$

Journal of Physics: Conference Series **1381** (2019) 012030 doi:10.1088/1742-6596/1381/1/012030

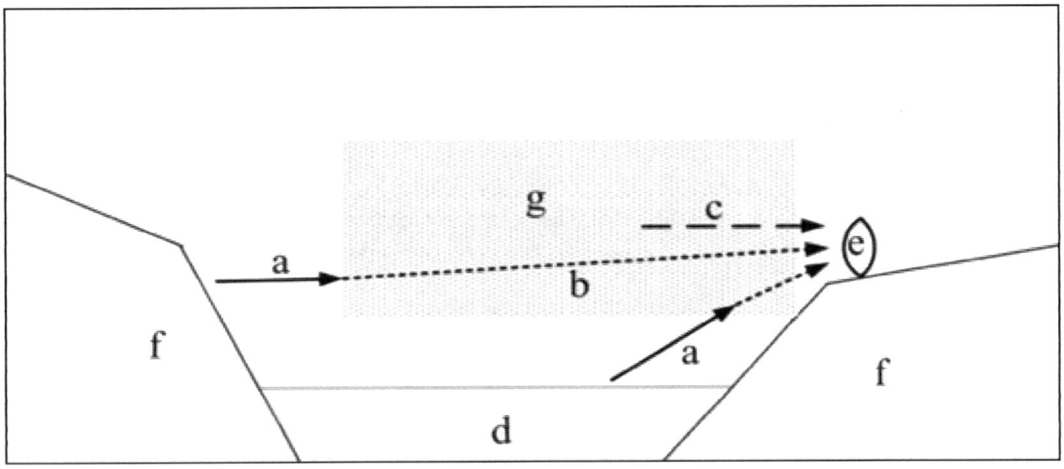


Figure 2. The physical model of image degradation. Where **a** is the propagating light, **b** is the attenuated light, **c** is the airlight, **d** is the lake water, **e** is the observer camera, **f** is the lakeside, and **g** is the haze above the lake

where $t_0$ is the transmisson map. It represents a portion of light that passes through particles in the air and finally reach the observer,$\beta$is the light scattering coefficient, and d represents the depth map. From equation **Error! Reference source not found.**, we get:

$$t_0(j) = e^{-\beta d(z)} \tag{3}$$

where the remaining problem in determining the transmission map is relied on the depth image d(x). The depth image is an image in graylevel which its pixels hold intensity value. In fact, each pixel represent the distance between point object or background to the observer.

Measuring the depth image is quite challenging [2,5,8]. Some tools such as Microsoft Kinect and stereo camera can be utilized for estimating the depth for measurable distance object. However, some outdoor scene it is troublesome to measure its depth.

A new outdoor depth model was introduced by [8]. It is formed from three coefficients which are positively correlated. They are brightness, saturation, and difference between them. The depth value is directly proportional to the difference between brightness and saturation. By this definition, the correlation can be modelled as:

$$d(z) \propto v(z) - s(z) \tag{4}$$

where d is the disparity, v,s are the value and saturation in HSV color domain respectively and z is the pixel position.

Since the equation **Error! Reference source not found.** is positive correlation, the mathematical representation can be defined as:

$$d(z) = \lambda_0 + \lambda_1 v(z) + \lambda_2 s(z) + r(z) \tag{5}$$

where d, v, and s are the depth, value, and saturation respectively. These are taken from equation **Error! Reference source not found.**. $\lambda_0$,$\lambda_1$, and $\lambda_2$ are the coefficients that must be estimated. And the last is residual r. In Zhu's works [8], the coefficients of this model was estimated using maximum likelihood estimation (MLE).

To simplify the calculation, the matrix form of **Error! Reference source not found.** are **K** represents the variables $v_0$ to $v_n$ and $s_0$to $s_n$, the vector **d** represents the depth $d_0$ to $d_n$, and the vector **p** represents $\lambda_0$ to $\lambda_n$. By this declaration, we have:

$$K = \begin{bmatrix} 1 & v_0 & s_0 \\ 1 & v_1 & s_1 \\ 1 & \vdots & \vdots \\ 1 & v_n & s_n \end{bmatrix}, p = \begin{bmatrix} \lambda_0 \\ \lambda_1 \\ \lambda_2 \end{bmatrix}, d = \begin{bmatrix} d_0 \\ d_1 \\ \vdots \\ d_n \end{bmatrix} \tag{6}$$

In order to estimate the coefficient **p**, the Least Squared Estimation (LSE) is proposed in this method. From **Error! Reference source not found.**, the estimated coefficients can be estimated by:

$$p = (K^T K)^{-1} V^T d \tag{7}$$

The data training are collected from [18]. The points are accumulated prior to data training. Once the coefficients are estimated, the disparity also can be estimated using equation **Error! Reference source not found.**.

*3.2. TME Module*

In case of dehazing, the condition of outdoor images varies. An image may composed of object which act as light source. This object can cause oversaturation for the restored scene. The other problem, global environment color also affects the restored image. This ill-posed nature needs different treatment [13].

A method based on gamma correction is adapted from [4]. This method classify the image intensity into bright and dark. Gamma correction not only is used to increase contrast but also to adjust the intensity of transformation function. Furthermore, this adaptive gamma correction (AGC) preserves the detailed edge. Thus, the modified TME can be formulated as

$$t_1(j) = V_m \left( \frac{t_0(j)}{V_m} \right)^{\gamma} \tag{8}$$

while $\gamma$ can be achieved from

$$\gamma = \begin{cases} 1 + \dfrac{z}{T_{thres}} & , z \geq T_{thres} \\ 1 & , z < T_{thres} \end{cases} \tag{9}$$

where$t_1$is the enhanced transmission, $V_m$is maximum intensity from grayscale, $\gamma$is the adaptive gamma value, $z$is intensity value when the cumulative distributed values reach certain threshold, and $T_{thres}$is the intesity constraint from $z$. We set the value of $T_{thres}$at 0.429 when $z = 0.1$. This value was obtained from our observation of crater images in various weather condition, such as clear, cloudy, and hazy.

Inspired from [4] in restoring bad weather images, the average of each RGB value is calculated. Therefore, we get

$$\mu_c = \frac{1}{s} \sum_{i=1}^{p} \sum_{j=1}^{q} H_c(i,j) \tag{10}$$

where $\mu$is the average intensity of each points, s is the number of pixel, p, q is the height and width of image respectively, and $H_c$is the intensity at i,j.

From equation **Error! Reference source not found.**, we get:

$$\delta^c = \mu_r - \mu_c \quad , \quad c \in \{r, g, b\} \tag{11}$$

*3.3. ImRec Module*

Finally, last but not least, this module plays as final role. When DispE and TME module has been achieved, the outdoor haze-free image can be obtained from

$$\Phi^c(j) = \frac{I^c(j) - (A^c - \delta^c)}{max(t_1(j), t_0)} + (A^c - \delta^c) \tag{12}$$

where$\Phi^c(j)$is the restored haze-free image, $I^c(j)$is the intensity of hazy image at$j$, A is the atmospheric light,$\delta$is a disparity from **Error! Reference source not found.**, and$t_0$and$t_1$are the boundaries. Typically, $t_0$is set at 0.1.

Subsequently, equation **Error! Reference source not found.** effectively manage to recover $\Phi^c(j)$from $I^c(j)$. To sum up, the proposed method is illustrated as in **Error! Reference source not found.**.

## 4. Experiments and Results

Our method were evaluated using generally used hazy images for instance: a hazy soccer field and a castle. This evalutaion is intended to measure the performance of our proposed method to other famous dehazing algorithms. They are He [2], Zhu [8], and Chen [19]. The evaluated parameters are fog density (FADE) and entropy.

This experiment was carried out using a PC with 16 GB RAM, Intel Core i7 8th gen, and the tested data was captured using CCTV camera model Axis Q1755-E with resolution 1280x720.

### *4.1. Disparity Estimation using LSE*

We gathered outdoor hazy images with their corresponding depth map from [18]the depth map parameters as in equation **Error! Reference source not found.** are estimated using Least-Squared Estimation (LSE). The colored images are converted into HSV color space. Subsequently, only saturation and value are extracted before they used to satisfy equation **Error! Reference source not found.**. With its depth images available, finally, the parameters were trained. As results, we found that$\lambda_0, \lambda_1, \lambda_2$are equal to 0.126111, -0.043755, and -0.117156 respectively. When these parameters obtained, the next step is TME module.

### *4.2. Transmission Map Enhancement Results*

In this part, a transmission map enhancement is proposed. This method is incorporated from dark channel for airlight estimation, color attenuation for its incredible outdoor depth map estimation, and finally the AGC from image restoration in a bad weather.

This task are quite challenging. In our previous work [5], it still has shortages. The restored image is influenced by global color. This global color causes obstruction to the scene. Futhermore, the haze-free image is leaving a rough amount of haze as well. Therefore, the gamma correction takes place in order to tackle these problems.

Since the condition of the lake suffers from uncertain weather condition, an adaptive parameter of gamma is deployed. This adaptive value relies on the cumulative density function (CDF). In order to optimize the AGC, color intensity distribution, as in equation **Error! Reference source not found.** and **Error! Reference source not found.**, is applied as well. Finally, the enhanced transmission map can be seen in Figure 3.

### *4.3. RecoveryResults*

In this part, we evaluated quantitatively using Fog-Awared Density Estimation (FADE) from [20]. As illustrated in Table 1, the proposed method is compared with previous famous dehazing algorithms in various hazy image conditions. In general, our method is able to tackle hazy condition. Even at certain condition, we fail to stand on top of all algorithms, however, the proposed method managed to overpowers the other methods. This can be seen that we achieve the average lowest value of FADE at 1.448 followed by He, Chen, and Zhu.

Generally speaking, most of hazy removal techniques are able to tackle such hazy condition. However, another experiment protocol must be used to measure the results in various conditions. Hence, all techniques are forced into these conditions.

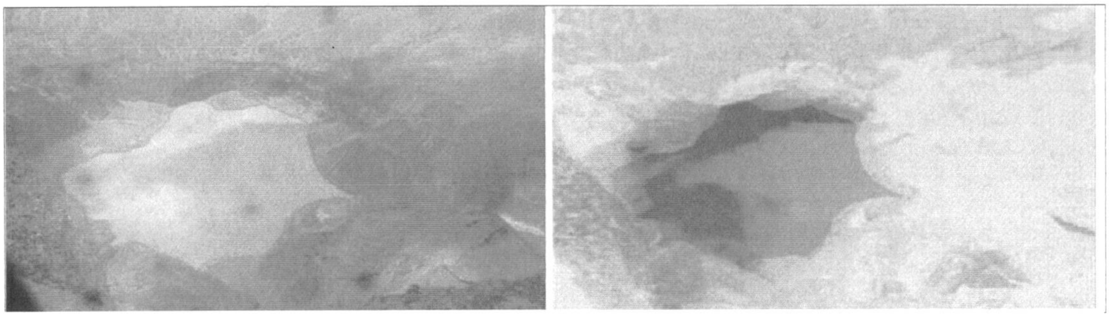

Figure 3: Original hazy image and its corresponding transmission map

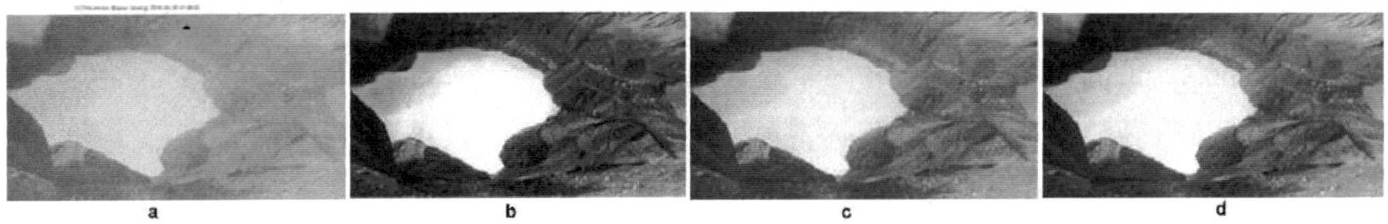


Figure 4: Comparison of proposed method and the others techniques. (a) is the hazy images, (b) He's result, (c) Zhu's result, and (d) our result.

As illustrated in Figure 4, the difference between the proposed method and the other three methods are quite distinguishable. It can be seen that the He's result contains less haze compared to others. However, even it is achieved such result, it tends to get affected by global blue color. Moreover, the lake color is oversaturated as well, not to mention the halo effects around the lake edge. This halo effects occurred due to the dark channel block patch artifacts. Even Meng's result has slight amount of haze, it is capable of tackling of halo effects, thanks to its adaptive feature of dark channel. Unfortunately, it still remain bluish. Finally, in our result, the proposed method is able to exhibit its ability to preserve edge and reduce oversaturation of lake color.

**Table 1.**FADE Results.

| | *He* *DCP* | *Zhu* *CAP* | Chen HGIR | *Our* *CAPAGC* |
|---|---|---|---|---|
| **N1** | 1.71 | 2.27 | 2.14 | **0.99** |
| **N2** | 0.29 | 2.56 | 2.29 | **1.26** |
| **N3** | 1.90 | 2.66 | 2.20 | **1.32** |
| **N4** | **2.17** | 3.26 | 3.02 | 2.45 |
| **N5** | 1.76 | 2.43 | 2.52 | **1.22** |

## 5. Conclusion and Discussion

In this paper, a hybrid method was proposed based on outdoor depth estimation from color attenuation and adaptive gamma correction. This method contains three main steps, which are the disparity estimation, enhanced transmission estimation, and finally the haze-free image recovery.

As a result, our method achived the lowest value of FADE at 1.448. Furthermore, the restored image is able to maintain edge and help to tackle oversaturation which is the shortage of dark channel prior. However, our result is still far from best condition. There are some issues that need to be put into concern such as accuracy and segmentation in hazy images, especially in using deep learning based algorithms. This has become our future works.

## 6. References

[1] Narasimhan S G and Nayar S K 2002 Vision and the Atmosphere *Int. J. Comput. Vis.* **48** 233–54
[2] He K, Sun J and Tang X 2011 Single image haze removal using dark channel prior *IEEE Trans. Pattern Anal. Mach. Intell.* **33** 2341–53
[3] Fattal R 2008 Single Image Dehazing *ACM Trans. Graph.* **27** 72:1--72:9
[4] Huang S C, Chen B H and Wang W J 2014 Visibility restoration of single hazy images captured in real-world weather conditions *IEEE Trans. Circuits Syst. Video Technol.* **24** 1814–24
[5] Putra O V, Prianto B, Yuniarno E M and Purnomo M H 2016 Visibility restoration of lake crater hazy image based on dark channel prior *2016 International Computer Science and Engineering Conference (ICSEC)* pp 1–6
[6] Hayer C S, Wadge G, Edmonds M and Christopher T 2016 Sensitivity of OMI SO2 measurements to variable eruptive behaviour at Soufrière Hills Volcano, Montserrat *J. Volcanol. Geotherm. Res.* **312** 1–10
[7] Gresse M, Vandemeulebrouck J, Byrdina S, Chiodini G and Bruno P P 2016 Changes in CO2 diffuse degassing induced by the passing of seismic waves *J. Volcanol. Geotherm. Res.* **320** 12–8
[8] Zhu Q, Mai J, Shao L and Member S 2015 A Fast Single Image Haze Removal Algorithm Using Color Attenuation Prior *IEEE Trans. Image Process.* **24** 3522–33
[9] Huang Z, Zhang T, Li Q and Fang H 2016 Adaptive gamma correction based on cumulative histogram for enhancing near-infrared images *Infrared Phys. Technol.* **79** 205–15
[10] Gao Y, Hu H M, Wang S and Li B 2014 A fast image dehazing algorithm based on negative correction *Signal Processing* **103** 380–98
[11] Chen B H and Huang S C 2016 Edge Collapse-Based Dehazing Algorithm for Visibility Restoration in Real Scenes *J. Disp. Technol.* **12** 964–70
[12] Tan R T 2008 Visibility in Bad Weather from a Single Image *2008 IEEE Conference on Computer Vision and Pattern Recognition* (IEEE) pp 1–8
[13] Lee S, Yun S, Nam J-H, Won C S and Jung S-W 2016 A review on dark channel prior based image dehazing algorithms *EURASIP J. Image Video Process.* **2016** 4
[14] He K, Sun J and Tang X 2013 Guided Image Filtering *IEEE Trans. Pattern Anal. Mach. Intell.* **35** 1397–409
[15] Meng G, Wang Y, Duan J, Xiang S and Pan C 2013 Efficient Image Dehazing with Boundary Constraint and Contextual Regularization *2013 IEEE International Conference on Computer Vision* pp 617–24
[16] McCartney J 1976 *Optics of the Atmosphere: Scattering by Molecules and Particles* (Wiley)
[17] Cai B, Xu X, Jia K, Qing C and Tao D 2016 DehazeNet : An End-to-End System for Single Image *IEEE Trans. Image Process.* **25** 5187–98
[18] Zhang Y, Ding L and Sharma G 2019 HazeRD: An Outdoor Scene Dataset and Benchmark for Single Image Dehazing
[19] Chen B H, Huang S C and Cheng F C 2016 A high-efficiency and high-speed gain intervention refinement filter for haze removal *J. Disp. Technol.* **12** 753–9
[20] Choi L K, You J and Bovik A C 2015 Referenceless Prediction of Perceptual Fog Density and Perceptual Image Defogging *IEEE Trans. Image Process.* **24** 3888–901

**Acknowledgments**
Authors would like to give regards to Indonesia Ministry of Higher Education (DIKTI) for this work and Ministry of Energy and Mineral Resources especially Pusat Vulkanologi dan Mitigasi Bencana Geologi Gunung Kelud for supplying data.

Journal of Physics: Conference Series **1381** (2019) 012031 doi:10.1088/1742-6596/1381/1/012031

# Use of Support Vector Machine to Classify Rhizomes Based on Color

**M Maimunah*, E R Arumi**

Department of Informatics Engineering, Engineering Faculty, Universitas Muhammadiyah Magelang, Jalan Mayjend Bambang Soegeng KM 5 Mertoyudan, Magelang, Indonesia

Corresponding author : maimunah@ummgl.ac.id

**Abstract**. Rhizome plants are one type of herbal medicinal plants which have various types and benefits. To find out the types of rhizome plants can be done by the classification process. Classification of rhizome plants can be done by identifying the color of the flesh of the rhizome. Identification technology in this study is needed to accelerate and facilitate the process of identifying rhizome species in the form of digital image data. Digital image processing can be used to identify rhizomes based on color using the support vector machine (SVM). In this study, the type of rhizome used as an input image was temulawak, temu hitam and temu mangga. Image preprocessing is done to get the color image of the rhizome flesh which is then used in the feature extraction stage. Image feature extraction process is done to get the color characteristics of the image by calculating the image RGB value. The image color value obtained has been used to identify rhizome species using SVM. The results obtained indicate that the identification of rhizome types using SVM provides an accuracy of 87.5%.

## 1. Introduction

Very famous medicinal plants are used as raw materials for herbal medicine and traditional medicine. Jamu is the original cultural heritage of the Indonesian nation which has been passed down from generation to generation. Also, it cannot be denied that herbal medicine is a very potential national asset so it should be developed into a superior and useful health commodity. To support herbal medicine as a health commodity, the government through the Minister of Health together with the Indonesian Coordinating Minister for Human Development and Culture inaugurated the Fit and Jamu Movement. The Ministry of Health supports and encourages the Herbal Medicine Movement for health and fitness. Besides, the Ministry of Health has developed a science of herbal medicine which is a service-based research program to obtain scientific evidence to get quality and nutritious herbs. A large number of herbal medicinal plants and the lack of public knowledge about the types and herbal medicinal plants make it difficult for the community in terms of distinguishing the types of herbal medicinal plants so that many people prefer to use chemical drugs. To provide information to the public, an herbal medicinal plant recognition system is needed that is capable of identifying and introducing herbal medicinal plants. The information obtained is in the form of a digital image which is then analyzed and processed by the system. The system identifies leaf images from herbal medicinal plants and performs recognition of a pattern or characteristics of the object [1]. Indonesia is very famous for its abundant natural wealth and very useful in pharmacotherapy. Several variables affect the decision to purchase herbal medicine in Indonesia. Social system and herbal characteristics showed a significant influence on the decision to purchase medicinal herbs, on the other hand, demographics and communication system shows no positive influence [2]. Identification of herbal plants based on leaf images using texture analysis has been carried out on 10 species. Texture analysis was performed using GLCM and KNN classification. The results showed the accuracy of identification using the 9-fold cross-validation method reached 83.33% using 9 subsets [3]. Classification of medicinal plants is not only based on color characteristics. Another feature that can be used for classification of medicinal plants is the texture feature. Classification of medicinal plant leaves based on texture feature using grey tone spatial dependency matrices method and Local Binary Pattern operators. The Classification that has been done obtained an accuracy of 94.7% [4]. Identification of medicinal plants in India has been carried out using neural networks. The features used for identification are the edge, area, and color of the leaf images of Indian medicinal plants. The classification result gives an accuracy of 75% [5]. Rhizome plants are very diverse and for each type has many varieties as well, an example is a turmeric. Classification of the type of turmeric has been carried out using neural networks. The types of turmeric studied were the samba refinery, samba verali and salmon verali. Classification of the type of turmeric based on histogram and morphology characteristics with an accuracy of 73.33% [6]. The characteristic texture of herbal plants can be classified by using support vector machines on herbal plants. Textural characteristics used include a histogram of oriented gradients (HOG), Local Binary Pattern (LBP) and Speed-Up Robust Features (SURF). The classification produces the best accuracy for HOG and LBP texture characteristics. SURF is more sensitive to rotation and illumination than HOG and LBP [7]. In this study, the classification of rhizome plants which includes temulawak, temu ireng, and temu mangga based on color feature.

## 2. Methodology

Details of the research stages are presented in Figure 1.

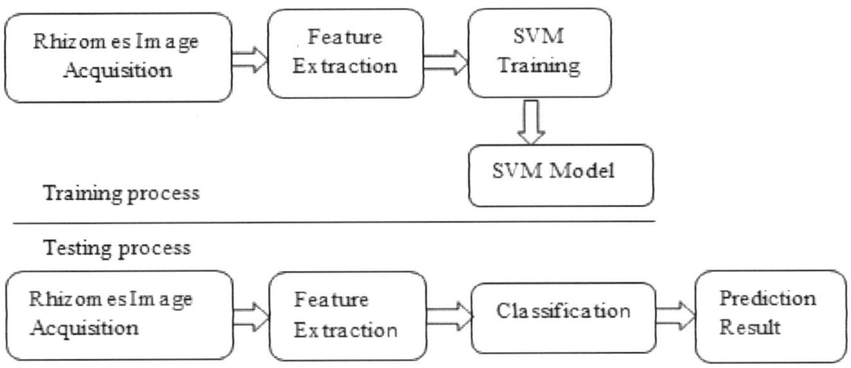


Figure 1. Detail of classification process

The stages of the research include :

### *2.1 Image acquisition*

The data used in this study were temulawak, temu putih and temu ireng. The image is taken using an 18MP DLSR camera with a distance of 30 cm. Image data is stored in the .jpeg extension format in the size 1296 x 864.

### *2.2 Feature Extraction*

In this stage, the RGB color feature extraction is carried out from the rhizome image. Before feature extraction is carried out, the image of the rhizome is first carried out by cutting the image of the rhizome by taking part in the object which states the color of the rhizome's flesh. Then the RGB color feature extraction from the rhizome image that has been extracted by calculating the average value of the entire pixel. The average RGB value for each component R, G, and B is obtained from the number of each component divided by the total number of pixels. The average RGB value of the entire pixel is normalized by dividing each value by the number 255 and stored as the RGB parameter which then becomes the input for the classification process.

*2.3 Classification*
At the classification stage, the method used is multiclass SVM. By using the tree method, first, compare temu ireng and temulawak. The winner between temu ireng and temulawak is compared to temu mangga. To solve this problem, an equation is made between the temu ireng and temulawak, temu ireng and temu mangga, temulawak and temu mangga.

## 3. Result and Discussion
The data used in this research are the image of temu ireng, temulawak and temu mangga obtained from the market with a total of 40 images each. The image is taken using a mini studio using a Canon EOS 600D camera and the image capture distance is 30 cm. The image capture process is carried out as in Figure 2 with the results of taking the image of the rhizome as in Figure 3.

Figure 2. Image acquisition process

(a) (b) (c)

Figure 3. Temu ireng(a), temulawak(b), temu mangga (c)

Rhizomes image used consisting of temu ireng,temulawak and temu mangga which has been obtained is then performed cropping to get the flesh of the findings only without any background objects. In this study, the classification process was carried out using Matlab software. The feature used for the classification of temu ireng, temulawak and temu mangga are color feature using RGB color space. Input images are calculated pixel values and normalized R, G and B values have a range of values between 0 and 1. Normalized values are expressed in variables r, g, and b which will then be used as input in the classification stage. The results of feature extraction are presented in table 1.

**Table 1**. Color feature extraction

| | r | g | b |
|---|---|---|---|
| Temu Ireng | 0.738852 | 0.384302 | 0.009729 |
| | 0.794535 | 0.436512 | 0.00824 |
| | 0.757932 | 0.409909 | 0.015808 |
| Temulawak | 0.768538 | 0.638383 | 0.075956 |
| | 0.713804 | 0.598794 | 0.043371 |
| | 0.773416 | 0.659666 | 0.059808 |
| Temu Mangga | 0.669368 | 0.670359 | 0.269494 |
| | 0.59195 | 0.600923 | 0.247807 |
| | 0.696839 | 0.681288 | 0.241195 |

To be able to classify temu ireng, temulawak and temu mangga, data learning is needed. In this study, training data are used in several of 32 pieces for each class. The multiclass SVM learning process is done using the tree method. After the training process is carried out, the testing process is done to get the classification results. The test data used were 8 pieces for each class. The classification process is presented in Figure 4.

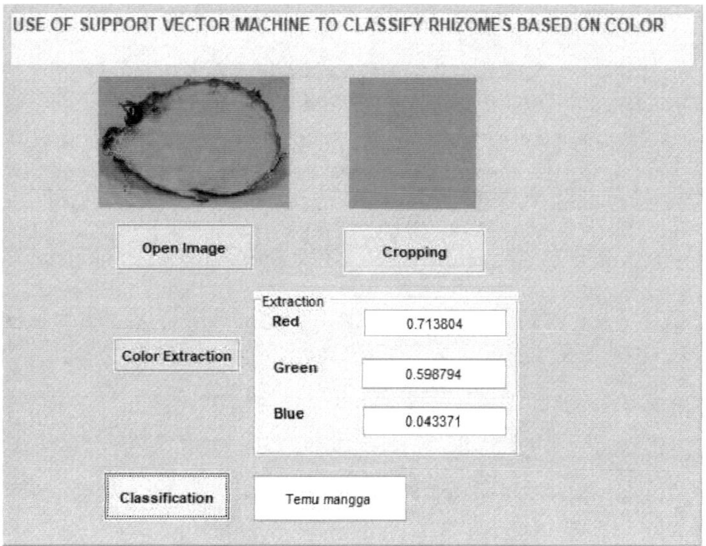


Figure 4. Rhizomes classification

The test results obtained are presented in the confusion matrix as in table 2.

**Table 2**. Confusion matrix

| | | Prediction | | |
|---|---|---|---|---|
| | | Temu Ireng | Temulawak | Temu Mangga |
| Actual | Temu Ireng | 7 | 1 | 0 |
| | Temulawak | 0 | 6 | 2 |
| | Temu Mangga | 0 | 0 | 8 |

## 4. Conclusion
Research has been carried out to classify rhizome plants which include temu ireng, temulawak and temu mangga. The classification of the types of findings is done because given the benefits of the types of findings are very many and on the other hand, have the same feature. One distinguishing characteristic is the color. The Classification has been carried out on types of rhizome plants using SVM based on RGB color. The test results obtained an accuracy of 87.5%. Further research is needed to identify temu ireng, temulawak and temu mangga based on odor feature.

**Acknowledgments**
The author of this paper would like to acknowledge great support and encouragement by LP3M Universitas Muhammadiyah Magelang for providing all the equipment used for Penelitian Revitalisasi Visi Institusi (PRVI) 2019 scheme.

## References
[1] Kementrian Kesehatan RI D J B K dan A K 2015 Informasi kefarmasian dan alat kesehatan **1**
[2] P. Kautsar A, Ayunovani F. S. M and Surahman E 2015 The Influence of Demographic, Social System, Communication System, and Herbal Characteristics on Purchase Decisions of Herbal Medicine in Indonesia *J. Econ. Bus. Manag.* **4** 235–8
[3] Ni'mah F S, Sutojo T and Setiadi D R I M 2018 Identifikasi Tumbuhan Obat Herbal Berdasarkan Citra Daun Menggunakan Algoritma Gray Level Co-occurence Matrix dan K-Nearest Neighbor *J. Teknol. dan Sist. Komput.* **6** 51
[4] Arun C H, Sam Emmanuel W R and Christopher Durairaj D 2013 Texture Feature Extraction for Identification of Medicinal Plants and Comparison of Different Classifiers *Int. J. Comput. Appl.* **62** 975–8887
[5] Aitwadkar P P, Deshpande S C and Savant A V 2018 Identification of Indian Medicinal Plant by using Artificial Neural Network *Int. Res. J. Eng. Technol.* **05** 1669–71
[6] Kaur A, Saini N, Kaur R and Das A 2016 Automatic classification of turmeric rhizomes using the external morphological characteristics *Int. Conf. Adv. Comput. Commun. Informatics, ICACCI 2016* 1507–10
[7] Ibrahim Z, Sabri N and Mangshor N N A 2018 Leaf recognition using texture features for herbal plant identification *Indones. J. Electr. Eng. Comput. Sci.* **9** 152–6

# Design Non-Invasive of Blood Sugar Detector Prototypes Using Cellular Technology GPS-Based

**G Santoso*, S Hani, S Kristiyana and Y A Saputra**
Electro Departement, IST Akprind Yogyakarta
[1]gatsan@akprind.ac.id*

**Abstract.** Patients with diabetes mellitus need regular blood sugar levels. This research aims to design a blood sugar level detector without taking blood samples so that it will improve the comfort of patients with diabetes mellitus as long as the measurement process has been completed and the measurement results can be sent to the doctor or nurse's cellular communication device and the ambulance unit using cellular technology.
The sensor used is the IR204 infrared LED to transmit infrared light with a wavelength of 850-940 nm and use PD204 photodiode for the receiver. Arduino Nano to be designed control all activities in the system, while for data communication using the GSM SIM800L module and the GPS module that is on a cell phone device. This tool is designed and calibrated to be able to measure blood glucose levels with a monitoring system that will later be measured and the location of the patient sent to the doctor's cell phone by SMS format, the location diabetes mellitus patients obtained by this tool has an average distance of 3,8 meters from five times testing.
The minimum measurements for all four samples with non-invasive techniques at 100 mg/dl blood sugar levels were measured with a minimum sensor reading of 65.600 and at a maximum sensor of 94.400, also at a blood sugar level of 102 mg/dl measured with a minimum sensor reading of 72.000 at a maximum sensor of 94.400.
The novelty of this research lies in the measurement of blood sugar levels carried out in a non-invasive manner and utilizing GPS-based cellular technology on the prototype that was developed. Based on the accuracy and sensitivity values it can be concluded that a non-invasive blood glucose detector using an infrared IR204 LED sensor has been made to be used as an alternative tool for measuring blood sugar levels for diabetics so that the patient can control himself.

## 1. Introduction

Facts show that up to now there are still many diabetics who are still unable to control their blood sugar levels. This problem denies more complicated effects. Therefore, diabetics are advised to always make an assessment of their blood sugar levels using invasive methods. Measurement of invasive blood sugar levels requires a blood sample in which diabetics will be pricked by one finger to agree to use the patient's blood sample and administration of blood sugar levels. This method of measurement causes sufferers to require uncomfortable treatment and do not use advice to measure their blood sugar levels routinely.

[1] Dino, S., in his research conducted measurements of blood glucose using a series of amplifiers and voltage dividers synchronized with LED (Light Emitting Diode) infrared and photodiode. Also

developing techniques to measure blood sugar levels in a non-invasive way using 940 nm infrared LED waves, 940 nm photodiode and Arduino nano. In this research, a set-up of non-invasive blood sugar level research will be developed using Labview-based microwaves sensors that are equipped with measurement data storage so that diabetics can control themselves.

[2] Syed M. Usman Ali et al in 2011 created a system that shows remote monitoring of glucose levels with the existing GPRS / GSM network structure using ZnO which has enabled the nanowire sensor array to be integrated with standard cellular phones available. Next research from [3] Sali et al in 2013 discussed microcontroller-based heart rate monitoring using an LCD, buzzer, keypad and GSM module as an interface. Reference is also found in a study conducted by [4] K.A Unnikrishna Menon in 2013, namely a survey on non-invasive blood glucose monitoring using NIR.

[5] Abidin, et al in 2013 also conducted a study using non-invasive methods to measure blood glucose concentration levels. By implementing two infrared rays with different wavelengths; 940nm and 950nm based on the use of light-emitting diodes and measuring transmittance through a solution of distilled water and d-glucose concentrations from 0 mg / dL to 200 mg / dL using a 1000 nm photodiode.

[6] R. A Buda and Addi, M.Mohd in 2014 made a blood glucose monitoring device which is done non-invasively by using an infrared sensor. In addition to being able to detect glucose concentrations in the blood, this device is also capable of displaying the required glucose levels and insulin doses, according to the body mass index of the user. The development of the tool was also made by [7] Boby Irfanudin Anwar, Raditiana Patmasari, and Hilman Fauzi in 2016 with the whole system controlled by the Wemos D1mini device. For IR, RED LED, and Photodetector all are integrated into the Nellcor DS-100a oxymeter sensor

[8] Junian Wicaksono in 2017 designed a tool for measuring blood glucose levels in a non-invasive way using a range of different sensors from previous studies based on near infrared, photodiode and arduino.

Continuing the previous research, a non-invasive blood sugar detector will be developed using IR204 infrared LED sensor, PD204 photodiode and GPS-based Cellular Technology so that diabetics can control themselves and the measurement results can be sent to the communication device. doctors or nurses' cellular phones and ambulance units using cellular technology.

## 2. Method

The system design includes the working principle of a non-invasive blood sugar detector that is equipped with GPS-based cellular technology can be described in Figure 1. First turn on the detector device by pressing the ON switch, and on the first LCD display the name will appear. After that the command will appear to check the glucose level in the blood indicating the device is ready. Then one of the fingers is inserted into the sensor that has been provided. The device will start detecting blood sugar levels after pressing the check button once. After the patient's GPS location and blood sugar levels are detected, the device sends information to the doctor's cell phone with the command to press the check button twice vulnerable at one pressure per second.

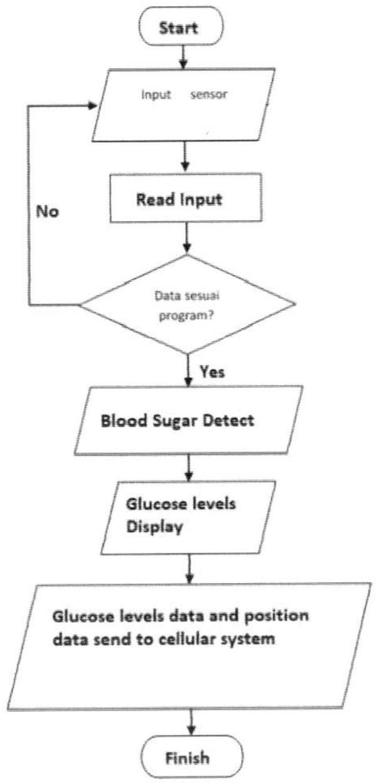


**Figure 1.** Flow diagram of non-invasive blood sugar level detector works using cellular technology based on GPS

The system designed in this research is a device that uses IR204 infrared LED sensor and PD204 photodiode, two Arduino nano, GPS module and GSM SIM800L module. The working principle of this tool can be shown by Figure 2.

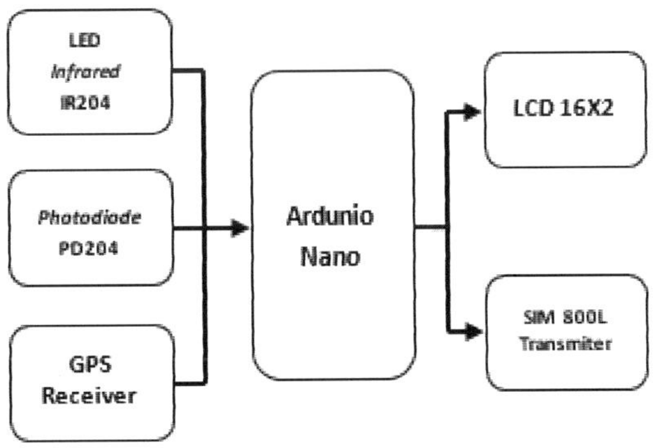


**Figure 2.** Flow diagram of the working principle of a non-invasive blood sugar level detector using GPS-based cellular technology

The design of a blood sugar level detector for diabetic patients is composed of several circuit blocks, namely Arduino as a controller, a sensor circuit as a detector of blood sugar levels and as a feedback value on the system, GPS module as a determinant of coordinates, and cellular transmitter module as an intermediary for communication to cellphone. From the block diagram the working principle of the tool will then be assembled in a schematic series of tools that will be designed using the proteus version 8.0 application. The schematic of the device made can be seen in Figure 3.

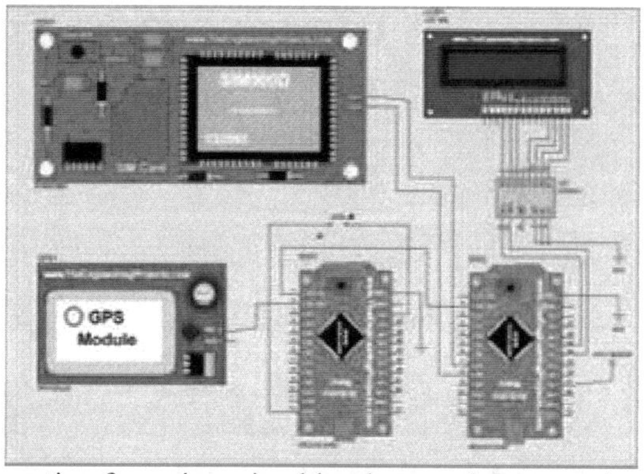


**Figure 3.** Circuit schematic of non-invasive blood sugar detector using a GPS-based cellular technology

## 3. Discussion

This sensor circuit analysis is intended to determine the working circuit by divided into three parts which are then assembled into a sensor circuit. A voltage divider circuit consisting of two 200 ohm and 56-ohm resistor components coupled with the basic theory of calculation with an input voltage of 5 volts, and the desired voltage to flow into the next circuit of 1 volt. so that for the calculation of the output voltage from the basic theory, a voltage of 1,094 volts is generated which flows in the next circuit to the LED photodiode. Low pass filter circuits filter high or high signals and pass low or low signals. Obtained the value of the frequency that is passed is 1592,35668 Hz according to the results of research conducted by Agung Junian Wicaksono, 2017. The non-inverting amplifier circuit uses OP07 which is an IC (Integrated Circuits) with low noise or low noise. the voltage entering the Op-Amp input V is 0.89 volts to 0.92 volts. Calculations of the Op-Amp circuit produce a measured voltage at the Op-Amp output of 0.40 volts to 0.45 volts.

In the analysis phase of the results of the fingers experiment aims to determine the value of the calibration results obtained from the invasive method and then do a comparison of quantities in accordance with the theoretical basis and direct application carried out on the voltage divider circuit, low pass filter, photodiode, infrared, and operational amplifier.

The data in the form of graphs were taken from the results of the trial of the first finger measurement on the fingers of a man aged 22 years with a height of 159 cm and a body weight of 72 kg. Readings from blood sugar measurement data that shows the results of 100 mg/dl as Figure 4.

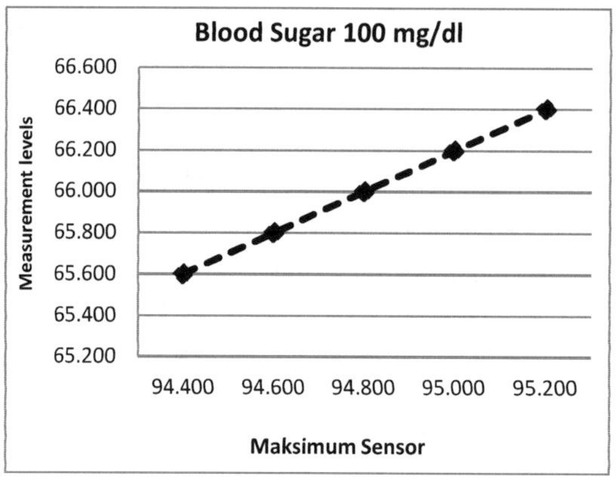


**Figure 4.** Experiment graph 1

Figure 4 it can be analyzed that measurement of blood sugar with an instrument using an invasive technique results in 100 mg / dL and if measured with a previously calibrated device using a non-invasive method with the same result that is 100 mg / dl the sensor parameter reading shows from the bottom rank determined from the maximum sensor value measurement of 94,400 obtained a sensor value of 65,600, while for the middle rank with a maximum sensor value of measurement of 94,800 obtained a value of 66,000, and the last in the upper rank with a maximum sensor value of measurement of 95,200 obtained a value of 66,400.

The results of the data in the form of graphs taken from the results of the experimental measurements of the 2nd finger experiment on the fingers of a man aged 28 years with a bandan height of 170 cm and weighing 98 kg. Readings from blood sugar measurement data that shows the results of 93 mg/dl as Figure 5.

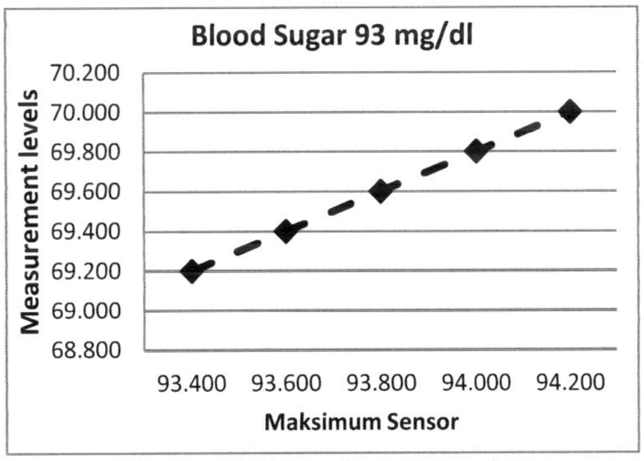


**Figure 5**. Experiment graph 2

Figure 5 it can be analyzed that measurement of blood sugar with an instrument using an invasive technique results in 93 mg / dl and if measured with a previously calibrated device using a non-invasive method with the same result that is 93 mg / dl the sensor parameter reading shows from the bottom rank determined from the maximum sensor value measurement of 93,400 obtained a sensor value of 69,200, while for the middle rank with a maximum sensor value of measurement of 93,800 obtained a value of 69,600, and finally in the top rank with a maximum sensor value of measurement of 94,200 obtained a value of 70,000.

The results of the data in the form of graphs taken from the results of measurements of the trial of the 3rd finger on the fingers of a 44-year-old woman with a bandan height of 158 cm and weighing 99 kg. Readings from blood sugar measurement data that show the results of 102 mg /dl as Figure 6.

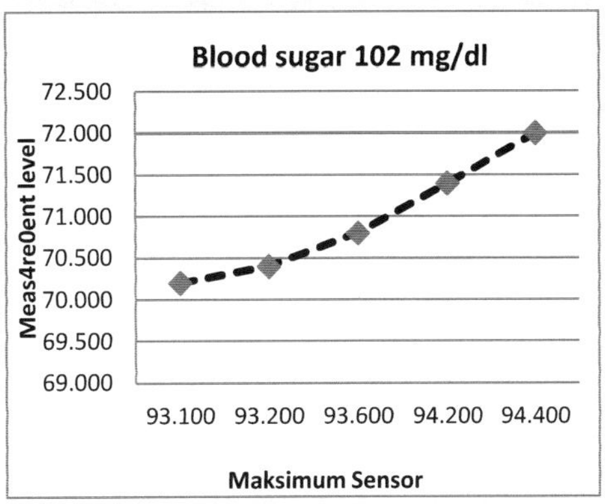


**Figure 6.** Experiment graph 3

Figure 6 it can be analyzed that the measurement of blood sugar using an instrument using an invasive technique yields a yield of 102 mg / dl and if measured with a previously calibrated device using a non-invasive method with the same result that is 102 mg / dl the sensor parameter reading shows from the bottom rank determined from the maximum sensor measurement value of 93,100 obtained a sensor value of 70,200, while for the middle rank with a maximum sensor value of measurement of 93,600 obtained a value of 70,800, and the last at the top rank with a maximum sensor measurement value of 94,400 obtained a value of 72,000.

The results of the data in the form of graphs taken from the results of measurements of the 4 finger experiments on the fingers of a woman aged 18 years with a bandan height of 165 cm and weighing 94 kg. Readings from the blood sugar measurement data showing 114 mg / dl as Figur 7.

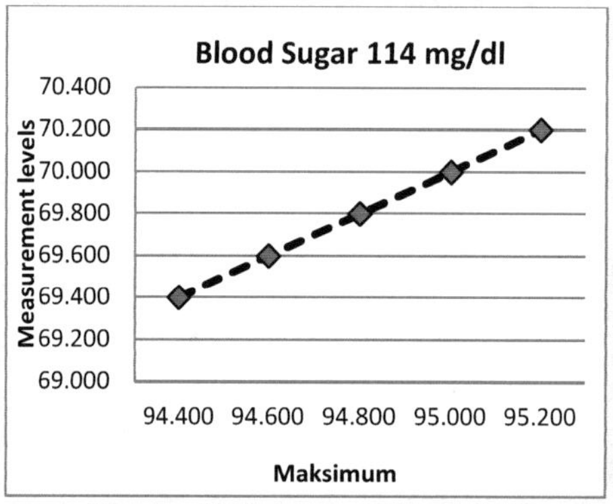


**Figure 7**. Experiment graph 4

Figure 7 it can be analyzed that the measurement of blood sugar with a device using an invasive technique results in 114 mg / dl and if measured with a previously calibrated device using a non-invasive method with the same result that is 114 mg / dl the sensor parameter reading shows from the bottom rank determined from the maximum sensor value measurement of 94,400 obtained a sensor value of 69,400, while for the middle rank with a maximum sensor value of measurement of 94,600 obtained a value of 69,600, and finally in the top rank with a maximum sensor value of measurement of 95,200 obtained a value of 70,200.

Minimum measurements for all four samples by comparison of invasive techniques with non-invasive techniques on blood sugar measured 100 mg / dl with a minimum sensor reading of 65,600 at a maximum of 94,400 sensors and 102 mg / dl with the highest reading of 72,000 sensors at a maximum of 94,400 sensors.

Testing of GPS data retrieval on this tool aims to retrieve coordinate or location data from the GPS module that is implanted in this blood sugar level detector. From the coordinate point that was previously determined using the google maps application then the results of the coordinates are compared with the results of the coordinates reading from the GPS device, following the results of the experiment.

**Table 1.** Difference in the reading distance of coordinates that have been determined with GPS data on the device

| Test | Fixed Coordinate | | Measurement | | Difference Distance (m) |
|---|---|---|---|---|---|
| | Long. | Lat. | Long. | Lat. | |
| 1. | 110.356098 | -7.788126 | 110.356101 | -7.788121 | 1 |
| 2. | 110.356094 | -7.788137 | 110.356101 | -7.788093 | 6 |
| 3. | 110.355138 | -7.779813 | 110.355085 | -7.779808 | 3 |
| 4. | 110.393616 | -7.763098 | 110.393637 | -7.763085 | 2 |
| 5. | 110.356353 | -7.787228 | 110.356356 | -7.787124 | 7 |
| | | Mean | | | 3.8 |

Testing of GPS data retrieval on this tool aims to retrieve coordinate or location data from the GPS module planted in this glucose detector tool. From the coordinates previously determined using the Google Maps application, the results of the coordinates are compared with the results of the coordinates from the GPS device. The results of the test as much as 5 times is getting an average of 3.8 meters from the tool location coordinates.

**Table 2.** Time of sending and receiving SMS

| Test | Time (s) | |
|---|---|---|
| | Sending | GPS Tracking |
| 1. | 10 | 41 |
| 2. | 8 | 6 |
| 3. | 9 | 6 |
| 4. | 8 | 7 |
| 5. | 8 | 6 |
| Mean | 8.6 | 13.2 |

Experiments above the GSM module can send SMS messages received by mobile phones in 10 seconds and the fastest time to send SMS messages is 8 seconds. However, the analysis conducted was in a good signal condition because the speed of sending messages depends on the signal quality of the sim card operator.

GPS module location tracking gets vulnerable for a long time only when the device is turned on for the first 41 seconds and for subsequent use at the same location GPS module location tracking gets the fastest time of 6 seconds, research is carried out in the open.

## 4. Conclusion

The design of the device consists of several circuits using GPS-based cellular technology which consists of several circuits using two Arduino Nano supported by a series of sensors (Photodiode and Infrared), SIM800l module, NEO-6M GPS module, and a power supply circuit (two lithium batteries) -ion) with an input voltage of 4.16 V DC changed by the step up module to 5.02 V DC and a step down module used to supply the transmitter driver (SIM800l) with a voltage of 3.72 V DC. This tool can provide measurement results in real time, at a minimum measurement for the four samples with a comparison of invasive and non-invasive techniques at a blood sugar level of 100 mg / dl measured with a minimum sensor reading of 65,600 at a maximum sensor of 94,400 while at a blood sugar level of 102 mg / dl the highest sensor reading is 72,000 at a maximum sensor of 94,400. This tool can provide coordinates or the location where the patient with diabetes is located that has an average distance difference of 3.8 meters from the location of the device with GPS tracking speed outside the room counted 14 seconds and the fastest counted 6 seconds, while the long GSM transmitter sends SMS messages received the cell phone counts 10 seconds and fast delivery counts 8 seconds. Based on the accuracy and sensitivity value, the non-invasive blood sugar level detector based on GPS technology that has been made can be used as an alternative tool for measuring blood sugar levels for diabetics so that patients can control themselves and can be monitored directly by a doctor.

## 5. Acknowledgment

On this occasion the researcher thanked LPPM Institut Sains & Teknologi AKPRIND Yogyakarta, so that this researcher could be completed well.

## 6. References

[1] P.Muhamaduyasic, K.M.Gopinath, K.Rohini and Sukanesh D (Mrs). R 2018 a Non-Invasive Blood Glucose Monitoring Device Using Red Laser Light 78–82

[2] Usman Ali S M, Aijazi T, Axelsson K, Nur O and Willander M 2011 Wireless remote monitoring of glucose using a functionalized ZnO nanowire arrays based sensor *Sensors* **11** 8485–96

[3] Sali S, Durge P, Pokar M and Kasge N 2017 Microcontroller Based Heart Rate Monitor *Int. J. Sci. Res.* 2–6

[4] Unnikrishna Menon K A, Hemachandran D and Abhishek T K 2013 A survey on non-invasive blood glucose monitoring using NIR *International Conference on Communication and Signal Processing, ICCSP 2013 - Proceedings* pp 1069–72

[5] Abidin M T B Z, Rosli M K R, Shamsuddin S A, Madzhi N K and Abdullah M F 2013 Initial quantitative comparison of 940nm and 950nm infrared sensor performance for measuring glucose non-invasively *2013 IEEE International Conference on Smart Instrumentation, Measurement and Applications, ICSIMA 2013*

[6] Buda R A and Addi M M 2014 A portable non-invasive blood glucose monitoring device *IECBES 2014, Conference Proceedings - 2014 IEEE Conference on Biomedical Engineering and Sciences: "Miri, Where Engineering in Medicine and Biology and Humanity Meet"* (Institute of Electrical and Electronics Engineers Inc.) pp 964–9

[7] Anwar B I and Patmasari R 2016 Perancangan dan Implementasi Alat Pengukur Kadar Glukosa dalam Darah Secara Non-Invasive Berbasis Arduio *e-Proceeding Eng. Vol.3, No.3 December 2016* **3** 4665–8

[8] Santoso G, Hani S and Wicaksono A 2018 Rancang Bangun Prototipe Detektor Glukosa Darah Secara Non-Invasive Menggunakan Near Infrared *Simp. Nas. RAPI XVII– 2018 FT UMS* 37–42

# The impact of geogebra classic application on learning geometry

**D Triwahyuningtyas*, S Rahayu, and W D Agustin**
Elementary School Teacher Education Study Program, Universitas Kanjuruhan Malang

*dyahtriwahyu@unikama.ac.id

**Abstract**
This study aimed to identify the impact of Geogebra Classic application towards geometry learning outcome of fifth graders. The method employed was experimental research using Non-equivalent Control Group Design. Serving as the subjects of this study were students of fifth grade in an elementary school in Malang, Indonesia. The data were required through statistical tests and analyses through SPSS 21.0 for Windows. The results of the study showed that the average difference of the score of fifth graders after having treatment of using the application were 66.4 for the pre-test and 81.2 for the post-test. In the meantime, the t-test on the study revealed that the value was $0.002 < 0.05$ meaning that the hypothesis was accepted. This indicated that there is an impact of the use of Geogebra Classic application on geometry learning outcome of the fifth graders. Therefore, it can be concluded that Geogebra Classic application was able to help students understand geometry-related learning materials.

## 1. Introduction

Geogebra is a software combining geometry, algebra, and calculus possibly representing a variety of mathematical things [1–3]. On Geogebra, such things as points, vectors, and lines of a cone, for instance, can be displayed in a dynamic graphical representation [4,5]. This characteristic enables various mathematical concepts to move back and forth [6]. The software also offers several interactive alternative applications for teaching mathematics [4].

There have been several studies discussing Geogebra software in learning mathematics. It has also been proven that the software is a good learning medium for effective learning [7]. In addition, it has been also found that Geogebra is able to help students understand basic concepts and contents of mathematics [8,9]. Geogebra has been proven effective in developing interactive learning texts (eBooks) with the most advanced technology [10] and integral topics [11]. In middle schools, Geogebra has been identified to be influential towards learning mathematics [12]. However, there is limited number of studies proving that Geogebra is effectively used in learning mathematics in elementary schools.

Journal of Physics: Conference Series **1381** (2019) 012033 doi:10.1088/1742-6596/1381/1/012033

In elementary schools, introduction to geometry usually starts with the characteristics of the important concepts and parts of a certain region of space [13]. A region of space consists of a couple of points, lines, and planes or surfaces [14]. Integrating Geogebra will possible encourage students to reflect how to find the parts of a geometry [15]. Geogebra is also able to help students investigate and represent geometry [16]. This study focuses on identifying the impact of the use of Geogebra Classic application towards geometry learning outcome of fifth graders. This is due to the fact that elementary school students still have difficulties in manually determining the parts of geometry.

## 2. Method

The study was conducted in an experimental method by collecting quantitative data using statistical analyses [17]. The subjects were 50 fifth graders in an elementary school in Malang, Indonesia, which were divided into two groups namely experimental and control groups. The experimental group received a treatment which was the use of Geogebra Classic application during learning sessions while the control group used Microsoft Power Point during the learning session. Thus, the research design used in this study was Nonequivalent Control Group Design in an attempt to find out the impact of the application on geometry learning outcome.

To collect the data needed, this study used a test as its research instrument. The type of test used was an achievement test where it investigated the learning outcome by comparing the results of pre-test and those of post-test. The pre-test was administered to show students' initial knowledge and ability and the post-test aimed to identify the results after they had received the treatment. The data collected were then analyzed through statistical analyses using SPSS 21.0 for Windows.

### Geogebra Classic

Geogebra Classic is used as a presentation media to illustrate various mathematical concepts [9]. Geogebra Classic provides a variety of mechanisms for learning mathematics that are taught and studied in class [1]. In elementary school students Geogebra Classic is presented in the form of videos
In this study, the Geogebra Classic application was implemented on learning a cube, a beam, and a triangular prism. Figure 1 shows how the application works on explaining a cube.

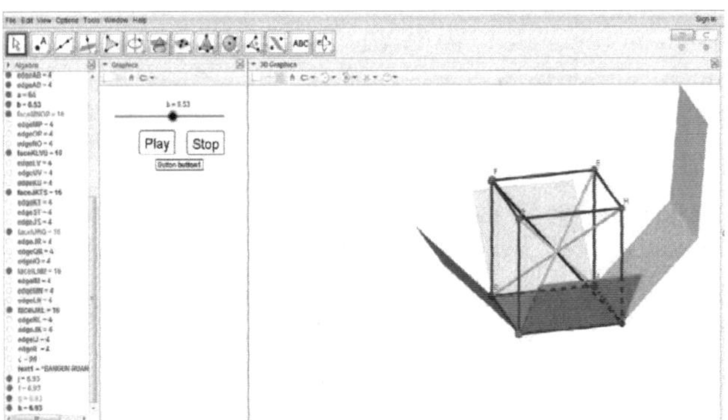


**Figure 1.** How a cube is formed using Geogebra Classic application

The fifth graders did not directly learn how to build a cube using Geogebra Classic, yet they had to identify how the cube is formed using the application. Presentation in the form of slow motion video displays the parts of the building that includes: vertex, ribs and diagonal space. So that through the Geogebra Classic application, students can describe the built webs of geometry.

## 3. Results and Discussion

The results of the study showed that the learning outcome of the experimental group was lower in the pre-test (66.4) in comparison with that of the control group (72.4). However, in the post-test, the experimental group outperformed the control group in terms of the learning outcome average score (81.2 for the experimental group and 78.4 for the control group). In the meantime, the results of the data normality distribution test of both the pre-test and post-test showed significant average (0.085) meaning that the data were normally distributed (> 0.05). The results of the data homogeneity test on both the pre-test and post-test also showed significant average (0.275) indicating that the data were homogeneous (>0.05). To find out how influential the treatment on both the groups, this study analyzed the results of the data distribution normality and data homogeneity tests using an independent sample t-test using SPSS 21.0 for Windows. The results are shown in Table 1.

**Table 1.** The results of hypothesis test using SPSS 21.0

| Independent Samples Test | | | | | | | |
|---|---|---|---|---|---|---|---|
| | | Levene's Test for Equality of Variances | | t-test for Equality of Means | | | |
| | | F | Sig. | t | Df | Sig. (2-tailed) | Mean Difference |
| Pretest Posttest | Equal variances assumed | 2,165 | ,148 | -3,304 | 48 | ,002 | -14,80000 |
| | Equal variances not assumed | | | -3,304 | 45,682 | ,002 | -14,80000 |

The results of the hypothesis test on Table 1 indicates that the significance of Levene's Test for Equality of Variances is 0.148 > 0.05 meaning that the variance between the pre-test and the post-test were homogeneous or similar. The significance value of the 2-tailed test also implied that the hypothesis is accepted (0.002 < 0.05). Decision making in the translation of the Independent Samples hypothesis test can be concluded that the hypothesis is accepted. This means that the use of Geogebra Classic application has a significant impact on learning geometry for fifth graders in Malang, Indonesia.
Based on the data analysis, Geogebra Classic encourages students to reflect on how a region of space is formed [15] and to thoroughly analyze a series of processes of forming a certain geometrical space [18] through an exploration phase [19]. The application also proves that learning geometry is fun [20].

## 4. Conclusion

Geogebra Classic was applicable in learning mathematics, particularly in learning geometry. The use of the application is also able to help students find the concepts of how a certain geometrical space is form. Based on the results of the Independent Samples t test or the average 2 difference test used to test two averages on two independent data groups using SPSS 21.0 For Windows, it was found that the use of Geogebra Classic application was influential towards fifth graders' learning outcome on geometry so that it is suggested that teachers use the application on learning geometry.

## References

[1] Hall J and Chamblee G 2013 Teaching Algebra and Geometry with GeoGebra: Preparing Pre-Service Teachers for Middle Grades/Secondary Mathematics Classrooms *Comput. Sch.* **30** 12–29

[2] Poon K K 2018 Learning fraction comparison by using a dynamic mathematics software–GeoGebra *Int. J. Math. Educ. Sci. Technol.* **49** 469–79

[3] Lavicza Z and Papp-Varga Z 2010 Integrating GeoGebra into IWB-equipped teaching environments: Preliminary results *Technol. Pedagog. Educ.* **19** 245–52

[4] Caligaris M G, Schivo M E and Romiti M R 2015 Calculus & GeoGebra, an Interesting Partnership *Procedia - Soc. Behav. Sci.* **174** 1183–8

[5] Hernández A, Perdomo-Díaz J and Camacho-Machín M 2019 Mathematical understanding in problem solving with GeoGebra: a case study in initial teacher education *Int. J. Math. Educ. Sci. Technol.* **5211**

[6] Attorps I, Björk K and Radic M 2016 Generating the patterns of variation with GeoGebra: the case of polynomial approximations *Int. J. Math. Educ. Sci. Technol.* **47** 45–57

[7] Arbain N and Shukor N A 2015 The Effects of GeoGebra on Students Achievement *Procedia - Soc. Behav. Sci.* **172** 208–14

[8] Caglayan G 2014 Static Versus Dynamic Disposition: The Role of GeoGebra in Representing Polynomial-Rational Inequalities and Exponential-Logarithmic Functions *Comput. Sch.* **31** 339–70

[9] Aktumen M and Bulut M 2013 Teacher candidates' opinions on real life problems designed in GeoGebra software *Anthropologist* **16** 167–76

[10] Radović S, Radojičić M, Veljković K and Marić M 2018 Examining the effects of Geogebra applets on mathematics learning using interactive mathematics textbook *Interact. Learn. Environ.* **0** 1–18

[11] Tatar E and Zengin Y 2016 Conceptual Understanding of Definite Integral with GeoGebra *Comput. Sch.* **33** 120–32

[12] Dockendorff M and Solar H 2018 ICT integration in mathematics initial teacher training and its impact on visualization: the case of GeoGebra *Int. J. Math. Educ. Sci. Technol.* **49** 66–84

[13] Bayrak N, Yüce S and Kaan M 2014 The Investigation of the Viewpoint of Academic Staff and Graduate Students in Teaching Geometry in Elementary School *Procedia - Soc. Behav. Sci.* **116** 2115–9

[14] Yang D-C and Wang T-L 2017 A Comparative Study of Geometry in Elementary School Mathematics Textbooks from Five Countries *Eur. J. STEM Educ.* **1** 1–10

[15] Verhoef N C, Coenders F, Pieters J M, van Smaalen D and Tall D O 2015 Professional development through lesson study: teaching the derivative using GeoGebra *Prof. Dev. Educ.* **41** 109–26

[16] Zengin Y 2018 Incorporating the dynamic mathematics software GeoGebra into a history of mathematics course *Int. J. Math. Educ. Sci. Technol.* **49** 1083–98

[17] McMillan J and Schumacher S 2014 Research in Education James McMillan Sally Schumacher **7**

[18] Takači D, Stankov G and Milanovic I 2015 Efficiency of learning environment using GeoGebra when calculus contents are learned in collaborative groups *Comput. Educ.* **82** 421–31

[19] Hähkiöniemi M 2017 Student teachers' types of probing questions in inquiry-based mathematics teaching with and without GeoGebra *Int. J. Math. Educ. Sci. Technol.* **48** 973–87

[20] Zengin Y 2017 The effects of GeoGebra software on pre-service mathematics teachers' attitudes and views toward proof and proving *Int. J. Math. Educ. Sci. Technol.* **48** 1002–22

Journal of Physics: Conference Series **1381** (2019) 012034 doi:10.1088/1742-6596/1381/1/012034

# Financial feasibility analysis of Pineapple carrot juice business

**M Sari, A Nurmaydha* and D U M Rohmah**
Agroindustrial Technology, University of Darussalam Gontor, Indonesia

*agensy@unida.gontor.ac.id

**Abstract.** Fruits and vegetables are horticultural commodities that have the potential to be developed. Pineapple and carrot are superior commodities that can be developed into juice drinks. The development of this juice product needs to pay attention to the financial feasibility of the business. This study will analyze the financial feasibility analysis using NPV, IRR, BC ratio, PBP, BEP methods. Financial feasibility analysis shows the Payback Period (PP) value of 2 years 7 months, Net Present Value (NPV) of IDR 102.128.626, Internal Rate of Return (IRR) of 17,29%, and B/C ratio of 1,04. Based on the results of the financial feasibility assessment, pineapple carrot juice business is feasible to run.

## 1. Introduction

Indonesia is an agricultural country that has great potential in agriculture development. One sub-sector of agriculture that has great potential to be developed is horticulture. Fruit is one of the horticultural commodity that has the potential to be developed. According to Central Bureau of Statistics Republic of Indonesia in 2010, the production of fruits especially pineapple, star fruit and guava has increased every year. Abundant fruit production is very unfortunate if traded in fresh fruit. Therefore, the horticulture sector needs to be developed into an industry to be able to produce products that have added value, for example, made various food and beverage products, such as jams, cocktails, fruit leather, syrups, fruit juices, chips to dried fruit preserves.

Pineapple is a tropical fruit that contains Potassium, Dextrose, Sucrose (sugar cane), and Bromelain Enzyme. Bromelain has anti-inflammatory properties, helps soften food in the stomach, interferes the growth of cancer cells, inhibits platelet aggregation, and has fibrinolytic activity[1]. The fiber content can help constipation. The pineapple leaves contain calcium oxalate and pectin substances[2]. Carrots contain anti-cancer nutrients, natural sugars, pectin, glutanion, minerals, calcium, phosphorus, iron, potassium, sodium, magnesium, vitamins (beta-carotene, B1, and C), and asparagine[3]. Pineapple and carrot are very good and beneficial for the health of the body and if combined as fruit juice. The abundant amount and affordable price make these two commodities very prospective to be developed into the fruit juice industry.

According to the quality requirements of fruit juice (SNI, 01-3719-1995), fruit juice beverage is drinks made from fruit juice and water with or without the addition of sugar and food additives that are permitted. The main purpose of making fruit juice is to increase the shelf life and diversification of horticultural products to give added value. Fruit juice is very suitable and has the opportunity to be developed in Indonesia which has a tropical climate. Many research has been done with various raw materials, methods and compositions to produce fruit juice that can be produced industrially. The process of making fruit juice is using pineapple and carrot with a ratio of 1: 3 through the Pulse Electric Field (PEF) method of low temperature 50° C, so the resulting fruit juice has the aroma and taste of the original fruit. This research will analyze the financial feasibility of the fruit juice industry. Expected the result of this research can be used to establish the industrial scale of pineapple carrot juice industry.

## 2. Methodology

This study uses secondary data from previous studies. This research will analyze the financial feasibility of establishing a fruit juice industry based on the investment criteria of NPV, Net B/C, IRR, and payback period. Financial feasibility analysis will provide information on whether the fruit juice industry is financially feasible or not to be undertaken.

## 3. Results

In corporate financial planning assumption is important because, it becomes the basic for making financial projection analysis. Assumptions can be based on several sources such as, agency, data and surveys[4]. The Assumptions used in this study are based on previous data, with a literature study of research results in the same field.

**Table 1.** Production Capacity Assumptions.

| No. | Description | Unit | Value |
|---|---|---|---|
| 1 | Production per year | L/year | 312.000 |
| 2 | Carrot | IDR/Kg | 10.000 |
| 3 | Pineapple | IDR/Kg | 5.400 |
| 4 | Sugar | Kg/year | 16.000 |
| 5 | Bottle @250 ml | Unit | 1.000 |
| 6 | Mineral water (16 L) | L | 6.000 |
| 7 | LPG | Kg | 7.000 |
| 8 | Machine PEF | IDR | 150.000.000 |
| 9 | Watt | KWh | 1.300 |
| 10 | IRR | % | 13 |

**Table 2.** Revenue Assumption.

| No. | Description | Unit | Volume |
|---|---|---|---|
| 1 | Day per month | day | 25 |
| 2 | Shift per day | shift | 2 |
| 3 | Month per year | month | 12 |
| 4 | Work hours per shift | Hour | 8 |
| 5 | Capacity production | | |
| | - first year | % | 80 |
| | - second year | % | 90 |
| | - third year and so on | % | 100 |
| 6 | Carrot pineapple juice | | |

| No. | Description | Unit | Volume |
|---|---|---|---|
| | - 15 kg carrot + 13,6 kg pineapple | L | 34 |
| 7 | Production fruit juice per shift | L | 136 |
| 8 | Production per day | bottle | 520 |
| 9 | Price fruit juice per bottle | IDR | 6000 |
| 10 | Revenue fruit juice | | |
| | First year ( 80 % ) | IDR/year | 748.800.000,00 |
| | Second year ( 90 % ) | IDR/year | 842.400.000,00 |
| | Third year and so on (100%) | IDR/year | 936.000.000,00 |

**Table 3**. Investment fund.

| Project cost | Total cost (IDR) |
|---|---|
| Financial capital | 858.994.200,00 |

**Table 4**. Raw material requirements per batch.

| No. | Particulars | Quantity | Rate (IDR/kg) | Total/Batch (IDR/kg) |
|---|---|---|---|---|
| 1 | Carrot | 15 | 10.000 | 150.000 |
| 2 | Pineapple | 13,6 | 5.400 | 73.440 |
| 3 | Sugar | 5 | 16.000 | 80.000 |
| 4 | Mineral water | 1 | 6.000 | 6.000 |
| 5 | Salt | 1 | 500 | 500 |
| | TOTAL | | | 309.940 |

**Table 5**. Raw material per year.

| No | Particulars | Volume | Price (IDR) | Value/year |
|---|---|---|---|---|
| 1 | Carrot and pineapple | 17.160 | 15.400 | 264.264.000 |
| 2 | Sugar | 3.000 | 16.000 | 48.000.000 |
| 3 | Mineral water | 600 | 6.000 | 3.600.000 |
| 4 | Salt | 600 | 500 | 300.000 |
| TOTAL | | | | 316.164.000 |

**Table 6**. Labor Requirements.

| No | Personnel | Required | Salary / month | Total (IDR)/month | Total (IDR)/ year |
|---|---|---|---|---|---|
| 1 | Production manager | 1 | 2.500.000 | 2.500.000 | 30.000.000 |
| 2 | Production | 6 | 1.250.000 | 7.500.000 | 90.000.000 |
| 3 | Marketing | 1 | 1.250.000 | 1.250.000 | 15.000.000 |
| | *TOTAL* | | | 11.250.000 | 135.000.000 |

**Table 7**. Investment costs.

| No. | Description | Volume | Unit | Unit Price | Value |
|---|---|---|---|---|---|
| 1 | Building | 4.000 | m2 | 50.000 | *200.000.000* |
| 2 | Machine and tool | | | | *153.330.000* |
| | a. Pulls Electric Field | 1 | Unit | 150.000.000 | 150.000.000 |
| | b. Blender | 4 | Unit | 500.000 | 2.000.000 |
| | c. measuring cylinder | 4 | Unit | 10.000 | 40.000 |
| | d. Scale | 2 | Unit | 250.000 | 500.000 |
| | e. Stirrer | 2 | Unit | 5.000 | 10.000 |
| | f. Mass colander | 2 | Unit | 25.000 | 50.000 |
| | g. Strainer filter | 1 | Meter | 25.000 | 25.000 |
| | h. Knife | 5 | Unit | 12.000 | 60.000 |
| | i. Press | 1 | Unit | 500.000 | 500.000 |
| | j. pH meter | 1 | Unit | 145.000 | 145.000 |
| 6 | Equipment | | | | *10.000.000* |
| | a. Office tools | 1 | Unit | 10.000.000 | 10.000.000 |
| 7 | Transport | | | | *50.000.000* |
| | a. Pick up | 1 | Unit | 50.000.000 | 50.000.000 |
| 8 | Pre-investment costs | | | | *15.000.000* |
| | a. Licensing fees | 1 | Unit | 10.000.000 | 10.000.000 |
| | b. AMDAL | 1 | Unit | 5.000.000 | 5.000.000 |
| | TOTAL INVESTASI | | | | *428.330.000,00* |

Journal of Physics: Conference Series **1381** (2019) 012034 doi:10.1088/1742-6596/1381/1/012034

**Table 8**. Cost account of depreciation, maintenance and land and building tax.

| No. | Description | Investment value | Economic life | Depreciation | Maintenance (1%) | Land and building tax (2.5%) |
|---|---|---|---|---|---|---|
| 1 | Building | 200.000.000 | 20 | 9.000.000 | 2.000.000 | 5.000.000 |
| 2 | Machine and tools | 153.330.000 | 10 | 13.799.700 | 1.533.300 | |
| 3 | Equipment | 10.000.000 | 5 | 1.800.000 | 100.000 | |
| 4 | Transport | 50.000.000 | 5 | 9.000.000 | 500.000 | |
| Total | | 413.330.000 | | 33.599.700 | 4.133.300 | |

**Table 9.** Operational cost analysis.

| No | Description | Year 1 | Year 2 | Year 3 | Year 4 | Year 5 | Year 6 | Year 7 | Year 8 | Year 9 | Year 10 |
|---|---|---|---|---|---|---|---|---|---|---|---|
| **I.** | **Variable Cost** | | | | | | | | | | |
| 1 | Direct labor cost | 90.000.000 | 90.000.000 | 90.000.000 | 90.000.000 | 90.000.000 | 90.000.000 | 90.000.000 | 90.000.000 | 90.000.000 | 90.000.000 |
| 2 | Raw material carrot and pineapple | 211.411.200 | 237.837.600 | 264.264.000 | 264.264.000 | 264.264.000 | 264.264.000 | 264.264.000 | 264.264.000 | 264.264.000 | 264.264.000 |
| 3 | Ingredient | 41.520.000 | 284.547.600 | 316.164.000 | 316.164.000 | 316.164.000 | 316.164.000 | 316.164.000 | 316.164.000 | 316.164.000 | 316.164.000 |
| ***Total Variable Cost*** | | **342.931.200** | **612.385.200** | **670.428.000** | **670.428.000** | **670.428.000** | **670.428.000** | **670.428.000** | **670.428.000** | **670.428.000** | **670.428.000** |
| **II.** | **Fixed Cost** | | | | | | | | | | |
| 1 | Indirect labor cost | 45.000.000 | 45.000.000 | 45.000.000 | 45.000.000 | 45.000.000 | 45.000.000 | 45.000.000 | 45.000.000 | 45.000.000 | 45.000.000 |
| 2 | Deprecation | 33.599.700 | 33.599.700 | 33.599.700 | 33.599.700 | 33.599.700 | 33.599.700 | 33.599.700 | 33.599.700 | 33.599.700 | 33.599.700 |
| 4 | Maintenance | 4.133.300 | 4.133.300 | 4.133.300 | 4.133.300 | 4.133.300 | 4.133.300 | 4.133.300 | 4.133.300 | 4.133.300 | 4.133.300 |
| 5 | Land and building tax | 5.000.000 | 5.000.000 | 5.000.000 | 5.000.000 | 5.000.000 | 5.000.000 | 5.000.000 | 5.000.000 | 5.000.000 | 5.000.000 |
| **Total Fixed Cost** | | **87.733.000** | **87.733.000** | **87.733.000** | **87.733.000** | **87.733.000** | **87.733.000** | **87.733.000** | **87.733.000** | **87.733.000** | **87.733.000** |
| **Total Operational Cost** | | 430.664.200 | 700.118.200 | 758.161.000 | 758.161.000 | 758.161.000 | 758.161.000 | 758.161.000 | 758.161.000 | 758.161.000 | 758.161.000 |

| | | | |
|---|---|---|---|
| Investment | : | 428.330.000,00 | |
| Operational cost for 1 year | : | *430.664.200,00* | (1 year = 240 days) |
| Capital Needed | : | *858.994.200,00* | |

**Table 10.** Sensitivity of increase in raw material prices

| No. | Description | Up 1% | Year 1 | Year 2 | Year 3 |
|---|---|---|---|---|---|
| I. | Variable Cost | | | | |
| 1 | Direct labor cost | 90.000.000 | 90.000.000 | 90.000.000 | 90.000.000 |
| 2 | Raw material carrot and pineapple | 264.264.000 | 211.411.200 | 237.837.600 | 264.264.000 |
| 3 | Ingredient | 51.900.000 | 41.520.000 | 46.710.000 | 51.900.000 |
| | ***Total Variable Cost*** | **406.164.000** | **342.931.200** | **374.547.600** | **406.164.000** |
| II. | Fixed Cost | | | | |
| 1 | Indirect labor cost | 45.000.000 | 30.000.000 | 30.000.000 | 30.000.000 |
| 2 | Deprecation | 33.599.700 | 33.599.700 | 33.599.700 | 33.599.700 |
| 3 | Maintenance | 4.133.300 | 4.133.300 | 4.133.300 | 4.133.300 |
| 4 | Land and building tax | 5.000.000 | 5.000.000 | 5.000.000 | 5.000.000 |
| | **Total Fixed Cost** | **87.733.000** | **72.733.000** | **72.733.000** | **72.733.000** |
| | **Total Operational Cost** | **493.897.000** | **415.664.200** | **7.280.600** | **478.897.000** |

| | |
|---|---|
| Operational cost for 1 year | 493.897.000,00 |
| Capital Needed | 922.227.000,00 |

Journal of Physics: Conference Series **1381** (2019) 012034 doi:10.1088/1742-6596/1381/1/012034

**Table 11**. Analysis cash flow

| No | Parameter | Year 0 | Year 1 | Year 2 | Year 3 | Year 4 | Year 5 | Year 6 | Year 7 | Year 8 | Year 9 | Year 10 |
|---|---|---|---|---|---|---|---|---|---|---|---|---|
| Sales | | | | | | | | | | | | |
| 1. | Fruit juice | | 748.800.000 | 842.400.000 | 936.000.000 | 936.000.000 | 936.000.000 | 936.000.000 | 936.000.000 | 936.000.000 | 936.000.000 | 936.000.000 |
| | Total sales | | *748.800.000* | *842.400.000* | *936.000.000* | *936.000.000* | *936.000.000* | *936.000.000* | *936.000.000* | 936.000.000 | 936.000.000 | 936.000.000 |
| | Investment | | | | | | | | | | | |
| 2. | Building | 200.000.000 | | | | | | | | | | |
| 3. | Machine and tools | 153.330.000 | | | | | | | | | | |
| 4. | Transport | 50.000.000 | | | | | | | | | | |
| 5. | Equipment | 10.000.000 | | | | | | | | | | |
| 6. | Pre-investment costs | 15.000.000 | | | | | | | | | | |
| Total Investment | | 428.330.000 | | | | | | | | | | |
| Operational cost | | | *430.664.200* | *700.118.200* | *758.161.000* | *758.161.000* | *758.161.000* | *758.161.000* | 758.161.000 | 758.161.000 | 758.161.000 | 758.161.000 |
| 1. | Fixed cost | | 87.733.000 | 87.733.000 | 87.733.000 | 87.733.000 | 87.733.000 | 87.733.000 | 87.733.000,00 | 87.733.000 | 87.733.000 | 87.733.000 |
| 2. | Variable cost | | 342.931.200 | 612.385.200 | 670.428.000 | 670.428.000 | 670.428.000 | 670.428.000 | 670.428.000,00 | 670.428.000 | 670.428.000 | 670.428.000 |
| Surplus (deficit) | | -428.330.000 | 318.135.800 | 142.281.800 | 177.839.000 | 177.839.000 | 177.839.000 | 177.839.000 | 177.839.000 | 177.839.000 | 177.839.000 | 177.839.000 |

| | | |
|---|---|---|
| Capital | : | 858.994.200,00 |
| Variable cost per bottle | : | 1.373,92 |
| Production per year | : | 249.600,00 |
| Price per bottle | : | 6.000,00 |
| Cost of goods sold | : | 1.725,42 |
| Profit percentage | : | 247,74% |
| Gross profit | : | 318.135.800,00 |

Journal of Physics: Conference Series **1381** (2019) 012034 doi:10.1088/1742-6596/1381/1/012034

**Table 12**. Income Statement

| No | Description | Year 1 | Year 2 | Year 3 | Year 4 | Year 5 | Year 6 | Year 7 | Year 8 | Year 9 | Year 10 |
|---|---|---|---|---|---|---|---|---|---|---|---|
| 1. | Sales | 748.800.000 | 842.400.000 | 936.000.000 | 936.000.000 | 936.000.000 | 936.000.000 | 936.000.000 | 936.000.000 | 936.000.000 | 936.000.000 |
| 2. | Cost | 478.384.570 | 721.460.470 | 784.836.850 | 784.836.850 | 784.836.850 | 784.836.850 | 784.836.850 | 784.836.850 | 784.836.850 | 784.836.850 |
| | Operational cost | 430.664.200 | 700.118.200 | 758.161.000 | 758.161.000 | 758.161.000 | 758.161.000 | 758.161.000 | 758.161.000 | 758.161.000 | 758.161.000 |
| | Total | 430.664.200 | 700.118.200 | 758.161.000 | 758.161.000 | 758.161.000 | 758.161.000 | 758.161.000 | 758.161.000 | 758.161.000 | 758.161.000 |
| 3. | Gross profit | 318.135.800 | 142.281.800 | 177.839.000 | 177.839.000 | 177.839.000 | 177.839.000 | 177.839.000 | 177.839.000 | 177.839.000 | 177.839.000 |
| 4. | Tax 15% | 47.720.370 | 21.342.270 | 26.675.850 | 26.675.850 | 26.675.850 | 26.675.850 | 26.675.850 | 26.675.850 | 26.675.850 | 26.675.850 |
| 5. | Net profit | 270.415.430 | 120.939.530 | 151.163.150 | 151.163.150 | 151.163.150 | 151.163.150 | 151.163.150 | 151.163.150 | 151.163.150 | 151.163.150 |

**Table 13.** Analysis for feasibility study

| **Year** | **Revenue** | **Cost** | **Net Benefit** | **DF 20%** | **NPV** | **PV (B)** | **PV (C)** |
|---|---|---|---|---|---|---|---|
| **0** | - | 428.330.000 | (428.330.000) | 1,000 | (428.330.000) | - | 428.330.000 |
| **1** | 748.800.000 | 478.384.570 | 270.415.430 | 0,833 | 225.346.192 | 624.000.000 | 398.653.808 |
| **2** | 842.400.000 | 721.460.470 | 120.939.530 | 0,694 | 83.985.785 | 585.000.000 | 501.014.215 |
| **3** | 936.000.000 | 784.836.850 | 151.163.150 | 0,579 | 87.478.675 | 541.666.667 | 454.187.992 |
| **4** | 936.000.000 | 784.836.850 | 151.163.150 | 0,482 | 72.898.896 | 451.388.889 | 378.489.993 |
| **5** | 936.000.000 | 784.836.850 | 151.163.150 | 0,402 | 60.749.080 | 376.157.407 | 315.408.328 |
| **6** | 936.000.000 | 784.836.850 | 151.163.150 | 0,335 | 50.624.233 | 313.464.506 | 262.840.273 |
| **7** | 936.000.000 | 784.836.850 | 151.163.150 | 0,279 | 42.186.861 | 261.220.422 | 219.033.561 |
| **8** | 936.000.000 | 784.836.850 | 151.163.150 | 0,233 | 35.155.717 | 217.683.685 | 182.527.967 |
| **9** | 936.000.000 | 784.836.850 | 151.163.150 | 0,194 | 29.296.431 | 181.403.071 | 152.106.640 |
| **10** | 936.000.000 | 784.836.850 | 151.163.150 | 0,162 | 24.413.693 | 151.169.226 | 126.755.533 |
| **Total** | **9.079.200.000** | **7.906.869.840** | **1.172.330.160** | | **102.128.626** | **2.578.212.963** | **2.476.084.336** |

Journal of Physics: Conference Series **1381** (2019) 012034 doi:10.1088/1742-6596/1381/1/012034

**Table 14.** Result financial feasibility analysis

| No. | Parameter | Value | Statement |
|---|---|---|---|
| 1 | BEP | 18.964,88 | ***Project feasible to run*** |
| 2 | PBP | 2,70 | |
| 3 | NPV | 102.128.626,50 | |
| 4 | IRR | 17,29% | |
| 5 | B/C ratio | 1,04 | |

Based on financial feasibility analysis that has been done, the results obtained if the Net Present Value (NPV) = IDR 102,128,626.50. The NPV value is positive or> 0 the investment is profitable or feasible. Thus the project is feasible to be developed and the industry is profitable if future revenues are greater than the investment value. Gross B/C ratio is the ratio of income (B = Benefit) compared to the cost (C = Cost) which the present value has been calculated (already in the discount factors). New investment projects are feasible to run if the B/C ratio is greater than 1 (one). The results of the gross B/C ratio analysis of this industry is feasible because it is more than one[5].

Internal Rate Return (IRR) calculates the interest rate when cash flow is equal to 0 (zero) or when the profit (revenue minus costs) which has been in the discount factor is equal to 0 (zero). The IRR is useful to know at what rate of interest investment projects continue to provide benefits[6]. The result shows the IRR value of 17.29% which is greater than the current interest, the project can be continued, whereas if the interest is more than the IRR, the investment project is better terminated. Based on the calculation of the discount factor in the table, a payback period value of 2.70 which means that in the 2nd year of the 7th month all investments will return.

## 4. Conclusions

From the results of the financial feasibility, the business of carrot pineapple juice is feasible to be developed. Evidenced by the results of the calculations that show the value of NPV, IRR, BC ratio is positive.

## 5. References

[1] Wiyati P I and Tjitraresmi A 2018 Karakterisasi, Aktivitas Dan Isolasi Enzim Bromelin Dari Tumbuhan Nanas (*Ananas sp.*) *Farmaka* **16** 179–85

[2] Herdyastuti N 2006 Isolasi Dan Karakterisasi Ekstrak Kasar Enzim Bromelin Dari Batang Nanas (*Ananas comusus L.merr*) *Berk. Penlitian Hayati* **12** 12–75

[3] Slamet A 2011 Fortifikasi Tepung Wortel Dalam Pembuatan Bubur Instan Untuk Peningkatan Provitamin A *AGROINTEK* **5** 1–8

[4] Gunawan M Y 2014 Business Plan Rosella and Fruit Juice UD. laberry *Ilm. Maahasiswa Univ. Surabaya* **3** 1–33

[5] Azis J A 2008 Analisis Prospek Pengembangan Industri Sirup Salak Bangkalan *Embryo* **5** 14–23

[6] Syafa'at I 2007 Peta Rakitan, Peta Proses Operasi Dan Diagram Tali pada Analisis Aliran Bahan Puller Jaws *J. Ilm. MOMENTUM* **3** 6–12

# Kahoot On Thematic Learning

**A R Hakim*, S Rahayu, R Affida**

Universitas Kanjuruhan Malang, Malang, Indonesia

*ariefrahman@unikama.ac.id

**Abstract**. Thematic learning, especially in elementary schools, requires a supporting component in the form of learning media. The applicable learning media needs are needed in an effort to facilitate the teaching and learning process. The purpose of this study was to describe students' responses to the Kahoot application for students at SD Kedungkandang 1 Malang.This research was a descriptive narrative study with a qualitative approach. The subjects in the study were all fourth grade students of Kedungkandang 1 Elementary School in Malang City, amounting to 32 people. The data collection procedure employed observation, interviews, questionnaires and documentation.The results showed that a positive response was given to the fourth grade students of Kedungkandang I SDN Malang City to the Kahoot application. This can be seen from the results of student questionnaire responses to aspects of expression by 70%, aspects of interest by 31%, active involvement aspects by 38%, accessbility aspects by 69%, and aspects of attractiveness of 82%

## 1. Introduction

Modern learning in this century cannot be separated from the internet connection. The use of social media and multimedia is crucial and essential for students and teachers lately. The balance between lifestyle and ways of learning is the biggest challenge for teachers in schools. Learning innovations need to be developed to obtain a better quality of education. In addition, innovation also aims to achieve learning objectives to the maximum. In the context of learning, innovation is a form of teacher creativity in managing learning. Innovation allows the transformation of monotonous, boring, saturating and conventional learning into fun, varied and meaningful learning. Digital games are one form of learning innovation that aims to balance these two things. Argues that the use of video games or digital video develops cognitive, motivational, emotional, and social aspects if the teacher is able to implement them well in the classroom [1].

Based on the results of interviews with fourth grade teachers, it was found that the use of media in Kedungkandang 1 Elementary School in Malang was still very limited. This is due to the limited time and cost to create media as innovations and variations of learning. Therefore, a learning innovation named Kahoot as one of the learning innovations is available.

Kahoot is one of the famous game-based learning platforms. Kahoot is a user-friendly learning platform, both for teachers and students. This platform was developed by Johan Brand, Jamie Brooker, and Morten Versvik in collaboration with the Norwegian University of Technology and Science[2].

Kahoot is an online application in the form of quizzes or practice questions developed and presented in the form of game. When accessing Kahoot, students get points for answering the quiz correctly and their names will be displayed on the player list when playing Kahoot. The Kahoot application makes it easy for teachers to evaluate students. Most teachers in France, design learning innovations by utilizing online-based technology in the classroom to create interesting learning[3].

The use of Kahoot in learning can create a fun and interactive atmosphere for students [4]. Positive enthusiasm is shown by students during the learning process. Based on the description, this research focuses on the responses given by fourth grade students of Kedungkandang 1 Elementary School in Malang to the use of Kahoot in the learning process.

## 2. Methodology

This paper employed descriptive design with qualitative approach. The researchers in this study took a role in planning, implementing, collecting data, analyzing, and reporting the research results. This study was employed in Kedungkandang 1 Elementary School Malang. This study took 32 students of the fourth graders as the subject. Data collection was done by implemeting observation, interview, documentation, and using questionnaires.

Data processing in this research was based on Miles and Huberman's Model which consisted of four stages, they were:

1. Data collection
2. Data reduction
3. Data presentation
4. Drawing conclusion

Checking the validity of the data in this study was employed triangulation of sources and triangulation of data collection techniques. Source triangulation was done by finding data from various sources that are still related to one another. While technical triangulation was done by checking the data to the same source with different techniques. In conducting this triangulation the researchers made interview sheets and conducted interviews with students and teachers. After that, the researchers gave questionnaires to students to determine student responses in the use of Kahoot. In addition, researchers conducted documentation during the learning process in class.

## 3. Findings and Discussions

The results of students' response on the application of Kahoot during Theme 7 lesson "*Indahnya Keberagaman di Negeriku*" (Unity in Diversity) with sub-theme of lesson 2 of The Diversity of Race and Religion in My Country fourth graders in Kedungkandang 1 Elementary School Malang is presented in the following Table 3.1.

**Table 3.1 Students' Response on The Application of Kahoot**

| No | Aspect | Percentage |
|---|---|---|
| 1 | Expression | 70% |
| 2 | Interest | 31% |
| 3 | Active involvement | 38% |
| 4 | Accessbility | 69% |
| 5 | Attractiveness | 82% |

Student responses regarding the learning by using the Kahoot Application at SDN Kedungkandang 1 Malang showed that each indicator that received an answer "strongly agree" was as follows: aspects of expression at 70%, aspects of interest at 31%, aspects of active involvement at 38%, aspects accessibility by 69%, and attractiveness by 82%.

The results of the questionnaire meant that students' responses to the use of the Kahoot application were very high. This can be seen from the percentage of questionnaires that have been obtained. The high response of students is because the Kahoot application is something new for students. This shows that applications that contain quizz in the form of games can prevent a boredom within students during the completion of quizz.

The first aspect, the aspect of expression, it obtained an average value of 70%. The number indicates that this aspect is categorized as high. The indicators of this aspect were pleasure and attentive in learning the content. The number signifies that the students enjoyed in learning the lesson content by using the application. It is further confirmed that the students were attentive in learning. Learning platform based on game is effective to be applied as a media to motivate students in learning process in classroom. It allows an enjoyable and delightful learning in classroom [5,6].

The second aspect, the aspect of interest, it obtained an average value of 31%. It correlates with students' experiences in attending the lesson. The number indicates that this aspect is categorized as moderate. The obtained number was due to the hindrance that students encountered during the learning process. Thus, when it comes to students' experience, students provided a negative response. One major reason of this issue was because students just knew the Kahoot Application and they were novice users. Learning activity using Kahoot creates a positive effect such as improving concetration, enhancing involvement and learning convenient instead of using conventional learning approach [7]. This application improves learning quality of students and minimizes obstacle during the learning process [8].

The third aspect, the aspect of active involvement, correlates with how students involve in learning and do not feel bored while using the application. It obtained a percentage of 38% and considered as moderate. The percentage signifies that students were actively involved in the learning. However, students also felt bored during the use of the application since they encountered several obastacle when the connection has been lose . Kahoot can be utilized as an ice breaker when the learning is too boring [9]. It can be used further to encourage students' involvement. The research results of Tivaraju that the entire subject were active in attending the learning process using Kahoot. The students enjoyed the learning activities which consequently it improved the language acquisition as the learning goal [10].

The fourth aspect is the aspect of accessbility. It deals with how do students easily access and use the media. It obtained a percentage of 69%. It is categorized as high. The percentage signifies that students found it easy when using Kahoot as learning media. Most students did not find it difficult during the application of Kahoot. Kahoot platform is a simple game which was developed to enable an easy access for the users [11].

The last but not least is the aspect of attractiveness. It deals with the students willingness, attention, and enthusiasm in learning by using Kahoot application. The percentage is 82%. It is considered as very high. It signifes that students are strongly willing, attentive, and enthusiastic on the learning by using Kahoot Application. Furthermore, it encourages students to be high achiever in school. Interestingly, some students tried to access the Kahoot but there was no admin which operated the application. It indicates that the students are strogly interested with the application. Kahoot application is a multiplayer-based game, it allows students to collaborate and compete with different user [12]. It further improves the interest of students in learning.

According to the data above, the highest response is the aspect of attractiveness which obtained the percentage of 82%. It is followed by the aspect of expression which obtained the percentage of 70%, the aspect of accessbility which obtained the percentage of 69%, the aspect of active involvement which obtained the percentage of 38%, and the last is the aspect of interest which obtained the percentage of 31%.

The obstacles found by students are the main reason of the low percentage in the aspect of interest. During the application of Kahoot, students were initially interested in using the application. But, after several errors of server and due to bad connection, it reduced students interest in using the application.

The findings of this research indicate that Kahoot application could stimulate students' involvement in learning through game quizzes. It further enables them to acquare further language effectively. Students can enjoy in learning English using this game quizzes and students expect a further development of game in the future.

According to the obtained data, students' responses regarding the application of Kahoot is moderately. The application of Kahoot provides several benefits to students for instance: stimulate students in remembering the past lesson content, encouraging the passion in learning, reducing boredom and saturation of learning, and enhançing the involvement [13]. Most students also consider that Kahoot allows them to be competitive and achive the best results in learning. The application of Kahoot could improve students interest and motivation in learning English, particularly in learning structure of English. It is in line with findings in this research.

## 4. Summary and Conclusions

The Positive responses have been given by fourth grade students of Kedungkandang 1 Elementary School in using Kahoot for learning Theme 7 lesson "*Indahnya Keberagaman di Negeriku*" (Unity in Diversity) with sub-theme of lesson 2 of The Diversity of Race and Religion in My Country obtained 70% for the aspect of expression, 31% for the aspect of interest, 38% for the aspect of active involvement, 69% for the accessbility aspect, and 82% for the aspect of attractiveness.

### References

[1] Siegle D 2015 Technology: Learning Can Be Fun and Games *Gift. Child Today* **38** 192–7
[2] Licorish S A, Owen H E, Daniel B and George J L 2018 Students' perception of Kahoot!'s influence on teaching and learning *Res. Pract. Technol. Enhanc. Learn.* **13**
[3] Darya M, Yulia S and Tamara T 2016 INTEGRATING ONLINE EDUCATIONAL APPLICATIONS IN THE CLASSROOM *Eur. Humanit. Stud. State Soc.* 145–56
[4] Sabandar G N C, Supit N R and Suryana E 2018 Kahoot!: Bring the Fun Into the Classroom! *IJIE (Indonesian J. Informatics Educ.* **2** 127
[5] Iaremenko N V. 2019 Enhancing English Language Learners' Motivation Through Online Games *Inf. Technol. Learn. Tools* **59** 126
[6] Nisa Omar N 2017 the Efffectiveness of Kahoot Application Towards Students' Good Feedback Practice *PEOPLE Int. J. Soc. Sci.* **3** 2551–62
[7] Wang A I and Lieberoth A 2016 The effect of points and audio on concentration, engagement, enjoyment, learning, motivation, and classroom dynamics using Kahoot! Game based Learning (EduApp4Syria) View project CODER-Center for Community Driven Research View project The effect of points
[8] Licorish S A, Owen H and Daniel B K 2017 "Go Kahoot!" Enriching Classroom Engagement, Motivation and Learning Experience with Games Requirements Engineering-Requirements

Prioritization View project BBN and virtual communities View project 755–64
[9] Damara G 2016 Students ' Perceptions on the Use of Kahoot! As An Ice Breaker in Movie Interpretation Class
[10] Tivaraju J, Md Yunus M and Badusah J 2006 Learning English Is Fun Via Kahoot : Students ' Attitude , Motivation and Perceptions *Semin. Pendidik. Transdisiplin (STEd 2017)* 218–29
[11] Graham K 2014 TechMatters : Getti ng i nt o K ah oo t !( s ): Exp lo ri ng a Ga m e - Based L ear ni ng S yst em t o En han ce S tu d en t Lear ni n g Creating a New kahoot Playing Your First kahoot Building a Better Kahoot ! **42** 6–7
[12] Rofiyarti F 2017 Penggunaan Platform "Kahoot!" Dalam Menumbuhkan Jiwa Kompetitif Dan Kolaboratif Anak *J. Anak Usia Dini dan Pendidik. Anak Usia Dini* **3** 227–39
[13] Barus I and Soedewo T 2013 Studi Kasus : Mahasiswa Sekolah Vokasi – Institut Pertanian Bogor *TEKNOLOGI TERAPAN BERBASIS KEARIFAN LOKAL PROSIDING* pp 589–96

Journal of Physics: Conference Series **1381** (2019) 012036 doi:10.1088/1742-6596/1381/1/012036

# Implementation of ID3 algorithm classification using web-based weka

**A S Fitrani*, M A Rosid, Y Findawati, Y Rahmawati, A K Anam**

Program Studi Informatika, Universitas Muhammadiyah Sidoarjo, Jl. Raya Gelam 250 Candi Sidoarjo, Jawa Timur, Indonesia

asfjim@umsida.ac.id

**Abstract**. The Bangil District Court is an IB class court that handles a large number of case cases. Every year more and more case cases are included in the Bangil District Court, but not all case cases are in a mutation status. By using classification techniques that can process large amounts of data to find patterns that occur in case data. Data processing is used to predict case minutation with the decision tree method using ID3 algorithm. Case data has 8 attributes and has been classified into 6 parts, namely division based on Case Type, Register, Case Classification, Length of Process, Public Prosecutor and Decision with a goal of Mutation Status. Weka 3.6 is an API that is used to build rules / rule bases. The rule that was formed was then implemented in the making of a case status prediction application in the web-based Bangil District Court.

## 1. Introduction

The Bangil District Court which is an IB class court that along with the increasing number of case cases in the Bangil District Court, the number of incoming case data automatically increases, but not all case cases are in a mutation status. This case data will be processed using data mining techniques that will generate new knowledge and information from patterns or models formed on large data.

By using classification techniques that can process large amounts of data[1][2][3] where it will be explained about the model produced by the data that has a data class on case data in Pengadialan Negeri Bangil. Data processing is used to predict case minutation with the decision tree method using ID3 algorithm.

Case data taken is data from the website of the Case Search Information System (SIPP) in the Bangil District Court. Data will be processed using the iterative classification method dichotomiser 3 (Id3) to find out predictions of case mutations from 2011 to 2015 in the Bangil District Court.

The impact of this application is that the IT Staff of the Bangil District Court is easier and more efficient in terms of completing the case data prediction process to find out the status of the mutation with ID3 algorithm.

## 2. Theoretical foundation

### *2.1 Definition of Case Search Information System*

The Case Search Information System (SIPP) is an official web-based application made by the Supreme Court of the Republic of Indonesia (MARI) to support and facilitate justice seekers in monitoring the process of handling criminal cases[4].

This application is intended for administration and tracking of first-degree court case data. Case Phase, Case Status, Case Costs, Session Schedule, Case Statistics, and History, including information services contained in this application and complete information, has been recorded.

All information can be accessed by the general public for free and realtime. This SIPP application allows access to case files. The community can directly access and search for the subject matter [5]

*2.2 Definition of Data Mining*

Data Mining is a process of collecting data that is processed with various methods. Another term for data mining is knowledge-discovery in a database (KDD). The purpose of data mining is to utilize data and process it to get new and useful information.

*2.3 Group Data Mining*

There are several types of data mining groupings, according to [6]:

1. Description, to describe the pattern in data.
2. Estimates, the goal is more numerical than category, and this estimate is almost similar to classification.
3. Prediction, the result of a prediction is to show something that hasn't happened yet
4. Classification, the purpose of this classification is categorical
5. Clustering, clustering is more towards grouping records or cases that have similar data
6. Association, to identify the relationship between events that occur at one time.

*2.4 Stages of Data Mining*

The data mining stage is interactive, and the stages of data mining are divided into several stages including. [6]

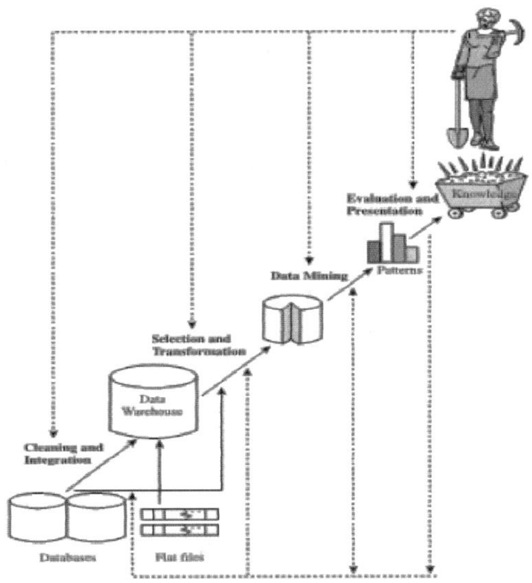


Figure 1. Stages of Data Mining

1. Cleaning (Cleaning)
2. Is the process of removing/cleaning inconsistent data.
3. Integration (integration)
4. Is the merging of data from various databases into one new database
5. Selection
6. Data selection (data selection), not all data contained in the database is used; the data taken is only suitable for analysis.
7. Transformation, data will be converted into a format that is compatible with the data mining process.
8. Data mining process, is the most important process in the method used to find knowledge.
9. Evaluate patterns (pattern evaluation), Identify a unique pattern and draw it into the knowledge base

10. Presentation of knowledge (knowledge presentation), is a presentation of knowledge from the method used to obtain knowledge obtained by the user.

### *2.5 Decision Tree Method*

Decision Tree is an approach that is widely used to solve clarification problems, and this method is used to estimate discrete values and target functions. The Decision Tree is a set of IF-THEN rules that each line or path is connected to the rule. In the decision, the tree can be used to map decision [7].

#### *2.5.1 ID3 algorithm*

ID3 algorithm is one method in the decision tree model, where estimating the discrete value of a function from a decision tree produces an if-then rule so that it is easier to understand. The following is an example of the Iterative Dichonomiser 3 or ID3 algorithm system [8][9].

#### *2.5.2 Entropy*

Entropy is a measure of data that has a measure of information theory that can determine the characteristics of data impurity and feature equations of members of an attribute[1][8]. To calculate the entropy value using the formula:

$$\textbf{\textit{Entropy}}\ (\mathrm{S}) = \sum_{i=1}^{n} -pi\ * \log_2 pi \tag{1}$$

Information:
S: a set of cases
n: number of S partitions
Pi: the proportion of Si to S

#### *2.5.3 Gain*

After the entropy value is known, the next step is to find the gain value of an attribute in classifying the data[1][10].
To calculate the Gain value using the formula:

$$\textbf{\textit{Gain(S,A)=Entropy}}\ (\mathrm{S})\ \text{-}\sum_{i=1}^{n} * \frac{|Si|}{S} * Entropy\ (Si) \tag{2}$$

Information:
S: a set of cases
A: Attribute
n: number of partition attributes A
| Si |: number of cases on a partition to i
| S |: number of cases in S

## 3. System Design

The design of the application of data mining prediction on minutation cases can be divided into the following stages, namely:

1. Classification Diagram
2. Data Attributes
3. Flowchart
4. Designing the User Interface

Journal of Physics: Conference Series **1381** (2019) 012036 doi:10.1088/1742-6596/1381/1/012036

### *3.1 Classification Diagram*

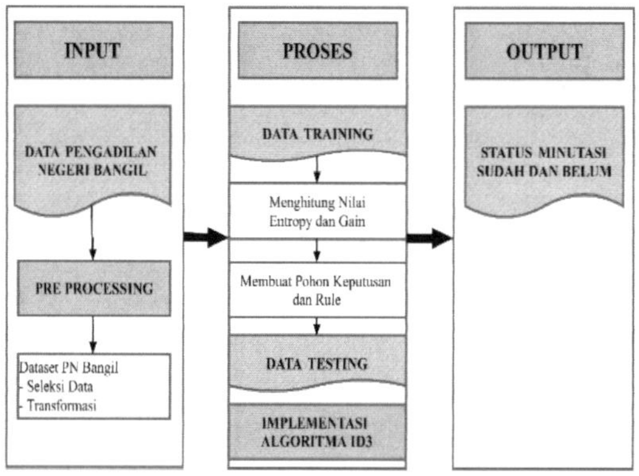


Figure 2. Classification Diagram

### *3.2 Flowchart*

The following is a description of the flowchart classification method with algorithm ID3 to predict case status minutation, namely:

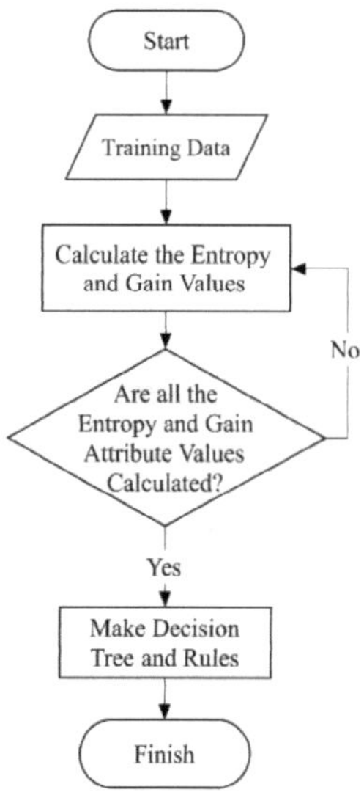


Figure 3. Flowchart Data Training

The description of figure 3 is:

1. Enter training data
2. Calculate the entropy and gain values
3. Create a root node (root node) based on the highest Gain value.
4. Calculate the Entropy value based on the attribute on the selected Gain.
5. Calculate all entropy values until you have one answer or leaf node if you have not found one answer, it needs to be recalculated by choosing the largest gain value remaining. After all, are counted, create a branch node (internal node) based on the results of the calculation of entropy.
6. After Gain and Entropy have calculated all, make a decision tree and Rule from the decision tree that has been made.

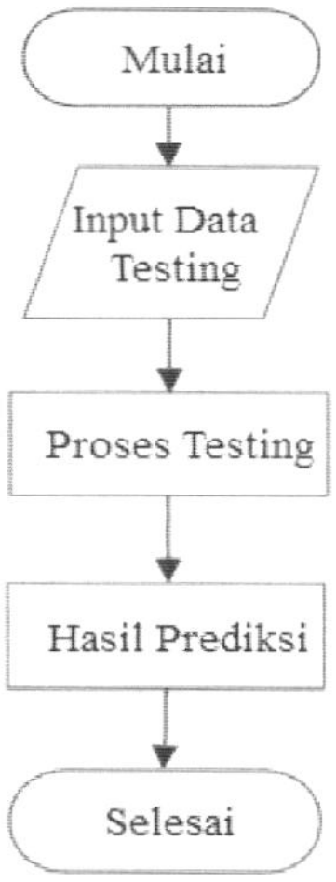


Figure 4. *Flowchart* Data *Testing*

The information from figure 4 is:

a. Input testing data
b. The testing process is based on the decision tree from the training data
c. Get the results of testing data

## 4. Discussion

### *4.1 Input*

Data obtained from the Bangil District Court totaled 998 case data. The data will be tested using the Decision Tree method with algorithm ID3 using WEKA tools and website[2]. Data collection was 65% of 649 training data and 35% of 349 testing data. Following are some examples of case data, namely:

Table1. Dataset

| No | Terdakwa/Pemohon | Jenis Perkara | Register | Klasifikasi Perkara | Lama Proses | Penuntut Umum | Putusan | Status Minutasi |
|---|---|---|---|---|---|---|---|---|
| 1 | IRA SUGIARTI | Pdt.G | A | Ganti Rugi | Sangat Lama | TD | Dikabulkan | Belum |
| 2 | PT. BPR PANDAAN ARTA JAYA | Pdt.G | A | Wanprestasi | Sangat Lama | TD | Dikabulkan | Belum |
| 3 | WARTI DAN PINI | Pdt.G | A | Wanprestasi | Sangat Lama | TD | Dicabut | Sudah |
| 4 | M. SYAIFUDDIN NUR | Pid.B | A | Pembunuhan | Lama | TD | Penjara | Belum |
| 5 | YOYOK SAPTOADI JOKO SUPRIOKO | Pid.B | A | Penggelapan | Cepat | TD | Penjara | Belum |
| 6 | DAIM | Pid.B | A | Narkotika | Lama | TD | Penjara | Belum |
| 7 | FATHUR ROHMAN | Pid.B | A | Penggelapan | Lama | TD | Penjara | Belum |
| 8 | RAKHMAT HIDAYAT | Pid.B | A | Narkotika | Lama | TD | Penjara | Belum |
| 9 | H. NUR HASAN | Pid.B | A | Pencurian | Lama | TD | Penjara | Belum |
| 10 | INUNG | Pid.B | A | Narkotika | Sangat Lama | TD | Penjara | Belum |
| 11 | IDA BAGUS KOKO | Pid.B | A | Lain-Lain | Sangat Lama | TD | Bebas | Belum |
| 12 | IRA SUGIARTI | Pdt.G | A | Ganti Rugi | Sangat Lama | TD | TDD | Belum |
| 13 | Disamarkan | Pdt.G | B | Penceraian | Cepat | D | Dikabulkan | Sudah |
| 14 | Disamarkan | Pdt.G | B | Penceraian | Sangat Cepat | D | Dikabulkan | Sudah |
| 15 | Disamarkan | Pdt.G | B | Penceraian | Cepat | D | Dikabulkan | Sudah |
| 16 | PT. BPR GUNUNG ADI DANA | Pdt.G | B | Lain-Lain | Sangat Lama | TD | Dikabulkan | Belum |
| 17 | PT. CIPTA MAKARYA INDAH | Pdt.G | B | Lain-Lain | Lama | TD | Dicabut | Sudah |
| 18 | GINTEN | Pdt.G | B | Lain-Lain | Sangat Lama | TD | Dikabulkan | Belum |
| 19 | Disamarkan | Pdt.G | B | Penceraian | Sangat Cepat | D | Dikabulkan | Sudah |
| 20 | LILA HERDIANTI | Pdt.G | B | Ganti Rugi | Cepat | TD | Dikabulkan | Sudah |

Table 2. Training Data

| No | Terdakwa/Pemohon | Jenis Perkara | Register | Klasifikasi Perkara | Lama Proses | Penuntut Umum | Putusan | Status Minutasi |
|---|---|---|---|---|---|---|---|---|
| 1 | IRA SUGIARTI | Pdt.G | A | Ganti Rugi | Sangat Lama | TD | Dikabulkan | Belum |
| 2 | PT. BPR PANDAAN ARTA JAYA | Pdt.G | A | Wanprestasi | Sangat Lama | TD | Dikabulkan | Belum |
| 3 | YOYOK SAPTOADI JOKO SUPRIOKO | Pid.B | A | Penggelapan | Cepat | TD | Penjara | Belum |
| 4 | DAIM | Pid.B | A | Narkotika | Lama | TD | Penjara | Belum |
| 5 | FATHUR ROHMAN | Pid.B | A | Penggelapan | Lama | TD | Penjara | Belum |
| 6 | RAKHMAT HIDAYAT | Pid.B | A | Narkotika | Lama | TD | Penjara | Belum |
| 7 | H. NUR HASAN | Pid.B | A | Pencurian | Lama | TD | Penjara | Belum |
| 8 | INUNG | Pid.B | A | Narkotika | Sangat Lama | TD | Penjara | Belum |
| 9 | IDA BAGUS KOKO | Pid.B | A | Lain-Lain | Sangat Lama | TD | Bebas | Belum |
| 10 | IRA SUGIARTI | Pdt.G | A | Ganti Rugi | Sangat Lama | TD | TDD | Belum |
| 11 | Disamarkan | Pdt.G | B | Penceraian | Cepat | D | Dikabulkan | Sudah |
| 12 | Disamarkan | Pdt.G | B | Penceraian | Sangat Cepat | D | Dikabulkan | Sudah |
| 13 | Disamarkan | Pdt.G | B | Penceraian | Cepat | D | Dikabulkan | Sudah |
| 14 | PT. BPR GUNUNG ADI DANA | Pdt.G | B | Lain-Lain | Sangat Lama | TD | Dikabulkan | Belum |
| 15 | PT. CIPTA MAKARYA INDAH | Pdt.G | B | Lain-Lain | Lama | TD | Dicabut | Sudah |
| 16 | GINTEN | Pdt.P | B | Lain-Lain | Sangat Lama | TD | Dikabulkan | Belum |
| 17 | Disamarkan | Pdt.G | B | Penceraian | Sangat Cepat | D | Dikabulkan | Sudah |
| 18 | LILA HERDIANTI | Pdt.G | B | Ganti Rugi | Cepat | TD | Dikabulkan | Sudah |
| 19 | LILA HERDIANTI | Pdt.G | B | Ganti Rugi | Lama | TD | Dikabulkan | Sudah |
| 20 | NINDIA AVISTA CITRAYANTI | Pdt.P | B | PAKT | Sangat Cepat | TAO | Dikabulkan | Sudah |

Table 3. Testing Data

| No | Terdakwa/Pemohon | Jenis Perkara | Register | Klasifikasi Perkara | Lama Proses | Penuntut Umum | Putusan | Status Minutasi |
|---|---|---|---|---|---|---|---|---|
| 1 | WARTI DAN PINI | Pdt.G | A | Wanprestasi | Sangat Lama | TD | Dicabut | Sudah |
| 2 | M. SYAIFUDDIN NUR | Pid.B | A | Pembunuhan | Lama | TD | Penjara | Belum |
| 3 | BAMBANG APRIYANTO | Pid.B | B | Penggelapan | Cepat | TD | Penjara | Sudah |
| 4 | KHOLIK | Pdt.G | B | Lain-Lain | Lama | TD | Dikabulkan | Sudah |
| 5 | SUNARKO Als REJES Bin DASIRUN | Pid.B | B | Penggelapan | Cepat | TD | Penjara | Sudah |
| 6 | H.HILMI | Pdt.P | B | PAKT | Sangat Cepat | TAO | Dikabulkan | Sudah |
| 7 | DIANTO | Pdt.P | B | PAKT | Sangat Cepat | TAO | Dikabulkan | Sudah |
| 8 | NURJANAH | Pdt.P | B | WDIJ | Sangat Cepat | TAO | Dikabulkan | Sudah |
| 9 | MAWARDI | Pdt.P | B | WDIJ | Sangat Cepat | TAO | Dikabulkan | Sudah |
| 10 | Disamarkan | Pdt.G | B | Perceraian | Sangat Cepat | D | Dikabulkan | Sudah |
| 11 | TRI SETYO JANUARIFIN | Pdt.P | B | PAKT | Sangat Cepat | TAO | Dikabulkan | Sudah |
| 12 | SRI RAHAYU YULIARTI | Pid.B | B | Pemalsuan | Cepat | TD | Penjara | Belum |
| 13 | Terdakwa | Pid.B | B | Lain-Lain | Cepat | TD | Penjara | Sudah |
| 14 | NURIYANTO | Pdt.P | B | PAKT | Sangat Cepat | TAO | Dikabulkan | Sudah |
| 15 | SAHUR BIN MARSAN | Pid.B | B | Penganiayaan | Sangat Cepat | TD | Penjara | Sudah |
| 16 | RUHANAH | Pdt.P | B | PAKT | Sangat Cepat | TAO | Dikabulkan | Sudah |
| 17 | YAYUK SUPRAPTI BROWNE | Pdt.P | B | PAKT | Cepat | TAO | Dikabulkan | Sudah |
| 18 | LOLOK WIJAYANTI | Pdt.G | B | Lain-Lain | Sangat Lama | TD | Dikabulkan | Sudah |
| 19 | MITA FITRIA | Pdt.P | B | PAKT | Sangat Cepat | TAO | Dikabulkan | Sudah |
| 20 | USMANTO BIN NGADIARTO | Pid.B | B | Pencurian | Cepat | TD | Penjara | Sudah |

### *4.1.1 Data*

The dataset used for the calculation process is 998 case data. The dataset is divided into two parts, namely training data and testing data. Training data is used to process a calculation by forming a classifier model. While data testing is used to measure the extent to which the classifier successfully classifies correctly.

### *4.1.2 Pre Processing Data*

The 1,000 datasets, which were the initial data of this study experienced pre-processing to 998 data. Where the pre-processing techniques in preparing data are discarding duplicate data, discarding data inconsistencies, and correcting data errors.

### *4.2 Proses*

The process used at this stage is to find out the results of the ID3 classification calculation on the Website.

#### *4.2.1 Website*

1. After that, take the dataset with the extension .csv / .arff

Figure 5. Display of Dataset Files

2. Select the classify method, select the ID3 algorithm and then select the dataset and Testing data and then select submit

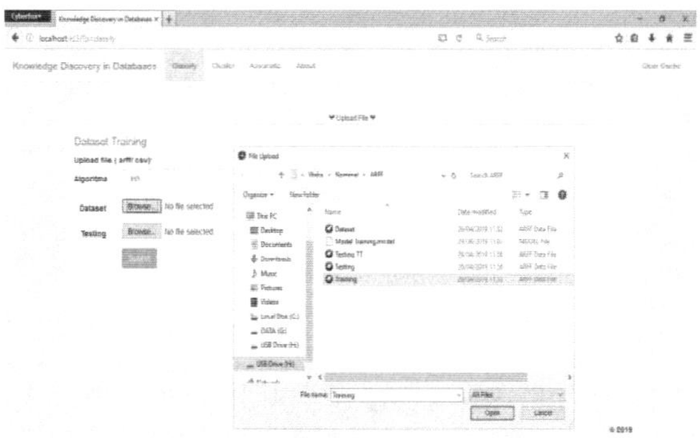

Figure 6. Classify Display

3. Wait for a while until the classification results appear

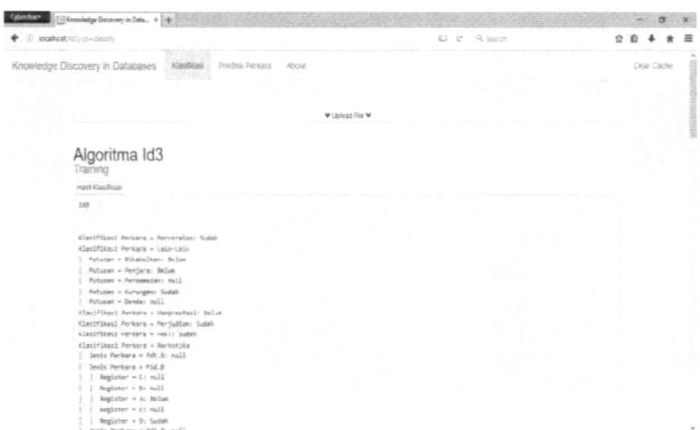


Figure 7. Display of ID3 Classification Results

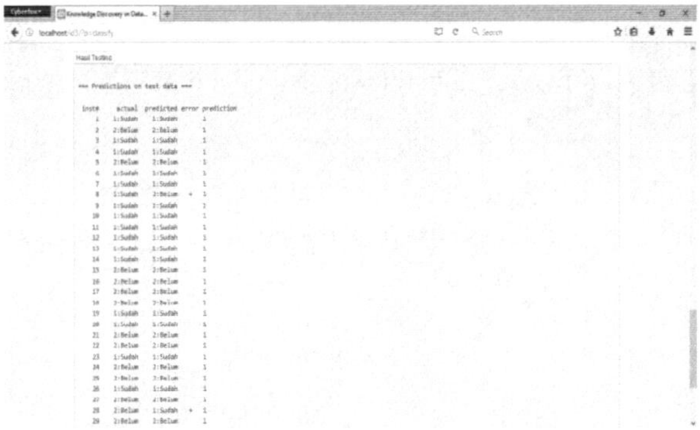

Figure 8. Display of Testing Results

4. Display prediction menu.

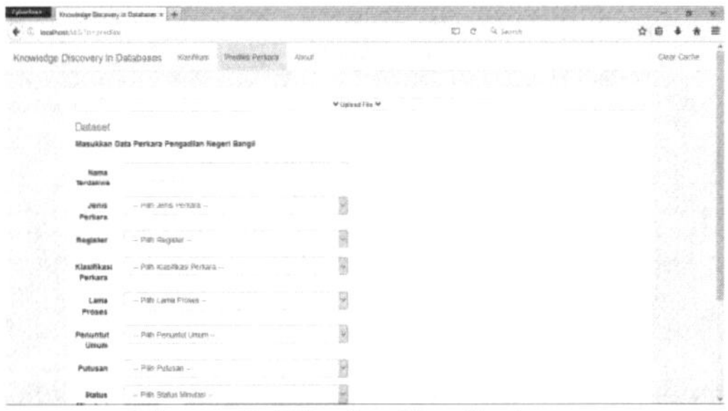


Figure 9. Display Prediction

5. Wait for a while until the prediction results appear

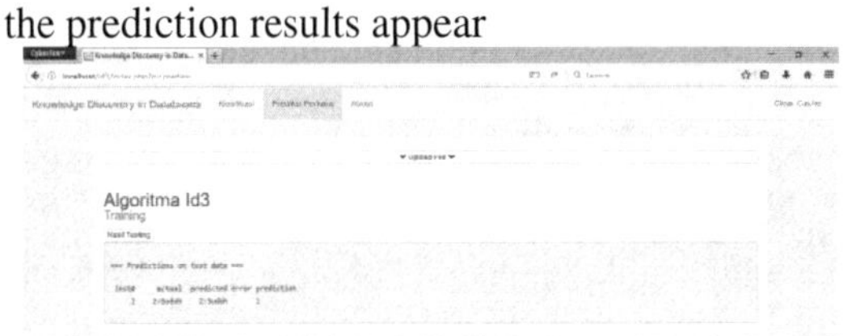


Figure 10. Display of Prediction Results

## 4.3 Output

### *4.3.1 Classification Results*

1. Predicted results on testing with a total testing data of 35% of a total dataset of 998.

=== Predictions on test data ===

| inst# | actual | predicted | error | prediction |
|---|---|---|---|---|
| 1 | 2:Sudah | 1:Belum | + | 1 |
| 2 | 1:Belum | ? | | ? |
| 3 | 2:Sudah | ? | | ? |
| 4 | 2:Sudah | 2:Sudah | | 1 |
| 5 | 2:Sudah | ? | | ? |
| 6 | 2:Sudah | 2:Sudah | | 1 |
| 7 | 2:Sudah | 2:Sudah | | 1 |
| 8 | 2:Sudah | 2:Sudah | | 1 |
| 9 | 2:Sudah | 2:Sudah | | 1 |
| 10 | 2:Sudah | 2:Sudah | | 1 |
| ... | ... | ... | | ... ... |

2. Prediction results on testing with single data testing.

=== Predictions on test data ===

| inst# | actual | predicted | error | prediction |
|---|---|---|---|---|
| 1 | 1:Belum | 2:Sudah | + | 1 |

### *4.3.2 Percentage Results*

Percentage of processes to determine accuracy. The following are the results of the percentage test accuracy Table 4 :

Table 4. Percentage Results

| **Data Testing** | **Result** | | | |
|---|---|---|---|---|
| | **True** | **%** | **False** | **%** |
| 988 | 326 | 93.41 % | 23 | 6.59 % |

## 5. Conclusion

Based on the results of the study, it can be concluded that the algorithm ID3 can be considered to be used in the case data prediction process to determine the mutation status. This is supported by the results of testing the system using algorithm ID3 in the Bangil District Court, which has an average accuracy of 93.41% and has an error rate of 6.59%.

Suggestions that can be given for this research are expected to be able to use other methods. So, can know the comparison between the percentage results in the calculation process according to the method used.

## 6. References

[1] Gupta B, Rawat A, Jain A, Arora A and Dhami N 2017 Analysis of Various Decision Tree Algorithms for Classification in Data Mining *Int. J. Comput. Appl.* **163** 15–9

[2] Guo H, Hong N, Shen Z, Duan W and Zhang Z 2018 Application of Classification Technique of Data Mining for Employee Management System **1** 12–22

[3] Sharma P, Singh D and Singh A 2015 Classification algorithms on a large continuous random dataset using rapid miner tool *2nd Int. Conf. Electron. Commun. Syst. ICECS 2015* 704–9

[4] Putra A 2015 PEMBAHARUAN SISTEM PERADILAN PIDANA MELALUI PENATAAN ADMINISTRASI PERADILAN **IV** 73–9

[5] Nisa K and Hany Fanida E 2000 PENERAPAN SISTEM INFORMASI ADMINISTRASI PERKARA PEGADILAN AGAMA ( SIADPAPLus ) MELALUI APLIKASI AUDIO TO TEXT RECORDING DI PENGADILAN AGAMA KABUPATEN MALANG Khoirun Nisa *Unasa* **1204067400** 1–11

[6] Ridwan M, Suyono H and Sarosa M 2013 Jurnal Pak Agung **7** 59–64

[7] B L K 2016 Algorithm Selection for Combinatorial Search Problems: A Survey **10101** 149–90

[8] Wang Y Y, Li Y Bin and Rong X W 2017 Improvement of ID3 algorithm based on simplified information entropy and coordination degree *Proc. - 2017 Chinese Autom. Congr. CAC 2017* **2017-Janua** 1526–30

[9] Jassim K S and Saleh H M 2018 Diabetes Classification Using ID3 and Naïve Bayes Algorithms **12**

[10] Kraidech S and Jearanaitanakij K 2018 Improving ID3 Algorithm by Combining Values from Equally Important Attributes *ICSEC 2017 - 21st Int. Comput. Sci. Eng. Conf. 2017, Proceeding* **6** 102–5

Journal of Physics: Conference Series **1381** (2019) 012037 doi:10.1088/1742-6596/1381/1/012037

# Epilepsi detection system based on EEG record using neural network backpropagation method

**Ade Eviyanti1, Hindarto hindarto[2], Sumarno[3], Herlian Aliyasa Alamj Duddin[4]**

Program Studi Informatika, Universitas Muhammadiyah Sidoarjo, Jl. Raya Gelam 250 Candi Sidoarjo

adeeviyanti@umsida.ac.id[1], hindarto@umsida.ac.id[2], sumarno@umsida.ac.id[3], aliyasa@umsida.ac.id[4]

**Abstract**. Epilepsy is a manifestation of brain disorders with a variety of etiologies, but with the typical single symptom, periodic and reversible attacks, Epilepsy is characterized by an excess amount of electricity coming out of the brain cells, which can cause seizures and abnormal movements. EEG signals on epilepsy attacks have a characteristic pattern that allows health professionals to distinguish them from normal conditions (nonseizure). There are many methods used by researchers to recognize patterns of EEG epilepsy and non epilepsy signals in this study using Discrete Cosinus Transform (DCT) to perform the extraction of EEG signal features and Backpropagation Neural Networks for identification of EEG signal patterns. This research data using five classes of data sets of digital EEG signal taken from clinic Epileptologie University of Bonn ie data set A normal open eye signal, set B normal eye closed signal, C set enter the epilepsy zone, set D enter epilepsy, set E epilepsy seizures. The five class data are processed using DCT to obtain feature extraction, so results from DCT are used to perform identification using the Backpropagation method. The results of this study indicate that with feature extracted using DCT and identification process using Artificial Neural Network Backpropagation got EEG signal identification obtained for data set A, B, C, D and E is 76%, for set AB and CDE class data 73 %.

## 1. Introduction

Epilepsy is a chronic neurological disease caused by abnormal electrical activity in the brain. As we know that our brain consists of about one hundred billion nerve cells called neurons, where they carry signals throughout the brain and between the brain to other parts of the body. Each neuron generates an electrical signal which is then spread by the neuro transmitter in the form of a nerve-conducting signal [1].

Electroencephalography (EEG) is a method used in measuring spontaneous electrical activity from the brain that is obtained by firing electrical signals into neurons in the brain[2]. The EEG signal recording process is carried out in a short time, usually for 20-40 minutes. Records are obtained by placing electrodes in various positions on the scalp[3]. There are two approaches to get EEG signals, namely invasive and non-invasive approaches. A noninvasive approach can be applied repeatedly for patients, normal adults, and children with almost no risk or limitation, so that almost all adult EEG recordings are done non-invasively [2]. EEG signal data used in this study is in the form of time series.

Transforming an EEG signal into a model, is a very effective way of assisting in the classification of EEG signals, identifying and estimating the spectrum of EEG signals. EEG signals contain certain components, known as alpha (8-13 Hz), beta (14-30 Hz), theta (4-7 Hz), and delta (0.5-3 Hz), so the transformation of EEG signals into frequency regions are very useful, especially in the identification of waves in the brain[4][5][6].

The algorithm used by Nigam and Graupe, 2004 uses a multistage nonlinear preprocessing filter combined with an artificial neural network (ANN) for automatic detection of epileptic seizures on EEG signals [7]. Güler et al., 2005 used recurrent neural networks (RNNs) and Lyapunov feature extraction which was trained with the Levenberg-Marquardt algorithm[3]. Übeyli, 2006 uses a multilayer perceptron neural network (MLPNN). Übeyli, 2010 uses the Least-Square Support Vector Machine (LS-SVM) and the Autoregressive (AR) coefficient [8].

In this study, the method used is using Discrete Cosine Transform (DCT) and Backpropagatioan artificial neural networks. The first step uses cosine transform to convert /extract a signal into its basic frequency component so that several features are obtained to capture the specific characteristics of the EEG signal and then these features are used as input to Backpropagation to obtain a classification of detected epilepsy or nonepilepsy.

## 2. Methodology

In the EEG signal classification system there are three functions, namely first to perform pre-processing (input processing) input signals which will be used for learning data, second to extract EEG signals and the third to classify EEG signals identified by epilepsy and non epilepsy

The epilesi and non-epilepsy identification system based on EEG signals in this study consisted of several stages of the process, the first taking EEG signals used as data for the learning process, the second stage was the process of extracting EEG signals using DCT[9] to obtain signal features that aim to bring out features and reduce the dimensions of the signal from the higher dimensions to the lower dimensions. The third stage of classification is to classify EEG signals detected by epilepsy and non-epilepsy based on similarity measurements using artificial neural network modeling Backpropagation method[10].

EEG signal data from the University of Bonn consists of five classes of datasets namely A, B, C, D, and E. Each dataset contains 100 single-channel EEG segments with a duration of 23.6 seconds[11]. Each segment is selected and cut from continuous multichannel EEG recordings after visual artifact inspection, such as eye movements or muscle activity. Sets A and B are signals taken from EEG records conducted on five healthy volunteers with a standard electrodes placement scheme (International 10-20 system). Volunteers are relaxed and awake with eyes open (for set A) and eyes closed (for set B). The C-E set is from the EEG archive for presurgical diagnosis. EEGs from five patients were selected, and all had achieved complete seizure control, after resection of one of the hippocampal formations, so that they were correctly diagnosed into the epileptogenic zone. Set D signals are recorded in the epileptogenic zone, and are at intervals without seizures and set C originates from the formation of the hippocampus in the opposite hemisphere of the brain. While sets C and D contain activities that are only measured during intervals without seizures, while sets E only contain seizure activities. The data set A and set E are used in this study. In accordance with existing references, all EEG signals are recorded with an amplifier system with 128 channels. Digitalization of data with a frequency of 173.61 samples per second using a 12-bit A / D converter. The passfilter band is set at 0.53 40 Hz (12 dB / oct). Each of the digital EEG signal data consists of 4097 discrete data. The EEG signal plots set A, B, C, D and set E used in this research are time series.

EEG signal data from the University of Bonn consists of five classes of datasets namely A, B, C, D, and E. Each dataset contains 100 single-channel EEG segments with a duration of 23.6 seconds[11]. Each segment is selected and cut from continuous multichannel EEG recordings after visual artifact inspection, such as eye movements or muscle activity. Sets A and B are signals taken from EEG records conducted on five healthy volunteers with a standard electrodes placement scheme (International 10-20 system). Volunteers are relaxed and awake with eyes open (for set A) and eyes closed (for set B). The

C-E set is from the EEG archive for presurgical diagnosis. EEGs from five patients were selected, and all had achieved complete seizure control, after resection of one of the hippocampal formations, so that they were correctly diagnosed into the epileptogenic zone. Set D signals are recorded in the epileptogenic zone, and are at intervals without seizures and set C originates from the formation of the hippocampus in the opposite hemisphere of the brain. While sets C and D contain activities that are only measured during intervals without seizures, while sets E only contain seizure activities. The data set A and set E are used in this study[11]. In accordance with existing references, all EEG signals are recorded with an amplifier system with 128 channels.

Digitalization of data with a frequency of 173.61 samples per second using a 12-bit A / D converter. The passfilter band is set at 0.53 40 Hz (12 dB / oct). Each of the digital EEG signal data consists of 4097 discrete data. The EEG signal plots set A, B, C, D and set E used in this research are time series.

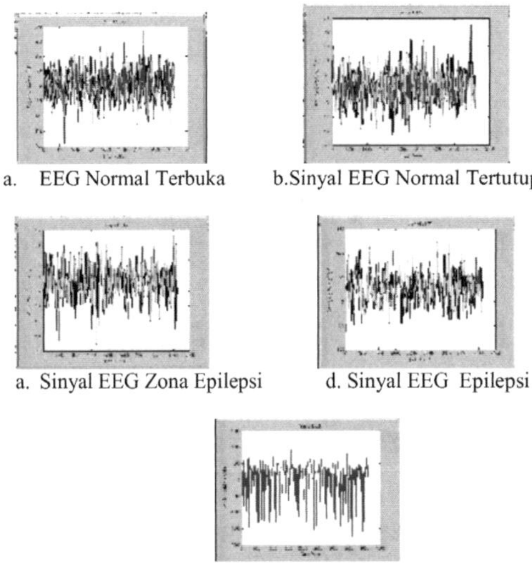

a. EEG Normal Terbuka b.Sinyal EEG Normal Tertutup

a. Sinyal EEG Zona Epilepsi d. Sinyal EEG Epilepsi

. e. Sinyal EEG Epilepsi Kejang

**Figure 1. Five Class EEG Signals (A, B, C, D, E)**

## 3. Extraction Process Using DCT

Discrete Cosinus Transform (DCT) is a technique used to convert signals into its forming frequency components. This transformation can be seen as a form of discrete time (discrete-time) of the fourier cosine transformation (Discrete Fourier Transform / DFT), unlike DFT, DCT only takes into account the real value of the transformed results and tends to have a fairly good approach to the original signal [9] .

Feature extraction is a stage to bring out features and reduce the dimensions of the image from the high dimensions to the lower dimensions. DCT can be used to carry out the EEG Signal extraction process which changes the function from the spatial domain to the basic frequency domain[9]. Discrete Cosine Transform represents an image of the sum of the sinusoid of varying magnitudes and frequencies. The nature of DCT is to change image information that is concentrated only on a few DCT coefficients, thus making the data represented in its frequency component[9].

Discrete Cosine Transform of a series of n real numbers s (x), x = 0, ..., n-1, formulated as Following.
(Watson1994):

Journal of Physics: Conference Series **1381** (2019) 012037 doi:10.1088/1742-6596/1381/1/012037

$$S(u)\sqrt{\frac{2}{n}C(u)}\sum_{x=0}^{n-1} s(x)\left(\cos\frac{(2x+1)u\pi}{2n}\right)$$

$$C_k = \begin{cases} \frac{1}{\sqrt{2}}, k = 0 \\ 1, k > 0 \end{cases}$$

If K = 0, the formula used is :

$$W_k = \sum\nolimits_{t=0}^{n-1} a_t \, cos\left[\frac{\pi}{n}\left(t+\frac{1}{2}\right)k\right], k = 0, \ldots, n-1$$

And if K> 0, then the formula used is :

$$W_k = C_k\sqrt{\frac{2}{n}}\sum\nolimits_{t=0}^{n-1} a_t \, cos\left[\frac{\pi}{n}\left(t+\frac{1}{2}\right)k\right] = \quad 0, \ldots, n-1$$

The purpose of EEG signal extraction is to facilitate input data in the training process on the network. Furthermore, this study takes the value of 4096 data points on each EEG signal, after the DCT process is obtained visualization patterns that can be read and still produce data point values, namely data points 256.

EEG signal data taken from data sets A, set B, set C, set D data set E each of 100 EEG signals using a signal length of 4096 points each time series data. The next process is the DCT process by cutting the signal with a signal length of 256 signals for all dataset segments. So in the process of classification of epilepsy and non-epilepsy, the data length is 256 for each of the five class EEG signal segments.

**Program Segment 3.1: The EEG Signal Extraction Process uses DCT for class A**

```
%Training data for class Z TO A
1.d01=load('E:\Tesis_Ade_Bab       3\ade\DATA SET\Z Set A\Z001.txt');
2.data_10=d01(1:256)
  figure(1)
  plot(data_10)
  % Transfer ke DCT
3.x1=dct(data_10)
  figure(2)
  plot(x1(:,1)); xlabel('Data Point'); ylabel('Amplitudo (Mikro Volt)');
  title('Sinyal EEG TerDCT')
  % Transfer ke iDCT
4.x2=x1(1:256)
  x3=idct(x2)
  figure(3)
  plot(x3)
  xlabel('DataPoint');ylabel('Amplitudo (Mikro Volt)');
```

Program segment 3.1 explains the DCT process on the EEG Z001 signal with a signal length of 1500 using the data set A. In line 1 it picks up the EEG Z001 signal in the file storage folder. The second line initializes the signal to be DCT. The third line is the process of transferring EEG signals with a length of 256 using DCT. The fourth line is the DCT inverse transfer process to prove that the initial signal after the DCT has not changed. Processes in the two segment program are applied to five classes of EEG signal data sets.

## 4. EEG Signal Identification Using Backpropagation Artificial Neural Networks

The data needed for the discussion of Backpropagation neural networks is the data from the feature extraction values from the DCT method. The feature extraction value data from the DCT method has variation data from five datasets taking a DCT value of 256 data points per EEG signal.

Network architecture that will be used in the learning process. The classification model is formed using Backpropagation Artificial Neural Network with Multi-layer architecture[12]. The number of input variables used in this study are 5 variables which are 5 different classes of data, where the researcher will make 3 structures of Artificial Neural Networks (ANN). The first ANN structure consists of two input variables, namely class A data of normal eye signal open) and class B (normal signal closed eyes), for the second ANN structure consisting of three input variables, namely class C data (entering epileptogenic zone), class D ( when the epileptogenic zone) and class E (seizure activity), for the third ANN structure consists of two input variables namely class A data (normal signal open eye) and Class E (seizure activity).

At each layer of the ANN architecture also set some parameters that will be given in the learning process and tested to form a classification model. The activation function used in the hidden layer uses the trainlm, while the output layer uses the purelin activation function. The minimum error tolerance (error) is 0.00001, the learning rate is 0.1, the maximum number of epochs in this study is 1000.

**Program segment 4.1: Training process (Learning)**

```
% Training Process (learning)
% DATA FROM DCT 256 CIRI
1. a1=load('D:\ade\DATA SET\Z Set A\Data_DCT_T_256_100.txt')'
   b1=load('D:\ade\DATA SET\O SET B\Data_DCT_T_256_100.txt')'
   c1=load('D:\ade\DATA SET\N SET C\Data_DCT_T_256_100.txt')'
   d1=load('D:\ade\DATA SET\F SET D\Data_DCT_T_256_100.txt')'
   e1=load('D:\ade\DATA SET\S SET E\Data_DCT_T_256_100.txt')'
2.%Define inputs (combine samples from all three classes)
   aa=a1(1:256,1:80)
   bb=b1(1:256,1:80)
   cc=c1(1:256,1:80)
   dd=d1(1:256,1:80)
   ee=e1(1:256,1:80)
3. % Data Target
   a =[1];
   b = [1];
   c = [-1];
   d = [-1];
   e = [-1]';
4. % Define Targets
   P = [aa,bb,cc,dd,ee];
   T    =    [repmat(a,1,80)    repmat(b,1,80)    repmat(c,1,80)    repmat(d,1,80)
   repmat(e,1,80)];
5. %Membuat jaringan
   net=newff(minmax(P),[30,20,1])
   net.trainParam.lr=0.1
   net.trainparam.show=100;
   net.trainParam.epoch=1000
   net.trainParam.goal=0.00001
6. %Proses Training
   net=train(net,P,T);
   %Hasil Setelah Pelatihan
   w1= net.IW{1,1}
   w2= net.LW{2,1}
   w3= net.LW{3,2}
   b1=net.b{1}
```

Journal of Physics: Conference Series **1381** (2019) 012037 doi:10.1088/1742-6596/1381/1/012037

```
  b2=net.b{2}
  b3=net.b{3}
7. save net.mat net
  y=sim(net,P)
   OutputLth=y'
```

Program Segment 4.1 explains the Training (learning) process in the ABCDE data set. In line 1 picks up the ABCDE data set signal in the file storage folder. The second line of signal initialization process will be in Training where the signal length used in each pattern is 256 and the data used in each class is 80 signal patterns. The third row is the initialization of the output layer value that will be generated, namely A (1), B (1), C (-1), D (-1) and E (-1). The fourth line is the expected target initialization process. The fifth line initializes the learning parameters used in the training process. The sixth line is the initialization of learning parameters used in the training process. Line 7 storage process value of the weights and bias values obtained in the training process. The process in segment one program is applied to the AB data set and CDE data set training processes.

**Segmen program 4.2 : Proses Testing**

%program Testing (uji coba)

```
1. %pemanggilan jaringan yang telah dilatih
   load net.mat
2. %proses membaca data yang diuji dari data set
   a1=load('D:\ade\DATA SET\Z Set A\Data_DCT_T_256_100.txt')'
   b1=load('D:\ade\DATA SET\O SET B\Data_DCT_T_256_100.txt')'
   c1=load('D:\ade\DATA SET\N SET C\Data_DCT_T_256_100.txt')'
   d1=load('D:\ade\DATA SET\F SET D\Data_DCT_T_256_100.txt')'
   e1=load('D:\ade\DATA SET\S SET E\Data_DCT_T_256_100.txt')'
3. %Define inputs (combine samples from all three classes)
   aaa=a1(1:256,81:100)
   bbb=b1(1:256,81:100)
   ccc=c1(1:256,81:100)
   ddd=d1(1:256,81:100)
   eee=e1(1:256,81:100)
   aaa1=aaa
   bbb1=bbb
   ccc1=ccc
   ddd1=ddd
   eee1=eee
   pa=[aaa1,bbb1,ccc1,ddd1,eee1]
4. % Define Output
   Y2=net(pa)
    out=Y2'
```

Program Segment 4.2 explains the Testing process in the ABCDE data set. In line 1 is the process of loading the weight and bias values of the training results. In the second line picks up the ABCDE data signal set in the file storage folder. In the third line the signal initialization process will be tested where the signal length used in each pattern is 256 and the data used in each class is 20 signal patterns. The fourth line is the process of testing output results. The process of the two segment program is applied to the process of testing AB data sets and CDE data sets.

Based on the testing process carried out with 2 Architecture namely the first architecture for ABCDE data sets, the second architecture for AB data sets and CDE sets, using a number of different hidden layers and layer nodes, the results of Identification of five classes of EEG signals can be seen as follows :

**Table 1**
**Results of ABCDE Data Set Identification**

| | *1 Hidden Layer* | *2 Hidden Layer* | *3 Hidden Layer* |
|---|---|---|---|
| Time | *5: 03minutes : 04second* | *08: 56minutes : 23second* | *0: 38minutes : 03second* |
| Iterasi | *709* | *1000* | *38* |
| *MSE* | *9,84x 10-6* | *0,00431* | *7,01x 10-6* |
| Ketepatan | *74%* | *75%* | *76%* |

**Table 2**
**Results of Identification of AB and CDE Data Sets**

| | *1 Hidden Layer* | *2 Hidden Layer* | *3 Hidden Layer* |
|---|---|---|---|
| Time | *0: 56minutes : 54second* | *2: 51minutes : 42second* | *14:11minut es : 14second* |
| Iterasi | *132* | *114* | *1000* |
| *MSE* | *7,47 x 10-6* | *6,04 x 10-6* | *0,0112* |
| Ketepatan | *72%* | *73%* | *79%* |

## 5. CONCLUSION

A careful analysis of the electroencephalogram (EEG) record can provide valuable information value to understand the mechanism behind epilepsy disorders. Since epileptic seizures occur irregularly and unexpectedly, automatic seizure detection in EEG records is needed. The results of this study indicate that with feature extraction using DCT and the identification process using Backpropagation Neural Networks, the test results obtained for the identification of EEG signal ABCDE class data sets by 75%, for AB and CDE class data sets by 73%.

## 6. BIBLIOGRAPHY

[1] Polat K and Güneş S 2007 Classification of epileptiform EEG using a hybrid system based on decision tree classifier and fast Fourier transform *Appl. Math. Comput.* **187** 1017–26

[2] Andrzejak R G, Lehnertz K, Mormann F, Rieke C, David P and Elger C E 2001 Indications of nonlinear deterministic and finite-dimensional structures in time series of brain electrical activity: Dependence on recording region and brain state *Phys. Rev. E - Stat. Physics, Plasmas, Fluids, Relat. Interdiscip. Top.* **64** 8

[3] Gu I 2005 Recurrent neural networks employing Lyapunov exponents for EEG signals classification **29** 506–14

[4] Adeli H, Ghosh-dastidar S and Dadmehr N 2006 A Wavelet-Chaos Methodology for Analysis of EEGs and EEG Sub-Bands to Detect Seizure and Epilepsy

[5] Subasi A 2007 EEG signal classification using wavelet feature extraction and a mixture of expert model **32** 1084–93

[6] Ataee P, Avanaki A N, Shariatpanahi H F and Khoee S M 2006 Ranking features of wavelet-decomposed EEG based on significance in epileptic seizure prediction *Eur. Signal Process. Conf.*

[7] Guo L, Rivero D, Dorado J, Rabuñal J R and Pazos A 2010 Automatic epileptic seizure detection in EEGs based on line length feature and artificial neural networks *J. Neurosci. Methods* **191** 101–9

[8] Karyawan M A, Zainal A, Saikhu A, Teknologi I and Nopember S 2011 Klasifikasi Sinyal Eeg Menggunakan Koefisien Autoregresif , **2** 1–13
[9] Khayam S A 2003 ( DCT ):
[10] Güler I and Übeyli E D 2005 Adaptive neuro-fuzzy inference system for classification of EEG signals using wavelet coefficients *J. Neurosci. Methods* **148** 113–21
[11] Gautama T, Mandic D P and Van Hulle M M 2003 Indications of nonlinear structures in brain electrical activity *Phys. Rev. E - Stat. Physics, Plasmas, Fluids, Relat. Interdiscip. Top.* **67** 5
[12] Handayani L and Adri M 2015 Penerapan JST ( Backpropagation ) untuk Prediksi Curah Hujan ( Studi Kasus : Kota Pekanbaru ) 238–47

Journal of Physics: Conference Series **1381** (2019) 012038 doi:10.1088/1742-6596/1381/1/012038

# Game education for child with disabled handle based on multimedia

**C Taurusta, A S Fitrani*, M Suryawinata, R Dijaya, and M W D Astutik**
Informatics Department, Universitas Muhammadiyah Sidoarjo, Sidoarjo, Indonesia

asfjim@umsida.ac.id

**Abstract.** Learning media is very influential on the interest of learning in children, especially in students. Many kinds of learning media, one of them using internet technology, computers and even mobile phones or smartphones that have a positive impact and have a negative impact. Learning media is also very influential in early childhood. Education at an early age is very important, especially for handicapped children who need more attention. Improving the effectiveness of learning should be improved to be easy to understand and not boring, then parents should improve in various ways. it includes academic education and creativity. The development of interactive multimedia learning technology allows the introduction of speech recognition technology into the application. Speech recognition is a voice command to operate a computer. Because speech recognition is a process for converting acoustic signals (sound) through a microphone as a computer operation command or writing a word (dictation). So in operation, users no longer need to use the cursor or press the keyboard. Through voice commands, users can operate the learning media..

## 1. Introduction

Early childhood is an individual who is experiencing a very rapid process of growth and development. It is even said to be a leap in development, because that is an early age said to be the golden age (golden age) which is a very valuable age compared to later ages. Early childhood (aged 0-6 years), it is easy to capture and remember the lessons taught by the teacher. Early childhood does not like the teacher who teaches the lesson with a serious method. Usually early childhood prefer to learn while playing (Wardhanie, 2011) That is very needed by early childhood especially for young children who suffer from hand disabilities. They more need special guidance for learning early childhood lessons.

Improving the effectiveness of learning should be improved so that it is easy to understand and not boring, so parents must improve in various ways. this includes academic education and creativity. The development of interactive learning multimedia technology has made it possible to incorporate speech recognition technology into the application. Speech recognition is a voice command to operate a computer, because speech recognition is a process to convert an acoustic signal (voice) through a microphone as a computer operating command or write words (dictation). So in operation, users no longer need to use the cursor or press the keyboard. Through voice commands, users can operate the learning media (widaryanto, 2015). For example in existing research, namely the use of speech recognition in spaceship counting games (hans alfon ericksoon, 2016).

Based on the problems and previous studies, what will be done is to build word recognition games for early childhood, especially children with hand disabilities using speech recognition technology to present learning and games using sounds from children. Children can learn the introduction of several words with English material, one of which is the introduction of color words (colors) and fruit (fruit) with sequential learning from letter recognition to several words. Accompanied by an innovative game that also uses sound as input from this game that is by inputting sound, to start learning and playing and also playing games. It aims to increase the effectiveness of learning by visualizing interactive learning so as to increase the enthusiasm of early childhood learning for people with disabilities.

## 2. Fundamental

There are many fundamental theory that supported this journal research. The theory are :

### *2.1 Early Childhood*

John Locke, a philosopher from England illustrates a small child as a person who is still clean and sensitive to various stimuli. Any stimulation, whether good or bad will be received by early childhood with a little filter, whether it is also received consciously or settles in the subconscious. That is why it is often said that children are accomplished learning. According to the 2003 National Education System Law, children aged 0-6 years are called early children. Age 0-6 years is an age called the golden age or golden age because at this time, children experience a process of rapid growth and rapid and rapid development that will not be replaced in the future. Early childhood is divided into 3 stages, namely:

a) The period of infancy from birth to 12 months (1 year);
b) Childhood or toddler age 1-3 years;
c) School period from 3-6 years old;

### *2.2 The Child with Special Needed*

In the big Indonesian dictionary, abnormal words are interpreted to be incompatible with unusual circumstances, have neglect and are not normal. In the RI law No.2 of 1989 concerning the national education system it is emphasized that children or students who have physical and mental neglect are referred to as extraordinary children. While in RI law No.20 of 2003 concerning the National Education System, children who have physical and mental neglect are referred to as children with special needs.

The development of early childhood there is a normal running both in terms of physical and psychological, there is also an abnormal going on in terms of physical and psychological terms that make them into the category of early childhood with special needs. Sevara is simple, early age children with special needs are children aged 0-6 years who experience developmental disorders that are significantly different from normal children so that in their daily lives and in various activities they require special treatment from others. (Novan Ardy Wiyani, 2014).

### *2.3 Games*

Games are a medium for playing activities. Play activity is an activity that includes solving problems that become challenges of the game, by following certain rules. Games become interesting because the challenges and rules in the game are packaged in a certain scenario. From one side, the activity of playing games is seen as an unproductive activity. However, on the other hand playing games can be seen as a learning activity. This happens because players are required to learn the ways that must be done to conquer the challenges given. Thus, by including learning content in it, the game can be used as an instructional system. (Affan Mahtarami, 2010).

#### 2.3.1 *The Types of Genre Game*

In a journal written by Putra (2016) states that according to Expro (2010) there are types of games as follows: (Vicky, 2012).

a. Action Games

Action game or action game involving human players who shoot a number of opponents or objects.

b. Fighting Games

Fighting games involving characters who fight are usually hand to hand, in one-on-one combat situations. The warriors are usually represented as human or animated characters.

c. First Person Shooter (FPS)

FPS (First Person Shooter) is a type of shooting game with the appearance of the game that is the character that we are playing, usually the perspective only shows the hands and weapons only. This fps game has missions for a specific purpose. The hallmark of this game is the use of long-range weapons.

d. Third Person Shooter

TPS (Third Person Shooter) is a game that has a characteristic similar to FPS that usually has a shooting gameplay except that the point of view used in this game is a third person looks half body.

e. Strategy

Strategy games emphasize the use of strategies that are contrary to fast action or the use of quick reflexes. In this game, players need to solve the problem of resource allocation, and organizing defenses and attacks.

f. Tycon

Tycoon is a game that makes us a businessman who will develop a property to be developed until it sells well.

g. Racing

Racing Game is a game or a type of vehicle racing game that is played by controlling our vehicle to win a race

h. Adventure

Adventure games are different from action games. They put more emphasis on stories, plots and puzzle solving than just capturing, shooting, capturing, or running away. Human players have to solve puzzles while adventuring. These game settings often lead to certain historical periods and places, such as the Middle Ages or Arthurian English, or thematically related to content. types like Science Fiction, Fantasy, or Espionage.

i. Sport

Sports games are adaptations of real world sports or variations of them.

j. Education

Educational games are games that are explicitly designed for educational purposes, or that have educational value. All types of games can be used in an educational environment. Educational games are games that are designed to help users learn about certain subjects, expand concepts, strengthen development, understand historical or cultural events, or assist users in learning skills when users play. Educational game is a game designed to teach humans about certain topics and teach them skills. Educational games are interactive games that teach users goals, rules, adaptations, problem solving, interactions, all represented as a story.

### *2.4 Speech recognition*

Automatic speech recognition (ASR) is a technology applied to software to receive input in the form of spoken words. This technology allows a device to recognize and understand a word spoken by digitizing words and matching the digital signal with a certain pattern stored in a device. The words spoken are converted into a group of numbers which are then adjusted to certain codes to identify those words (hans alfon ericksoon, 2016). Speech recognition used in this word recognition game is a plugin from Unity. In use speech recognition will send the sound received to the server to be processed into text. (hans alfon ericksoon, 2016).

## 3. Metodology

Materials used by the author to conduct research are several literary readings including, reference books about color recognition, fruit for children, journals and scientific articles relating to early childhood abilities, children with hand disabilities, about educational games and about Automatic Speech technology Recognition (ASR).

### *3.1. Research Framework*

The framework of this study will be presented in the form of a flowchart below:

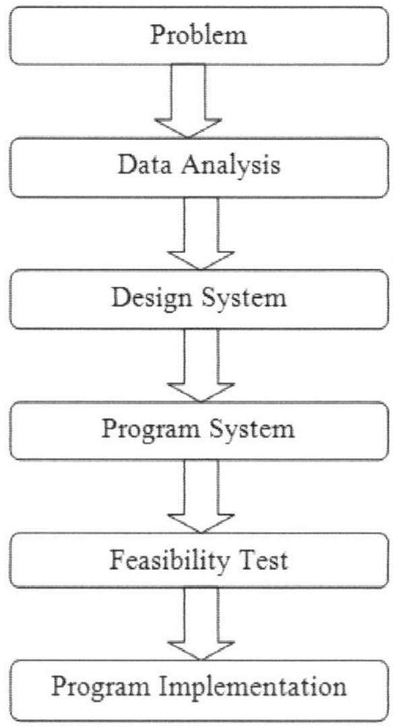


**Figure 1.** Research Framework

*3.2. Design System*
There are several design systems in this picture, and the namely are : Flowchart, Use Case, and Finite State Machine (FSM).

*3.2.1 Flowchart*
Following the General flowchart system design of this research :

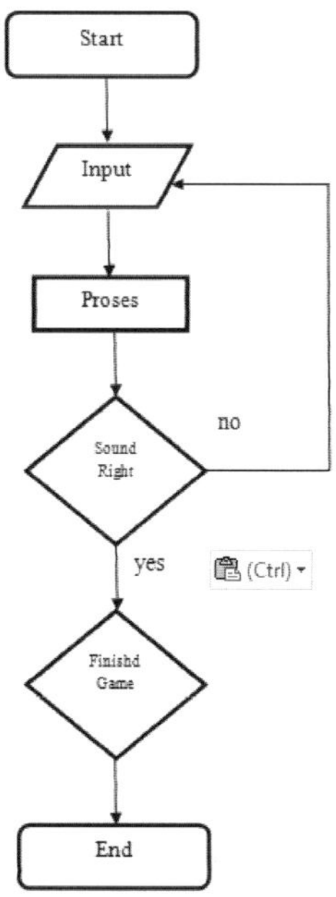


**Figure 2.** Flowchart Game

Figure 2 shows the game starts by selecting the game according to the material that has been studied first. To starst the game, which is using voice input. Likewise with how to play that is by inputting the sound from the user and the sound will be processed to match whether it is correct. If true, it will go to the next game and vice versa if it is wrong, there will be a reminding sound and the user must input the new sura until completing the game. This General Flowchart have many part, there are :

**a. Flowchart Learn and Submenu Learn**

**Figure 3.** Flowchart Learn

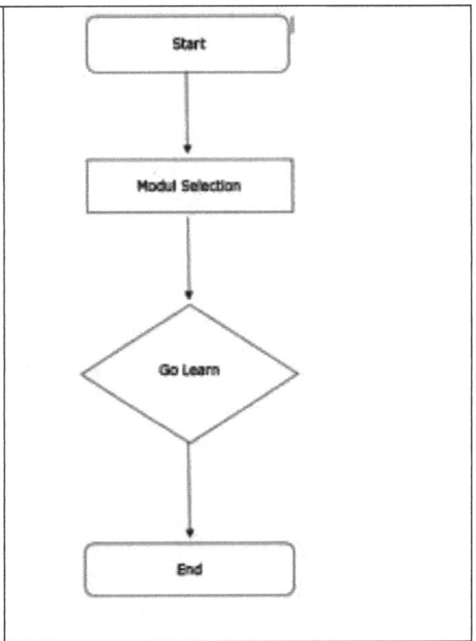


**Figure 4.** Flowchart Submenu Learn

**b. Flowchart Play and Submenu Play**

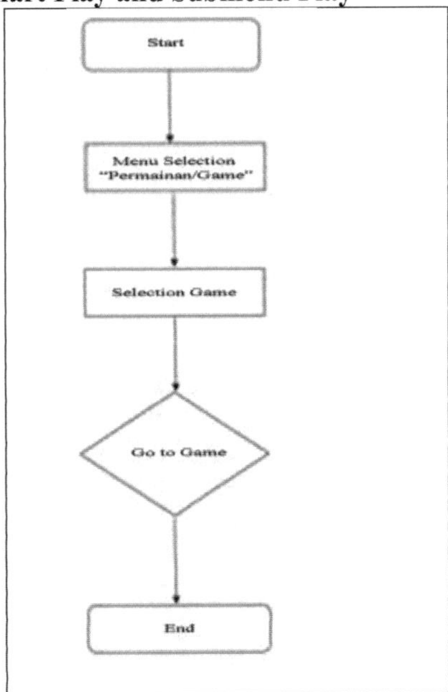


**Figure 5.** Flowchart Play

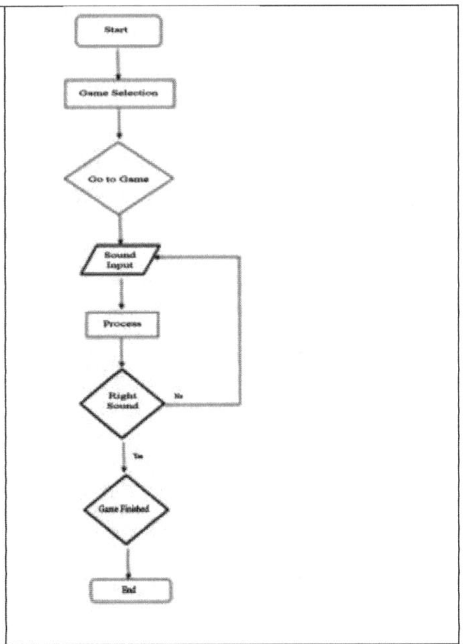


**Figure 6.** Flowchart Submenu Play

Figure 5 shows the selection of the game menu, namely by selecting a menu and selecting submenu which later contains a game that is available to play according to the learning material. And Figure 6 shows how to play in the game menu, which is by choosing a game that is suitable with the material that has been studied, then the user will play the game using sound

### *3.3. Use Case*

This is use case for this game :

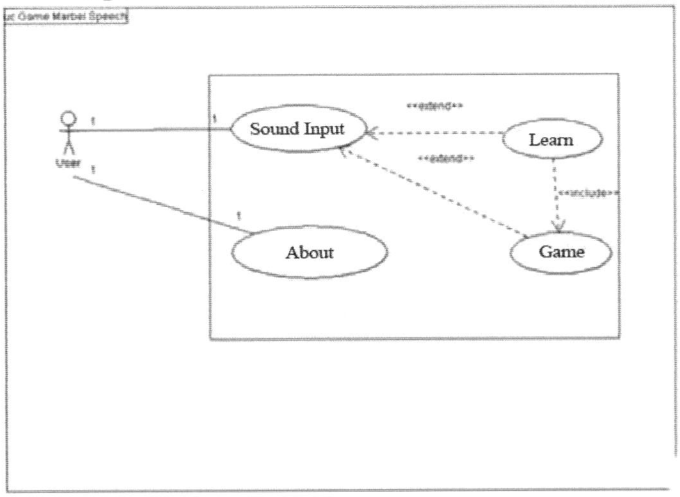


**Figure 7.** Use Case Game

Figure 7 shows how users input words using sound, then the sound will be processed and recognized, if appropriate, it will bring up the results and move to the next level.

### 3.4. *Finite State Machine (FSM)*

*1. Scene Learn*

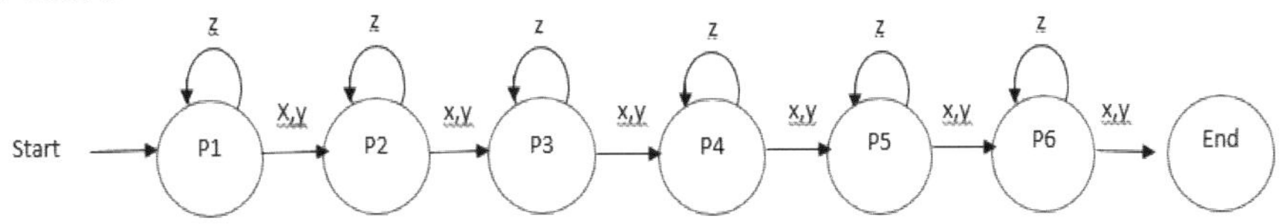


**Figure 8.** FSM Scene Learn

Q = {P1, P2, P3,P4,P5,P6 End}
Σ = {x, y, z}
S = {P1}
F = {End}
δ = {((P1, x), P2), ((P1, y), P2), ((P1, z), P1), ((P2, x), P3), ((P2, y), P3), ((P2, z), P2), ((P3, x), P4), ((P3, y), P4), ((P3, z), P3), ((P4, x), P5)), ((P4, y), P5)), ((P4, z), P4)), ((P5, x), P6)), ((P5, y), P6)), ((P5, z), P5)), ((P6, x),End), ((P6, y), End)), ((P6, z), P6))}

Figure 8 shows the flow in the learning scene. So the user inputs their sound (x), and gestures / clicks on the button (y) in the learning menu to open or study some material and if it fails (z) recognize the sound entered, it will re-enter the sound. Likewise in learning the second material and so on.

Journal of Physics: Conference Series **1381** (2019) 012038 doi:10.1088/1742-6596/1381/1/012038

2. *Scene Play*

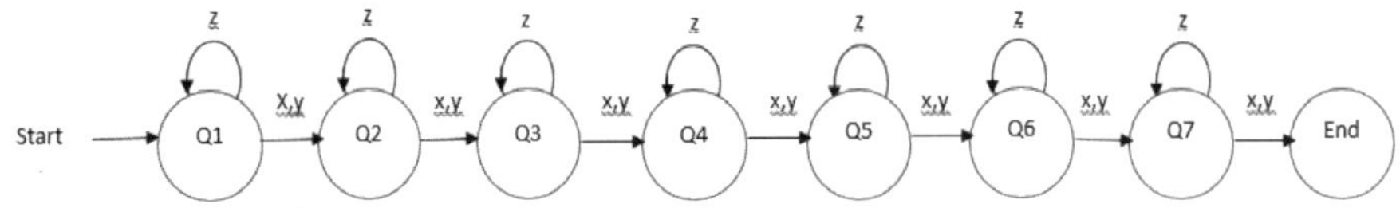


**Figure 9.** FSM Scene Play

{Q1, Q2, Q3,Q4,Q5,Q6 ,Q7, End}
{x, y, z}
{Q1}
{End}
{((Q1, x), Q2), ((Q1, y), Q2), ((Q1, z), Q1), ((Q2, x), Q3), ((Q2, y), Q3), ((Q2, z), Q2), ((Q3, x), Q4), ((Q3, y), Q4), ((Q3, z), Q3), ((Q4, x), Q5)), ((Q4, y), Q5)), ((Q4, z), Q4)), ((Q5, x), Q6)), ((Q5, y), Q6)), ((Q5, z), Q5)), ((Q6, x),Q7), ((Q6, y), Q7)), ((Q6, z), Q6)), ((Q7,x),End),((Q7,y),End),((Q7,z),Q7)}

Figure 9 shows the plot in the game scene. So the user inputs their sound (x), and moves / clicks on the button (y) in the learning menu to open or play a game or quiz in the game menu and if it fails (z) recognize the sound entered, it will re-enter the sound. Likewise in the second game and so on.

3. *Scene Word Recognition*
There are 2 word recognition in this game, letter and fruit. Figure 9 shows FSM Letter recognition and Figure 10 shows FSM fruit recognition.

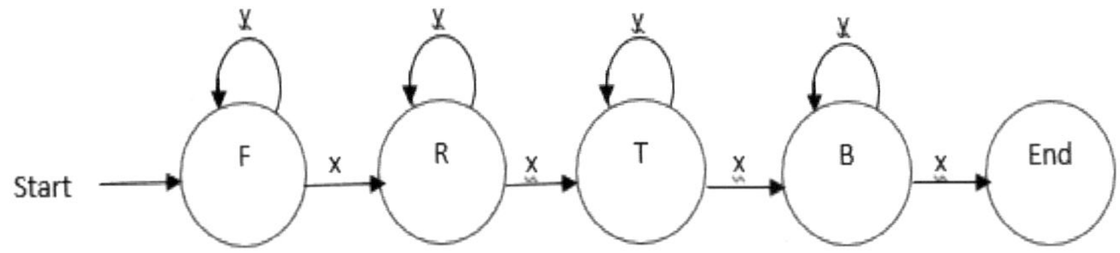


**Figure 10.** FSM Scene Letter Recognition

Q = {F,R,T,B, End}

Σ = {x, y}

S = {F}

F = {End}

δ = {((F, x), R), ((F, y), F), ((R, x), T), ((R, y), R), ((T, x), B), ((T, y), T), ((B, x), End), ((B, y), B)}

Figure 10 shows the speech recognition flow. So the user inputs his voice (x), without clicking to open guessing the letters that are already on the display and if it fails (y) recognize the sound entered, it will re-enter the sound. Likewise with the second question and so on until the game ends.

Journal of Physics: Conference Series **1381** (2019) 012038 doi:10.1088/1742-6596/1381/1/012038

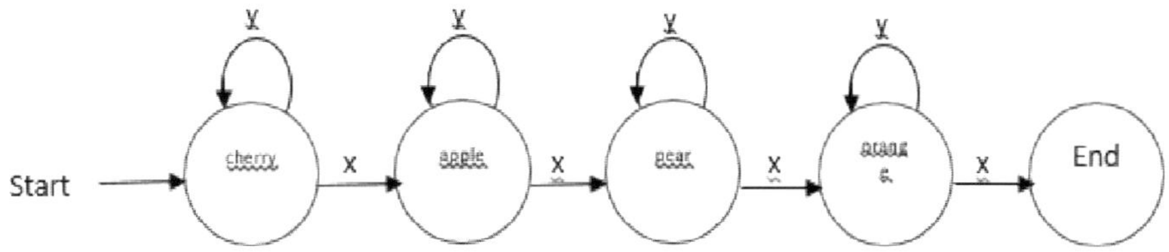


**Figure 11.** FSM Scene Fruit Recognition

Q = {cherry,apple,pear,orange, End}

Σ = {x, y}

S = {cherry}

F = {End}

Figure 11 shows the speech recognition flow. So the user inputs his voice (x), with no click to open guessing the wxcord on the display and if it fails (y) recognizes the sound entered, it will re-enter the sound. Likewise with the second question and so on until the game ends.

## 4. Result

This word recognition game in English is a learning application that is suitable for early childhood accompanied by parents to get more insight into learning English that is packaged interactively by displaying learning starting from the basic recognition of letters and how to read it, writing in small letters and there are some games that are suitable for children with special needs, especially children with disabilities.

a. Interface Main Menu

**Figure 12.** Main Menu

The main display of this application which consists of learning menus, games, about, help and close.

This button is used to enter the learning menu which contains learning animations with interactive sounds from each material.

This button is used to enter the game menu which contains a game that uses Speech Recognition to play it.

This button contains information about the application and the application maker.

This button is used to exit and close the application.

b. Interface Menu Learn

**Figure 13.** Interface Menu Learn

c. Interface Learn Letter

**Figure 14.** Interface Learn Letter

Figur 14 displays several examples of consonant letters with writing how to read them and lowercase letters to differentiate the writing of uppercase and lowercase letters alternately accompanied by sound to find out how to read them.

d. Interface Vokal Letter

**Figure 15.** Interface Vokal Letter

e. Interface Playing Guess the Letter

**Figure 16.** Interface Guess the Letter

f. Interface Playing Guess the Fruit

**Figure 17.** Interface Guess the Fruit

g. User Testing

User testing is done by testing the application with the user directly, this is done to find out whether the application is feasible to use. The main purpose of this trial is to find out the problems that exist in the application when used by users. The author will summarize the problem by asking 3 questions with a score of 1-25 with 4 respondents. With the score description as follows:

**5 - 10 = Less**
**11 - 15 = Enough**
**16-20 = Good**
**21 - 25 = Very Good**

**The questions that will be asked are as follows:**

- Is this application easy enough to use?
- To what extent does this application help early childhood sufferers with hand disabilities be able to learn to recognize English words?
- Do the results displayed by the application match your needs / desires?

**Table 1. Evaluation Result**

| | | Question | | |
|---|---|---|---|---|
| | | A | B | C |
| Respondent | 1 | 20 | 25 | 25 |
| | 2 | 25 | 20 | 20 |
| | 3 | 24 | 20 | 24 |
| | 4 | 25 | 25 | 25 |
| Total | | 94 | 85 | 94 |

**Average percentage = (number of grades) / (many scores) x 100%**
**Average percentage** = 273/300 x 100% = **91%**

Based on the table above, it can be concluded that the trial application runs smoothly with a success of 91%, it can be concluded that the Alfa Speech game application has decent quality.

## 5. Summary

In accordance with the results of the discussion that has been done, it can be concluded that:

1) With the English word recognition application, it can help early childhood learn English and help add vocabulary around them in English.
2) With the input in the form of sound to help early childhood sufferers with hand disabilities in playing games that are in the application.

## References

Wardhanie, R. I. 2011. Game Pengenal Nama Hewan Dan Nama Buah Untuk Anak Playgroup.

Widaryanto, D. P. 2015. Media Pembelajaran Interaktif Berbasis Voice recognition.

Hans Alfon Ericksoon, I. K. 2016. Rancang Bangun Game Berhitung Spaceship Dengan Pengendali Suara Menggunakan Speech recognition Plugin Pada Unity.

Novan Ardy Wiyani, M. 2014. *penanganan Anak Usia Dini Berkebutuhan Khusus.* Yogyakarta: Ar-Ruzz Media.

Affan Mahtarami, M. N. 2010. Pengembangan Game Pembelajaran Otomata Finit

Putra, G. F. 2016. Pembuatan Game Tentang Permainan Tradisional Patil Lele Bergenre Sports.

Vicky. (2012, September 9). *belajar-komputer-mu.com*. Retrieved 12 15, 2017, from http://belajar-komputer-mu.com/pengertian-dan-kegunaan-program-corel-draw.

# Classification Of Dengue Hemorrhagic Disease Using Decision Tree With Id3 Algorithm

**M A Rosid*, A S Fitrani, Y Findawati, S Winata, and V A Firmansyah**

Program Studi Informatika, Universitas Muhammadiyah Sidoarjo, Jl. Raya Gelam 250 Candi Sidoarjo, Jawa Timur, Indonesia

*alfanrosid@umsida.ac.id

**Abstract**. Dengue Hemorrhagic Fever (DHF) is a disease caused by the dengue virus which is transmitted by Aedes Aegypty, and Aedes Albopictus (DHF) mosquitoes is a disease caused by the dengue virus transmitted by mosquitoes Aedes Aegypty and Aedes Albopictus. There are three categories of DHF with different symptoms; the first category is called Dengue Fever (DF), the symptoms of DF are characterized by fever, these symptoms are almost the same as symptoms of typhoid or Typhoid Fever (TF). The second category is Dengue Hemorrhagic Fever (DHF) symptoms of DHF characterized by fever, nausea, vomiting, red spots, and nosebleeds. The last category is dengue shock syndrome (DSS) the advanced category of DHF. To determine the level of DHF disease experienced by patients with a background of various symptoms, the DHF disease classification study was conducted using the ID3 algorithm. It is hoped that this study can help doctors diagnose DHF disease. The achievement of predictions from research using the ID3 algorithm can produce an approach in the form of symptoms that affect the level of DHF disease with an accuracy rate of 82%.

## 1. Intoduction

Dengue hemorrhagic fever, also known as DHF, is a disease caused by dengue virus through the bite of the Aedes Aegypty and Aedes Albopictus mosquitoes. Dengue hemorrhagic fever is characterized by sudden fever and bleeding either in the skin or elsewhere in the body which can cause shock and death [1]. according to Data from the Directorate for Prevention and Control of Zoonotic and Zoonotic Infection Disease, in Indonesia cases of dengue still occur every year[2], the Ministry of Health (Ministry of Health) released data on the development of dengue fever (DHF) from 2014 to January 2019. Director Vector and Zoonotic Infectious Diseases Ministry of Health Siti Nadia Tarmizi claimed to have received reports on DHF the last five years in a row namely since 2014, 2015, 2016, 2017, 2018, until 2019.

The number of DHF sufferers in 2014 was 100,347 people, then in 2015 it was 129,650, then in 2016 it was 204,171[2]. Then in 2017 as many as 68,407, then in 2018 as many as 53,075, and in 2019 as many as 13,683 people, from these data the number of dengue patients who died in 2014 were 907 people, the following year 1,071 people, then in 2016 as many as 1,598 people, and 2017 as many as 493 Then during 2018 there were 344 people and in 2019 (until 29 January 2019) there were 133 people[3]. Frans Yosep Sitepu et al conducted research on the Epidemiology and Entomology of Extraordinary Dengue Fever in South Nias Regency, from the results of his research An outbreak of DF was confirmed in Teluk Dalam Subdistrict, South Nias District, North Sumatra. Not eliminating mosquito breeding sites routinely and having the habit of hanging worn clothes are the most potential risk factors associated with the outbreak[4]. Hina Gul et al. Made a study to identify the findings of ultrasound in severe dengue infection and the results of this study will assist doctors in developing treatment options strategies[5]. To predict DHF, the diagnosis of symptoms experienced by patients requires a method. ID3 algorithm is one of the decision making methods in data mining techniques[6][7][8] which can be applied to make it easier to predict DHF because the attributes of the dataset will be managed categorically. Based on the background above, we apply Data Mining for Dengue Hemorrhagic Fever Classification Using the Decision Tree Method With ID3 Algorithm.

## 2. Dangue Hemorrhagic Fever

Dangue hemorrhagic fever or known as DHF is a disease caused by dengue virus through the bite of Aedes Aegypty and Aedes Albopictus mosquitoes. DHF can be classified according to the severity of the patient. Infants and adults who have been infected for the first time can usually be characterized by fever, skin rashes, respiratory symptoms and digestion. The following picture of the phase of dengue fever:

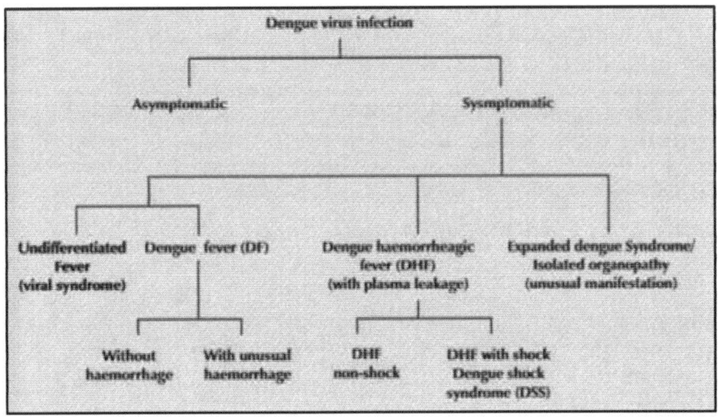


Figure 1. Dengue Fever Phase (World Health Organization , 2011)

At the initial level called Dengue Fever (DF), DF often occurs in children, adolescents and adults. DF can be characterized by acute fever with headaches, muscle aches and rashes on the skin, although DF is not dangerous, in some cases, DF can also cause nosebleeds.

At the second level, it is called Dengue Haemorrhagic Fever (DHF), at the age of under 15 years, it is often infected with DHF. In the initial phase the symptoms are similar to DF symptoms while in the final phase, fever tends to experience hypovolemic shock due to plasma leakage or called Dengue Shock Syndrome (DSS). Before the final phase there are symptoms that can be characterized by nausea, vomiting, abdominal pain, body weakness and nosebleeds. At the final level or DSS, it can involve vital organs such as the heart, kidneys, brain, heart and lungs. Symptoms include unstable breathing, fainting and can cause death[9].

## 3. ID3 algorithm

ID3 algorithm is a classic algorithm that is often used in data mining technology, which is mainly applied to data mining implementation[8], classification algorithm for data mainly includes decision tree induction classification, rule-based classification methods, Byes, artificial neural networks, K- nearest neighbor , support vector machines and associations rules classification. In this data classification method, the classification process can be concluded as follows: the first is the learning of classification rules, that is, the learning phase of rules is to establish the classification stage; the second stage is the stage of test data, treatment of data classified using class analysis, and treat classification of data to be classified[8][10]. Following are the ID3 algorithm steps in building a decision tree:

1 Select the attribute as the root node from the highest gain value.
2 Create a branch of each variable based on the selected attribute.

3 Calculate each variable until all cases have the same class.

To calculate the gain can be seen from equation 1 as follows:

$$\text{Gain}(S,A) = \text{Entropy}(S) - \sum_{i=1}^{n} * \frac{|Si|}{S} * \text{Entropy}(Si) \quad (1)$$

Information:

S : case set
A : attribute
n : Number of attribute attributes A
|Si| : Number of cases on partitions
|S| : Number of cases in S

Meanwhile entropy value calculation can be seen from equation 2 below:

$$\text{Entropy}(S) = \sum_{i=1}^{n} -pi * \log_2 pi \quad (2)$$

Information:

S : case set
A : the features
n : number of partitions S
Pi : the proportion of Si with respect to S

**4. Research methodology**

The system design that will be made, based on data mining[11]classification techniques using the Decision Tree method with ID3 algorithm to classify dengue fever (DHF). There are several steps that will be carried out, namely:

a. Classification Diagram ID3

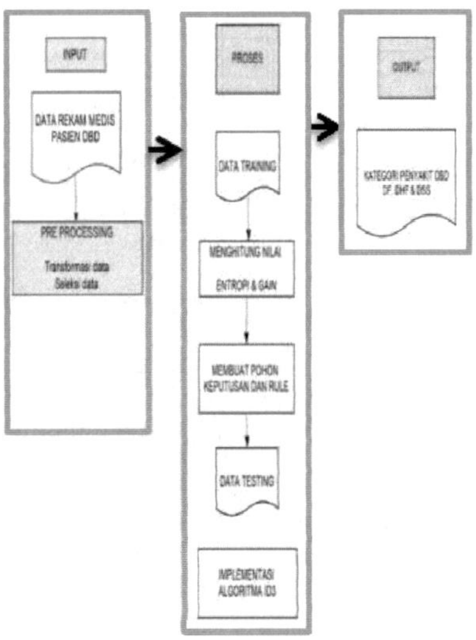


Figure 2. Classification Diagram ID3

The description of the classification diagram is as follows:

1. INPUT Phase, at this stage the data will be pre-processed where the data will be selected and the data will be transformed according to the method to be used.
2. PROCESS Phase, after the input stage is completed, the data will be divided into two namely training data and testing data. at the training data stage, the data will be calculated to find a rule and will be tested with data testing.
3. Output stage, at this stage the results of the data that has been processed include DHF disease categories such as DF, DHF and DSS.

## 5. Result

Data of DHF patient medical records obtained from the Bhayangkara Pusdik Brimob Watukosek Hospital were 481 data. The data will be tested using the Decision Tree method with ID3 algorithm using WEKA tools and website. Here are some examples of DHF patient medical record data, namely:

Table 1. Dataset

| No | Nama | Umur | Demam | Badan_ny | Mual | Diare | Pusing | Nafsu_ma | Badan_le | Mimisan | Gusi_berd | Bintik_me | Trombosit | Diagnosis_utama |
|---|---|---|---|---|---|---|---|---|---|---|---|---|---|---|
| 1 | Didik Hen | lansia | Ya | Ya | Ya | Tidak | Ya | Tidak | Tidak | Tidak | Tidak | Tidak | pre-syok | DF |
| 73 | Weni Kus | kanak-kar | Ya | Ya | Tidak | Tidak | Tidak | Tidak | Tidak | Tidak | Tidak | Tidak | pre-syok | DF |
| 164 | Lintang N | kanak-kar | Ya | Tidak | Ya | Tidak | Tidak | Tidak | Tidak | Tidak | Tidak | Tidak | ambang-dengue | DF |
| 10 | Kadek Yur | remaja | Ya | Ya | | | | | | | | | | |

Table 2. Training Data

| No | Umur | Demam | Badan_ny | Mual | Diare | Pusing | Nafsu_ma | Badan_le | Mimisan | Gusi_berd | Bintik_me | Trombosit | Diagnosis_utama |
|---|---|---|---|---|---|---|---|---|---|---|---|---|---|
| 1 | lansia | Ya | Ya | Ya | Tidak | Ya | Tidak | Tidak | Tidak | Tidak | Tidak | pre-syok | DF |
| 73 | kanak-kar | Ya | Ya | Tidak | Tidak | Tidak | Tidak | Tidak | Tidak | Tidak | Tidak | pre-syok | DF |
| 164 | kanak-kar | Ya | Tidak | Ya | Tidak | Tidak | Tidak | Tidak | Tidak | Tidak | Tidak | ambang-dengue | DF |

Table 3. Testing Data

| No | Umur | Demam | Badan_ny | Mual | Diare | Mual | Diare | Pusing | Nafsu_ma | Badan_le | Mimisan | Gusi_berd | Bintik_me | Trombosi | Diagnosis_ |
|---|---|---|---|---|---|---|---|---|---|---|---|---|---|---|---|
| 290 | remaja | Ya | Tidak | Ya | Tidak | Ya | Tidak | Ya | Tidak | Tidak | Tidak | Tidak | Tidak | ambang-d | DF |
| 298 | remaja | Ya | Tidak | Ya | Tidak | Tidak | Tidak | Tidak | Tidak | Tidak | Tidak | Tidak | Tidak | ambang-d | DF |
| 317 | kanak-kar | Ya | Tidak | Ya | Tidak | Ya | Tidak | Tidak | Tidak | Tidak | Tidak | Tidak | Tidak | ambang-d | DF |

The dataset used for the calculation process was 481 case data. The dataset is divided into 2 parts, training data and testing data. Training data is used to process a calculation by forming a classifier

Journal of Physics: Conference Series **1381** (2019) 012039 doi:10.1088/1742-6596/1381/1/012039

model. While testing data is used to measure the extent to which the classifier managed to classify correctly. Variables used for the calculation process are age, fever, body aches, nausea, diarrhea, dizziness, decreased appetite, body weakness, nosebleeds, bleeding gums, platelets and the main diagnosis.

### *5.1 Preprocessing*

Unprocessed data is raw data, so it needs to be prepared in advance so that it can be used in the data mining process. The techniques contained in pre-processing in preparing data are removing data duplication, removing data inconsistency, correcting data errors and transforming data according to the method used. The amount of data after the pre-processing is 339 data that will be divided into two, 65% training data with 220 data and 35% testing data with 119 data.

### *5.2 Process*

The process carried out at this stage is to determine the results of the classification calculation with the ID3 algorithm. The use of Weka tools in implementing classification methods to predict dengue fever. From the results of the process using Weka, it is obtained Re Evaluated Model, from the testing data as many as 5,298 data have been obtained as predicted results as Figure 3.

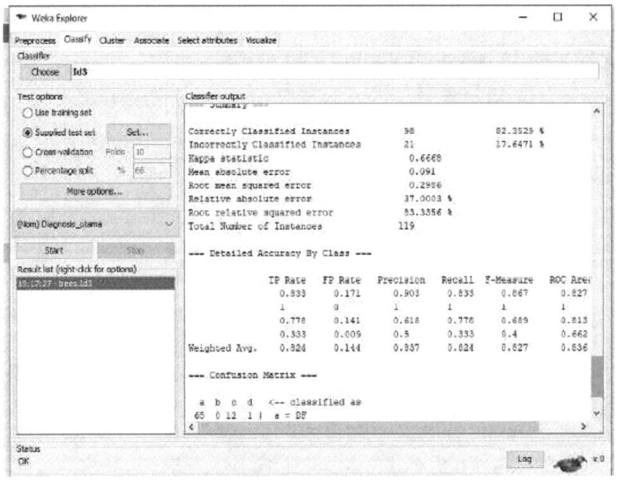


Figure 3. Display of Re-Evaluated Model Results

### *5.3 Cross Validation*

To verify the results of the experiment, cross validation is carried out, the experiment will be carried out 10 times, A 10-fold cross-validation procedure suggests a quite stable selection[12], the dataset of this experiment is divided into two pieces namely training data and testing data, determination of training data and testing data is carried out randomly and evenly so that no grouping of documents occurs[10]. documents from one particular category in an experiment. The training data is taken as much as 277 of the dataset randomly and evenly for each category, while the rest is used as testing data, this data sharing will be done 10 times for ten experiments. The results of the experiments will find the average accuracy value. The results of cross validation from 10 experiments are shown in table 4.

Table 4. Cross Validation Result

| Cross Validation | | Benar | Salah |
|---|---|---|---|
| Split 1 | Hasil | 188 | 32 |
| | % | 85,45 | 14,54 |
| Split 2 | Hasil | 188 | 31 |
| | % | 85,45 | 14,09 |
| Split 3 | Hasil | 193 | 27 |
| | % | 87,72 | 12,27 |
| Split 4 | Hasil | 191 | 27 |
| | % | 86,81 | 12,27 |
| Split 5 | Hasil | 187 | 30 |
| | % | 85 | 13,64 |
| Split 6 | Hasil | 188 | 30 |
| | % | 85,45 | 13,63 |
| Split 7 | Hasil | 188 | 32 |
| | % | 85,45 | 14,54 |
| Split 8 | Hasil | 187 | 33 |
| | % | 85 | 15 |
| Split 9 | Hasil | 189 | 29 |
| | % | 85,9 | 13,18 |
| Split 10 | Hasil | 190 | 29 |
| | % | 86,39 | 13,18 |
| Rata-Rata | Hasil | 189 | 30 |
| | % | 85,87 | 13,71 |

## 6. Conclusion

From the research that has been done, it can be concluded:

1. DHF patient data obtained from the Bhayangkara Hospital Pusdik Brimob Watukosek as many as 481 data using 13 variables that have been determined include age, fever, dizziness, nausea, diarrhea, body pain, body weakness, weakness, decreased appetite, red spots, nosebleeds, bleeding gums, platelets and the main diagnosis. Data will pass the pre-processing stage to get rid of data duplication, remove data inconsistencies, correct data errors and transform data according to the method used. The amount of data after the pre-processing is 339 data that will be divided into two, 65% training data with 220 data and 35% testing data with 119 data
2. In evaluating the test using 10-fold cross validation, it produces the highest correct accuracy value (Correctly Classified Instances) of 87.72% in split to 3 with an average accuracy of 85.87%.
3. The accuracy value by using the classification method on the ID3 algorithm produces a truth level of 82% with an error rate of 18%.

**Acknowledgements**

We would like to thank the University Muhammadiyah Sidoarjo for the support provided in the form of funds and laboratory facilities.

**References**

[1] Tosepu R, Tantrakarnapa K, Nakhapakorn K and Worakhunpiset S 2018 Climate variability and dengue hemorrhagic fever in Southeast Sulawesi Province, Indonesia *Environ. Sci. Pollut. Res.* **25** 14944–52
[2] Kementrian Kesehatan RI 2018 InfoDatin Situas Demam Berdarah Dengue
[3] Damanik C (Kompas. com. 13.683 Kasus DBD di Indonesia dalam Sebulan, 133 Orang Meninggal Dunia
[4] District S N, Province N S, Sitepu F Y, Nasution H, Supriyadi T and Depari E 2018 Epidemiological and Entomological Investigation of Dengue Fever Outbreak in **11** 8–12
[5] Tufan Z K, Yigit H, Kacar M, Bulut C, Canpolat G, Hatipoglu C A, Kinikli S, Kosar P and Demiroz A P 2014 Sonographic findings in patients with crimean-congo hemorrhagic fever *J. Ultrasound Med.* **33** 1999–2003
[6] Kraidech S and Jearanaitanakij K 2018 Improving ID3 Algorithm by Combining Values from

Journal of Physics: Conference Series **1381** (2019) 012039 doi:10.1088/1742-6596/1381/1/012039

Equally Important Attributes *ICSEC 2017 - 21st Int. Comput. Sci. Eng. Conf. 2017, Proceeding* **6** 102–5

[7] Wang Y Y, Li Y Bin and Rong X W 2017 Improvement of ID3 algorithm based on simplified information entropy and coordination degree *Proc. - 2017 Chinese Autom. Congr. CAC 2017* **2017-January** 1526–30

[8] Feng W and Cai X 2016 The application and improvement of ID3 algorithm in WEB log data mining *Lect. Notes Electr. Eng.* **393** 577–83

[9] WHO 2011 *Comprehensive guidelines for prevention and control of dengue and dengue haemorrhagic fever*

[10] Rosid M A, Gunawan G and Pramana E 2015 Centroid Based Classifier With TF – IDF – ICF for Classfication of Student's Complaint at Appliation E-Complaint in Muhammadiyah University of Sidoarjo **1**

[11] Linoff G S and Berry M J A 2011 Data Mining Techniques, Third Edition 847

[12] Steyerberg E W 2009 Validation of Prediction Models 299–311

# Practicing Energy Saving Habits of Elementary Students Through Development of Lectora Inspire Software Based Instructional Media

**P Sulistyowati[1], D W Utomo[1], JR Batlolona[2], A Saregar[3], M N Hudha[1*], A C Yusro[4]**
[1]Universitas Kanjuruhan Malang, Jl. S. Supriadi no 48, Malang 65148, Indonesia
[2]Universitas Pattimura, Jl. Dr. Tamaela, Unpatti Campus B Ambon 97223, Indonesia
[3]Universitas Islam Negeri Raden Intan Lampung. Jl. Endro Suratmin, No.1 Sukarame Bandar Lampung, Lampung, Indonesia
[4]Universitas PGRI Madiun, Jl Setiabudi No. 85 Madiun 63118, Madiun, Indoneisa

*muhammadnurhudha@unikama.ac.id

**Abstract**. This paper intends to develop a software-based intructional media, Lectora Inspire. The data were obtained through experts validation on the developed instructional media. It took one media expert, one material expert, and 11 students. The data were then analyzed descriptively and quantitatively. The results and discussion confirm that the instructional media based on software, Lectora Inspire, is valid and feasible to be used in the learning process. The instructional media has met both media and materal requirements. It offers a convenient approach for students to understand the lesson content since it provides a real-life example which encourages them to imitate the habits. However, this instructional media needs a supporting facility such as computer which not all schools already have. More importantly, this software-based instructional media can be used as an effective learning resources.

## 1. Introduction

Instructional media is the important instrument in the learning process [1]. Elementary school student's development has reached a stage of concrete operational development [2]. The presence of instructional media plays a role as a media to introduce students a real form of certain object or plays a role as an object imitation to support lesson content delivery. Effective learning is a learning process which reach certain objective in a relatively short time [3]. An interesting instructional media promotes student's motivation in the learning process [4]. Lectira inspire is an interactive instructional media which is based on computer software [5] and is able to be operated in an android-based device [6][4]. Lectora inspire is an instructional media based on slide presentation which presents an image, text, and video [7]. The presentations promote student's attention in observing lesson content [5]. Furthermore, by observing to the slide presentation, students are invited to imitatet what they observe.

Technological development in the context of pedagogy takes a great role [6]. A technological-based media currently is well-developed, from a computer-based to an android-based media. Android-based smartphone nowadays is the main necessity in which almost all people have it in their hand, including students and the parents [8]. The android-based instructional media allows students to access a lesson content in anywhere; allows the to learn in their home or in anywhere. As a results it promotes a better understanding and learning outcomes [6]. Supplementary, the parents are also able to assist their kids in learning by watching the lesson content in the media.

An positive student's behavior is one of the goals of the kearning process [9]. This instructional media develop a lesson content regarding the energy resources. The energy resources lesson also discusses an understanding on how to preserve energy resources [10]. After learning and discussing the energy resources by observing, understanding, and comprehending, it is further expected that students could implement an energy-saving habits in their daily life [11][12]. Recently, the reserve of energy resource gradually decreased, thus people are expected to be wise in using any kind of available energy [12]. Energy-saving habit is the important key to maintain the availablity of energy resources, particularly the availability of non-renewable energy resources [13]. Students as the next generation must be educated and knowladgeable regarding the environment preservation, partiuclary in the context of energy [14][13]. Energy resources play a vital role in the earth, therefore people must be able to preserve it well for the future.

This research aims at developing interactive instructional media based on Lecora Inspire software which intends to enhance student's understanding about energy resource and thus promotes an energy-saving habits. The development product in this research aims at facilitating teacher in explaining and delivering the lesson content as well as assisting students in the learning process. The lesson contents contained in the instructional media are based on the applied curriculum; it focuses on students' active involvement and promotes a meaningful learning process.

## 2. Methodology

The stages of this research and development are presented in the following Figure 1.

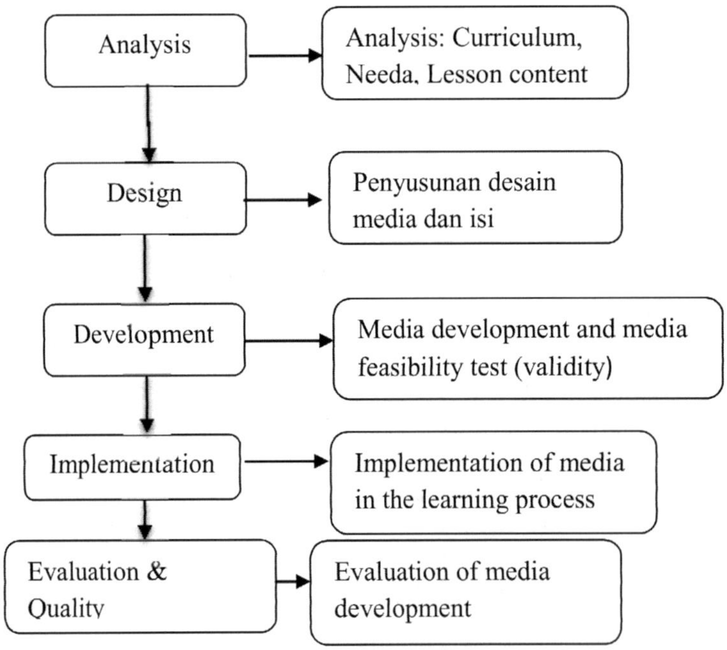


Figure 1 : Media Development Stages [15]

The stages of development [16] consisted of: 1) Analysis. This stage began by analyzing the needs in the field based on problems related to the learning process. The information obtained was based on observations and interviews with teachers and students relating to learning resources and instructional media used in school. The second analysis was an analysis of the curriculum used in learning in primary schools. The third analysis was an analysis of learning material that needs to use learning resources and interesting media. 2) The design stage (media design) dealt with media framework compilation and its contents. The process of compiling the media included: determining the material, determining the components in the media, and determining the evaluation. 3) Development stage was the process of making media through the stage of pre-writing (story board), darft writing, and editing process (edit, review, and edit). 4) In the implementation phase, the developed media was given to 11 students of elementary schools for the purpose of field trials. 5) Evaluation stage was the stage of evaluating each stage of media development as a reference for revision and feedback.

To test the validity of the developed instructional media, it was assessed by media expert, a lecturer of Information technology Kanjuruhan University, Malang; material expert to assess the content, a lecturer of Elementary School education; and field trials by teacher and 11 students of IV graders in SD YBPK Wonorejo. After being assessed in terms of its validity, revisions were made to improve the product based on the suggestions obtained.

The data obtained were in the form of qualitative and quantitative data. Qualitative data is the validators suggestions and recommendations. While, the quantitative data is the questionnaires scores and students' evaluation results. Based on the data obtained, it was then analyzed to acquire a valid research results.

The development of Lectora Inspire media is presented in teh following Figure 2.

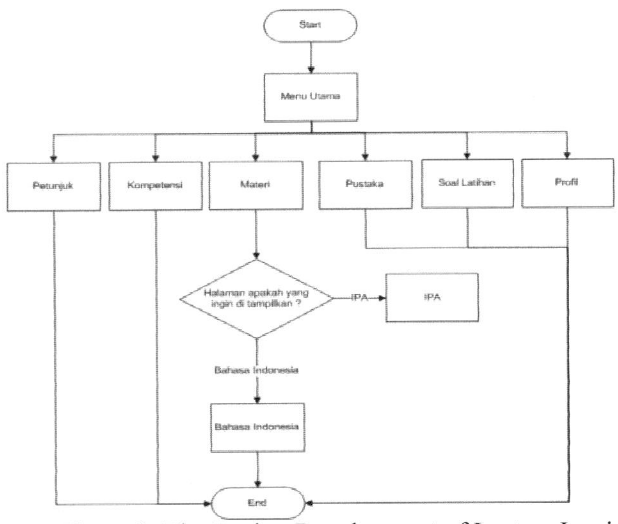


Figure 2: The Design Development of Lectora Inspire

## 3. Findings and Discussion

The results of this research development was in the form of interactive instructional media based on Lectora Inspire software. Lectora inspire software can be used as an interactive learning media that makes it easy for teachers to convey and delivery messages of learning contents for students to understand [4]. The use of interactive learning media lectora inspire software requires a supporting instrument such as a computer or laptop. Thus this media relies on the availability of these facilities. The use of lectora inspire interactive learning media is very easy to operate since it is simialr as when we operate slides on powerpoint.

The software-based instructional media Lectora Inspire provides a learning instruction as follows: user instruction, basic competences of lesson, lesson contents in the form of text, images, and videos, it is followed by learning evaluation. At the end of the part, it presents a bibliography and references. The following Figures 3 is the user interface of Lectora Inspire.

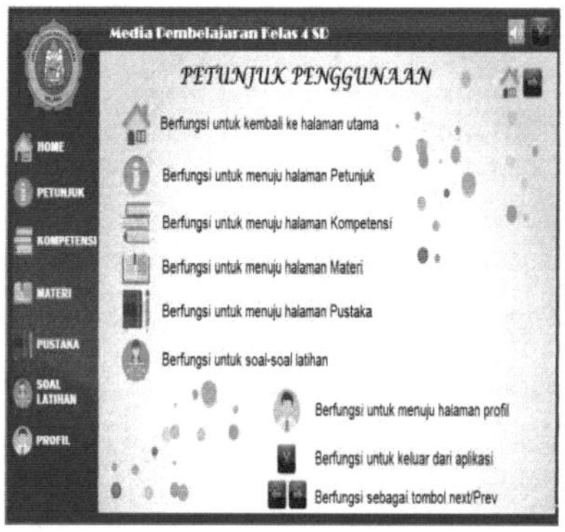


Figure 3. Lesson Instruction

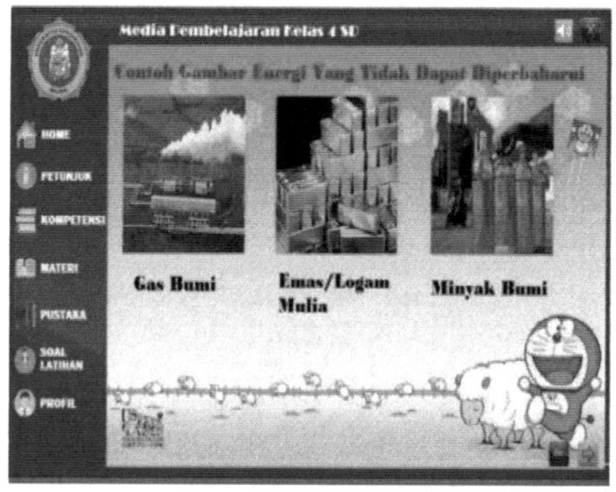


Figure 4. Content in the form of images

Figure 5. An example of student's exercise

According to the Figure 3, it presents an initial part of the media—the application instruction of media Lectora Inspire which consists of home button to return to the homepage; I button to go to instructions page; Competence button to go to competence page; Content button to load a page of lesson content; Reference button to go to the references page; and Exercise button to load a page of evaluation exercises. Figure 2 also presents an example of lesson content with the animation which explains the benefits of preserving energy resoiurces. Figure 5 presents the learning activties in the form of interview which discusses energy resource and how to preserve.

The validation results of Lectora Inspire instructional media can be seen in the following Table 1:

**Table 1: The validation results of Lectora Inspire**

| Validators | Results | Criteria |
|---|---|---|
| Material expert | 83% | Valid |
| Media expert | 76 % | Valid |

According to Table 1 above, it can be seen that the interactive instructional media based on software Lectore Inspire obtained scores which were categorized as valid from both material and media validators. Then, the instructional media was tested to the teacher and 11 students. The results of field trial confirmed that the instructional media is interesting and accessible. It offers students a new learning experience since it presents a real-life context of habit to preserve energy resource.

The results of the development of Lectora Inspire, an interactive software-based instructional media, are expected to help teachers deliver learning content and its messages to students effectively and pleasantly [7][6]. In addition, students can motivate their learning and imitate energy-saving habits into daily habits [11][10][14]. The rapid development of technology, especially the development of computer technology, is expected to support the learning process which is not only related to instructional media but also other components in learning such as the availability of digital teaching materials, as well as the digital instrument of assessment systems [7][4][5].

## 4. Summary and Conclusion

This paper concludes that Lectora Inspire, an interactive software-based instructional media, is valid to be used in the learning process; it is seen from the validation results by two experts. This instructional media offers teacher an accessible approach to deliver a learning content and its messages to students. In addition, this instructional media offers students an easy way to understand the learning content and encourage them to imitate a positive habit related to energy-saving. According to the findings and discussion, thus, it is expected for teacher to use this instructional media to deliver the learning content. More importantly, this paper encourages the teachers to also develop a varied instructional media with a different lesson content.

### Reference

[1] M. Ali, "Pengembangan Media Pembelajaran Interaktif Mata Kuliah Medan Magnet Elektromagnetik," *J. Edukasi@Elektro*, vol. 5, no. 1, pp. 11–18, 2009.
[2] Mukhlisah AM, "Pengembangan Kognitif Jean Piaget Dan Peningkatan Belajar Anak Diskalkulia," vol. 6, pp. 118–143, 2015.
[3] S. Prasetyo, "Pengembangan Media Lectora Inspire dalam Pembelajaran

Journal of Physics: Conference Series **1381** (2019) 012040 doi:10.1088/1742-6596/1381/1/012040

Sains di Madrasah Ibtidaiyah," *J. Pendidik. Islam*, vol. IV, no. 2, 2015.

[4] A. Faruk, "Development of Interactive Learning Media Based Lectora Inspire in Discrete Method Course," *Proceeding Int. Conf. Res. Implement. Educ. Math. Sci.*, pp. 69–77, 2014.

[5] S. C. Wibawa, R. Harimurti, Y. Anistyasari, and M. S. Sumbawati, "The Design And Implementation Of An Educational Multimedia Interactive Operation System Using Lectora Inspire," *Electron. Informatics, Vocat. Educ.*, vol. 2, no. 1, pp. 74–79, 2017.

[6] N. D. Shalikhah, A. Primadewi, and M. S. Iman, "Media Pembelajaran Interaktif Lectora Inspire Sebagai Inovasi Pembelajaran," *War. LPM*, vol. 20, no. 1, pp. 9–16, 2017.

[7] N. Shalikah, "Pemanfaatan Aplikasi Lectora Inspire Sebagai Media Pembelajaran Interaktif," *Cakrawala*, vol. XI, no. 1, pp. 101–115, 2016.

[8] F. N. K. and M. N. H. 1 P Sulistyowati1*, L Setyaningrum1, "Android- based monitoring applications of students ' learning outcomes," *Ser. I O P Conf. Sci. Mater.*, vol. 3, 2018.

[9] M. Matsuo, "Empowerment through self-improvement skills : The role of learning goals and personal growth initiative," *J. Vocat. Behav.*, vol. 115, no. June, p. 103311, 2019.

[10] P. Ferrão, "Understanding Resident ' s Perception of Energy Habits in Households in Bangkok," *Energy Procedia*, vol. 138, pp. 247–252, 2017.

[11] Z. Tang, M. Warkentin, and L. Wu, "Resources , Conservation & Recycling Understanding employees ' energy saving behavior from the perspective of," vol. 140, no. July 2018, pp. 216–223, 2019.

[12] E. Van Der Wer, D. Tau, and L. Venhoeven, "Energy Research & Social Science Pull the plug : How private commitment strategies can strengthen personal norms and promote energy-saving in the Netherlands," *Energy Res. Soc. Sci.*, vol. 54, no. January 2018, pp. 26–33, 2019.

[13] F. Mehmood, N. Fatima, and K. Ullah, "Impact of China-Pakistan economic corridor on Pakistan ' s future energy consumption and energy saving potential : Evidence from sectoral time series analysis," *Energy Strateg. Rev.*, vol. 25, no. March, pp. 34–46, 2019.

[14] M. Pérez-urdiales and M. Á. García-valiñas, "Efficient water-using technologies and habits : A disaggregated analysis in the water sector," *Ecol. Econ.*, vol. 128, pp. 117–129, 2016.

[15] R. Sharifah and S. Faaizah, "The Development of Online Project Based Collaborative Learning using ADDIE Model," *Procedia - Soc. Behav. Sci.*, vol. 195, pp. 1803–1812, 2015.

[16] F. Schott, D. Psychology, and T. U. Dresden, *Instructional Design*, Second Edition., vol. 11, no. 1922. Elsevier, 2015.

Journal of Physics: Conference Series **1381** (2019) 012041 doi:10.1088/1742-6596/1381/1/012041

# A Review on Cryptography Protocol for Securing Data

**Robbi Rahim***
Sekolah Tinggi Ilmu Manajemen Sukma, Medan, Indonesia

*usurobbi85@zoho.com

**Abstract.** Cryptography is one type of protection that is widely used to secure information, encryption and decryption and the use of keys is a common process used to secure information; and also the cryptographic processes have a classic problem that is the distribution of keys that are vulnerable to interception when the key is sent by the sender to the receiver, the cryptographic protocol (Authentication) could be used to minimize key distribution problem because there is no need for key exchanges. Shamir's Three-Pass Protocol, Secret Splitting, Bit-Commitment Protocol and Blind Signature are few cryptographic protocols that can be used to help with key distribution issues, this article performs systematic approach to protocol cryptography for security level and the using other algorithms to combine with protocol.

## 1. Introduction

Security [1], [2] of information, data, and files or in a communication is very important in the digital era today[3], cryptographic techniques [4], [5] is one technique that could be used to secure information or communication. The AES, DES, MMB, GOST, WAKE and RSA [6], [7] algorithms are some of the modern cryptographic algorithms that can be used to secure information, data or communications. Cryptography algorithms generally require a key (private or public) for encryption and decryption process and the main thing to do is key distribution to decryption process, and this is one of the major problems in cryptography algorithm [1], [8], [9].

A cryptographic protocols [8]–[10] a technique that can be used to accommodate key distribution issues (key exchanges), cryptography protocols typically used by participating parties to share secret components, generate random number sets [11], assure the identity of others (authentication) and most importantly cryptography protocols involving 2 or more users and involves cryptographic algorithms in its security process [12].

The discussion on this article will overview the differences between Shamir's Three-Pass Protocol [9], [13], Secret Splitting [14], Bit-Commitment Protocol [15] and Blind Signature [16], [17] on encryption and decryption without the key exchanges mechanism.

## 2. Methodology

The cryptography protocol is one of the mechanisms that can be used to anticipate key distributions in the encryption and decryption process, generally protocols are used with cryptographic algorithms for security process and using good algorithms like AES, DES, RC4 and RSA.

Sui [13] applies Shamir's Three-Pass to a double image sharing scheme in the sender and receiver communication process, 2 (two) images and may be more combined using chaotic permutation random methods with phase mask keys; and obtained new cipher image, communication using Shamir's Three-Pass Protocol is quite effective because it does not need to provide the key (phase mask key) encryption and decryption because each sender or receiver uses each key and can interact with each other, Shamir's Three-Pass The protocol on double-image sharing scheme is quite good but requires a considerable computation process (time efficiency).

Another research performed by Nithya [18] analyzed the use of Shamir's Three-Pass Protocol on routing protocols, the experiment performed to acquire better security results on routing protocols by applying Shamir's Three-Pass Protocol especially on IoT communications, and also avoid some of the possible security attacks.

Secret Splitting Protocol differs from Shamir's Three-Pass Protocol (TPP) from user or communications participation, TPP is only done in two directions (1 Sender and 1 Receiver) and this is quite effective because there is no third party communication and each can using private key. The Secret Splitting Protocol involves two or more recipients in which secret messages and keys are distributed to all recipients, Secret Splitting Protocol research conducted by Shah [14] on Cloud of Thing by applying XOR and Diffie-Hellman operations as a key exchange process, from experiment XOR and Diffie-Hellman shows resistance to various types of attacks, theoretically the use of Diffie-Hellman algorithm with XOR will not increase the level of security but will ease brute force attack and get the information transmitted.

Another protocol that can be used is the Bit-Commitment Protocol, this protocol has differences from the two protocols that have been discussed, the Bit-Commitment Protocol allows to identify the legitimacy of messages from changes made until commitment occurs, some articles about the Bit-Commitment Protocol, and especially on quantum is still a problem today. Arash [15] implemented a quantum bit-commitment protocol to anticipate attacks on security and from experiment result proved successful, Arash's research results were different from those of Chau and Chiribella [19], [20] where they argue that quantum bit-commitment protocols are impossible to implement because the possibility of cheating by the sender of the message may occur.

Blind Signature Protocol has a different concept from Shamir's Three-Pass Protocol, Secret Splitting Protocol and Bit-Commitment Protocol, research done by Tianjie [17], Wang [21], Sun [22], Chow [23], Wu [23] and Hua [16] provide similar results that Blind Signature Protocol is safe to use in communications network and banking if the recipient believes that communication is secure, the blind signature protocol requires high trust of information received and this is one of the weaknesses of this protocol (mistrust communication).

The results of research on cryptographic protocols indicate that security communications without key distributions can be done well and in terms of security are also good, and most importantly that cryptographic protocols are possible to be used with another cryptography algorithms.

## 3. Results and Discussion

The use of cryptographic protocols as communication between the sender and receiver without the key distribution or key exchange process is possible and the security is also quite good, the use of cryptographic algorithms in cryptographic protocols is not all feasible.

An experiment Pohlig-Hellman algorithm on Shamir's Three-Pass Protocol (TPP), Secret Splitting Protocol (SSP), Bit-Commitment Protocol (BCP) and Blind Signature (BS) do not work well, table 1 describes the Pohlig-Hellman algorithm component used on those protocols.

**Table 1.** Component Pohlig-Hellman

| | Key Generation | XOR Function | Extended Euclidean |
|---|---|---|---|
| TPP | ✓ | ✓ | ✓ |
| SSP | ✓ | ✓ | ✓ |
| BCP | ✓ | | |
| BS | ✓ | ✓ | |

Table 1 above is an encryption and decryption component of the Pohlig-Hellman algorithm. The selection of Pohlig-Hellman algorithm in this exeperiment because the keys on Pohlig-Hellman are not distributed, but the sender and receiver make their own keys, based on experiments conducted on Shamir's Three Pass Protocol and Secret Splitting Protocol, both cryptography protocols can be used well with the Pohlig-Hellman algorithm.

## 4. Conclusion

The cryptography protocol can be used to solve key distribution problems between the sender and receiver, the use of Shamir's Three-Pass Protocol with the Pohlig-Hellman algorithm can work well and with a sufficient level of security, cryptography protocol development is possible by improvisation using combination cryptography algorithms.

## REFERENCES

[1] R. Rahim, "Man-in-the-middle-attack prevention using interlock protocol method," *ARPN J. Eng. Appl. Sci.*, vol. 12, no. 22, pp. 6483–6487, 2017.

[2] R. Rahim, M. Dahria, M. Syahril, and B. Anwar, "Combination of the Blowfish and Lempel-Ziv-Welch algorithms for text compression," *World Trans. Eng. Technol. Educ.*, vol. 15, no. 3, pp. 292–297, 2017.

[3] Iswanto, O. Wahyunggoro, and A. I. Cahyadi, "3D object modeling using data fusion from laser sensor on quadrotor," in *AIP Conference Proceedings*, 2016, vol. 1755.

[4] A. Putera, U. Siahaan, and R. Rahim, "Dynamic Key Matrix of Hill Cipher Using Genetic Algorithm," *Int. J. Secur. Its Appl.*, vol. 10, no. 8, pp. 173–180, Aug. 2016.

[5] R. Rahim and A. Ikhwan, "Cryptography Technique with Modular Multiplication Block Cipher and Playfair Cipher," *Int. J. Sci. Res. Sci. Technol.*, vol. 2, no. 6, pp. 71–78, 2016.

[6] R. Rahim, "128 Bit Hash of Variable Length in Short Message Service Security," *Int. J. Secur. Its Appl.*, vol. 11, no. 1, pp. 45–58, Jan. 2017.

[7] H. Nurdiyanto, R. Rahim, and N. Wulan, "Symmetric Stream Cipher using Triple Transposition Key Method and Base64 Algorithm for Security Improvement," *J. Phys. Conf. Ser.*, vol. 930, no. 1, p. 012005, Dec. 2017.

[8] A. G. D. Uchoa, M. E. Pellenz, A. O. Santin, and C. A. Maziero, "A Three-Pass Protocol for Cryptography Based on Padding for Wireless Networks," in *2007 4th IEEE Consumer Communications and Networking Conference*, 2007, pp. 287–291.

[9] R. Rahim and A. Ikhwan, "Study of Three Pass Protocol on Data Security," *Int. J. Sci. Res.*, vol. 5, no. 11, pp. 102–104, Nov. 2016.

[10] Y. Yin, Y. Gan, H. Wen, and T. Li, "A symmetric key exchange protocol bsaed on virtual S-box," *China Commun.*, vol. 11, no. 14, pp. 46–52, 2014.

[11] R. Rahim, H. Winata, I. Zulkarnain, and H. Jaya, "Prime Number: an Experiment Rabin-Miller and Fast Exponentiation," *J. Phys. Conf. Ser.*, vol. 930, no. 1, p. 012032, Dec. 2017.

[12] H. Li and P. Liu, "An Identification System Combined with Fingerprint and Cryptography," in *First International Multi-Symposiums on Computer and Computational Sciences (IMSCCS'06)*, 2006, pp. 105–108.

[13] L. Sui, K. Duan, and J. Liang, "A secure double-image sharing scheme based on Shamir's three-pass protocol and 2D Sine Logistic modulation map in discrete multiple-parameter fractional angular transform domain," *Opt. Lasers Eng.*, vol. 80, pp. 52–62, May 2016.

[14] R. H. Shah and D. P. Salapurkar, "A multifactor authentication system using secret splitting in the perspective of Cloud of Things," in *2017 International Conference on Emerging Trends & Innovation in ICT (ICEI)*, 2017, pp. 1–4.

[15] S. Arash Sheikholeslam and T. Aaron Gulliver, “A practical quantum bit commitment protocol,” *Results Phys.*, vol. 2, pp. 97–99, Jan. 2012.

[16] J. Hua and H. Ran-dong, “Online/Offline Blind Signature,” *Procedia Eng.*, vol. 29, pp. 2214–2218, Jan. 2012.

[17] T. Cao, D. Lin, and R. Xue, “A randomized RSA-based partially blind signature scheme for electronic cash,” *Comput. Secur.*, vol. 24, no. 1, pp. 44–49, Feb. 2005.

[18] S. Nithya and K. M. alias Jeyanthi, “Genetic algorithm based bacterial foraging optimization with three-pass protocol concept for heterogeneous network security enhancement,” *J. Comput. Sci.*, vol. 21, pp. 275–282, Jul. 2017.

[19] H.-K. Lo and H. F. Chau, “Why quantum bit commitment and ideal quantum coin tossing are impossible,” *Phys. D Nonlinear Phenom.*, vol. 120, no. 1–2, pp. 177–187, Sep. 1998.

[20] G. Chiribella, G. Mauro, D. ’ Ariano, P. Perinotti, D. Schlingemann, and R. Werner, “A short impossibility proof of quantum bit commitment,” *Phys. Lett. A*, vol. 377, pp. 1076–1087, 2013.

[21] J. hui WANG, J. wei LIU, X. hui LI, and W. dong KOU, “Fair e-payment protocol based on blind signature,” *J. China Univ. Posts Telecommun.*, vol. 16, no. 5, pp. 114–118, Oct. 2009.

[22] H.-M. Sun, B.-T. Hsieh, and S.-M. Tseng, “On the security of some proxy blind signature schemes,” *J. Syst. Softw.*, vol. 74, no. 3, pp. 297–302, Feb. 2005.

[23] S. S. M. Chow, “Blind signature and ring signature schemes: Rehabilitation and attack,” *Comput. Stand. Interfaces*, vol. 31, no. 4, pp. 707–712, Jun. 2009.

# Data Security On RFID Information Using Word Auto Key Encryption Algorithm

**Robbi Rahim[1*], Syamsuddin Lubis[2], N Nurmalini[2], Haida Dafitri[3]**
[1]Sekolah Tinggi Ilmu Manajemen Sukma, Medan, Indonesia
[2]Department of Informatics, AMIK Intel Com Global Indo, Kisaran, Indonesia
[3]Department of Informatics, Universitas Harapan Medan, Medan, Indonesia

*usurobbi85@zoho.com

**Abstract.** RFID technology is a technology that is commonly used, one of the problems that exist in RFID is the information contained in RFID Tags are not encrypted and easily read by others. Word Auto Key Encryption (WAKE) is one of the cryptographic algorithms that can be used to secure data on RFID Tag. Experiments resulted in the process of reading and writing encrypted data on RFID Tags can be done well and data more secure than before.

## 1. Introduction

Development of communications technology in the digital age is currently vulnerable to security attack [1], [2] One way to secure data is to change the data into another form that cannot be known by others [3], encryption using cryptography is one of solution that can be used.

Data on Radio Frequency Identification (RFID)[4], [5] are in the form of text and this can be read by transponders remotely without having direct contact [6], the data on the RFID label is generally not encrypted and can be read or modified by irresponsible parties, thus action can cause deprivation to the owner. Securing data on RFID by using a cryptography algorithm is chosen because it takes time for irresponsible parties to read and if data on RFID modified with plain text it will not be read by transponders due to different data (Non Encryption), Word Auto Key Encryption (WAKE) algorithm [7] is used on RFID data security.

Security testing on RFID is perform by using RFID Card, RFID Reader and Arduino as experiment material, RFID Card contains encrypted text using WAKE algorithm. An experiment is done to prove that data on RFID is safe and cannot be easily read when data encrypted using WAKE algorithm.

## 2. Methodology

### *2.1 Cryptography*

Cryptography is the study of mathematical techniques related to aspects of information security such as confidentiality, data integrity, sender/recipient authentication, and data authentication so messages or documents are secure and unreadable by unauthorized parties [3], [8]–[12]. General purpose cryptography [3], [9]:

A. Confidentiality
Ensure that data or information can only be accessed by certain parties only.
B. Authentication
the sender and the recipient of the message know that the message came from the actual party.

C. Integrity
This issue relates to the assurance that the data or messages sent are not modified to the recipient.

D. Nonrepudiation
Nonrepudiation prevents both senders and receivers from denying that they have sent or received a message / information. If a message is sent, the recipient can prove that the message was indeed sent by the sender listed. Conversely, if a message is received, the sender can prove that the message has been received by the intended party.

E. Access Control
Limit data sources only to specific people.

F. Availability
Required all information on a computer system and must be available to all parties entitled to such information.

The six aspects of data security, four of them can be overcome by using cryptography that is confidentiality, integrity, authentication, and nonrepudiation.

### *2.2 Word Auto Key Encryption*

Word Auto Key Encryption (WAKE) is a stream cipher algorithm that works using 128 bit keys and a 256 x 32 bit table [7], [13]. The WAKE algorithm uses XOR, AND, OR and Shift Right operations in the encryption and decryption process. The main process of WAKE algorithm such as [7]:

a. Formation of S-Box table.
b. Key generation.
c. Encryption and Decryption.

The core of the WAKE algorithm exists in the process of forming the S-Box table and the key generation. The S-Box table of the WAKE algorithm is flexible and different for each round.

Program listing of the WAKE algorithm used for the encryption and decryption process [7].

a. Wake Encryption

```
strPlain = pcEnkripsi
X = ""
For N = 1 To Len(strPlain)
  X=X&FormatS(FDecToBiner(Asc(Mid(strPlain, N, 1))), "0", 8)
Next N
X1 = "": X2 = "": cHasil = ""
For N = 1 To Len(X) Step 8
   X1=Mid(strKunciBiner,(N Mod 32),8)
   X2=FOpBiner("XOR",Mid(X,N,8),X1,8)
   cHasil=cHasil&Chr(FBinerToDec(X2))
Next N
strCipher = cHasil
```

b. Wake Decryption

```
strCipher = pcDekripsi
X = ""
For N = 1 To Len(strCipher)
  X=X&FormatS(FDecToBiner(Asc(Mid(strCipher, N, 1))), "0", 8)
Next N
X1 = "": X2 = "": cHasil = ""
For N = 1 To Len(X) Step 8
  X1=Mid(strKunciBiner,(N Mod 32),8)
  X2=FOpBiner("XOR",Mid(X,N,8),X1,8)
  cHasil=cHasil&Chr(FBinerToDec(X2))
Next N
strPlain = cHasil
```

c. Key Generatin

```
X = ""
For N = 1 To Len(pcKunci)
```

```
  X=X&FormatS(FDecToBiner(Asc(Mid(pcKunci, N, 1))), "0", 8)
Next N

A(0) = Mid(X, 1, 32)
B(0) = Mid(X, 33, 32)
C(0) = Mid(X, 65, 32)
D(0) = Mid(X, 97, 32)

For N = 1 To pnPutaran
  A(N) = FungsiM(A(N - 1), D(N - 1))
  B(N) = FungsiM(B(N - 1), A(N))
  C(N) = FungsiM(C(N - 1), B(N))
  D(N) = FungsiM(D(N - 1), C(N))
Next N

strKunciBiner = D(N - 1)
```

### *2.3 Radio Frequency Identification*

Radio Frequency Identification (RFID)[4] is an object identification method by utilizing radio waves. The identification process is done by RFID Reader and RFID Transponder (RFID tag). The system works by placing RFID tags on an object or object to be identified. Each RFID tag has a unique identification number (ID number), so no RFID tags have the same ID number [4], [14]. RFID technology can be used to describe a system capable of transmitting identity data from an object wirelessly using radio waves[14].

Generally RFID system consists of 4 components[15]–[17], such as:

a. Tags
   Tags are devices that store information for object identification. RFID tags are often also referred to as transponders.
b. Antenna
   A device used to transmit radio frequency signals between RFID readers and RFID tags.
c. RFID Reader
   It is a device compatible with RFID tags that will communicate wirelessly with tags.
d. Application Software
   An application on a workstation or PC that can read data from tags through RFID reader. Both the RFID tag and reader are equipped with an antenna so that it can receive and emit electromagnetic waves

RFID reader is a connection between the application software with the antenna that will radiate radio waves to RFID tags [14]. Radio waves emitted by the antenna operate in the surrounding space. Consequently data can move wirelessly to RFID tags that are located adjacent to the antenna. In this research used RFID reader RC522 as reader/writer and using tags with Mifare S50 RFID card / Contactless card 13.56 Mhz HF.

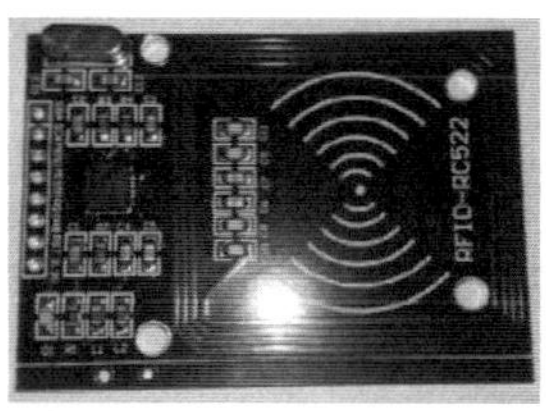


**Figure 1.** RFID Reader RC522

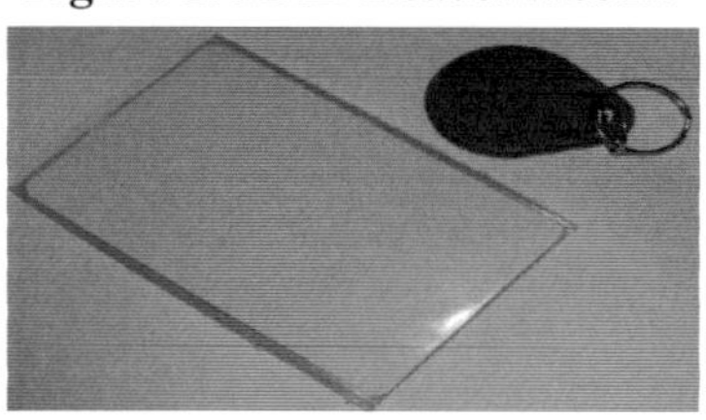

**Figure 2.** Mifare S50

Security system application architecture for RFID [15], [18], [19] that can be used there are 4 (four) types:

a. Fixed Code System
This system is the simplest and most used system. Fixed code stored in the RFID tag is read and compared to the code stored in the database. This system does not allow copying of RFID tags. This simple system has the lowest security level.

b. Rolling Code System
Operates in the same way as the Fixed Code system, but the secret code in the RFID tag is only valid for a certain period of time. RFID readers on this system must have the ability to write RFID tags. The RFID tags used must be programmed multiple times. This system provides better security, but that should be considered is the secret code synchronization process.

c. Password Protection System
Simple authentication systems can be provided by RFID systems with password protection. Confidential data on RFID tags will only be transmitted after the RFID reader sends the appropriate password data to prove the validity of the RFID reader. The length of the password can vary between several minutes to several years depending on the length of the password. For security systems with multiple users and with different passwords, the limitations of total communication time are very long, because RFID readers should expect passwords from the available databases.

d. Combination System Rolling Code with Password
It is an aggregated system with a revamped secret code facility and password to protect secret codes stored in RFID tags. The critical issue of this system is the timing of communication and password synchronization. With this system will provide a high level of security.

Weaknesses in the fixed code system that the data stored in the text and must be compared in the database, to overcome the weakness of fixed code system can be done with the encryption before to write in RFID Tag and the key used is Serial Number in RFID Tags that would not be possible same all over RFID Tag.

## 3. Results and Discussion

The mechanism of data security on RFID Tag consists of writing and reading process on RFID Tags, all communication process is done by using Arduino UNO as communication between RFID Reader, RFID Tag and computer as host, see figure 3 below for thus communication.

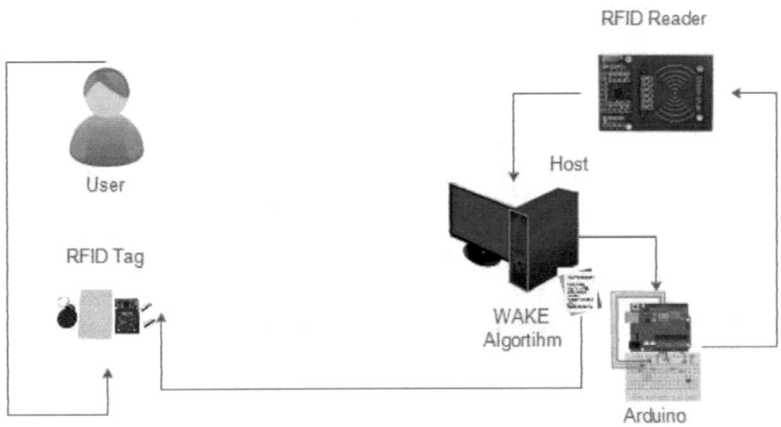

**Figure 3.** WAKE and RFID Mechanism

The above mechanism describes the information security process in RFID Tag, where the process of writing and reading is done on the host.

An experiment of data security on RFID is accomplished by determining plaintext, key and number of key rotation (S-Box), as previously explained under the formation of key and S-Box is very dynamic.
Plaintext = SCITEPRESS
Key = 000009002E4EA2CB
Rotation = 3
In addition the above key is obtained from the RFID Tag serial number by adding 4 digits 0000 before "09002E4EA2CB" so that the key is equal to 2 bytes. The result of the plaintext encryption process is as follows:
Plaintext : 'SCITEPRESS'
ASCII from 'S' = 53
ASCII from 'C' = 43
ASCII from 'I' = 49
ASCII from 'T' = 54
ASCII from 'E' = 45
ASCII from 'P' = 50
ASCII from 'R' = 52
ASCII from 'E' = 45
ASCII from 'S' = 53
ASCII from 'S' = 53
Plaintext (hexadecimal) = 53434954455052455353

Key in hexadecimal = 2B1C125A

Ciphertext = Plaintext XOR Key
53 XOR 2B = 78 = 'x'
43 XOR 1C = 5F = '_'
49 XOR 12 = 5B = '['
54 XOR 5A = 0E = '_'
45 XOR 2B = 6E = 'n'
50 XOR 1C = 4C = 'L'
52 XOR 12 = 40 = '@'
45 XOR 5A = 1F = '_'
53 XOR 2B = 78 = 'x'
53 XOR 1C = 4F = 'O'

Ciphertext = x_[_nL@_xO
The next test is ready with same plaintext and key but with different rotation number, S-Box rotation is used 6 rounds to get the result below.
Key generation from s-box process
Key = A3D23A2A

Cipher text = Plaintext XOR Key
53 XOR A3 = F0 = 'ð'
43 XOR D2 = 91 = '‘'
49 XOR 3A = 73 = 's'

54 XOR 2A = 7E = '~'
45 XOR A3 = E6 = 'æ'
50 XOR D2 = 82 = ','
52 XOR 3A = 68 = 'h'
45 XOR 2A = 6F = 'o'
53 XOR A3 = F0 = 'ð'
53 XOR D2 = 81 = '• '

Cipher text = ð's~æ,hoð•

Ciphertext result then written to RFID Tag using arduino as the medium. Based on the above two test results obtained different ciphertext when the number of key rotation with S-Box function made 3 rounds and 6 round, ciphertext results can be decrypted using the same step where the key and the number of key rounds should be same, reading of information (ciphertext) on RFID Tag with RFID Reader will activate the decryption process automatically to know data in RFID Tag were true.

## 4. Conclusion

Data security on RFID Tags by using WAKE algorithm can be done well, the data available on RFID Tag will also be difficult to read by unauthorized if doesn't not know the algorithm, key and key rotation count. Improvisation from this research can be developed chip or RFID Tag which have component of cryptography which by default can be read and written by system and easy in synchronization.

## REFERENCES

[1] H. Nurdiyanto, R. Rahim, and N. Wulan, "Symmetric Stream Cipher using Triple Transposition Key Method and Base64 Algorithm for Security Improvement," *J. Phys. Conf. Ser.*, vol. 930, no. 1, p. 012005, Dec. 2017.

[2] H. Arrosida and M. E. Echsony, "The Design of Optimal PID Control Method for Quadcopter Movement Control," *Int. J. Artif. Intell. Res.*, vol. 2, no. 1, p. 1, Feb. 2018.

[3] R. Rahim, "Man-in-the-middle-attack prevention using interlock protocol method," *ARPN J. Eng. Appl. Sci.*, vol. 12, no. 22, pp. 6483–6487, 2017.

[4] R. Want, "An introduction to RFID technology," *IEEE Pervasive Computing*, vol. 5, no. 1. pp. 25–33, 2006.

[5] Iswanto, O. Wahyunggoro, and A. I. Cahyadi, "3D object modeling using data fusion from laser sensor on quadrotor," in *AIP Conference Proceedings*, 2016, vol. 1755.

[6] A. Randall, "RFID Can Be Hacked: Here's How, & What You Can Do To Stay Safe," 2017. [Online]. Available: http://www.makeuseof.com/tag/rfid-hacked-stay-safe/. [Accessed: 03-Sep-2017].

[7] L. Legito and R. Rahim, "SMS Encryption Using Word Auto Key Encryption," *Int. J. Recent Trends Eng. Res.*, vol. 3, no. 1, pp. 251–256, 2017.

[8] S. Manna and S. Dutta, "A Stream Cipher based Bit-Level Symmetric Key Cryptographic Technique using Chen Prime Number," *Int. J. Comput. Appl.*, vol. 107, no. 12, pp. 975–8887, 2014.

[9] R. Rahim, M. Dahria, M. Syahril, and B. Anwar, "Combination of the Blowfish and Lempel-Ziv-Welch algorithms for text compression," *World Trans. Eng. Technol. Educ.*, vol. 15, no. 3, pp. 292–297, 2017.

[10] R. Rahim and A. Ikhwan, "Cryptography Technique with Modular Multiplication Block Cipher and Playfair Cipher," *Int. J. Sci. Res. Sci. Technol.*, vol. 2, no. 6, pp. 71–78, 2016.
[11] R. Rahim, "128 Bit Hash of Variable Length in Short Message Service Security," *Int. J. Secur. Its Appl.*, vol. 11, no. 1, pp. 45–58, Jan. 2017.
[12] H. Wandani, M. Budiman, and A. Sharif, "Implementasi Sistem Keamanan Data dengan Menggunakan Teknik Steganografi End of File (EOF) dan Rabin Public Key Cryptosystem," *Alkhawarizmi*, 2012.
[13] H. Gultom, "Penyandian Email Menggunakan Algoritma Kriptografi WAKE( Word Auto Key Encryption)," *Pelita Inform. Budi Darma*, vol. 4, no. 1, pp. 107–111, 2013.
[14] M. Kaur, M. Sandhu, N. Mohan, and P. S. Sandhu, "RFID Technology Principles, Advantages, Limitations & Its Applications," *Int. J. Comput. Electr. Eng.*, vol. 3, no. 1, pp. 1793–8163, 2011.
[15] A. N. Nambiar, "RFID technology: a review of its applications," *Proc. World Congr. Eng. Comput. Sci.*, vol. 2, pp. 20–22, 2009.
[16] S. Garrido Azevedo and H. Carvalho, "Contribution of RFID technology to better management of fashion supply chains," *Int. J. Retail Distrib. Manag.*, vol. 40, no. 2, pp. 128–156, 2012.
[17] C. M. Roberts, "Radio frequency identification (RFID)," *Computers and Security*, vol. 25, no. 1. pp. 18–26, 2006.
[18] S. Spiekermann and S. Evdokimov, "Critical RFID privacy-enhancing technologies," *IEEE Secur. Priv.*, vol. 7, no. 2, pp. 56–62, 2009.
[19] H. Knospe and H. Pohl, "RFID security," *Inf. Secur. Tech. Rep.*, vol. 9, no. 4, pp. 39–50, 2004.

Journal of Physics: Conference Series **1381** (2019) 012043 doi:10.1088/1742-6596/1381/1/012043

# Enhancement IDEA Algorithm with Digital Image as Key Encryption and Decryption

**Robbi Rahim[1*], Helmy Fauzi Siregar[2], and Delima Sitanggang[3]**
[1]Sekolah Tinggi Ilmu Manajemen Sukma, Medan, Indonesia
[2]Department of Informatics, Universitas Asahan, Kisaran, Indonesia
[3]Department of Informatics, Universitas Prima Indonesia, Medan, Indonesia

*usurobbi85@zoho.com

**Abstract.** Cryptography is one type of protection that is widely used to secure information, encryption, and decryption and the use of keys is a common process used to secure information; and also the cryptographic processes have a classic problem that is the distribution of keys that are vulnerable to interception when the sender sends the key to the receiver, the cryptographic protocol (Authentication) could be used to minimize key distribution problem because there is no need for key exchanges. Shamir's Three-Pass Protocol, Secret Splitting, Bit-Commitment Protocol and Blind Signature are few cryptographic protocols that can be used to help with key distribution issues, this article performs systematic approach to protocol cryptography for security level and the using other algorithms to combine with protocol.

## 1. Introduction

Data security [1]–[5] is still a very serious concern until today, with many technological developments and algorithms being undertaken to secure data from irresponsible parties [6]–[9]. Security can be done in various ways such as cryptography and steganography [10], [11], e.g. cryptography using AES, RSA, MMB, GOST[12] and IDEA algorithms [13], while LSB [14], EOF, Pixel Value Differencing for steganography [11]. Utilization of the algorithm can be done independently or can be combined, or also could be applied to certain processes such as on network communications.s

International Data Encryption Algorithm or IDEA is a block cipher algorithm with 128 bit key length and 64 bit encryption block [13], [15], [16]. IDEA was originally a powerful and reliable algorithm for securing data, and like an algorithm if it has advantages it also has weaknesses and on IDEA algorithm the weakness lies in the key rotation especially linear factor [15], [17], this weakness can be overcome by modifying the key scheduling from IDEA algorithm [13], [17]. Another technique that can be used is a key that is used instead of a common text but an image that has the composition of pixel values [8], [18], [19], hashing values and also byte code so that have different variables to be key.

Digital image used as a key in the process of encryption and decryption is done by reading the value of hashing as the key [18], [20], this process would make it difficult for cryptanalyst to know the keys used. One of the reason because the keys that do not have a pattern, in addition to perform the decryption process also requires a digital image used. Digital image as a key is expected to be one of the solutions to improve the security of the IDEA algorithm.

## 2. Methodology

### *2.1 Cryptography IDEA*

Cryptography can be defined as a science that studies mathematical techniques related to aspects of information security such as confidentiality, data integrity, sender/data receiver authentication, and data authentication. Cryptanalysis is the study of how to solve cryptographic mechanisms[21]–[24].

The IDEA algorithm used in this study has several cryptographic mathematical foundations used for the process of encryption and decryption [15], [16], [25], [26]:

a. Modular Arithmetic
Modular arithmetic is a mathematical operation that is widely implemented in cryptographic methods. In the IDEA method, the modular arithmetic operation used is the modulo $2^{16}$ addition operation and the modulo $2^{16} + 1$ multiplication operation. Modulo operation involves numbers 0 and 1 only so that it is identical to the bits on the computer. Example,
$(65530 + 10) \bmod 2^{16} = 65540 \bmod 65536 = 4$
$(32675 * 4) \bmod (2^{16} + 1) = 131060 \bmod 65537 = 65523$

b. Invers Multiplication
Inverse multiplication used on the IDEA method is not like the inverse of the multiplication operation in mathematics. This multiplication cannot be explained mathematically, but by using the following algorithm:

```
Function InverseA As Double) As Double
    n = 65537
    G0 = n
    G1 = A
    V0 = 0
    V1 = 1
    While (G1 <> 0)
        Y = IntG0 / G1)
        G2 = G0 - Y * G1
        G0 = G1
        G1 = G2
        V2 = V0 - Y * V1
        V0 = V1
        V1 = V2
    Wend
    If (V0 >= 0) Then
        Inverse = V0
    Else
        Inverse = V0 + n
    End If
End Function
```

c. Inverse Addition
Inverse addition in the IDEA method using the following algorithm:
Inverse sum = 65536 – pnBil
Example: Inverse sum of 32654 is 65536 – 32654 = 32882

d. XOR Operation
XOR is an Exclusive-OR operation denoted by a "⊕" sign. The XOR operation will return a bit value of "0" (zero) if XOR two bits of the same value and will result in a "1" bit value if XOR two bits of a different bit value.

e. Permutation
Permutations in cryptography are often used to move the position of a number of bits to a predetermined position in the permutation table. Permutations in the IDEA algorithm are used as the process of establishing sub keys.

f. Bit Shift
The bit shift is a shift operation against a sequence of bits as much as desired.

g. Bit Rotation
Rotation of bits is the operation of rotation to a sequence of bits as much as desired, there are 2 operations that can be used left rotation and right rotation.

h. Modulo Multiplication
Multiplication with zero always produces zero and has no inverse. Multiplication modulo *n* also has no inverse if the number multiplied is not relatively prime to *n*. In IDEA, for multiplication operations, a 16 bit number consisting of zeros is all considered a number 65536, while other numbers remain in accordance with the unidentified numbers it represents, example:
(32542 * 10) mod 65537 = 325420 mod 65537 =
63272
(3154 * 25) mod 65537 = 78850 mod 65537 =
13313

### *2.2 Digital Image*

Digital image is a two-dimensional image that is displayed on certain media as a set of digital values called pixels [18], [27]–[29]. In a mathematical overview, the image is a continuous function of the intensity light in a two-dimensional plane.

Digital image is the image f (x, y) where discrete spatial coordinates are coordinated and quantization level discretization. Digital image is a function of light intensity f (x, y), where x and y are spatial coordinates. The function at each point (x, y) represents the brightness level of the image at that point. The digital image is a matrix in which the row and column indices represent a point on the image and the matrix element (pixel element) represents gray level at that point [28], [30], [31].

The pixel value form digital image used as key for encryption and decryption process has a 3x3 matrix size, see example of the following RGB image pixel.

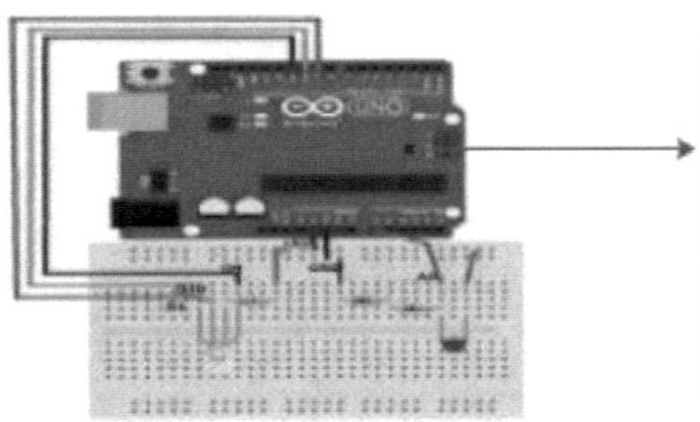

| | | |
|---|---|---|
| 67 | 69 | 75 |
| 110 | 111 | 111 |
| 90 | 99 | 89 |

**Figure 1.** Pixel Image Sample

The pixel value in Figure 1 will be the key to the encryption and decryption process performed on the IDEA algorithm, the pixel value will be taken randomly for the encryption process and when the decryption process also requires the same image.

## 3. Results and Discussion

Security analysis on IDEA algorithms using images as key requires several sets of rules that must be met as follows:

a. Maximized digital image size of 200x200
b. The key for encryption and decryption is obtained from randomly generated RGB image pixels by 3x3 pixel reading from left to right horizontally.

c. Key authentication is derived from digital images or not before encryption and decryption processes
d. Encryption and decryption process is completed by using IDEA algorithm.

The key used from digital image with 3x3 matrix as follows:

**Table 1.** 3x3 Matrix Key

| | | |
|---|---|---|
| 78 | 77 | 89 |
| 25 | 55 | 56 |
| 56 | 55 | 00 |

RGB Pixel value is then read from the left (78) to the right (00) horizontally and obtained the following result: 787789255556565500. THOMPSON is a message to be secured using IDEA algorithm, the process as below:

**ROUND 1**
01) L#1 = (X1 * K1) mod (2^16 + 1) = 0101010001001000 * 0011011100111000 mod (2^16 + 1) = 1101010110010011

02) L#2 = (X2 + K2) mod 2^16 = 0100111101001101 + 0011011100110111 mod 2^16 = 1000011010000100

03) L#3 = (X3 + K3) mod 2^16 = 0101000001010011 + 0011100000111001 mod 2^16 = 1000100010001100

04) L#4 = (X4 * K4) mod (2^16 + 1) = 0100111101001110 * 0011001000110101 mod (2^16 + 1) = 1001011110011001

05) L#5 = L#1 XOR L#3 = 1101010110010011 XOR 1000100010001100 = 0101110100011111

06) L#6 = L#2 XOR L#4 = 1000011010000100 XOR 1001011110011001 = 0001000100011101

07) L#7 = (L#5 * K5) mod (2^16 + 1) = 0101110100011111 * 0011010100110101 mod (2^16 + 1) = 1001111100010001

08) L#8 = (L#6 + L#7)) mod 2^16 = 0001000100011101 + 1001111100010001 mod 2^16 = 1011000000101110

09) L#9 = (L#8 * K6) mod (2^16 + 1) = 1011000000101110 * 0011010100110110 mod (2^16 + 1) = 1000101100010110

10) L#10 = (L#7 + L#9)) mod 2^16 = 1001111100010001 + 1000101100010110 mod 2^16 = 0010101000100111

11) L#11 = L#1 XOR L#9 = 1101010110010011 XOR 1000101100010110 = 0101111010000101

12) L#12 = L#3 XOR L#9 = 1000100010001100 XOR 1000101100010110 = 0000001110011010

13) L#13 = L#2 XOR L#10 = 1000011010000100 XOR 0010101000100111 = 1010110010100011

14) L#14 = L#4 XOR L#10 = 1001011110011001 XOR 0010101000100111 = 1011110110111110

Next encryption rotation:
X1 = L#11 = 0101111010000101
X2 = L#12 = 0000001110011010
X3 = L#13 = 1010110010100011
X4 = L#14 = 1011110110111110

**ROUND 2**
01) L#1 = (X1 * K1) mod (2^16 + 1) = 0101010001001000 * 0011011100111000 mod (2^16 + 1) = 1101010110010011

02) L#2 = (X2 + K2) mod 2^16 = 0100111101001101 + 0011011100110111 mod 2^16 = 1000011010000100

03) L#3 = (X3 + K3) mod 2^16 = 0101000001010011 + 0011100000111001 mod 2^16 = 1000100010001100

04) L#4 = (X4 * K4) mod (2^16 + 1) = 0100111101001110 * 0011001000110101 mod (2^16 + 1) = 1001011110011001

05) L#5 = L#1 XOR L#3 = 1101010110010011 XOR 1000100010001100 = 0101110100011111

06) L#6 = L#2 XOR L#4 = 1000011010000100 XOR 1001011110011001 = 0001000100011101

07) L#7 = (L#5 * K5) mod (2^16 + 1) = 0101110100011111 * 0011010100110101 mod (2^16 + 1) = 1001111100010001

08) L#8 = (L#6 + L#7)) mod 2^16 = 0001000100011101 + 1001111100010001 mod 2^16 = 1011000000101110

09) L#9 = (L#8 * K6) mod (2^16 + 1) = 1011000000101110 * 0011010100110110 mod (2^16 + 1) = 1000101100010110

10) L#10 = (L#7 + L#9)) mod 2^16 = 1001111100010001 + 1000101100010110 mod 2^16 = 0010101000100111

11) L#11 = L#1 XOR L#9 = 1101010110010011 XOR 1000101100010110 = 0101111010000101

12) L#12 = L#3 XOR L#9 = 1000100010001100 XOR 1000101100010110 = 0000001110011010

13) L#13 = L#2 XOR L#10 = 1000011010000100 XOR 0010101000100111 = 1010110010100011

14) L#14 = L#4 XOR L#10 = 1001011110011001 XOR 0010101000100111 = 1011110110111110

Next encryption rotation:
X1 = L#11 = 0101111010000101
X2 = L#12 = 0000001110011010
X3 = L#13 = 1010110010100011
X4 = L#14 = 1011110110111110

**ROUND 3**
01) L#1 = (X1 * K1) mod (2^16 + 1) = 0101010001001000 * 0011011100111000 mod (2^16 + 1) = 1101010110010011

02) L#2 = (X2 + K2) mod 2^16 = 0100111101001101 + 0011011100110111 mod 2^16 = 1000011010000100

03) L#3 = (X3 + K3) mod 2^16 = 0101000001010011 + 0011100000111001 mod 2^16 = 1000100010001100

04) L#4 = (X4 * K4) mod (2^16 + 1) = 0100111101001110 * 0011001000110101 mod (2^16 + 1) = 1001011110011001

05) L#5 = L#1 XOR L#3 = 1101010110010011 XOR 1000100010001100 = 0101110100011111

06) L#6 = L#2 XOR L#4 = 1000011010000100 XOR 1001011110011001 = 0001000100011101

07) L#7 = (L#5 * K5) mod (2^16 + 1) = 0101110100011111 * 0011010100110101 mod (2^16 + 1) = 1001111100010001

08) L#8 = (L#6 + L#7)) mod 2^16 = 0001000100011101 + 1001111100010001 mod 2^16 = 1011000000101110

09) L#9 = (L#8 * K6) mod (2^16 + 1) = 1011000000101110 * 0011010100110110 mod (2^16 + 1) = 1000101100010110

10) L#10 = (L#7 + L#9)) mod 2^16 = 1001111100010001 + 1000101100010110 mod 2^16 = 0010101000100111

11) L#11 = L#1 XOR L#9 = 1101010110010011 XOR 1000101100010110 = 0101111010000101

12) L#12 = L#3 XOR L#9 = 1000100010001100 XOR 1000101100010110 = 0000001110011010

13) L#13 = L#2 XOR L#10 = 1000011010000100 XOR 0010101000100111 = 1010110010100011

14) L#14 = L#4 XOR L#10 = 1001011110011001 XOR 0010101000100111 = 1011110110111110

Next encryption rotation:
X1 = L#11 = 0101111010000101
X2 = L#12 = 0000001110011010
X3 = L#13 = 1010110010100011
X4 = L#14 = 1011110110111110

Encryption rotation will loop until eight rotation and the last result as below:

01) Y1 = (X1 * K1) mod (2^16 + 1) = 0111011100010010 * 1100110111001110 mod (2^16 + 1) = 1101101011000100

02) Y2 = (X2 + K2) mod 2^16 = 1001000100111101 + 0000111001001100 mod 2^16 = 1001111110001001

03) Y3 = (X3 + K3) mod 2^16 = 1101100001010000 + 1000110101001101 mod 2^16 = 0110010110011101

04) Y4 = (X4 * K4) mod (2^16 + 1) = 0110101101001110 * 0100110101001101 mod (2^16 + 1) = 1001110000010000

Encryption Result:
Y1 = 1101101011000100 = ÚÄ
Y2 = 1001111110001001 = Ÿ‰
Y3 = 0110010110011101 = e•
Y4 = 1001110000010000 = œ

Cipher text = ÚÄŸ‰e□œ

The chipertext result is then decrypted by following predefined rules, here is the decryption process for round 1

01) L#1 = (X1 * K1) mod (2^16 + 1) = 1101101011000100 * 1000110001100010 mod (2^16 + 1) = 0111011100010010

02) L#2 = (X2 + K2) mod 2^16 = 1001111110001001 + 1111000110110100 mod 2^16 = 1001000100111101

03) L#3 = (X3 + K3) mod 2^16 = 0110010110011101 + 0111001010110011 mod 2^16 = 1101100001010000

04) L#4 = (X4 * K4) mod (2^16 + 1) = 1001110000010000 * 1110101001100011 mod (2^16 + 1) = 0110101101001110

05) L#5 = L#1 XOR L#3 = 0111011100010010 XOR 1101100001010000 = 1010111101000010

06) L#6 = L#2 XOR L#4 = 1001000100111101 XOR 0110101101001110 = 1111101001110011

07) L#7 = (L#5 * K5) mod (2^16 + 1) = 1010111101000010 * 1100011010100110 mod (2^16 + 1) = 0010100011001110

08) L#8 = (L#6 + L#7)) mod 2^16 = 1111101001110011 + 0010100011001110 mod 2^16 = 0010001101000001

09) L#9 = (L#8 * K6) mod (2^16 + 1) = 0010001101000001 * 1100011010100110 mod (2^16 + 1) = 0000011011001011

10) L#10 = (L#7 + L#9)) mod 2^16 = 0010100011001110 + 0000011011001011 mod 2^16 = 0010111110011001

11) L#11 = L#1 XOR L#9 = 0111011100010010 XOR 0000011011001011 = 0111000111011001

12) L#12 = L#3 XOR L#9 = 1101100001010000 XOR 0000011011001011 = 1101111010011011

13) L#13 = L#2 XOR L#10 = 1001000100111101 XOR 0010111110011001 = 1011111010100100

14) L#14 = L#4 XOR L#10 = 0110101101001110 XOR 0010111110011001 = 0100010011010111

Next decryption rotation
X1 = L#11 = 0111000111011001
X2 = L#12 = 1101111010011011
X3 = L#13 = 1011111010100100
X4 = L#14 = 0100010011010111

Rotation is done until round 8 and produces the following output transformation:

01) Y1 = (X1 * K1) mod (2^16 + 1) = 1101010110010011 * 1001011000010100 mod (2^16 + 1) = 0101010001001000

02) Y2 = (X2 + K2) mod 2^16 = 1000011010000100 + 1100100011001001 mod 2^16 = 0100111101001101

03) Y3 = (X3 + K3) mod 2^16 = 1000100010001100 + 1100011111000111 mod 2^16 = 0101000001010011

04) Y4 = (X4 * K4) mod (2^16 + 1) = 1001011110011001 * 0100101100010010 mod (2^16 + 1) = 0100111101001110

Decryption result
Y1 = 0101010001001000 = TH
Y2 = 0100111101001101 = OM
Y3 = 0101000001010011 = PS
Y4 = 0100111101001110 = ON

Plain text = THOMPSON

## 4. Conclusion

The experiment performed to improve the encryption and decryption of the IDEA algorithm use the random Pixel value of the image, and from the examination it is found that the process of decryption with different images cannot be done due to the different pixel image's, in the experiment at key generator it requires a lot of time to read pixel images randomly, for future development can be done optimization by using a special algorithm image processing so that the combination of algorithms make the key generation much faster with maximum results.

## REFERENCES

[1] A. Putera, U. Siahaan, and R. Rahim, "Dynamic Key Matrix of Hill Cipher Using Genetic Algorithm," *Int. J. Secur. Its Appl.*, vol. 10, no. 8, pp. 173–180, Aug. 2016.
[2] R. Rahim, "Man-in-the-middle-attack prevention using interlock protocol method," *ARPN J. Eng. Appl. Sci.*, vol. 12, no. 22, pp. 6483–6487, 2017.
[3] R. Rahim, M. Dahria, M. Syahril, and B. Anwar, "Combination of the Blowfish and Lempel-Ziv-Welch algorithms for text compression," *World Trans. Eng. Technol. Educ.*, vol. 15, no. 3, pp. 292–297, 2017.
[4] H. Nurdiyanto, R. Rahim, and N. Wulan, "Symmetric Stream Cipher using Triple Transposition Key Method and Base64 Algorithm for Security Improvement," *J. Phys. Conf. Ser.*, vol. 930, no. 1, p. 012005, Dec. 2017.
[5] R. F. Marta, G. S. Achmad Daengs, A. Daniar, W. O. Seprina, and A. P. Menayang, "Information security risk and management in organizational network," *Int. J. Eng. Adv. Technol.*, vol. 8, no. 6 Special Issue 2, pp. 1152–1156, Aug. 2019.
[6] K. J. Fitzgerald, "Security and data integrity for LANs and WANs," *Inf. Manag. Comput. Secur.*, vol. 3, no. 4, pp. 27–33, 1995.
[7] R. J. Boyle, C. D. Challa, and J. A. Clements, "Valuing Information Security: A Look at the Influence of User Engagement on Information Security Strength," *J. Inf. Priv. Secur.*, vol. 13, no. 3, pp. 137–156, Jul. 2017.
[8] Q. Kester, "A Visual Cryptographic Encryption Technique for Securing Medical Images," *arXiv Prepr. arXiv1307.7791*, vol. 3, no. 6, pp. 3–7, 2013.
[9] R. Rahim, A. Indahingwati, I. K. Sudarsana, G. S. Achmad Daengs, and Y. Yuniningsih, "Caesar cipher and end of file combination algorithm for steganography," *J. Adv. Res. Dyn. Control Syst.*, vol. 10, no. 10 Special Issue, pp. 947–951, 2018.
[10] R. Rahim and A. Ikhwan, "Cryptography Technique with Modular Multiplication Block Cipher and Playfair Cipher," *Int. J. Sci. Res. Sci. Technol.*, vol. 2, no. 6, pp. 71–78, 2016.
[11] D. Nofriansyah and R. Rahim, "Combination of Pixel Value Differencing Algorithm with Caesar Cipher for Steganography," *Int. J. Res. Sci. Eng.*, vol. 2, no. 6, pp. 153–159, 2016.
[12] H. Nurdiyanto and R. Rahim, "Enhanced pixel value differencing steganography with government standard algorithm," in *2017 3rd International Conference on Science in Information Technology (ICSITech)*, 2017, pp. 366–371.
[13] O. Almasri and H. M. Jani, "Introducing an Encryption Algorithm based on IDEA," vol. 2, no. 9, pp. 334–339, 2013.
[14] B. S. Champakamala, K. Padmini, R. D. K. A. Professors, and D. Bosco, "Least Significant Bit algorithm for image steganography Overview of Steganography," *Int. J. Adv. Comput. Technol.*, vol. 3, no. 4, p. 5, 2014.
[15] J. Chen, D. Xue, and X. Lai, "An analysis of international data encryption algorithm(IDEA) security against differential cryptanalysis," *Wuhan Univ. J. Nat. Sci.*, vol. 13, no. 6, pp. 697–701, 2008.
[16] G. A. Ludwig and P. Alegre, "International Data Encryption Algorithm (IDEA)," vol. 8, no. 1, pp. 6–11, 2007.
[17] J. Daemen, R. Govaerts, and J. Vandewalle, "Weak Keys for IDEA," in *Advances in Cryptology*

— *CRYPTO' 93*, Berlin, Heidelberg: Springer Berlin Heidelberg, 1993, pp. 224–231.

[18] R. I. Al-Khalid, R. A. Al-Dallah, A. M. Al-Anani, R. M. Barham, and S. I. Hajir, "A Secure Visual Cryptography Scheme Using Private Key with Invariant Share Sizes," *J. Softw. Eng. Appl.*, vol. 10, no. 01, pp. 1–10, Jan. 2017.

[19] L. Sui, K. Duan, and J. Liang, "A secure double-image sharing scheme based on Shamir's three-pass protocol and 2D Sine Logistic modulation map in discrete multiple-parameter fractional angular transform domain," *Opt. Lasers Eng.*, vol. 80, pp. 52–62, May 2016.

[20] Iswanto, O. Wahyunggoro, and A. I. Cahyadi, "3D object modeling using data fusion from laser sensor on quadrotor," in *AIP Conference Proceedings*, 2016, vol. 1755.

[21] M. B. Pramanik, "Implementation of Cryptography Technique using Columnar Transposition," *Int. J. Comput. Appl.*, pp. 19–23, 2014.

[22] Q.-A. Kester, "A HYBRID CRYPTOSYSTEM BASED ON VIGENÈRE CIPHER AND COLUMNAR TRANSPOSITION CIPHER," *Int. J. Adv. Technol. Eng. Res.*, vol. 3, no. 1, pp. 141–147, 2013.

[23] S. Bruce, *Applied cryptography*. 1996.

[24] S. Manna and S. Dutta, "A Stream Cipher based Bit-Level Symmetric Key Cryptographic Technique using Chen Prime Number," *Int. J. Comput. Appl.*, vol. 107, no. 12, pp. 975–8887, 2014.

[25] H. P. Singh, S. Verma, and S. Mishra, "Secure-International Data Encryption Algorithm," *Int. J. Adv. Res. Electr. Electron. Instrum. Eng.*, vol. 2, no. 2, pp. 780–792, 2013.

[26] H. K. Sahu, V. Jadhav, S. Sonavane, and R. K. Sharma, "Cryptanalytic attacks on international data encryption algorithm block cipher," *Def. Sci. J.*, vol. 66, no. 6, pp. 582–589, 2016.

[27] E. Hariyanto and R. Rahim, "Arnold's Cat Map Algorithm in Digital Image Encryption," *Int. J. Sci. Res.*, vol. 5, no. 10, pp. 1363–1365, Oct. 2016.

[28] X. Li, "Watermarking in secure image retrieval," *Pattern Recognit. Lett.*, vol. 24, no. 14, pp. 2431–2434, 2003.

[29] A. M. Bagade and S. N. Talbar, "Image morphing concept for secure transmission of image data contents over internet," *J. Comput. Sci.*, vol. 6, no. 9, pp. 987–992, 2010.

[30] R. Gonzalez and R. Woods, *Digital image processing*. 2002.

[31] A. Swaminathan, Y. Mao, and M. Wu, "Robust and secure image hashing," *IEEE Trans. Inf. Forensics Secur.*, vol. 1, no. 2, pp. 215–230, 2006.

# Online Expert System for Diagnosis Psychological Disorders Using Case-Based Reasoning Method

**Robbi Rahim[1*], Windania Purba[2], Mufida Khairani[3], and R Rosmawati[4]**

[1]Sekolah Tinggi Ilmu Manajemen Sukma, Medan, Indonesia
[2]Department of Information System, Universitas Prima Indonesia, Medan, Indonesia
[3]Department of Informatics, Universitas Harapan Medan, Medan, Indonesia
[4]Department of Industrial Engineering, Institut Teknologi dan Sains TD Pardede, Medan, Indonesia

*usurobbi85@zoho.com

**Abstract.** Expert system of diagnosis of psychological disorders is a system designed to diagnose early symptoms of psychological disorders that may be suffered by individuals, online-based expert systems by applying the method of Case-Based Reasoning allows users to check the psychological condition directly with the results that can be used as a basis for more examination continue. The use of Case-Based Reasoning method facilitates the examination process for each user consultation because the result of past examination data can be used to compare with consultation data, psychological disorder examination with Case-Based Reasoning method can be done well.

## 1. Introduction

There are many areas outside the computer field that utilize computer technology to facilitate their work, for example, is the field of health, physical infrastructure, psychology, communications, Virtual Reality, Modelling and many other areas[1]. The field of health is one field that has been widely used computer applications to help work efficiency. One of the forms is an expert system [2], [3]. Expert system application is one of the branches of AI (Artificial Intelligence) which can be described as a computer device that has a knowledge base for a particular field that uses inference reasoning resembles an expert in solving a problem [4], [5].

Psychological disorder [6] is a disease that may be experienced by some people due to various factors such as trauma, stress, environmental conditions and hereditary diseases. Depression, Phobia, Bipolar Disorder are some types of psychological disorder that can be suffered by anyone, to know psychological self can be done in consultation with experts in the field of psychology or by using an expert system.

Expert systems are not a substitute for experts in a particular field, but a method used to adopt specialist knowledge into systems that can be accessed through computers, smartphones or other devices [3], [7], [8]. Case-Based Reasoning (CBR) [9]–[11] Method is one method of solving problems by using a solution that has been used previously against similar issues [11], the use of CBR methods on the diagnosis of psychological disorders will facilitate the process of early examination before a direct consultation to a specialist.

## 2. Methodology

An expert system is a branch of Artificial Intelligence (AI) developed in mid-1960. Expert system first time appeared was General Purpose Problem Solver (BPS). An expert system is a system that uses human knowledge in which knowledge is incorporated into a computer and then used to solve problems that usually require human skills [2], [12].

### *2.1 Expert System*

Expert systems are designed on a specific knowledge area for expertise to approaching human ability in one field [7]. The expert system was trying to find a satisfactory solution as one expert does. Also, expert systems can also explain each step was taken and provide reasons for suggestions or conclusions that have been found [8], [13], [14].

An expert is a person who has expertise in a particular field, an expert who has the knowledge or unique ability that others do not know or are capable of in the area it has [14]. Knowledge[15], [16] in expert systems may be an expert, or knowledge commonly found in books, magazines and people who know a field In the expert system, the user delivers the facts or information to the system and then accepts the advice of the expert or the expert answer. The inside of the expert system consists of two main components, the knowledge base which contains the knowledge and the inference engine that illustrates the conclusion. The conclusion is the response of the expert system at the request of the user [12], [14].

### *2.2 Case-Based Reasoning*

Case-Based Reasoning [9] method is one of the ways to build expert systems with decision-making from new cases based on solutions from previous cases. There are four process steps in the Case-Based Reasoning method. Retrieve the most similar case, Reuse (using) information and knowledge of the case to solve the problem, Revise (repair) the proposed solution, and Retain (s) part of the experience that might be useful for solving problems in the future [5], [9], [11], see figure 1 for diagram process Case-Based Reasoning.

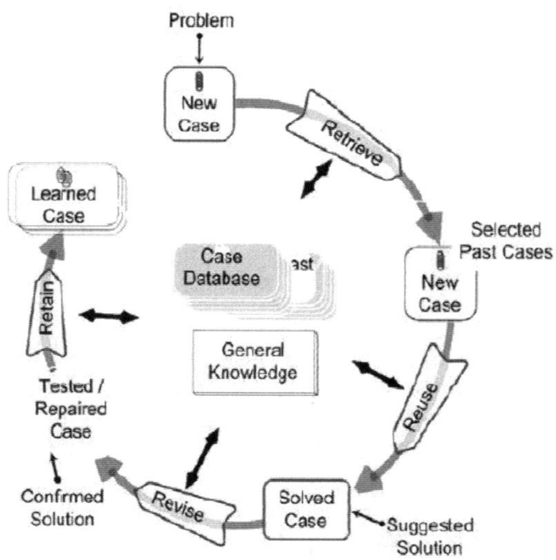


Figure 1. Case-Based Reasoning Method

Case-Based Reasoning process can use several techniques, one of which is the Nearest Neighbor method [11], [17] using equations.

$$Similiarity\ (S,T) = \sum_{t-1}^{n} f(T_S, S_t)\ x\ wt \qquad (1)$$

The closeness of the case is determined by the value of 0 to 1, because of Nearest Neighbor calculations using the real calculation is between the values (0.1). A value of 0 means that two similar cases are not absolute and the absolute value of 1 case is similar [5], [9], [18].

## 3. Results And Discussion

Application of Case-Based Reasoning to a diagnosis of psychological disorder requires several data objects such as symptom data, symptom categories, solutions, and diseases, here are the tables of each data used.

**Table 1.** Symptom Categories

| No | Code | Name |
|---|---|---|
| 1 | S1 | Emotional Symptom |
| 2 | S2 | Physical Symptom |
| 3 | S3 | Mind Symptom |
| 4 | S4 | Attitude Symptom |

**Table 2.** Symptom Data

| No | ID | Name |
|---|---|---|
| 1 | G01 | Speaking less smoothly |
| 2 | G02 | Rigid and inflexible |
| 3 | G03 | Selfish |
| 4 | G04 | Unstable emotional disturbance |
| 5 | G05 | Respiratory disturbances, shortness of breath |
| 6 | G06 | Diet disorders |
| 7 | G07 | Mindset disorder |
| 8 | G08 | Insomnia |
| 9 | G09 | Want to kill yourself |
| 10 | G10 | Loss of memory |
| 11 | G11 | Antisocial |
| 12 | G12 | Doing a job/activity over and over again |
| 13 | G13 | Experiencing convulsions |
| 14 | G14 | Maintain excessive hygiene |
| 15 | G15 | Indigestion |
| 16 | G16 | Feeling heavy on the limbs |
| 17 | G17 | Excessive suspicion |
| 18 | G18 | Feeling low self-esteem |
| 19 | G19 | Feeling worried about excessive disease |
| 20 | G20 | Feeling weak and tired easily |
| 21 | G21 | Feeling suffering from a severe illness |
| 22 | G22 | Feeling body parts pain |
| 23 | G23 | Feeling hopeless |
| 24 | G24 | Feel very sinful |
| 25 | G25 | Feeling very guilty / very afraid to make mistakes |
| 26 | G26 | Feel very excited |

| | | |
|---|---|---|
| 27 | G27 | Feel very smart in a particular field |
| 28 | G28 | Feel very afraid of certain objects/conditions |
| 29 | G29 | Excessive sadness |
| 30 | G30 | Feel happy after pulling hair |
| 31 | G31 | Feel happy to grab an eye-catching object |
| 32 | G32 | Fear of excessive |
| 33 | G33 | Fearing excessive dirty |
| 34 | G34 | Feeling not confident |
| 35 | G35 | Feeling never guilty |
| 36 | G36 | Easy to sweat cold |
| 37 | G37 | Easy to panic |
| 38 | G38 | Easily offended |
| 39 | G39 | Underwent traumatic / violent |
| 40 | G40 | Feeling stressed for one thing |
| 41 | G41 | Always think badly |
| 42 | G42 | Nice to quarrel, hostile and often screwed up |
| 43 | G43 | Nice to interfere in other people's business |
| 44 | G44 | Often lie |
| 45 | G45 | Often counts in the mind |
| 46 | G46 | Often thinking is unrealistic |
| 47 | G47 | often being attentive |
| 48 | G48 | Frequent delusions |
| 49 | G49 | Often shaking |
| 50 | G50 | Frequent hallucinations |
| 51 | G51 | Often angry / irritable |
| 52 | G52 | Often check the body for mild problems |
| 53 | G53 | Often crying hysterically |
| 54 | G54 | Often pull hair until it fall out |
| 55 | G55 | Often experience of loss of physical function, suddenly paralyzed, feeling deaf, blind, numb but only briefly |
| 56 | G56 | Often have severe cramps |
| 57 | G57 | Often experience significant mood swings, very happy to be very sad / vice versa |
| 58 | G58 | Often experience pain / health problems |
| 59 | G59 | Often hurt yourself |
| 60 | G60 | Often feel neglected and inattentive |
| 61 | G61 | Often feel uneasy |
| 62 | G62 | Often feel anxious, restless and daydreaming |
| 63 | G63 | Often feel heart palpitations |
| 64 | G64 | Often feel headache |
| 65 | G65 | Often happy, excessively cheerful |
| 66 | G66 | Often feel tense and anxious |
| 67 | G67 | Often feel depressed with the circumstances |
| 68 | G68 | lost consciousness |
| 69 | G69 | impatient |
| 70 | G70 | difficult to speak normally |

| | | |
|---|---|---|
| 71 | G71 | Difficult concentration and decreased memory |
| 72 | G72 | Unable to adjust / difficult to adjust |
| 73 | G73 | Do not have a sense of humor |
| 74 | G74 | Never fear |

**Table 3.** Solution

| No | Code | Name |
|---|---|---|
| 1 | T01 | Hospitalization |
| 2 | T02 | Psychoanalysis psychotherapy approach Cognitive Behavioral Therapy (CBT) |
| 3 | T03 | Individual Psychotherapy |
| 4 | T04 | Group psychotherapy |
| 5 | T05 | Environmental psychotherapy |
| 6 | T06 | Supportive psychotherapy |
| 7 | T07 | Relaxing training |
| 8 | T08 | PHBS Technique |
| 9 | T09 | Sports Therapy |
| 10 | T10 | Cognitive behavioral therapy |

**Table 4.** Psychological Disorders Disease

| No | Code | Name |
|---|---|---|
| 1 | P01 | Ansietas Cemas |
| 2 | P02 | Anorexia Nervosa |
| 3 | P03 | Bipolar Disorder |
| 4 | P04 | Conversion Disorder |
| 5 | P05 | Depresi |
| 6 | P06 | Enosimania |
| 7 | P07 | Fobia |
| 8 | P08 | Hipokondria |
| 9 | P09 | Hysteria |
| 10 | P10 | Multiple personality |

A simple case example of using Case-Based Reasoning method can see in table 5.

**Table 5.** Example Case Based Reasoning

| No | Disease | Symptom | Initial Case |
|---|---|---|---|
| 1 | Anorexia Nervosa | [G06]<br>[G32]<br>[G34]<br>[G58] | Anorexia |
| 2 | Anxiety | [G15]<br>[G36]<br>[G37]<br>[G40]<br>[G49]<br>[G61] | Anxiety |

| | | | |
|---|---|---|---|
| | | [G64] | |

The consultation process obtained symptoms as follows:
1. [G06] Diet disorders
2. [G34] Feeling not confident
3. [G58] Often experience pain/health problems
4. [G15] Indigestion
5. [G49] Often shaking

To obtain a diagnostic result, the system compares the symptoms chosen in consultation with the existing symptoms of the previously stored case.

**Table 6.** Case-Based Reasoning Process

| No | Diseases | Case | Weight | Match | Weight |
|---|---|---|---|---|---|
| 1 | Anorexia Nervosa | [G06]<br>[G32]<br>[G34]<br>[G58] | 3<br>3<br>3<br>3 | [G06]<br>[G34]<br>[G58] | 3<br>3<br>3 |
| 2 | Anxiety | [G15]<br>[G36]<br>[G37]<br>[G40]<br>[G49]<br>[G61]<br>[G64] | 3<br>1<br>5<br>1<br>1<br>3<br>1 | [G15]<br>[G49] | 3<br>1 |

Based on table 6 data, it means the most likely disease experienced is Anorexia Nervosa with a percentage of 75%, from this data an expert can add the case to a new case on a system that can later be reused or change the existing case data (revision).

The use of case-based reasoning methods in Table 6 is completed in an online expert system.

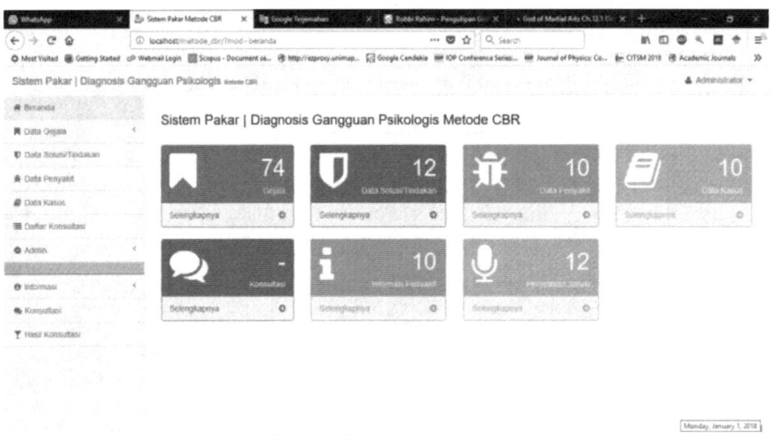


**Figure 2.** Main Display

Figure 1 displays information on symptoms, categories of symptoms, diseases, and solutions. The central section also displays the exact amount of data contained in the database; Figure 2 shows another example of the system.

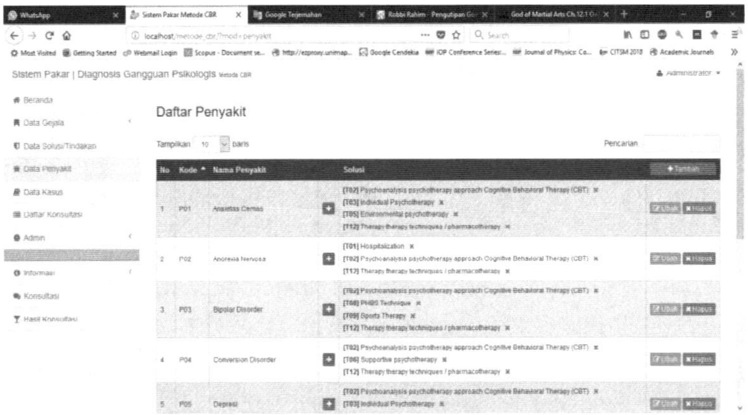


**Figure 3.** List of Psychology Disorder Disease

Testing and implementation done by using case-based reasoning method in the expert system can help to know the psychological disorder.

## 4. Conclusion

Expert system with Case-Based Reasoning method can be applied well for cases of psychological disorders and can also be used for other cases. Development of this method can also be done by combining with other methods or also compare the results with other expert system methods.

## References

[1] Iswanto, O. Wahyunggoro, and A. I. Cahyadi, "3D object modeling using data fusion from laser sensor on quadrotor," in *AIP Conference Proceedings*, 2016, vol. 1755.

[2] B. Ruiz-Mezcua, A. Garcia-Crespo, J. L. Lopez-Cuadrado, and I. Gonzalez-Carrasco, "An expert system development tool for non AI experts," *Expert Syst. Appl.*, vol. 38, no. 1, pp. 597–609, 2011.

[3] N. A. Hasibuan, K. Yusmiarti, F. T. Waruwu, and R. Rahim, "Expert systems with genetics probability," *Int. J. Res. Sci. Eng.*, vol. 3, no. 2, pp. 112–116, 2017.

[4] K. C. Lee and S. Lee, "A causal knowledge-based expert system for planning an Internet-based stock trading system," *Expert Syst. Appl.*, vol. 39, no. 10, pp. 8626–8635, 2012.

[5] G. H. Lee, "Rule-based and case-based reasoning approach for internal audit of bank," *Knowledge-Based Syst.*, vol. 21, no. 2, pp. 140–147, 2008.

[6] R. Retnowati and A. Pujiyanta, "Implementasi Case Base Reasoning Pada Sistem Pakar Dalam Menentukan Jenis Gangguan Kejiwaan," *J. Sarj. Tek. Inform.*, vol. 1, no. 1, pp. 69–78, 2013.

[7] L. A. Latumakulita, "Sistem Pakar Pendiagnosa Penyakit Anak Menggunakan Certainty Factor (Cf)," *J. Ilm. Sains*, vol. 12, no. 2, p. 120, 2012.

[8] D. T. Yuwono, A. Fadlil, M. T. Informatika, U. Ahmad, and D. Yogyakarta, "PENERAPAN METODE FORWARD CHAINING DAN CERTAINTY FACTOR PADA SISTEM PAKAR," vol. 04, no. 02, pp. 136–145, 2017.

[9] I. Watson and F. Marir, "Case-Based Reasoning: A Review," *Knowl. Eng. Rev.*, vol. 9, no. 4, pp. 327–354, 1994.

[10] C. Diki Andita Kusuma, "Rancang Bangun Sistem Pakar Pendiagnosa Penyakit Paru-Paru Menggunakan Metode Case Based Reasoning," *Infotel*, vol. 6, no. 2, pp. 57–62, 2014.

[11] W. Yulianti, "Aptitude Testing Berbasis Case-Based Reasoning Dalam Sistem Pakar Untuk Menentukan Minat Dan Bakat Siswa Sekolah Dasar," *J. Teknol. dan Sist. Inf. UNIVRAB*, vol. 1, no. 2, pp. 104–118, 2016.

[12] K. P. Tripathi, "A Review on Knowledge-based Expert System : Concept and Architecture," *Artif. Intell. Tech. - Nov. Approaches Pract. Appl.*, vol. 4, no. 4, pp. 19–23, 2011.

[13] Roni Pambudi & Sumarno, "Aplikasi Sistem Pakar Diagnosa Penyakit Kanker Menggunakan Metode Certainty Factor," *J. Ilm. Tek. Inf.*, no. Sistem Pakar, p. IF-64-IF-65, 2015.

[14] A. Keleş, "Expert Doctor Verdis: Integrated medical expert system," *Turkish J. Electr. Eng. Comput. Sci.*, vol. 22, no. 4, pp. 1032–1043, 2014.

[15] D. Siregar, D. Arisandi, A. Usman, D. Irwan, and R. Rahim, "Research of Simple Multi-Attribute Rating Technique for Decision Support," *J. Phys. Conf. Ser.*, vol. 930, no. 1, p. 012015, Dec. 2017.

[16] F. Haswan, "Decision Support System For Election Of Members Unit Patients Pamong Praja," *Int. J. Artif. Intell. Res.*, vol. 1, no. 1, p. 21, Jun. 2017.

[17] J. Papathanasiou, N. P. B, T. Bournaris, and B. Manos, "A Decision Support System for Multiple Criteria Alternative Ranking Using TOPSIS and VIKOR : A Case Study on Social Sustainability in Agriculture," *ICDSST*, vol. 2, pp. 3–15, 2016.

[18] J. Sebestyénová, "Case-based reasoning in agent-based decision support system," *Acta Polytech. Hungarica*, vol. 4, no. 1, pp. 127–138, 2007.

Journal of Physics: Conference Series **1381** (2019) 012045 doi:10.1088/1742-6596/1381/1/012045

# Development of physics learning e-module based on local culture wisdom in Pontianak,West Kalimantan

**Matsun[1], V S Andrini[2], T W Maduretno[2], A C Yusro[4]**

[1] IKIP PGRI Pontianak, Indonesia
[2.2] STKIP PGRI Nganjuk, Indonesia
[4]Universitas PGRI Madiun, Indonesia

E-mail: matsunpfisika@ikippgriptk.ac.id

**Abstrac**, This study aims to determine the feasibility and effectiveness of physics learning e-module based on local culture wisdom. This study is a development using ADDIE model of study design comprised five steps of analysis, design, development, implementation and evaluation. The sample used by 70 students with one group pretest-posttest design. Based on the research results, e-module based on local wisdom that is validated by three experts and three experts media material showed 83% in the category of very viable for use in learning. The effectiveness of e-module was tested using paired sample t test. The results showed no effect of the use-based e-modules of local wisdom to the learning process of physics. Values of N-Gain score indicates the percentage of 81% (in the effective category). E-modules used in this study utilizes the android based smartphone technology. Local wisdom that becomes a source of learning is a game "carbide cannon" who came fromPontianak, West Kalimantan, the terms of the concepts of physics and chemistry. Students' response to the media developed showed an average score of 87% ratings with very good category.

**Keyword:** E-Module, Physics, Wisdom of Local Culture, Research and Development

## Introduction

Learning media is everything that used to deliver the message and to stimulate thoughts, feelings, concerns and volition of the study so as to encourage the process of learning a deliberate aim and control. Utilization of instructional media can be applied wherever possible in all subjects, including physics. The media should be packed interesting learning so that students can linger study material. If media be understood, the meaning is human, material, or events that establish conditions under which students are able to acquire the knowledge, skills or attitudes [1]. In this sense, teachers, textbooks and school environment is a medium. More specifically, the notion of media in learning processes tend to be interpreted as graphics tools,

Advancement of Information and Communication Technology (ICT) in Indonesia is growing very rapidly. ICT advances that can be felt in many areas, one in the field of education. The development of ICT in education will make education in Indonesia today can be more advanced and developed. The development of ICT can be used by teachers to create learning interesting and more colorful in the classroom. Regular instructional media packaged in the form of a power point, video learning physics textbooks and worksheets but over the times, learning media is also provided in electronic form, or better known as Mobile Learning. M-Learning is able to create a conducive academic atmosphere for students to learn anywhere and anytime.

There is a close relationship between the quality of learning resources (learning media) with an understanding of the nature of science, such as empirical aspects, methods, theories, creative, social, and cultural [3]. Student culture has changed drastically as the era of the industrial revolution 4.0. shows that the technology has important and positive if managed properly on science learning [4]. Especially physics science education should involve technology, innovation in learning in school settings, and accelerate technological innovation [5]. The motivational characteristics of learning resources (learning media) will increase interest in learning, especially involving issues that emerge at the international level or even interdisciplinary [6].

Instructional media developed in this study was associated with a typical local knowledge of West Kalimantan. Supposedly learning involves aspects of art, science and technology and value / value, and the implementation of local knowledge [7]. Local knowledge by utilizing traditional technologies as a source of learning in the learning of physics. Physics learning resources (learning media) based on the study of the traditional technology when implemented in the classroom from the perspective of sustainability is expected to bring new innovations example of environmentally friendly technologies.

## Method

The method used in this research is the method of Development (Research and Development / R & D). Method R & D is the research methods used to produce a specific product and test the effectiveness of such products [8]. The study design used in the study is the ADDIE model of research design consisting of five stages consisting of analysis, design, development, implementation and evaluation [9]. The sample used in this study consisted of 70 students in Physical Education IKIP PGRI Pontianak, Indonesia. Data collection techniques used in this study in the form of interviews, questionnaires, tests of learning outcomes, and documentation. The study design using one group pretest posttest.

## Result

Research development of e-module based on local knowledge is done by utilizing the android application. The research activities started from the observation phase of the study (research) and development (development), the design and testing of products. Based on the research goal is to determine the effectiveness of the e-module according to subject matter experts, media experts, to determine the student's response. Standardized assessment of subject matter experts and media experts are used as a reference for whether or not the media used. While the effectiveness of instructional media depend on the results of trials that have been given to the study of media users. In summary the data results of research and development are as follows:

1. **According to the feasibility Learning Media Expert Content**

   Before the product tested to the range, products are validated in advance by subject matter experts. Validation is done so that the media developed products is guaranteed that the initial product developed decent tested on students. Additionally validation of subject matter experts is useful to anticipate material errors, material deprivation, anticipation as range trials and so on. Expert validation is done so that the products developed are not having a lot of mistakes and in accordance with the needs of students in the range. Once the completed product design, product design was submitted to the validator to be validated and assessed its feasibility.

   Validator matter experts in the validation of the e-module is based on local wisdom one physics education study program lecturers and one teacher subject matter experts in the range of physics. Validation is done by giving subject matter experts along with the assessment sheet products. Assessment sheet form validation sheet with seventeen grains of assessment of the matter, matter / quizzes, language and enforceability as Table 1.

   The process of assessment of the subject matter expert validation performed twice matter experts. After validation of the product can be seen there are deficiencies and necessary improvements to the Android-based learning media products that will be developed. Then to subject matter experts in the assessment did consist of 4 aspects of materials, Questions / Quiz, Linguistic and Keterlaksanaan which translated into 17 statement. The following Table 4.1 the results of expert assessment materials based on every aspect that is in use.

   **Table 1. Acquisition of Material Aspects Validation Sheet**

| Aspect | Percentage | Criteria |
|---|---|---|
| Matter | 81% | Very Decent |
| Problem / Quiz | 84% | Very Decent |
| Linguistic | 85% | Very Decent |
| keterlaksanaan | 82% | Very Decent |
| **Average Percentage** | **83%** | **Very Decent** |

Based on Table 1, it can be concluded that based on aspects of material to get an average score of 83% with the criteria very decent, aspects of a matter / quiz to get an average score of 86% with the criteria very decent, aspects of language to get an average score of 87% with the criteria very decent and adherence to get an average score of 85% with a very decent criteria. Based on the above data that the development of e-module based on local knowledge has an average score of 85% and in the category of very decent.

**2. According Feasibility Study Media Expert Media**

Before the products tested to the range, products are validated in advance by media experts. At the stage of design validation is performed by submitting and presenting the initial product design to be validated by a team of experts. Media validation assessment based media perspective contained in the e-module. Validation is done so that the products developed can be said is reasonable and tested on students. Besides expert validation is useful to anticipate material errors, material deprivation, anticipation is currently in trials in the range and others. In addition to a completed product design, product design can be submitted to the validator for validation and in the value of its feasibility.

Validator media experts carried out by one person expert lecturers in the field of media. Media expert validation done by providing products along with the assessment sheet. Assessment sheet form validation sheet with 18 point ratings on software engineering and visual communication.

The process of assessment of media experts do one media expert validation. After validation of the product can be known nothing too significant revisions in the media developed. Then for media expert assessment will be undertaken consists of two aspects: Software Engineering and Visual Communication were translated into 18 statements such as Table 2.

Table 2 Acquisition Aspects Validation Sheet Media

| Aspect | Percentage | Criteria |
|---|---|---|
| Software engineering | 80% | very Decent |
| Visual communications | 84% | very Decent |
| **Average Percentage** | **82%** | **very Decent** |

Based on Table 2, the media expert's assessment of the aspects of software engineering is 80% and visual communication is 84%. The overall rating is 82% and the very feasible category is used to support the learning process.

**3. Respond of The Students**

After the media is said to be appropriate by material and media experts, the researchers then test the learning media with the aim of knowing students' responses to the products developed. Trials are conducted in the classroom using e-modules for each student with permission from the school and subject teachers. Furthermore, the learning process ends with knowing the responses of students using questionnaires. The results of student responses are classified into 7 aspects, namely content quality, pleasure, character, evaluation, grammar, motivation and use of illustrations which are then described into 18 statements such as Table 3.

Table 3. Obtaining Student Response Aspects

| Aspect | Percentage | Criteria |
|---|---|---|
| Quality | 90% | Very Good |
| Pleasure | 86% | Very Good |
| Character | 85% | Very Good |
| Evaluation | 83% | Very Good |
| Grammar | 88% | Very Good |
| Motivation | 86% | Very Good |
| use of illustrations | 91% | Very Good |
| **Mean** | **87%** | **Very Good** |

Based on Table 3, the percentage of the content quality aspect is 90%, the pleasure aspect is 86%, the character aspect is 85%, the evaluation aspect is 83%, the grammar aspect is 88%, the motivation aspect is 86%, and the illustration use aspect is 91 %. Overall, from 7 aspects in the category it is very good.

**4. Effectiveness Test**

Based on statistical tests using SPSS, the output table "Paired Sample Test" is known as the Sig. (2-tailed) of 0.008. So it can be concluded that there are differences in the average between the pretest learning outcomes with the posttest value. This means that there is the influence of the use of e-modules based on local cultural wisdom on student learning outcomes. The next step, the N-Gain test was carried out to determine the effectiveness of the e-module developed. The result is a Gain Score of 81%, which means that the product is developed effectively to support the learning process.

**Discussion**

The stages of e-module development start from (1) the analysis phase of the achievement of competencies and characteristics of students about their learning capacity, knowledge, attitudes, and skills; (2) the design stage is carried out with the following terms of reference: (a) for whom is learning designed? (learners); (b) what abilities do you want to learn? (competence); (c) how can subject matter or skills be learned? (learning strategies); (e) how do you determine the level of mastery of the lessons already achieved? (assessment and evaluation); (3) the development stage in the form of activities to translate design specifications into physical forms, so that this activity produces a development product prototype. Everything that is done at the design stage, namely the selection of material in accordance with the characteristics of students and the demands and evaluations used are realized in the prototype; (4) the stage of implementation of the development results applied in learning to determine the effect on the quality of learning which includes effectiveness, attractiveness and efficiency of learning; and (5) stages which include formative evaluation and summative evaluation. Formative evaluation is done to collect data at each stage used for refinement and summative evaluation at the end of the program to determine the effect on student learning outcomes and the quality of learning widely. The final product of this development is an e-module that integrates local cultural wisdom.

Increasing students' understanding of the material learned can be measured using a score post test. Measurement of this understanding is done by essay test with 5 items. From the measurement of students' understanding of the material obtained a classical completeness score of 70% with very good criteria. The use of learning media in the teaching and learning process can generate new desires and interests, generate motivation and stimulation of learning activities and even bring psychological influences to students. The use of learning media at the learning orientation stage will greatly help the effectiveness of the learning process and the delivery of messages and content of the lesson at that time. In addition to arousing students' motivation and interests, learning media can also help students improve understanding, present data interestingly and reliably, facilitate interpretation of data and obtain new information ([10].

The use of e-modules in this learning is significantly effective and can improve student learning outcomes. The integration of technology and information in the learning process actually depends on the probability of each student. However, if the lecturer is able to present an interesting e-learning concept it will be able to motivate students to use it. The potential of ICT is very useful to improve the skills of each individual. Students will be trained to use electronic media to learn and communicate positively [11]. Other research also mentions that the use of e-learning is more effective than learning with a regular face in a conventional manner and is able to improve learning outcomes [12]. E-learning in fact reduces interaction between individuals directly, resulting in good communication between students through the internet network. Collaboration between students also increases [13].

Local wisdom is part of the culture that exists in the community. One of the local wisdom taken from Pontianak, West Kalimantan is the Karbit Cannon. Karbit Cannon is the main tool that is played as a marker for the arrival of the month of Ramadan until the eid feast of Eid or Eid for the Malaya who live in Pontianak. This tradition began in 1771 and has become the identity of the Malaya people of Pontianak. The history of carbide cannons associated with the establishment of the city of Pontianak in the form of the Pontianak Sultanate with its first king was Syarif Abdrrahman Al Qadrie. The sound of the explosion of the carbide cannon is very hard to vibrate the nearby building. Carbide cannons are made of large wood with a diameter of 50 cm - 80 cm and a length between 4-7 meters perforated in the middle. The wooden bar which has been hollow in the middle is filled with water and carbide is inserted as the machine. Carbide which reacts with water will produce gas which if ignited with fire will produce an explosion. If the cannon is 50-80 cm in diameter, then one explosion requires 3-5 ounces of carbide. The sound will be louder if the carbide mass of carbide used is getting bigger.

Integration of the values of local wisdom in the learning process associated with the cultivation of the spirit of nationalism and the character of students. It is intended that regional cultures is not lost and can be preserved as any part of the national culture. Lecturers can take the example of the cultural wisdom but remain linked and discussed in science. As an example of the use of carbide cannon who discussed the concept of physics andchemistry. Another goal of planting the value of local wisdom capable of educating students to always be close to the environment and the daily problems [14].

The times, technology and information are the main factors in the loss of local culture. Many young people are beginning to forget the culture of the region and move to modern culture and technology such as smartphone and others. Based on this, the role of the teacher must be able to facilitate between technological developments while not leaving the regional culture [15]. The learning process becomes more varied and not boring. So that the material provided will be absorbed and easily understood by students. There are many ways that lecturers can take in integrating the content of local cultural wisdom in learning. One of them is through the curriculum. The opening of pedagogic innovation based on local wisdom can motivate students to learn according to their own traditions, so that they are not free from the prevailing culture in their social systems) [16].

The chemical concepts contained in carbide cannons can be explained as follows. Chemically, carbide in the form of grey crystal rhombies is used for welding, asitelin gas factory with CaC2 chemical formula. In solid form it is not dangerous, but if mixed with water / steam it will cause ASETYLEN gas which is flammable and explosive. Chemically the reaction occurs as follows:

$CaC_2\ (s) + 2H_2O\ (l) \rightarrow Ca\ (OH)_2\ (s) + C_2H_2\ (g)$ ......(1)

and at the time of the asetylen gas eruption there was a reaction

$2\ C_2H_2\ (g) + CO_2\ (g) \rightarrow 4CO_2\ (g) + 2H_2O(g)$ + Panas (2) ......

Based on reaction (1), calcium carbide (CaC2) before reacting with water (H2O) does not produce any reaction. But when calcium carbide is reacted with water, expansion will occur which produces asetylen gas (C2H2). Production of acetylene gas by mixing $C_2H_2$. In water that is put in to the cannon made of bamboo in the cap, causing the development of the bamboo inside because there is gas that has reacted in it. The gas molecules are moving in all directions at high speed and collide with the walls of the cannon so that the pressure in the larger cannon. When using a fire lit, the reaction that occurs as the equation (2). An eruption would form karbondiokasida, water, and heat. The more a mixture of calcium carbide and water was added to the cannon made of wood, the resulting explosion will be even greater.

In physics, students can learn about concept of the collision. In the carbide cannon blast, including the concept of the collision is not elastic at all ($e = 0$). Then apply the law of conservation of momentum:

$$\mathrm{mv} = m_1 v_1' + m_2 v_2' \qquad (3)$$

Based on the equation 3, when the carbide cannon blast the mass of matter before the collision is greater than after the collision. While the speed of the substance before the explosion was smaller than the speed of substance after the collision. Therefore, the kinetic energy of the substance after the explosion is also greater. T The kinetic energy of the substance produced causes vibrations of matter particles in the air. So that the energy of these vibrations cause sound energy is greater. If the variable mass is increased, then the sound energy generated is also getting bigger.

**Conclusion**

This study reports important findings about the concept of using e-learning which is integrated with local cultural wisdom. The research was conducted by developing technology-based modules. As the final product in the form of an electronic module or e-module. The concept of local wisdom values taken is adjusted to the material given and studied in physics and chemistry. Based on the results of product trials and discussion, it can be generally concluded that the use of e-module based on local cultural wisdom is appropriate to be used and applied in the learning process. The feasibility aspect of the media developed can be seen from the results of the assessment by material experts with an average score of 83% with very feasible criteria and the results of the assessment from media experts get an average score of 82% with very feasible criteria. The responses of students obtain an average score of 87% with very good criteria. The conclusion of this study is that e-module has a significant impact on improving learning outcomes. Material can be learned anytime by students. However, the weakness in this study is that there are still some students who have not been able to make maximum use of the e-module. Students get tired faster when reading modules in electronic form. There are even some students who cannot open e-modules via their own smartphone.

**Reference**

[1] Tabor, S. W., & Minch, R. P. (2013). Student Adoption & Development of Digital Learning Media: Action Research and Recommended Practices. Journal of Information Technology Education, 12.
[2] Pratama, H., & Prastyaningrum, I. (2019, February). Effectiveness of the use of Integrated Project Based Learning model, Telegram messenger, and plagiarism checker on learning outcomes. In Journal of Physics: Conference Series (Vol. 1171, No. 1, p. 012033). IOP Publishing.
[3] Ramnarain, U. D., & Chanetsa, T. (2016). An analysis of South African Grade 9 natural sciences textbooks for their representation of nature of science. International Journal of Science Education, 38(6), 922-933.
[4] Rehmat, A. P., & Bailey, J. M. (2014). Technology Integration in a Science Classroom: Preservice Teachers' Perceptions. J. Sci. Educ. Technol., 23: 744–755.
[5] Sotiriou, S., Riviou, K., Cherouvis, S., Chelioti, E., & Bogner, F. X. (2016). Introducing large-scale innovation in schools. Journal of Science Education and Technology, 25(4), 541-549.
[6] Morris, H. (2014). Socioscientific Issues and Multidisciplinarity in School Science Textbooks. International Journal of Science Education, 36, 7, 1137–1158.
[7] Kun, P. Z. (2013, September). Pembelajaran Sains Berbasis Kearifan Lokal. In PROSIDING: Seminar Nasional Fisika Dan Pendidikan Fisika (Vol. 4, No. 1).
[8] Sugiyono. (2015). Metode Penelitian Pendidikan (Pendekatan Kuantitatif, Kualitatif, dan R&D). Bandung: Alfabeta.
[9] Tegeh, Made., I Nyoman Jampel dan Ketut Pudjawan. Model Penelitian Pengembangan. Yogyakarta: Graha Ilmu.
[10] El-Seoud, S., Seddiek, N., Taj-Eddin, I., Ghenghesh, P., & El-Khouly, M. (2013). The Effect of E-learning on Learner's Motivation: A Case Study on Evaluating E-Learning and its Effect On Egyptian Higher Education. In The International Conference on E-Learning in the Workplace (pp. 12-14).

[11] Abou El-Seoud, M. S., Taj-Eddin, I. A., Seddiek, N., El-Khouly, M. M., & Nosseir, A. (2014). E-Learning and Students' Motivation: A Research Study on the Effect of E-Learning on Higher Education. International Journal of Emerging Technologies in Learning, 9(4), 20-26.
[12] Martiningsih, I., Lisdiana, L., & Susilowati, SME (2019). Development of Module Based on Scientific Contextual Additives Material to Increase Learning Outcomes and Science Process Skills in Junior High School. Journal of Innovative Science Education, 8 (1), 372-381.
[13] Getuno, DM, Kiboss, JK, Changeiywo, JM, & Ogola, LB (2015). Effects of an E-Learning Module on Students' Attitudes in an Electronics Class. Journal of Education and Practice, 6 (36), 80-86.
[14] Albantani, AM, & Madkur, A. (2018). Think globally, act locally: the strategy of incorporating local wisdom in foreign language teaching in Indonesia. International Journal of Applied Linguistics and English Literature, 7 (2), 1-8.
[15] Darojah, R., Winarni, R., & Murwaningsh, T. (2018). The Local Culture Values of "Perdikan Cahyana" as a Source of Instructional Material for Elementary Students. International Journal of Multicultural and Multireligious Understanding, 5(2), 69-78.
[16] Sulistyo, E. T., Prayitno, B. A., & Pratama, H. (2014). Integrasi Budaya Jawa Pada Pengembangan Bahan Ajar Bumi Dan Alam Semesta. Jurnal Pendidikan Fisika Indonesia, 10(1).

# Effect of materials character on the design of the 12 slot 8 pole generators on power efficiency

**A T A Salim[1*], N Romandoni[1], A Aminudin[1], D W S Wardana[1], E Sudibyo[2]**

[1,]Politeknik Negeri Madiun, Kota Madiun, Jawa Timur, Indonesia.
[2] Universitas Negeri Surabaya, Kota Surabaya, Jawa Timur, Indonesia.

[*]alfitranggono@pnm.ac.id

**Abstract.** The development of renewable energy as electrical energy needs to be followed by the provision of high-efficiency generators, especially for small-scale alternative energy uses. This study discusses the effect of the character of materials on the design of 12 slots 8 pole generator on power efficiency. the material used for the discussion of this research is copper and aluminum as materials for the coil. comparison of these two material to find the coil forming materials with the highest efficiency. As a result of the experiments described above, the use of aluminium material as a coil can reduce efficiency by 0.1% from the use of copper material as a coil.

## 1. Introduction

The use of electricity in Indonesia in 2017 reached 114,347.64 MVA [1]. Electricity can be generated from the power plant, one of which is the Electric Power Bank (EPB). EPB utilizes wind energy and utilizes electrical energy. The EPB component consists of wind turbines, generators, and batteries [2].

The working principle of the generator is mechanical energy into electrical energy. The generator consists of two main parts, namely the rotor and stator. The rotor is a moving part and produces a magnetic field. The stator is the stationary part and receives a flux magnet [3,4]. Faraday's law explains that a wire or coil of the conductor is in a magnetic field that changes with time so that at the ends of the wire or conductor it will cause induction voltage or electromotive force. Changes in the magnetic field produce electric current or electromotive force [5].

This research will find the highest efficiency from the comparison of the use of windings from copper and aluminum for generators.

## 2. Research Methods

a. Modeling 12 Slot 8 Pole generators with MagNet Software. In figure 1 and 2 illustrate the generator model with MagNet Software.

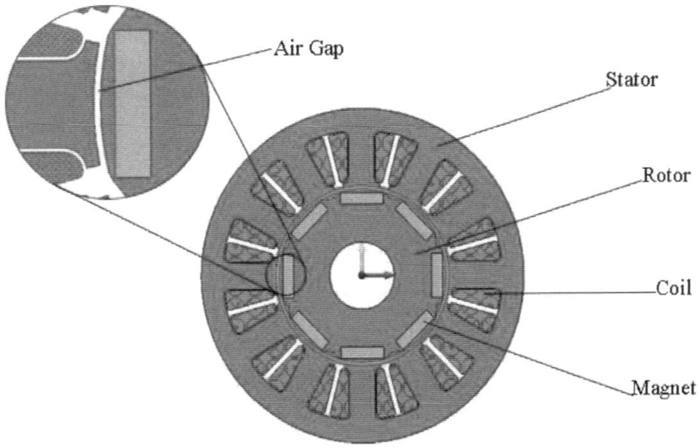


**Figure 1.** the first model of 12 Slot 8 Pole generator using copper wire

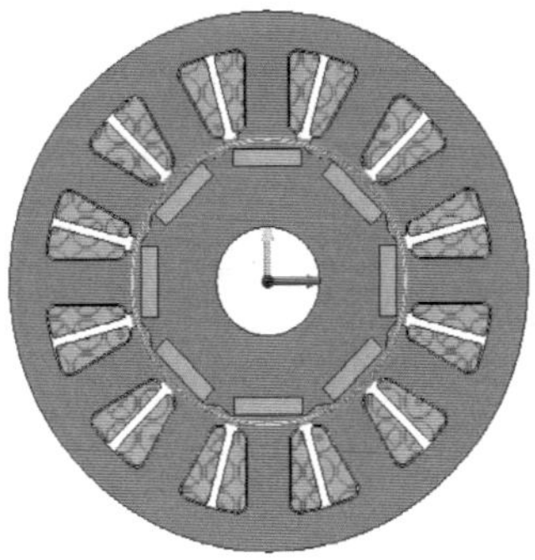

**Figure 2.** Model of the two 12 Slot 8 Pole generator using aluminum wire

Material model Figure 1 of the 12 Slot 8 Pole generator is shown in Table 1:

**Table 1.** Material 12 Slot 8 Pole generator in figure 1

| No | Name | Material |
|---|---|---|
| **1.** | Stator | Carpenter: Silicon steel |
| **2.** | Coil | Copper: 5.77e7 Siemens/meter |
| **3.** | Magnet | Neodymium Iron Boron: 48/11 |
| **4.** | Rotor | Carpenter: Silicon steel |
| **5.** | Air Gap | Air |

Material model Figure 2 of the 12 Slot 8 Pole generator is shown in Table 2:

**Table 2.** Material 12 Slot 8 Pole generator in figure 1

| No | Name | Material |
|---|---|---|
| **1.** | Stator | Carpenter: Silicon steel |
| **2.** | Coil | Aluminum: 3.8e7 Siemens/meter |
| **3.** | Magnet | Neodymium Iron Boron: 48/11 |
| **4.** | Rotor | Carpenter: Silicon steel |
| **5.** | Air Gap | Air |

b. The next process is making rectifier circuits as shown in Figure 3. To be able to simulate and get results from the simulation.

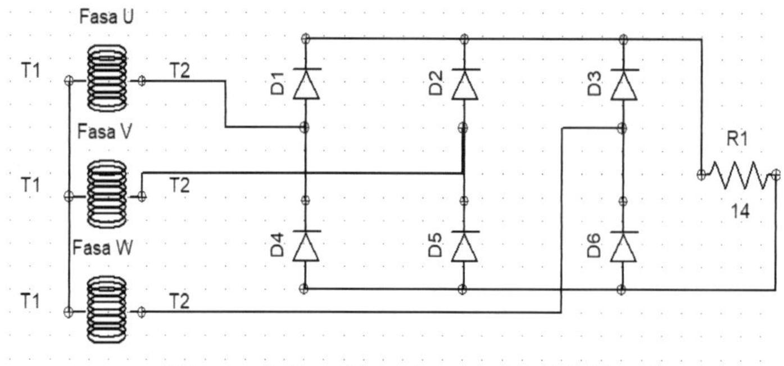


**Figure 3**. Rectifier circuit with 14 Ohm load

## 3. Result

This simulation aims to find the efficiency of a permanent magnet generator with two different materials and in loading 14 ohms at a speed of 1000 RPM. The part of the circuit is composed of 3 parallel, rectifier, and load phases. shown in Figure 3.

After adding a rectifier and load circuit, then the generator is solved / re-simulated so that some data is obtained, namely voltage, current, and average torque. Simulation data are listed in Table 3 below.

**Table 3.** Average data of simulation results

| Coil Material | Voltage (V) | Current (A) | Torque (Nm) |
|---|---|---|---|
| Copper | 23.042 | 1.664 | 0.44 |
| Aluminum | 23.029 | 1.645 | 0.44 |

To determine the efficiency of the generator, can then be calculated the value of the input power, output, and efficiency is by the formula [6, 7]:

$$P_{In} = \tau \,.\, \upsilon \,.\, \frac{2\pi}{60} \tag{1}$$

Explanation: $P_{In}$ : Entry Power (W)
$\tau$ : Torque (Nm)
$\upsilon$ : Speed (RPM)

$$P_{Out} = I \text{ x } V \tag{2}$$

Explanation: $P_{Out}$ : Power Out (W)
I : Curret (A)
V : Voltage (V)

$$\eta = \frac{POut}{PIn} \text{ x } 100\% \tag{3}$$

Explanation: $\eta$ : Efficiency (%)
$P_{Out}$ : Power Out (W)
$P_{In}$ : Entry Power (W)

Calculation of input and output power in the simulation results is shown in Table 4. From Table 4 we get a graph of the comparison of input and output power in the variation of material used on the coil as in Figure 4.

Calculation of input and output power efficiency in the simulation results is shown in Table 5. From Table 5 we get a graph of efficiency comparison on the variation of material used on the coil as shown in Figure 4.

**Table 4.** Comparison data from input and output power from simulation results

| **Coil Material** | **Power Input (W)** | **Power Output (W)** |
|---|---|---|
| Copper | 48.004 | 38.90 |
| Aluminum | 48.004 | 38.83 |

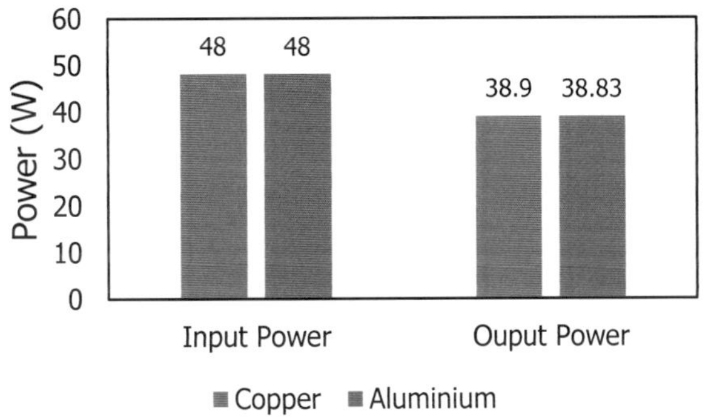


**Figure 4**. Graph of comparison of input and output power from the simulation results

**Table 4.** the comparison data of efficiency between copper and aluminum

| **Coil Material** | **Efficiency** |
|---|---|
| Copper | 81 |
| Aluminum | 80.9 |

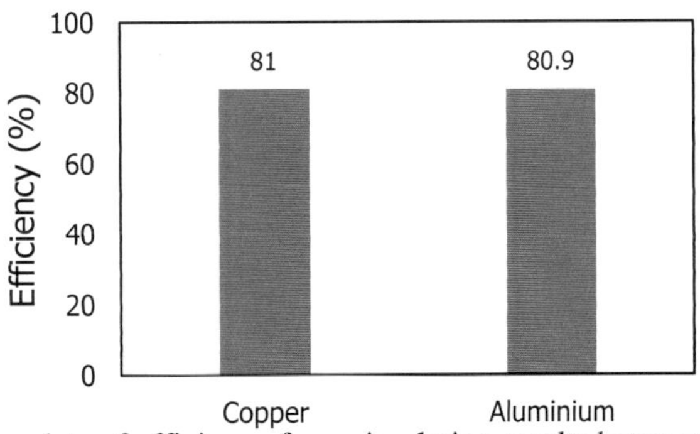


**Figure 5**. the comparison data of efficiency from simulation results between copper and aluminum

## 4. CONCLUSION

This paper has discussed the comparison of copper and aluminum material as a material for a coil with high efficiency from the 12 slots 8 pole generator model using software from infolytica.
As a result of the experiments described above, the use of aluminum material as a coil can reduce efficiency by 0.1% from the use of copper material as a coil.

## References

[1] Direktorat Jendral Ketenagalistrikan KESDM. 2018. *Statistik Ketenagalistrikan.*

[2] LAN. 2014. Pengenalan Teknologi Pemanfaatan Eenergi Angin. Tasikmalaya.

[3] Kundur, Prabha. 1994. Power System Stability and Control. *McGraw-Hill, Inc*. New York

[4] Akbar, M. 2012. Rancang Bangun Generator Turbin Angin Axial Tiga Fasa untuk Kecepatan Angin Rendah. *Skripsi. Program Studi Teknik Elektro.Universitas Indonesia.*

[5] Issetyorini, Apik dan Antono, Djodi. 2012. Gaya Gerak Listrik Pada Motor AC. *Jurusan Teknik Elektro. Politeknik Negeri Semarang.*

[6] Lestari, Ayu. 2018. Analisis Efisiensi pada Generator 12 Slot 8 Pole. *Program Studi Teknik Mesin. Universitas Jember.*

[7] Windarto J*, Sudjadi, Karnoto, Sukmadi T, Santoso I and Desmiarti A. 2018. Effect of Geometry Generator Variation Design 12 Slot 8 Pole on Power Efficiency Design. *Electrical Engineering. Diponegoro University.*

Journal of Physics: Conference Series **1381** (2019) 012047 doi:10.1088/1742-6596/1381/1/012047

# Providing different type of bracing in the existing building structure under seismic load

**R K Rohman[1], S D Cahyono[1], L Fatmawati[1]**
[1] Civil Engineering Department, Faculty of Engineering, Merdeka University of Madiun, Indonesia

E-mail:rosyid@unmer-madiun.ac.id

**Abstract.** This paper presented the influence of the providing different types of the bracing system in the existing building structure. The selected object of analysis is Timbul Jaya Plaza building that located in Madiun City Indonesia. The system of building structure is Moment Resisting Frame (MRF). Seismic loading analysis use equivalent static analysis base on SNI 1726 2012. Modelling and analysis of structure use computer software ETABS version 9.00. Aim of this research to compare the use of different types of the bracing system under seismic load. Type of bracing that analyzed is X-bracing, V-bracing, and Inverted V-bracing. Model of structure without bracing used as control. Result analysis known that the providing of X-bracing will decrease maximum displacement until 36.28 % compared with existing structure without bracing and increase the stiffness of structure.

## 1. Introduction

Indonesian country is located in an active seismic region. Indonesia has experienced several major earthquakes. Some of the earthquake events struck modern buildings have been constructed and were expected to resist the earthquake effects. Recognizing this fact, the government revised Code on Seismic Resistant Design for Buildings, SNI-03-1726-2002 by SNI 1726 2012. This code based on the ASCE 7-10 Standard [1].

Rohman (2014) compared the using of SNI 1726 2002 and RSNI 1726 201X for analyzing the model of building structure in Madiun city. Results of the analysis known that earthquake load base on SNI 1726 201X bigger 27.5% compared with SNI 1726 2002 and the lateral load on each floor bigger too. Using SNI 1726 2012 in structural analysis due to increasing of earthquake loading [2].The existing building that constructed base on SNI 1726 2002 must be analyzed to know its structural stability. If the structural stability cannot fulfill the requirement of the new code, existing building structure must be retrofitted.

One of retrofitting techniques is providing of the bracing system. The bracing system can be provided to enhance stability of building structure[3]. Aim of retrofitting is to increase the stiffness of the structure and the lateral strength. Retrofitting also expected to increase the ductility and to enhance the capacity of energy dissipation and it should be resulted in effective cost. The application of bracing system will enhance the capacity of building structure under seismic load.

There are some research about the using of the bracing system to building structure, such as Barat Patel, et. al[4], Swetha Sunil and Sujith PS[5], Kulkarni and Kore[6], Nauman Mohammed and Islam Nazrul[7], Karthik K and Vidyashree D[8], Muhammad Taher Khaleel and Dilep Kumar U[9], Rishi Mishra, Sharma and Garg[10] and etc. From their study can be concluded that the providing of bracing will decrease the value of displacement at the top story of model structure.

This paper will analyze Timbul Jaya Plaza building located at Pahlawan street, Madiun city. Timbul Jaya Plaza building constructed in 2007. The reference for the planning of this building was SNI 1726 2002. Modelling of structure and calculation of the base shear and displacement value of building structure use ETABS software version 9.00. Structural analysis will be done to an existing structure without bracing and after retrofitted by the bracing system.

## 2. Methodology

The method that used in this research is modeling and analysis of structure. The bracing is provided at the possible position of the model structure. The position of the bracing on the corner of the building. There are four models of the structure are being analyzed using ETABS software.

Model 1 is Momen Resisting Frame (MRF) non bracing. The structure of existing building was analyzed and its result used as control. The first model of structure shown in figure 1.

Model 2 is Momen Resisting Frame with X bracing type. For the second model, the building structure is provided with X bracing type to enhance the stability of structure under lateral force. The second model is shown in Figure 2.

Model 3 is Momen resisting Frame with V bracing type . The existing building structure is provided with the V bracing type as shown in Figure 3.

Model 4 is Momen Resisting Frame with Inverted V bracing as shown in figure 4.

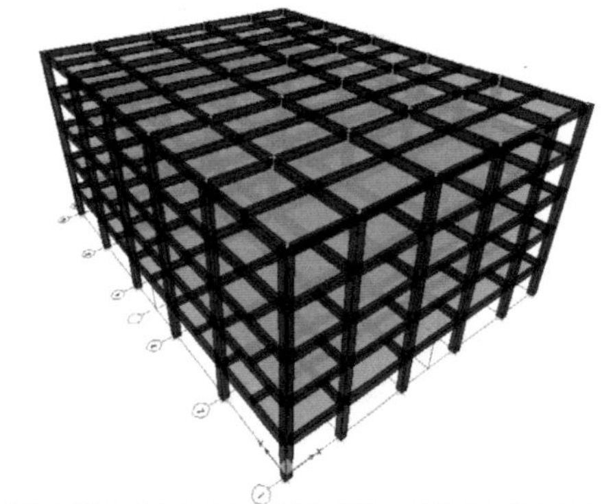

**Figure 1.** MRF non bracing (NB)

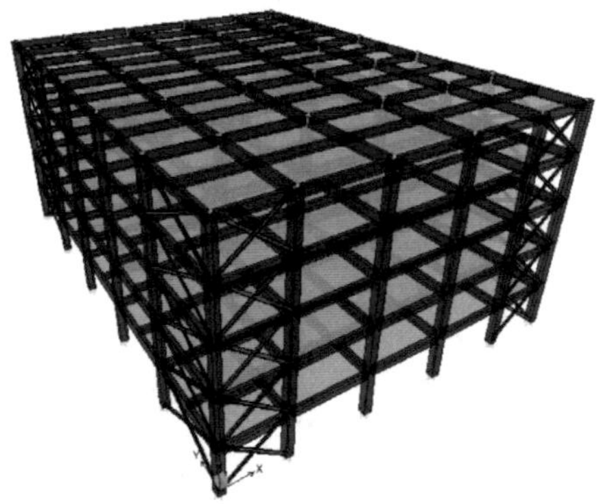

**Figure 2.** MRF with X bracing (XB)

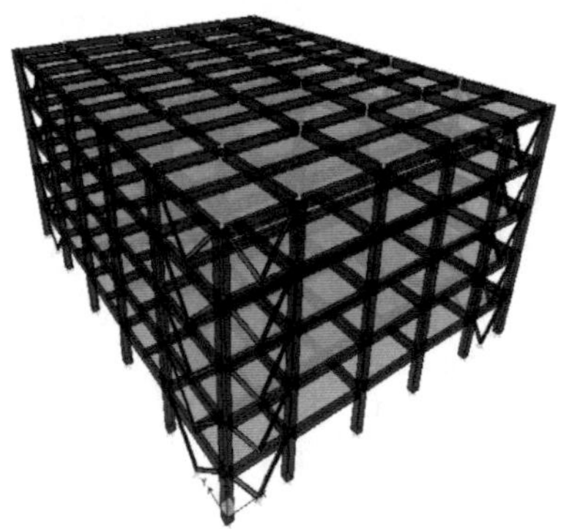

**Figure 3.** MRF with V bracing (VB)

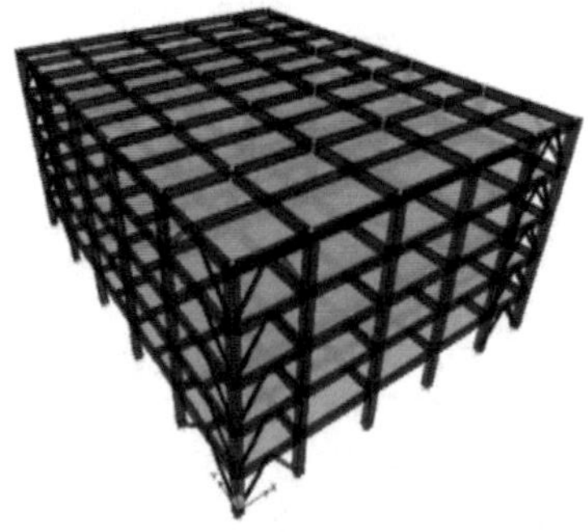

**Figure 4.**MRF with Inverted Vbracing (IVB)

### *2.1. Modeling and Analysis of Structure*

Modeling of building structure use ETABS 9.0 computer software. The selected object of this modeling was Timbul Jaya Plaza (TJP) building. The location of TJP building at Pahlawan street, Madiun city, East Java Indonesia. The structure of Timbul Jaya Plaza is reinforced concrete structure. The number of stories is 5 and high of each column 4.0 m. The grade of concrete is 25 MPa and the grade of steel is 400 MPa. The bracing system uses steel which grades 240 MPa. The building data used for modelling of building structure was shown in table 1 below.

**Table 1.** Building Property

| Property | Dimensions |
|---|---|
| Length | 40 m |
| Width | 30 m |
| Thickness of slab | 120 mm |
| Beam B1 | 300 x 800 |
| Beam B2 | 300 x 700 |
| Beam B3 | 300 x 600 |
| Columns | 600 x 600 |
| Bracing | WF 250 x 125 x 6 x 9 |

*2.2. Equivalent Static Analysis*

The seismic load can be calculated use equivalent static analysis for regular building[11]. The calculation of base shear (V) uses the following formula base on SNI 1726 2012[12]:

$$V = Cs.\ W$$

where Cs is coefficient of seismic response, and W is the total weight of the structure. The base shear is distributed as a lateral force which assigns at the joint of each story level of the building. The calculation of lateral force (Fx) each story was done by using the following equation below:

$$Fx = Cvx.V$$

$$Cvx = \frac{wi.hi^k}{\sum_{i=1}^{n} wi.hi^k}$$

where wi = weight of story i
hi = column height of story i
n = number of story
k = exponent related to period of structure

## 3. Result and Discussion

Structural analysis has been conducted by using ETABS software version 9.0. Output analysis is the lateral displacement value on the top level of the building structure. After get analysis output then determine the displacement value and base shear on each type of the provided bracings. The result was analyzed under equivalent static analysis by different type of the bracing. Then maximum displacement and base shear will be compared to get effective of the bracing system. The location of maximum displacement selected at joint of the top story level.

*3.1. Displacement value for different types of MRF model*

Displacement value (x-axis) was shown in figure 5. From that figure shown that maximum value of displacement at the top story. The value of displacement will decrease due to the using of the bracing system. The using of X-bracing influence displacement value decrease by 36.28 % compared with MRF without bracing (NB). Then displacement value decreases 30.95 % at the existing building which retrofitted by Inverted-V bracing and decreases 26.99 % after retrofitted by V-bracing system.

Displacement value for y-axis can be shown in figure 6. The displacement value after providing by X-bracing at existing building more effective to improve structural stability. The decreasing of displacement value 42.85 % for X-bracing, 42.43 % for Inverted V-bracing and 39.78 % for V-bracing compared with MRF without bracing. It is also conclused that providing bracing enhance of the siffness of the structure.

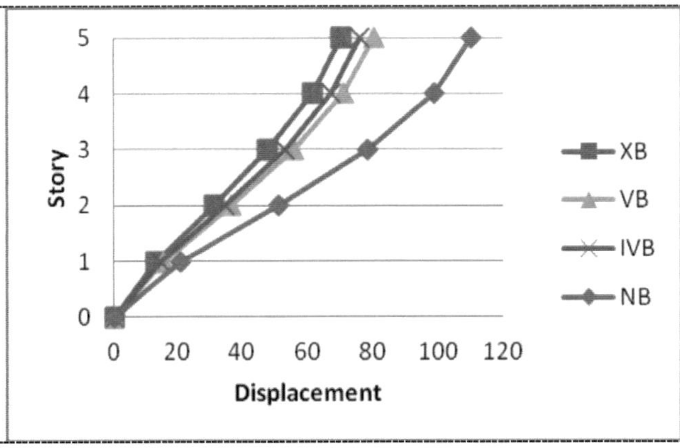


**Figure 5.** Displacement of each story (x-axis)

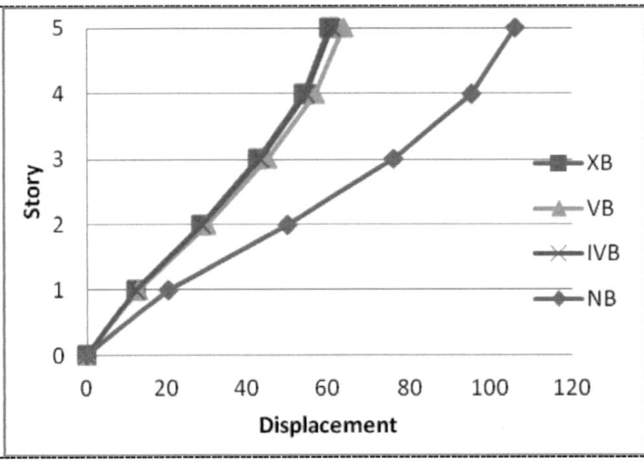


**Figure 6.** Displacement of each story (y-axis)

*3.2. Base shear on the different type of MRF model*

Base shear of each model can be shown in table 2. The minimum value of base shear is Momen Resisting Frame non bracing (NB), 18368.59 KN. The maximum base shear is in second model Momen Resisting Frame (MRF) with providing by X bracing type (XB), 18429.43 KN. It is also observed that base shear is higher in X bracing type. It also indicated that providing of bracing system will enhance value of the base shear compared with control (NB).

**Table 2.** Base Shear

| Type of Model structure | Base Shear (KN) |
|---|---|
| Model NB | 18368.59 |
| Model XB | 18429.43 |

| | |
|---|---|
| Model IVB | 18401.40 |
| Model VB | 18399.32 |

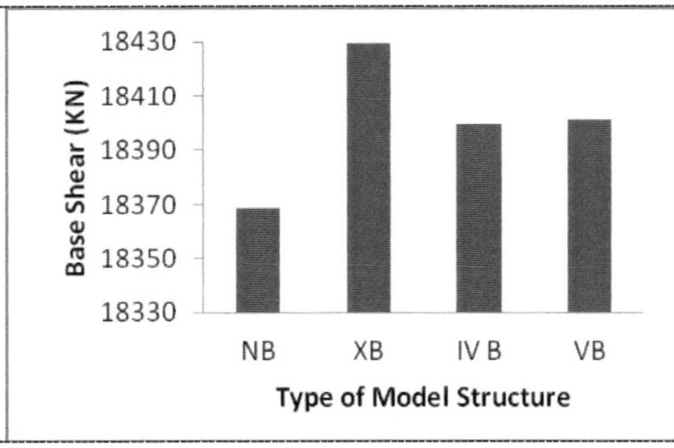


**Figure 7.** Base Shear of each Bracing System

## 4. Conclusions

Providing of the steel bracing in the Timbul Jaya Plasa (TJP) building can enhance its structural stability. Application of the bracing system in the existing structure will increase base shear and decrease the maximum displacement value. Providing of the X bracing type will reduce the displacement until 36.28 % and more effective compared with Inverted-V bracing and V-bracing.

**Acknowledgments**

Authors gratefully acknowledge to the Merdeka University of Madiun that supported this research. Thanks to all staff in the Civil Engineering Department, Engineering Faculty and management of TJP Madiun city.

## References

[1] Achmad Fauzi, Usama Juniansyah Fauzi, Hendriyawan, Wayan Sengara and Masyhur Irsyam 2015 Proposal of Indonesia Seismic Hazard Deaggregation Maps for Sumatra, Indonesia *J. Civ. Eng. Archit.* **9** 410–8

[2] Rohman R K, Fakultas D, Universitas T and Madiun M 2014 Studi Perbandingan Analisis Gaya Gempa Terhadap Struktur Gedung Di Kota Madiun Berdasar SNI 1726 2002 dan RSNI 201X **15** 46–56

[3] Ismail R, Ismail K and Ishak I S 2018 Retrofitting of soft storey building by using different bracing system due to earthquake load *IOP Conf. Ser. Mater. Sci. Eng.* **431** 8–15

[4] Patel B, Mali R, Jadhav P and Ganesh G M 2017 Seismic behavior of different bracing systems in high rise RCC buildings *Int. J. Civ. Eng. Technol.* **8** 973–81

[5] Sunil S and Sujith P S 2017 Seismic Study of Multistorey RC Building With Different Bracings 8667–73

[6] Kapur V and Jain A 1982 Seismic Response of Reinforced Concrete Braced Frames. *Indian Concr. J.* **56** 242–52

[7] Mohammed N and Nazrul I 2013 Behaviour of Multistorey RCC Structure withDifferent Type of Bracing System (A Software Approach) *Int. J. Innov. Res. Sci. Eng. Technol.* **2** 7465–78

[8] . K K M 2015 Effect of Steel Bracing on Vertically Irregular R.C.C Building Frames Under Seismic Loads *Int. J. Res. Eng. Technol.* **04** 90–6

[9] Khaleel M T, U D K and Tech M 2016 Seismic Analysis of Steel Frames with Different Bracings using ETSBS Software . 2081–7

[10] Mishra R, Sharma A and Garg V 2014 Analysis of RC Building Frames for Seismic Forces Using Different Types of Bracing Systems **3** 1135–40

[11] SNI-03-1726 2002 Tata Cara Perencanaan Ketahanan Gempa untuk Bangunan Gedung 03–

1726
[12] Badan Standardisasi Nasional 2003 SNI 1726:2012 Tata cara perencanaan ketahanan gempa untuk struktur bangunan gedung dan non gedung *Rethink. Marx.* **15** 316–25

# DC-DC Hi-Voltage Converter in application of multi-stage coil gun power charger

**B Winarno[1*], R G Putra[1], I Yuwono[1], B Sumantri[2], and A I Gunawan[2]**

[1]Politeknik Negeri Madiun, Indonesia
[2]Politeknik Elektronika Negeri Surabaya, Indonesia

basuki@pnm.ac.id[1],gusta@pnm.ac.id[1], indarto@pnm.ac.id[1], bambang@pens.ac.id[2], agus_ig@pens.ac.id[2]

**Abstract**. The multi-stage coil gun is a coilgun with many coils as launchers. Coil gun requires a lot of power to work. High voltage and large currents are needed. Sources of energy to support the coil gun arelimited. Alternatives that are often used by using high voltage capacitors bank. To store energy in a capacitor bank from a battery, a charging device is needed. High voltage DC to DC converter is required. In this article, a high voltage capacitor bank charging device is discussed to supply a multi-stage coil gun. The device has been successfully implemented to support six coilgun levels with the ability to increase the voltage from 12 V to 1.1 kV DC.

## 1. Introduction

Many researchers have studied coilgun, starting from system modeling simulations, calculating each force and energy on a single level coilgun[1], construction and simulation of coilgun parts with capacitors as energy storage[2], coilgun current injection systems using genetics algorithmic methods to get pulse parameters to activate multi-stagecoilgun[3][4]. Comparison of simulations and experiments on the three-level coilgun has also been done[5]. The weakness of the coilgun is its relatively lower power compared to a conventional firearm. Multi coils used to increase the throwing power on the coilgun[6][7][8][9][10].

Most of the research is about simulation and modeling. The problem with real applications from coilgun is more complex. Coilgun requires a lot of energy to work.Taking energy directly from the battery is almost impossible because it requires very high current and voltage.CapacitorBank is used to store energy. Capacitors with large capacity will be charged to the required limit. The required voltage can reach of kilovolts (kV). This capacitor charging system is very rarely resolved. This is a very vital system. Therefore, in this article about the application of DC-DC Hi-Voltage Converter for the application of capacitor bank charging on multistagecoilgun. The results of this study are that the coilgun energy charging system model can be used on real devices on multilevel coilgun as an alternative weapon to replace conventional firearms.

## 2. Research Method

To realize a multistage coilgun requires several parts including power supply, DC-DC Hi-Voltage Converter, Charging Controller, Bank Capacitor, Coils, Switching Circuit, and Control Unit. The complete system of the multistage coilgun is shown in Figure 1. In this study, 12V batteries are used as a power supply. The charger is a High-Voltage DC / DC converter consisting of a DC / AC converter,

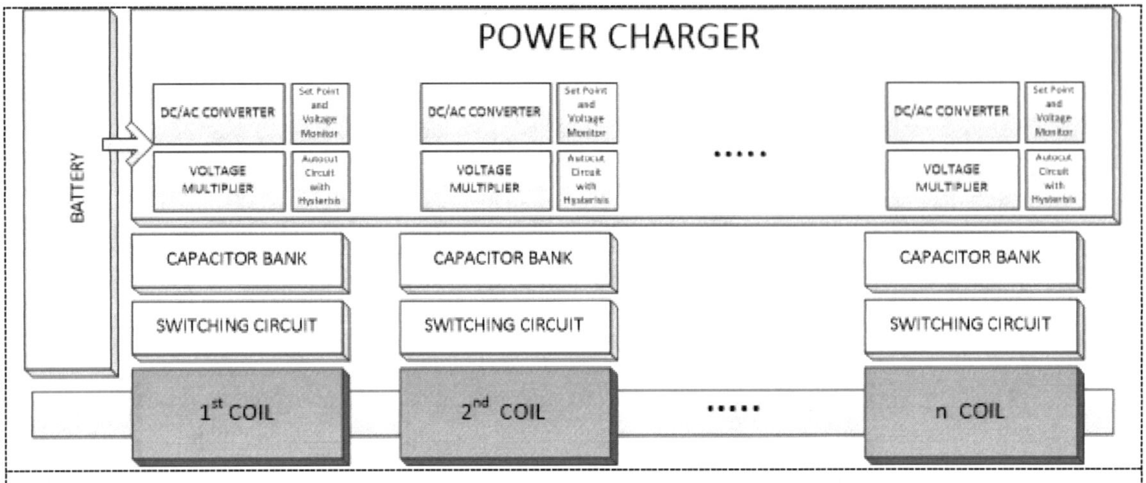


**Figure 1.** Block System Diagram.

voltage multiplier, set point and voltage monitor and charging control circuit consisting of an autocut circuit with hysteresis.

In designing devices that will be realized several important parameters have been established. Some of them are battery capacity, DC / AC converter capability, number of multipliers, switching capabilities and charging controllers.The parameters in the realization of the device are shown in Table 1.

**Table 1.**Parameters of the device

| No | Parameters | Value |
|---|---|---|
| 1. | Battery | 12 V 5000mAH |
| 2. | DC/AC Converter | 12 V to 380V / 50Hz/ 500 Watt |
| 3. | Voltage Multiplier | 10 Stage |
| 4. | Switching Circuit | IGBT 1500 V/ 300 A Max |
| 5. | Charging Control Circuit | Comparator w/ hysteresis |

The capacitor bank specifications used are shown in table 2. In this study, the capacitors bank can supply 6 stages of coilgun with the energy of each stage displayed in the table. The energy displayed is the maximum energy that capacitors can store. Energy (Joule) can be calculated from the maximum capacitance and voltage values with equation 1.

$$W = \frac{1}{2} CV^2 \quad (1)$$

**Table 2.**Specification of Capacitor Bank

| Stage | 1 | 2 | 3 | 4 | 5 | 6 |
|---|---|---|---|---|---|---|
| Voltage Max (V) | 100 | 100 | 300 | 600 | 900 | 1500 |
| Capacity (uF) | 1000 | 1000 | 333.33 | 166.66 | 111.11 | 66.66 |
| Energy Max (J) | 5 | 5 | 15 | 30 | 45 | 75 |
| Total Energy Storage Max (J) | | | | 175 | | |

## 3. Result and Analysis

Journal of Physics: Conference Series **1381** (2019) 012048 doi:10.1088/1742-6596/1381/1/012048

To design and realize the device can be divided into several parts. In the discussion, the design and results of each part will be reviewed as a representation of the whole system.The characteristics of the charging for each stage will also be displayed here.

### *3.1.1. DC/AC Converter*

The DC / AC converter circuit is built from a pulse generator mechanism, switching devices from MOSFETs and step-up transformers. Part of this system is shown in Figure 2. The output of the DC / AC is at the transformer output. The circuit afterward is only for monitoring the peak voltage of the system and the selection device ON / OFF. In testing, the system can provide 380 V output with a frequency of 50 Hz.

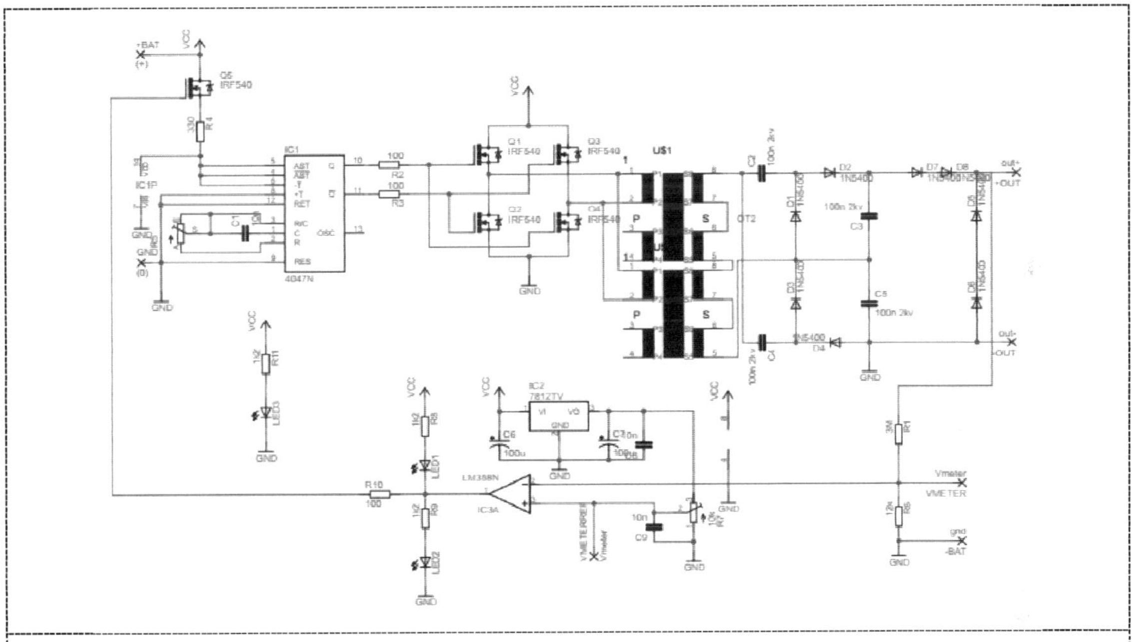


**Figure 2.** Part of DC/AC Converter.

### *3.1.2. Voltage Multiplier*

The voltage multiplier design can be adjusted as needed. In this study using 10 levels of voltage multiplier for all stages. This is done by taking into account the maximum voltage needed from all stages. At lower levels, the voltage multiplier here will not have a significant effect. By making all the stages the same it will make it easier in the process of implementing the tool. Design 10 voltage multiplier levels shown in Figure 3.

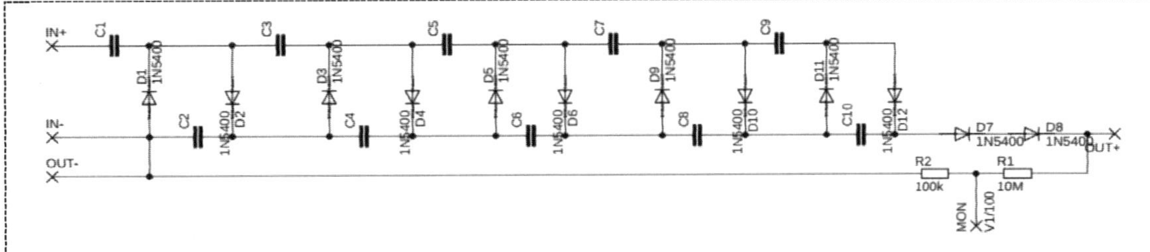


**Figure 3.** 10 Levels Voltage Multiplier.

Journal of Physics: Conference Series **1381** (2019) 012048 doi:10.1088/1742-6596/1381/1/012048

### *3.1.3. Capacitor Bank*

The capacitors bank used are in accordance with the specifications in the previous section. The capacitor bank is composed of 1000uF / 100V electrolytic capacitors. To get the desired voltage, capacitors are arranged in series. In Figure 4 a series of 10 capacitors is displayed. Each capacitor is coupled with resistors in parallel. This is done as a voltage divider so that the capacitor is balanced. The imbalance of charging capacitors is dangerous because there is a risk that capacitors are overcharged.

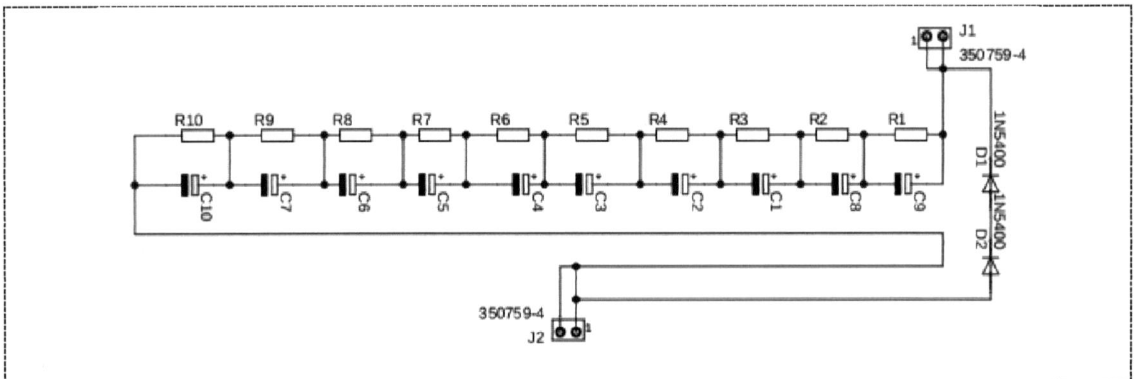


**Figure 4.** Capacitor Bank Circuit.

### *3.1.4. Charging Control Circuit*

Charging control circuit consists of a comparator with hysteresis and relay circuit. This circuit is used as a voltage monitor on the capacitor bank. When the voltage on the capacitor bank exceeds the set-point voltage, the relay will turn off the charger circuit. The comparator uses hysteresis in order to avoid oscillation in the relay circuit due to the rise and fall of the voltage of the capacitor bank. The comparator provides a lower tolerance than the upper tolerance. The comparator has a stable reference voltage from the voltage regulator. The setpoint voltage can be adjusted by changing the existing variable resistor. This circuit is shown in Figure 5.

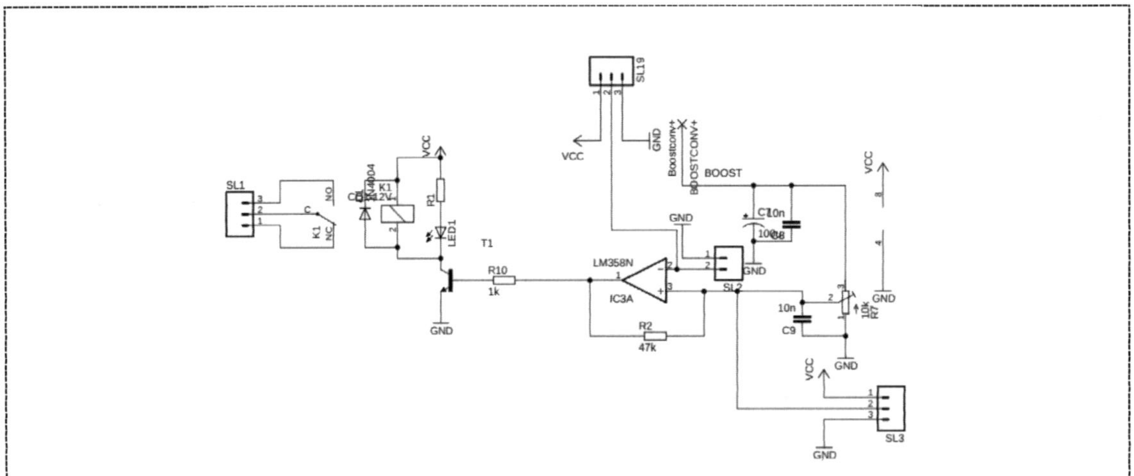


**Figure 5.** Charging Control Circuit.

### *3.1.5. Switching Circuit*

The switching section is the part that regulates the discharge time from the capacitor bank to the coil. With a large capacity capacitor bank, the current that can be passed is also very large. The choice of electronic components for high voltage and large currents is very limited. One of the main choices is IGBT (Insulated Gate Bipolar Transistor). IGBT provides maximum specifications of relatively large voltages and currents. In this study used specifications IGBT 1300V and 60 Ampere. To get a maximum current of 300 Ampere, IGBT can be arranged in parallel. To connect the control unit to the IGBT circuit is isolated using an optocoupler. This is done so that there is no voltage or reverse flow from the high voltage section to the low voltage section. The circuit used in this study is shown in Figure 6.

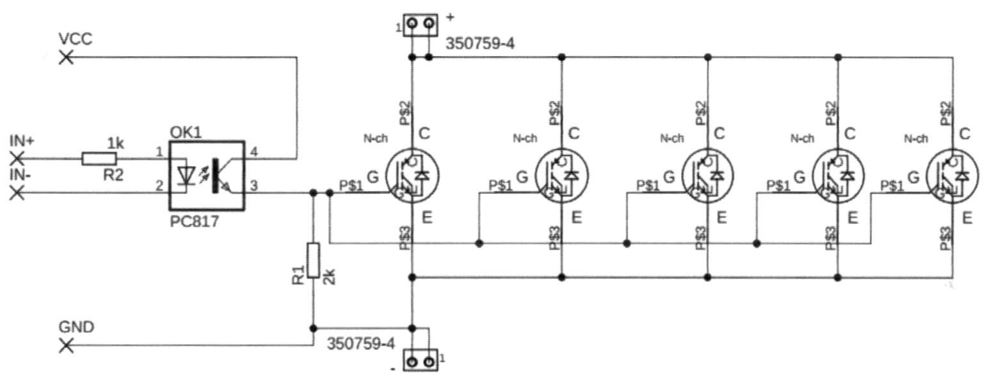


**Figure 6.** Switching Circuit.

### *3.1.6. Charging Characteristic*

Charging characteristics are observed by monitoring the voltage of the capacitor bank. With the six stages available, the capacitor banks used vary. The set-points used are in accordance with the table in the previous section. The charging time needed to fill the capacitor is shown in Figure 7. Based on the experiment, the charging time varies according to the value of the capacitor used. The greater the value of the capacitor, the charging time will be longer. The consecutive times needed for stage one to stage six are shown in Table 3. The stage one takes 25 seconds while sixstagesrequire a longer time which is greater than 400 seconds.

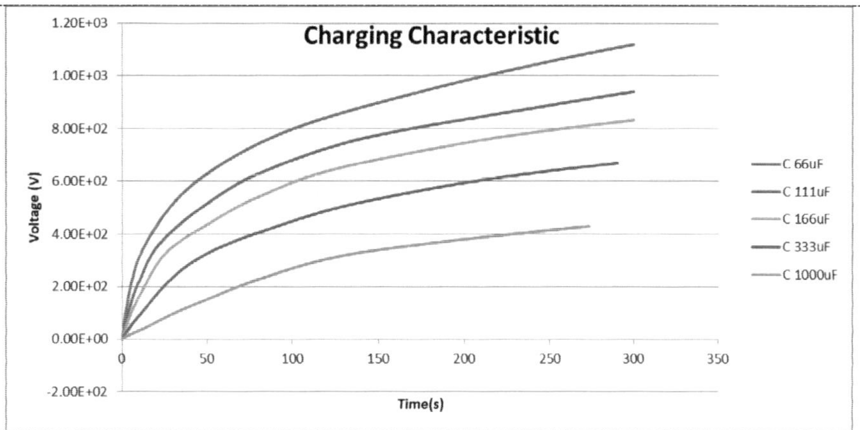


**Figure 7.** Charging Characteristic.

**Table 3.**Charging Time each Stage Capacitors Bank

Journal of Physics: Conference Series **1381** (2019) 012048 doi:10.1088/1742-6596/1381/1/012048

| Stage | 1 | 2 | 3 | 4 | 5 | 6 |
|---|---|---|---|---|---|---|
| Voltage set-point (V) | 100 | 100 | 300 | 600 | 900 | 1500 |
| Time (s) | 25 | 25 | 45 | 100 | 300 | >400 |

All the circuits discussed have been implemented. The implementation documentation of the DC / AC converter system and the charging control circuit is shown in Figure 8. Implementation of the system from bank capacitors and switching circuits is shown in Figure 9. All circuits are packaged in a metal box 50 x 50 x 12 cm.

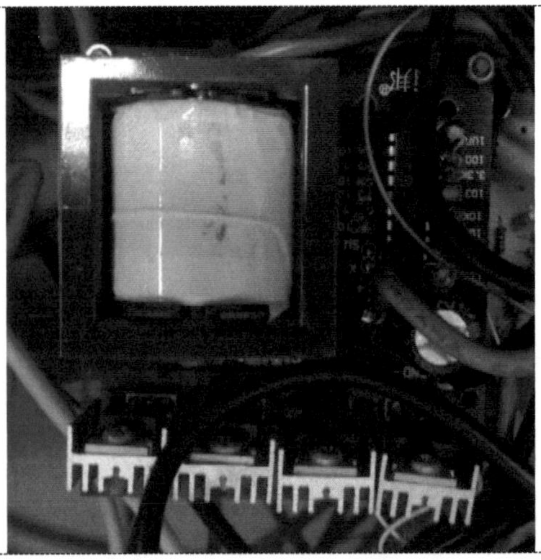

**Figure 8a.** DC/AC Converter.

**Figure 8b.** Charging Control Circuit.

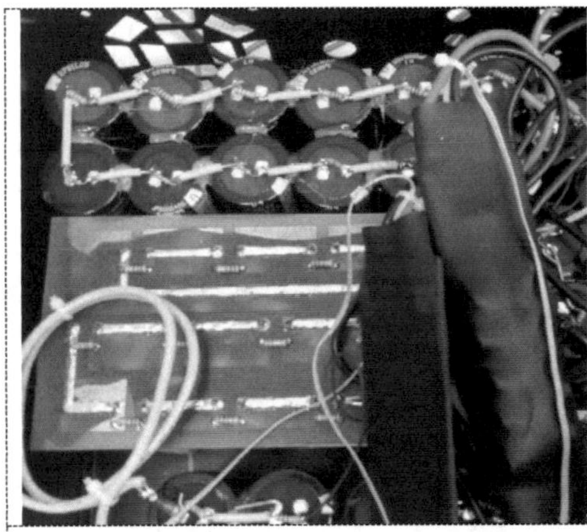

**Figure 9a.** Capacitor Bank.

**Figure 9b.** Switching Circuit.

## 4. Conclusion

Hi-Voltage Converter in Application DC-DC Multi-Stage Coil Gun Power Charger has been implemented. Based on the experiments carried out each part of the system functions properly. The charging time of the capacitor bank varies according to its capacitance. The greater the capacitance, the charging time will be longer. The higher the set-point voltage also requires more time. With a limited capacitor, the higher the capacitor voltage, the lower the capacitance value. Capacitor charging time for the first stage reaches 25 seconds while the last stage reaches> 400 seconds. So that the time of the six-stagecoilgun to be used requires the longest time, which is> 400 seconds with the energy charging ability reaching 175 Joules. The charging time produced until now is still too long, so it is still less effective. further research can be made of a mechanism for charging capacitor banks faster. Another alternative is to use other alternative sources that are more efficient.

**Acknowledgements**
The author would like to thank the Ministry Of Research, Technology And Higher Education, Ristekdikti for financial support to this research.

**References**

[1] A. W. Horodeński and C. Pochrybniak 2018 A Digital Model of the Electrodynamic Acceleration Process in a Coil Gun Device *Probl. Mechatronics Armament, Aviat. Saf. Eng.* **4** 113–126
[2] Skala B and Kindl V 2014 Electromagnetic Coil Gun – Construction and Basic Simulation *Mechatronics Springer International Publishing Switzerland* pp 87–8
[3] Winarno B, Wijono W and Hasanah R N 2015 Pengaturan Injeksi Arus pada Coilgun dengan PWM menggunakan Metode Algoritma Genetika *J. Arus Elektro Indones.* **1** 13–8
[4] Winarno B and Yuwono I 2017 Optimalisasi Gaya pada Coil dengan Modifikasi Dimensi menggunakan Metoda Algoritma Genetika *JEECAE (Journal Electr. Electron. Control. Automot. Eng.* **2** 77–82
[5] Yadong Z, Ying W and Jiangjun R 2011 Capacitor-Driven Coil-Gun Scaling Relationships *IEEE Trans. Plasma Sci.* **39** 220–4
[6] Lee S J, Kim J H, Song B S and Kim J H 2013 Coil gun electromagnetic launcher (EML) system with multi-stage electromagnetic coils *J. Magn.* **18** 481–6
[7] Zhang T, Guo W, Fan W, Zhang H hai, Liu Y and Su Z zhou 2018 An efficiency-improved method of multi-stage induction coilgun based on magnetic field arrangement *Def. Technol.* 6–13
[8] Lee S, Kim J H and Kim S 2016 Design and experiments of multi-stage coil gun system *J. Vibroengineering* **18** 2053–60
[9] Sandhya T, Chandan K S and Rao P M 2015 Modelling and Critical Analysis of a Two Stage Tubular Coil Launcher *Int. Journals Power Electron. Control. Convert.* **1** 17–24
[10] Winarno B, Gust Putra R, Yuwono I, Gunawan A I and Sumantri B 2018 Control of Velocity Projectile on Multistage Coil gun (Manado, Indonesia: IEEE) pp 24–9

# Ecological function of green open space as water infiltration: study in kalijodo green open space, north jakarta

**D P P Mbarep[1], H Herdiansyah[1*]**

School of Environmental Science, University of Indonesia, Jakarta, Indonesia

*herdis@ui.ac.id

**Abstract.** Kalijodo Green Open Space is a green open space set by the DKI Jakarta Provincial Government in 2017. Related to the ecological function of green open space as water infiltration, a literature study was conducted to analyze the condition of Kalijodo green open space based on supporting data sources using descriptive analysis techniques. The results of the analysis show that the Kalijodo green open space has not been able to carry out ecological functions as well as water absorption. Based on the imagery in Google Earth, it is seen that Kalijodo's green open space is very minimal vegetation and looks barren. This condition has implications for hardening the soil, so that water is difficult to seep into the soil. In the last 5 years, the average rainfall intensity from October 2013 to March 2014 was 134.97 mmph and was a very high level of rainfall intensity. Green open space is included in flood-prone areas. With the condition of the Kalijodo green open space and the high intensity of rainfall, the ecological function of green open space as water infiltration has not been maximized.

## 1. Introduction

Green Open Space is an important element in urban areas because it forms an environment that is comfortable and healthy. The availability of green open space in Jakarta is 9.98% and has not met the ideal requirements of 30% [1]. Kalijodo is an area around the scouting road II, Pejagalan Village, Penjaringan District, along with the eastern bank of the flood canal that used to be a place of night entertainment and prostitution. Then in 2017, the DKI Jakarta Provincial Government sets the Kalijodo area as Green Open Space and Integrated Child-Friendly Public Space. The one of the ecological functions of green open space is as water infiltration [2]. Green Open Space (GOS) can help absorb water so that the supply of water in the soil increases and saves the cost of building Stormwater Drainage which must be provided by big cities to overcome flooding [3]. Regarding the ecological function of green open space as an absorbent of rainwater, the condition of Kalijodo Green Open Space received little attention. This can be seen from the poorly maintained vegetation resulting in soil surface hardening. Some factors that influence the rate of infiltration of water into the soil, namely depth detention on the surface (surface detention) and saturated layer thickness, soil moisture, compression by rainwater, vegetation (cover crops), topography, and intensity of rain [4]. As a result of the lack of vegetation available, soil surface permeability decreases, thereby reducing water infiltration and increasing surface water flow which can lead to flooding [5].

These problems can disturb people's comfort because it can cause flooding. With the occurrence of floods, it can disrupt the economic activities of the community and the damage caused by floods can damage the physical condition of the Kalijodo Green Open Space so that its aesthetic value decreases. Therefore, this literature study was carried out to analyze the ecological function of the Kalijodo green open space as rainwater infiltration and to be an evaluation material for the improvement of the condition of the sustainable Kalijodo Green Space.

## 2. Methodology

### *2.1. Place*

The study was conducted on the Kalijodo Green Open Space located on the scouting road II, Pejagalan Village, Penjaringan District, North Jakarta Administrative City. Kalijodo Green Open Space was chosen as the location of the literature study because it was only determined by the DKI Jakarta Provincial Government and is located east of Banjir Kanal, so it is necessary to conduct a study and analysis to find out the current conditions.

### *2.2. Source of data*

The data sources used came from various agencies such as the Badan Penanggulangan Bencana Daerah (BPBD) of DKI Jakarta province which had data in the form of data on flood-prone areas, Badan Pusat Statistik (BPS) which had data on the intensity of rainfall in the last 5 years in DKI Jakarta Province from the year 2013 to 2014. Then a literature review of several previous research journals and books relating to the ecological function of green open space as rainwater infiltration.

### *2.3. Data analysis technique*

The data analysis technique used is descriptive analysis, which analyzes and concludes the current condition of Kalijodo green open space based on the data sources obtained.

## 3. Result and Discussion

### *3.1. The Condition*

Based on observations made through google earth, the Kalijodo green open space looks barren and has minimal vegetation. The new Kalijodo green open space was inaugurated in 2017 and of course it is still in the process of developing towards ideal conditions, in accordance with its functions, however, the development of green open space at this time does not seem to be going well. The condition of vegetation in the Kalijodo green open space has not been well organized and tends to be a lot of grasses that cover damaged soil. Grass in green open space Many times are withered, so they look very dry due to poorly maintained conditions. The condition of the unkempt green open space can be seen in Figure 1.

**Figure 1**. Area of Kalijodo Green Open Space

The one of the factors influencing the rate of infiltration of water into the soil is the presence of plants (cover crops) [4]. The number of plants that cover the surface of the soil, such as grass or forests, can increase the infiltration capacity of the land. Trees as important elements in urban green open space have ecological functions in improving environmental quality, among others in storing and trapping carbon, reducing the rate of runoff water, absorbing solar radiation and reducing energy use in urban environments [6].

### *3.2. The Location*

The location of Kalijodo green open space is located in Pejagalan Village, Penjaringan District, North Jakarta Administrative The administrative city of North Jakarta is a flood-prone area [7]. Areas prone to flooding in North Jakarta are scattered throughout sub-districts and villages. One of the flood-prone areas is Penjaringan District, Pejagalan Village. Data on flood-prone areas, especially in Penjaringan District, North Jakarta can be seen in table 1.

**Table 1.** List of Flood-Prone Areas in Penjaringan District, North Jakarta.

| No | Districts | Village |
|---|---|---|
| 1 | Penjaringan | Kamal Muara |
| | | Kapuk Muara |
| | | Pejagalan |
| | | Penjaringan |
| | | Pluit |

(Source: Badan Penanggulangan Bencana Daerah DKI Jakarta, 2018)

Flood-prone areas in DKI Jakarta Province, generally are areas with drainage conditions and high population densities. Therefore, one of the drainage systems that can be used is a natural drainage system. The natural drainage system is by providing water absorption space with green land where there are various variations of plants. Plants (grasses and trees) have the benefit of increasing soil permeability so that the soil can absorb surface water optimally, according to the standard infiltration rate which is good. Water infiltration spaces such as forests / urban parks and green open spaces are needed in large cities with high density. In addition to one of the functions, namely as an absorbent of rainwater, another function of the forest/city park and green open space is that it can be used as a means of socialization and means for economic development.

### *3.3. The Intensity of Rainfall*

Based on the rainfall data of the last 5 years in DKI Jakarta Province and the calculation of the intensity of the rain carried out, especially during the rainy season between October 2013 and March 2014, it was noted that the average rainfall intensity reached 134.97 mmph [8]. This average rainfall intensity is at the level of rainstorms [9]. According to Badan Meteorologi Klimatologi Geofisika (BMKG), the intensity of rainfall is at a very dense level. The intensity of rainfall at this level can result in floods in the DKI Jakarta area. The graph of rainfall intensity can be seen in Figure 2.

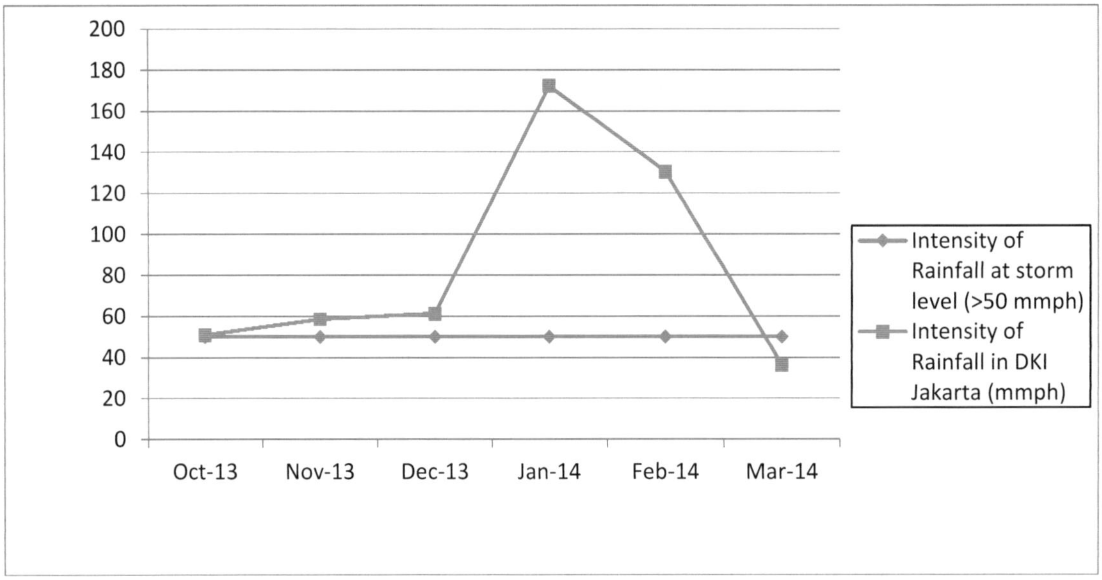


**Figure 2.** DKI Jakarta Province Rainfall Intensity Chart.

The intensity of rainfall can affect the rate of infiltration. The large volume of surface water that continues to inundate the surface of the land, without the presence of media or locations that can accommodate and absorb into the soil can cause floods. The hardened soil conditions cannot be used to carry out the infiltration process properly so that the surface water flow continues to increase and can move towards the lower area. Soil hardening occurs due to lack of cover plants. Plants play a role in increasing soil permeability so that the soil pores become wider and can absorb surface water flow. Therefore, the flow of surface water that flows continuously with large volumes is destructive so that it can hamper all economic activities carried out and disrupt the social activities of the community.

## 4. Conclusion

The results of the analysis of Kalijodo green open space are based on the data sources obtained, concluding that:

1. The condition of Kalijodo green open space has the potential to produce a low infiltration rate due to a lack of vegetation (grass and trees) as a land cover so that the soil experiences hardening and difficult water seeps into the soil.
2. Apart from the lack of plants that make the soil hardened, the intensity of very high rainfall can increase runoff /surface water flow that cannot be absorbed by the soil, causing flooding.
3. The condition of green open space which has a decrease in ecological functions, can have a negative impact on other functions such as social functions where people who are unable to carry out activities such as sports and others properly, and bad economic impacts will occur.

**Acknowledgment**

This research is funded by the Grant of Indexed International Publication for Master Student/ *Publikasi Internasional Terindeks Mahasiswa Magister*-PITMA B, Universitas Indonesia 2019 with contract number NKB-1023/UN2.R3.1/HKP.05.00/2019

## Reference

[1] Dinas Kehutanan 2018 *The Amount of Green Open Space in Jakarta* (Jakarta)
[2] Sekretariat Negara (Republic of Indonesia) 2007 *Penataan Ruang* vol 7 (Republic of Indonesia)
[3] Zhang B, Xie G, Zhang C and Zhang J 2012 The economic benefits of rainwater-runoff reduction

by urban green spaces: A case study in Beijing, China *Journal of Environmental Management* **100** 65–71
[4] Vijay P. Singh 1991 *Elementary Hydrology* (New Jersey: Pearson)
[5] Bai T, Mayer A L, Shuster W D and Tian G 2018 The hydrologic role of urban green space in mitigating flooding (Luohe, China) *Sustainability (Switzerland)* **10** 1–13
[6] Fatimah I S, Sinukaban N, Munandar A and Kholil K 2013 Valuasi Manfaat Ekologis Ruang Terbuka Hijau (Rth) Di Kota Bogor Dengan Aplikasi Citygreen 5.4 *Jurnal Pengelolaan Sumberdaya Alam dan Lingkungan (Journal of Natural Resources and Environmental Management)* **3** 31
[7] Badan Penanggulangan Bencana Daerah DKI Jakarta 2018 *The data of flood-prone areas in 2014* (Jakarta)
[8] Badan Pusat Statistik 2018 *The data of rainfall intensity and rainy day in 2011 to 2014* (Jakarta)
[9] Chay Asdak 2007 *Hidrologi dan Pengelolaan Daerah Aliran Sungai* (Yogyakarta: Universitas Gadjah Mada Press)

# Application of waste bank use in reducing household waste in sub-urban area?

**S Susilowati[1] and H Herdiansyah[1*]**

[1]School of Environmental Science University of Indonesia, Jl. Salemba Raya No. 4 Central Jakarta 10430, Indonesia

[*]herdis@ui.ac.id

**Abstract.** Household waste becomes a problem in sub-urban area like Bekasi City. The quantity of waste to landfill increased from 2011-2015. To improve the situation the local government initiatives to reduce it by establishing Bank Sampah Induk Patriot (BSIP-waste bank in the city level) with the responsibilities to improve the participation of citizens on 3R (reduce, reuse, recycle) by educating and supporting them to build a waste bank in each rukun warga (RW-informal unit under village). This action is followed by issuing the new regulation concerning waste management based on community participation. This study aims to analyse the correlation between waste bank and household waste in term on reduction to landfill and financial benefit using Pearson correlation analysis. The results show issuing BSIP and new regulation impact on increasing the number of waste banks significantly. However, it is found the number of active waste bank is only 18.5%. The results show there is very strong correlation between the number of waste banks and the amount of waste reduction. It means the waste banks is effective to reduce waste to landfill. It's recommended to BSIP to encourage the inactive waste banks and coordinate with the local authorities.

## 1. Introduction

The Bekasi City as a sub-urban area faces the problem of waste. The quantity of waste increases in line with population growth and urbanization. Data from the Central Statistics Agency [1] shows that the volume of waste sent to landfills or Final Processing Site rises from 576,479 $m^3$ in 2011 to 1,311,822 $m^3$ in 2015, equivalent to 395.01 to 898.54 tons per day. To overcome the waste problem, the local government formed the Patriot Main Waste Bank (BSIP) which was formalized through a mayor regulation which functioned as a government partner and as the parent of the waste bank at the RW, village and sub-district levels. The waste management can be successful with community participation and government involvement [2] [3].

The familiar method of waste management carried out by several countries is the 3Rs (reduce, reuse, recycle) such as in China [4], in Ghana [5], Bangladesh [6], and Tehran [7]. In Indonesia, waste banks are the place to implement the 3Rs. The application of the waste bank adopts the 3R method which refers to environmental regulations [8]. The more waste banks are expected to have an impact on reducing the amount of waste to landfill, changing people's behaviour in sorting waste and getting economic benefits from the waste business. This study aims to determine 1) the correlation between the number of waste banks and the amount of waste reduction to landfills, 2) the correlation between the number of waste banks and the value of selling waste, as an impact of the implementation of local regulations. It uses The Pearson correlation analysis. It is found there is very strong correlation between the number of waste banks and the amount of waste reduction to landfill.

## 2. Method

*2.1 Method of Data Collection*

This study was carried out through literature studies to collect and examine the results of previous studies. The data of the number of waste banks and the waste reduction are obtained from the Patriot Main Waste Bank (BSIP). Data is analysed using Pearson correlation through excel. The variables tested are the number of waste banks, the amount of waste reduction and the value of selling waste. The number of garbage banks uses two data, namely the number of active waste banks (only) and the number of all waste banks (active and passive). The correlation tests carried out are:

- The number of waste bank operates vs the number of waste reduction.
- The number of waste bank operates vs the value of selling waste.
- The number of active waste banks vs the number of waste reduction.
- The number of active waste banks vs the value of selling.
- The number of waste reduction vs the value of selling waste.

## 3. Result and discussion

*3.1 Amount of Waste in Bekasi City*

Bekasi City is a suburban area of Jakarta as the capital city of Indonesia. It has a population of 2,402,465 people in 2016. The area consists of 12 sub-districts, most of which are residential areas, trade areas, and services. The number of population and the development activities of the city are in line with the increasing amount of waste. As shown in table 1, there is a significant rising volume of waste to landfill in the period of 2011-2015.

**Table 1.** The tonnage and volume of waste sent to the Bekasi Sumur Batu TPA Waste from 2011-2015.

| Year | Routes (times) | Weight (Ton) | Volume (M3) | Average (Ton/day) |
|---|---|---|---|---|
| 2015 | 123.398,00 | 327.970,59 | 1.311.882,37 | 898,54 |
| 2014 | 115.338,00 | 201.841,50 | 807.366,00 | 552,99 |
| 2013 | 107.543,00 | 188.200,25 | 752.801,00 | 6.185,22 |
| 2012 | 82.683,00 | 144.695,25 | 578.781,00 | 395,64 |
| 2011 | 82.355,00 | 144.121,25 | 576.479,00 | 395,01 |

Source: [1]

Sumur Batu TPA locates on an area 20 hectares, where 17 hectares occupied or 85%. The data in Table 1 describes the behaviour of waste producer that does not treat the waste at the source properly. The increase in waste disposal into landfills will accelerate the full landfill capacity, so that new land for waste disposal is needed. Therefore, steps are needed to slow down the full landfill facility by implementing waste management at the source. The household must sort out the waste it produces. Waste that has economic value sorted can then be deposited into the waste bank or sold directly to collectors. The local government issued a regional regulation on community-based waste management to overcome the issues. The government expects more communities to be involved in waste management at the source. The more garbage can be collected for recycling or reused will help to reduce waste sent to landfills. The authority recommends waste banks as community-based waste management sites. The administration expects each RW initiates to establish a waste bank as a place for managing inorganic waste which has economic value. The number of RWs are 1013 with the assumption that each family consists of 4 people in one RW an average of 600 families.

*3.2 The Patriot Main Waste Bank (Bank Sampah Induk Patriot)*

The Patriot Main Waste Bank established on October 27, 2016, based on the Bekasi City Mayor Regulation No. 10A of 2017 concerning Position, organizational structure, main tasks and functions and work procedures for the Patriot Main Waste Bank of Bekasi City. The purpose is to organize part of the technical affairs of environmental management, especially handling waste at its source. BSIP tasks of helping the government increase community participation in waste management through the 3R, additional the number of environmental volunteers, provide awareness on waste to cash and opportunity business from waste banks.

BSIP has a vision of "realizing a healthy and productive waste bank throughout Bekasi City" with a mission: 1) Increasing the quality and quantity of waste banks, 2) Increasing the human resource, 3) Additional the facilities and infrastructure, and 4) Improving relationship with partnerships. The impact of the establishment of the BSIP and the mayor's regulation significantly increased the number of waste banks in all sub-districts in 2017 as listed in Table 2.

**Table 2.** Data on the number of waste banks (WB) in 2016 and 2017 per sub-district.

| No | Sub-district | No of village | No of RW | No of waste bank 2016 | | No of waste bank 2017 | |
|---|---|---|---|---|---|---|---|
| | | | | Listed | Active | Listed | Active |
| 1 | Medan Satria | 4 | 72 | 16 | 5 | 69 | 16 |
| 2 | Bekasi Utara | 6 | 144 | 10 | 4 | 144 | 22 |
| 3 | Bekasi Barat | 5 | 91 | 4 | 3 | 92 | 8 |
| 4 | Bekasi Selatan | 5 | 94 | 94 | 11 | 96 | 24 |
| 5 | Bekasi Timur | 4 | 83 | 7 | 6 | 84 | 17 |
| 6 | Rawalumbu | 4 | 97 | 35 | 6 | 48 | 9 |
| 7 | Mustika Jaya | 4 | 92 | 13 | 6 | 89 | 31 |
| 8 | Bantar Gebang | 4 | 37 | 4 | 2 | 37 | 3 |
| 9 | Pondok Melati | 4 | 57 | 15 | 3 | 59 | 14 |
| 10 | Pondok Gede | 5 | 78 | 5 | 3 | 24 | 4 |
| 11 | Jatiasih | 6 | 100 | 10 | 5 | 94 | 15 |
| 12 | Jati Sampurna | 5 | 68 | 55 | 3 | 68 | 5 |
| | Total | 58 | 1013 | 268 | 57 | 904 | 168 |

Source: [8]

`The data in Table 2 shows the number of waste banks increase dramatically growing up 337% from 268 in 2016 to 904 in December 2017. However, the percentage of the active waste bank, unfortunately is very low 21% (2016) and 18.5% (2017).

The first garbage bank was first established in Thailand in 2006, while in Indonesia it was first established in Bantul in 2008 [9]. The waste bank adopts the 3Rs system and conventional banks where customers deposit waste that has been sorted and handed over to the waste bank. Customers are given a savings book to record the amount of waste saved which has been converted into a currency amount based on the calculation of the determined waste price. Savers can also withdraw their savings if needed. The principle of activities in the waste bank is to follow the 3Rs, which is to collect, sort, recycle, and reuse waste that has economic value. The 3Rs gives positive results in protecting the environment and improving environmental behaviour, reducing air, water and soil pollution including reducing the incidence of greenhouse gases, while also providing economic benefits and can stimulate the growth of micro-recycling businesses [4] [6]. Also stated by [10], garbage banks provide financial benefits that have helped the learning process in an Al-Quran Education Park.

Although the benefits of the 3Rs are known, it does not automatically inspire people to implement it. The participation of community members influences the success of the 3Rs implementation. In the housing areas, the household has significant roles to participate in the 3Rs. As mentioned by [8] the appropriate support in the term of training, adequate information dissemination, the availability of environmental cadres and the existence of a waste bank influence the participation of households. And the authority holds an important role in issuing regulations and policies [7]. The Indonesian Government Regulation states that everyone is mandatory to reduce and handle waste. However, this regulation is not well socialized and promoted to the citizens. The administrator of the Bekasi City establishes a local regulation with the aims to encourage community participation and changing people's behaviour. The people to have a view to transform the waste into the new valued products. People can create money from processing waste. It will support to improve their finances. The other benefit of this practice is better cleanliness of the area and decrease the quantity of waste sent to landfill. Community-based waste management is practiced as well in other cities such as Makassar [2], Surabaya [11], Bandung and Yogyakarta [12].

*3.3 Correlation Test*

The correlation coefficient has a score between -1 and 1. If the score is <0 (negative) means the variable 1 has opposite direction with the variable 2. It means if the variable 1 increases, the variable 2 decreases. If the value > 0 (positive) means the variable 1 and variable 2 have the same directions. If the variable 1 increases, the variable 2 increases too. The score of correlation coefficient shows the strength of the relationship. It is justified based on the specified criteria [13].

| | |
|---|---|
| 0 | : there is no correlation |
| > 0-0,25 | : very weak correlation |
| > 0,25-0,5 | : fair correlation |
| > 0,5-0,75 | : strong correlation |
| > 0,75-0,99 | : very strong correlation |
| 1 | : a perfect correlation |

Table 3 shows the data on waste reduction and selling waste, and table 4 displays the Pearson correlation test results.

**Table 3.** waste reduction data in garbage banks from December 2016 to December 2017

| Month | No of active waste bank | Waste reduction (kg) | Selling waste value (IDR) |
|---|---|---|---|
| December 16 | 33 | 17923 | 24.085.320 |
| January 2017 | 43 | 25485 | 36.077.637 |
| February 2017 | 56 | 30226 | 35.170.793 |
| March 2017 | 69 | 38810 | 25.775.957 |
| April 2017 | 82 | 40222 | 31.471.057 |
| May 2017 | 82 | 59644 | 53.952.378 |
| June 2017 | 48 | 20286 | 26.857.154 |
| July 2017 | 93 | 49372 | 86.113.671 |
| August 2017 | 93 | 43265 | 75.096.727 |
| September 2017 | 103 | 53365 | 89.027.219 |
| October 2017 | 90 | 46976 | 77.933.227 |
| November 2017 | 99 | 46749 | 80.674.768 |
| December 2017 | 103 | 46610 | 87.919.139 |

**Table 4.** Result of Pearson Correlation Test

| Correlation name | Correlation coefficient | Criteria |
|---|---|---|
| No of active waste bank vs waste reduction | 0.876 | Very strong, positive |
| No of active waste bank vs selling waste | 0.868 | Very strong, positive |
| No of all waste bank vs waste reduction | 0.808 | Very strong, positive |
| No of all waste bank vs selling waste | 0.826 | Very strong, positive |
| Waste reduction vs selling | 0.871 | Very strong, positive |

The results of the correlation test conclude the number of active waste banks very strongly correlated with the quantity of waste reduction, as well as the value of selling waste. It shows increasing the number of active waste banks in line with the increment of waste reduction. It means that the more waste banks the load of TPA become lighter, it will help to lengthen the TPA life. Table 2 shows the number of the active waste bank is only 18.5%. Furthermore, BSIP needs to pay attention to the passive waste bank and to find the reasons why the waste bank is not active. It is necessary for BSIP to collaborate with the other parties, such as the head of village, the head of RW and the organization of Empowerment and Family Welfare (*Pemberdayaan dan Kesejahteraan Keluarga* – PKK). The chief of village could motivate the chief of each RW and the PKK caders at the RW level to actively initiate the building of waste bank.

## 4. Conclusion

One of the household waste management in Bekasi City is a waste bank. The government takes some actions to reduce waste sent to TPA through establishing a master waste bank as a managing partner, updating the previous regulations by involving the community in waste management at its source, encouraging each RW to form a waste bank and providing incentives people who innovate in waste management. These steps show positive results with the indicators as increasing the number of waste banks in each sub-district, increasing the amount of waste collected in the waste bank to be recycled or reused, decreasing the amount of waste sent to the landfill, increasing the value of selling waste. Correlation test results show that the number of waste banks correlates very strongly with the reduction of waste to landfill, so that the more waste banks, the higher reduction waste to landfill. The high number of passive waste banks is a challenge for BSIP and local governments. It needs appropriate strategies and policies to encourage the activity of waste banks. BSIP is recommended to coordinate with the chief of village and PKK as a community empowerment organization to increase community participation in waste management.

## Acknowledgment

This research is funded by *Masters Thesis Research*/ Penelitian Tesis Magister Universitas Indonesia-DRPM Ministry Research, Technology and Higher Education 2019 with contract number NKB-1901/UN2.R3.1/HKP.05.00/2019.

## References

[1] Bekasi B P S K Data sampah Kota Bekasi 2011-2015

[2] Towolioe, S., Permana, A.S., Aziz, N.A., Ho, C.S., Pampanga D G 2016 THE RUKUN WARGA-BASED 3RS AND WASTE BANK AS SUSTAINABLE SOLID WASTE MANAGEMENT STRATEGY *Planning Malaysia* 135–46

[3] Xiao L, Zhang G, Zhu Y and Lin T 2017 Promoting public participation in household waste management: A survey based method and case study in Xiamen city, China *Journal of Cleaner Production* **144** 313–22

Journal of Physics: Conference Series **1381** (2019) 012050 doi:10.1088/1742-6596/1381/1/012050

[4] Samiha B 2013 The importance of the 3R principle of municipal solid waste management for achieving sustainable development *Mediterranean Journal of Social Sciences* **4** 129–35
[5] Peprah K, Amoah S T, Thomas G and Achana W 2015 Assessing '3Rs' Model in Relation to Municipal Solid Waste Management in Wa, Ghana *World Environment* **5** 112–20
[6] Hye Chowdhury D A, Mohammad N, Haque M R U and Hossain D T 2014 Developing 3Rs (Reduce, Reuse And Recycle) Strategy for Waste Management in the Urban Areas of Bangladesh: Socioeconomic and Climate Adoption Mitigation Option *IOSR Journal of Environmental Science, Toxicology and Food Technology* **8** 09–18
[7] Ahmadi M 2017 Evaluating the Performance of 3Rs Waste Practices: Case Study-Region One Municipality of Tehran *Advances in Recycling & Waste Management* **02**
[8] Patriot B S I 2018 Data reduksi sampah di bank sampah Kota Bekasi 2017
[9] Wijayanti D R and Suryani S 2015 Waste Bank as Community-based Environmental Governance: A Lesson Learned from Surabaya *Procedia - Social and Behavioral Sciences* **184** 171–9
[10] Indrianti N 2016 Community-based Solid Waste Bank Model for Sustainable Education *Procedia - Social and Behavioral Sciences* **224** 158–66
[11] Dhokhikah Y, Trihadiningrum Y and Sunaryo S 2015 Community participation in household solid waste reduction in Surabaya, Indonesia *Resources, Conservation and Recycling* **102** 153–62
[12] Ramadhan M A 2016 Perbandingan Efektivitas Bank Sampah Di Kota Bandung Dan Kota Yogyakarta *Inersia* **XII** 85–90
[13] Sarwono J Buku lepas Korelasi Jonathan Sarwono

# A new model of open end yarn twist using torus coordinate based on dynamical mechanics

V G V Putra*, S Rohmah

Physics and Mechatronics Laboratory, Politeknik STTT Bandung, Jalan Jakarta No.31 Bandung, Jawa Barat, Indonesia. 40272

a)Corresponding author: valentinus@kemenperin.go.id

**Abstract.** Yarn structure and its relationship with the properties have attracted many researchers and scientist in textile science. The importance of this subject has increased due to the need of yarn with best possible quality. In this study, we have established a new equation to relate open end (OE) yarn count number in metric count ($N_m$) toward yarn twist on OE spinning. Based on this research, twist has been determined and formulated by analyzing the movement of fibre inside yarn using tensor calculus on torus coordinate based on dynamical mechanics.

Keywords: Yarn Twist, Torus Coordinate, Fibre Movement

## 1. Introduction

Yarn structure and its relationship with the properties have attracted many researchers and scientist in textile science. The importance of this subject has increased due to the need of yarn with best possible quality. Fibre migration or fibre movement inside of OE yarn is the change in the range of a fibre (along it's length) from the axis of a yarn, which happens during the movement of fibre inside OE yarn. The research of yarn and fibre equation as well as its influence of characteristics of yarn have been studied by some scientists both experimentally as well as theoretically [1-16]. According to Putra, et al [1-3] the fibre movement of yarn can be analyzed by classical mechanics in a certain coordinate. As stated by Lawrence [7], the characteristic of spun yarn can be explained by the fibre movement inside yarn in certain coordinate and also yarn structure as shown in Figure 1 below.

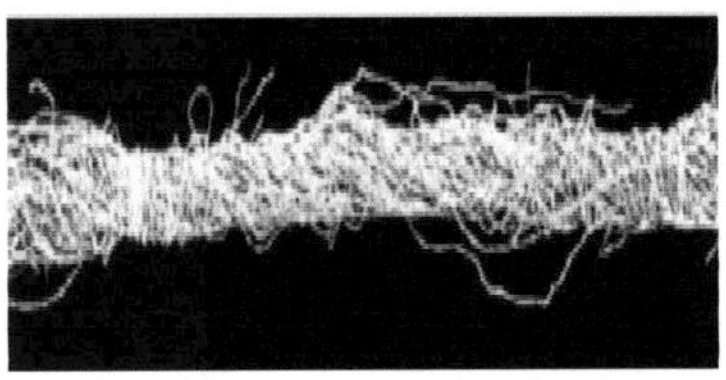

Rotor spinning

**Figure 1.** OE Yarn Properties [7]

As reported by Backer, et al [5], the tenacity or strength of amount of fibre is determined by the rate of twist and the relation can be exposed as the lower of twist, the higher is the tenacity or strength of yarn per tex and vice versa. Some researchers, such as Rohlena [11] reported that breakage rate of yarn is affected by the yarn twist. The lower the twist, the lower is the breakage rate. According to Musa and Ayse [8], Prendzova [10] and Penava & Oreskovic [9], the yarn diameter influenced the yarn strength and yarn count number in indirect system (Nm). Musa and Ayse [8] reported that the wider yarn radius, the stronger is the tenacity or strength of yarn. Putra, et al.[1] composed the relationship between twist and yarn count number in metric count as Eq.(1) which is related by twist multiplier ($\alpha_m$). Rohlena [11] and Backer, et al [5] formulated yarn twist as a function of yarn count number as Equation (2) and Equation (3), whereas Lawrence [7] and Putra et al.[3] composed the relationship between twist and yarn delivery speed, $v_d$, with rotor diameter, $d_{rotor}$, as well as rotor speed, $n_{rotor}$, as Equation (4) and Equation (5)

$$T_{solenoid} = \frac{\sqrt{F_o}}{n_r \pi^2 d_{yarn} d} \sqrt{N_m}, \tag{1}$$

$$T = \frac{\tan\theta}{2\pi r_{yarn}} \tag{2}$$

$$T = \alpha_m \sqrt{N_m} \tag{3}$$

$$T = \frac{n_{rotor}}{v_d} \tag{4}$$

$$T = \frac{n_{rotor}}{v_d} + \frac{1}{\pi d_{rotor}} \tag{5}$$

In this study, the relationship of yarn twist to yarn count number would be derived by looking the movement of fibre inside yarn using torus coordinate. Tensor calculus has a main role to solve the movement of fibre inside yarn and to get the relationship of it. According to many researchers [15-19] mathematical models have their basis in applied physics and can be used to determine the motion of object. Since the subject of mechanics deals with the motion of objects in space, fibre equation inside of yarn and its characteristic of analytical mechanics can be formulated better with a certain coordinate, such as torus coordinate. An n-dimensional Euclidean space is defined to consist of vectors S whose components are $r, u, v$, now assumed to be real. In this study, we used geometric mechanics to measure and to derive the equation of fibre in curve linear coordinate on torus coordinate.

## 2. Research Method

### Development of Fibre Equation on Torus

Fibre movement on torus coordinate (caused by the external force ($F_o$) and the internal force) can be obtained using analytical mechanics model. By knowing the certain coordinate system, the fibre movement inside a certain yarn can be formulated. In particular, a fibre inside a yarn with certain length ($dS$) travels in a certain coordinate, whose length is $du$ during a time ($dt$) and radius of yarn cross section ($dv$) . A yarn can be pretended to be formed in torus coordinate, whose radius of yarn ($r$) and the length of gap ($b$), are shown in Figure 2.

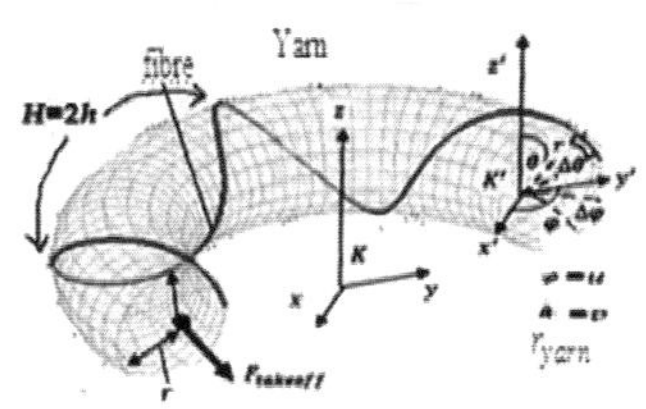


**Figure 2**. Model of OE yarn on torus coordinate

A coordinate system, such as a torus, is kind of the curvilinear type. Consider the three quantities of $V_m$ $(m = 1,2,3)$ which relate to the rectangular coordinates and the three transformed quantities (torus coordinate) of $\widetilde{V}_\mu$ ($\mu$ =1,2,3) related to $V_m$, as:

$$S = (x, y, z) = (r, u, v) \tag{6}$$

$$S = (\rho + r\cos v)\cos u, (\rho + r\cos v)\sin u, r\sin v. \tag{7}$$

Equation (7) can be used to get the fibre movement on torus coordinate, the square of line element ($\boldsymbol{dS^2}$) can be given below

$$\boldsymbol{dS^2 = (b + r\cos v)^2 du^2 + r^2 dv^2 + dr^2}, \tag{8}$$

The metric element ($\boldsymbol{g_{mn}}$) can be demonstrated as shown below:

$$\boldsymbol{g_{mn}} = \begin{pmatrix} \boldsymbol{b + r\cos v} & \boldsymbol{0} & \boldsymbol{0} \\ \boldsymbol{0} & \boldsymbol{r^2} & \boldsymbol{0} \\ \boldsymbol{0} & \boldsymbol{0} & \boldsymbol{1} \end{pmatrix}. \tag{9}$$

The components of a unit vector can be written in matrix format, as illustrated below

$$\begin{pmatrix} \hat{u} \\ \hat{v} \\ \hat{r} \end{pmatrix} = \begin{pmatrix} -\sin u & \cos u & 0 \\ -\sin v\cos u & -\sin v\sin u & \cos v \\ \cos v\cos u & \cos v\sin u & \sin v \end{pmatrix} \begin{pmatrix} \hat{i} \\ \hat{j} \\ \hat{k} \end{pmatrix}. \tag{10}$$

The acceleration of fibre on each axis ($\boldsymbol{a^u, a^v, a^r}$) on torus coordinate can be written as (after some calculations):

$$\frac{d^2x^1}{dt^2} + \Gamma^1_{\alpha\beta}\frac{dx^\alpha}{dt}\frac{dx^\beta}{dt} = \frac{d^2u}{dt^2} - \frac{2r\sin v}{b + r\cos v}\frac{du}{dt}\frac{dv}{dt} + \frac{2\cos v}{b + r\cos v}\frac{du}{dt}\frac{dr}{dt} = a^u, \tag{11}$$

$$\frac{d^2x^2}{dt^2} + \Gamma^2_{\alpha\beta}\frac{dx^\alpha}{dt}\frac{dx^\beta}{dt} = \frac{d^2v}{dt^2} + \frac{1}{r}\sin v(b + r\cos v)\frac{du}{dt}\frac{du}{dt} + \frac{2}{r}\frac{dv}{dt}\frac{dr}{dt} = a^v, \tag{12}$$

$$\frac{d^2x^3}{dt^2} + \Gamma^3_{\alpha\beta}\frac{dx^\alpha}{dt}\frac{dx^\beta}{dt} = \frac{d^2r}{dt^2} - \cos v\,(b + r\cos v)\,\frac{du}{dt}\frac{du}{dt} - r\frac{dv}{dt}\frac{dv}{dt} = a^r. \tag{13}$$

The general acceleration ($\boldsymbol{a^i}$) for a certain mass ($\boldsymbol{M}$), including the stress tensor ($\boldsymbol{\sigma^{ij}}$) and the external force ($\boldsymbol{F^i}$), can be written as shown below:

$$\sigma^{ij}{}_{,j} + F^i = Ma^i \tag{14}$$

to find the general equation of fibre movement influenced by the stress tensor as traction tensor ($\boldsymbol{t^r, t^u, t^v}$) caused by the external force, hence we can substitute Equation (11), Equation (12) and Equation (13) to Equation (14) and after some calculations, it can be shown as below:

$$\frac{\partial t^r}{\partial r} + \frac{1}{b + r\cos v}\frac{\partial t^u}{\partial u} + \frac{1}{r}\frac{\partial t^v}{\partial v} + \frac{\cos v t^r - \sin v t^v}{b + r\cos v} + \frac{t^r}{r} + \bar{F} = M\bar{a}. \tag{15}$$

Equation (15) can be derived as

$$\frac{\partial\sigma^{rr}}{\partial r}\hat{r}+\frac{\partial\sigma^{rr}}{\partial r}\frac{\partial\sigma^{ru}}{\partial r}\hat{u}+\frac{\partial\sigma^{rv}}{\partial r}\hat{v}+\frac{1}{b+r\cos v}\left(\frac{\partial\sigma^{ur}}{\partial u}\hat{r}+\frac{\partial\sigma^{uu}}{\partial u}\hat{u}+\frac{\partial\sigma^{uv}}{\partial u}\hat{v}\right)+\frac{1}{r}\left(\frac{\partial\sigma^{vr}}{\partial v}\hat{r}+\frac{\partial\sigma^{vu}}{\partial v}\hat{u}+\frac{\partial\sigma^{vv}}{\partial v}\hat{v}\right)+\frac{\cos v(\sigma^{rr}\hat{r}+\sigma^{ru}\hat{u}+\sigma^{rv}\hat{v})-\sin v(\sigma^{vr}\hat{r}+\sigma^{vu}\hat{u}+\sigma^{vv}\hat{v})}{b+r\cos v}+\frac{(\sigma^{rr}\hat{r}+\sigma^{ru}\hat{u}+\sigma^{rv}\hat{v})}{r}+\bar{F}=M\bar{a}, \quad (16)$$

If we take $\hat{\boldsymbol{r}}$ axis and analyze the fibre movement inside yarn which is influenced by the external force from machine ($\boldsymbol{F_o}=\boldsymbol{F_{takeoff}}$) as well as internal force (caused by deformation force), also substitute Equation (13) to Equation(16), therefore we get Equation (17)

$$\frac{\partial\sigma^{rr}}{\partial r}+\frac{1}{b+r\cos v}\left(\frac{\partial\sigma^{ur}}{\partial u}\right)+\frac{1}{r}\left(\frac{\partial\sigma^{vr}}{\partial v}\right)+\frac{\sigma^{rr}\cos v-\sigma^{vr}\sin v}{b+r\cos v}+\frac{\sigma^{rr}}{r}-F_o=m\left(\frac{d^2r}{dt^2}-\cos v\,(b+r\cos v)\dot{u}^2-r\dot{v}^2\right), \quad (17)$$

If $\boldsymbol{v}$ is kept constant and the speed of $\dot{\boldsymbol{u}}\approx\dot{\boldsymbol{v}}$ and there's no fibre migration on $\hat{\boldsymbol{r}}$ axis, $\frac{\boldsymbol{d^2r}}{\boldsymbol{dt^2}}=\boldsymbol{0}$, hence it gives

$$\frac{\partial\sigma^{rr}}{\partial r}+\frac{1}{b+r\cos v}\left(\frac{\partial\sigma^{ur}}{\partial u}\right)+\frac{1}{r}\left(\frac{\partial\sigma^{vr}}{\partial v}\right)+\frac{\sigma^{rr}\cos v-\sigma^{vr}\sin v}{b+r\cos v}+\frac{\sigma^{rr}}{r}-F_o=-M\cos v(b+r\cos v)\dot{u}^2-Mr\dot{v}^2\approx-M\dot{u}^2(\cos v(b+r\cos v)+r), \quad (18)$$

In case $\boldsymbol{r}\ll\boldsymbol{b}$ and $\dot{\boldsymbol{u}}\approx\dot{\boldsymbol{v}}$ **or** $\boldsymbol{n_r}\approx\boldsymbol{n_d}$, hence

$$\boldsymbol{\frac{\partial\sigma^{rr}}{\partial r}+\frac{1}{b+r\cos v}\left(\frac{\partial\sigma^{ur}}{\partial u}\right)+\frac{1}{r}\left(\frac{\partial\sigma^{vr}}{\partial v}\right)+\frac{\sigma^{rr}\cos v-\sigma^{vr}\sin v}{b+r\cos v}+\frac{\sigma^{rr}}{r}-F_o=-M\cos v(b+r\cos v)\dot{u}^2}, \quad (19)$$

For a case, if twist angle,$\boldsymbol{\theta}$, is kept constant, $\boldsymbol{\sigma^{ii}}$**is constant** value and $\boldsymbol{r}\ll\boldsymbol{b}$, $\boldsymbol{\sigma^{ij}=0}$, we can formulate

$$F_o-\sigma^{rr}\left(\frac{\cos v}{b+r\cos v}+\frac{1}{r}\right)=M\sin\theta(b+r\sin\theta)\dot{u}^2, \quad (20)$$

Now by Equation (20) , we have

$$F_o-\frac{\sigma^{rr}}{r}\left(\frac{2r\cos v+b}{b+r\cos v}\right)=F_o-\frac{\sigma^{rr}}{r}\approx M\sin\theta b\dot{u}^2. \quad (21)$$

$$M(R_r\sin^2\theta)\dot{u}^2=F_o-\frac{\sigma^{rr}}{r}, \quad (22)$$

After some calculations, we have

$$\boldsymbol{\tan\theta=0.7\frac{\sqrt{F_{take-off}}}{n_r\pi d}\sqrt{N_m}}. \quad (23)$$

Substitute Equation (23) to Equation (2), hence

$$T = \frac{\tan\theta}{2\pi r_{yarn}} = \frac{0.7}{2\pi r_{yarn}} \frac{\sqrt{F_{take-off}}}{n_r \pi d} \sqrt{N_m}, \tag{24}$$

According to Backer et al. [5] and Putra et al.[1-2] , the relation among yarn angle, twist and yarn count number can be related by Equation (24) as shown below:

$$T = \frac{0.7}{2\pi r_{yarn}} \frac{\sqrt{F_{take-off}}}{n_r \pi d} \sqrt{N_m} = \alpha_m \sqrt{N_m} \tag{25}$$

Equation (25) can be drawn the relationship of twist and yarn count in metric as Figure 3 below (for a special case that the yarn constant, $\alpha_m = 136$)

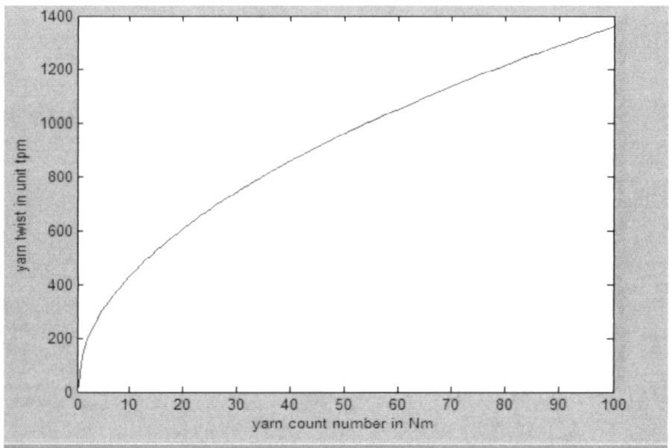


**Figure 3.** Relationship of Twist and Yarn Count in Metric

According to the experimental result in Industry by following data below (Table 1)

**Table 1.** Experimental result

| *rotor speed* (rpm) | *tenacity of yarn* (cN/tex) | $a_m$ | *Nm yarn count* | *Tt*(tex) | $v_d$ (m/min) | $T_{machine}$ (tpm) |
|---|---|---|---|---|---|---|
| 72000 | 15.51 | 136 | 67 | 14.79 | 77.7 | 1118.18 |
| 72000 | 15.13 | 136 | 83 | 11.83 | 69.2 | 1250 |
| 72000 | 14.56 | 136 | 100 | 9.86 | 63.1 | 1369,31 |

By using Table 1, the relationship of twist and yarn count in metric $N_m$ as shown in Figure 4 below can be created

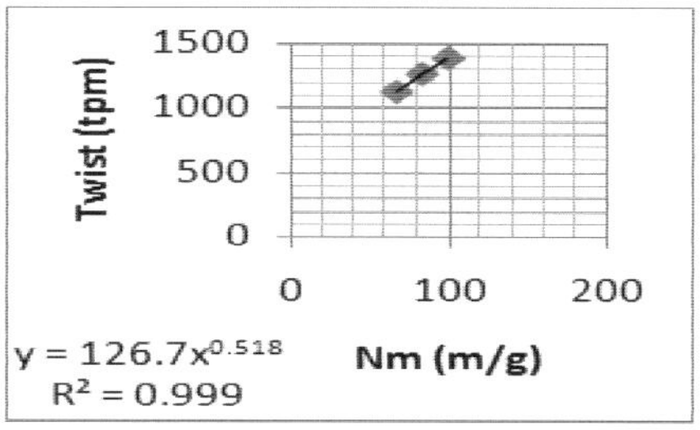


**Figure 4.** Relationship of Twist and Yarn Count in Metric as $T_m = Const\sqrt{N_m}$

## 3. Result and Discussion

The equation of fibre movement traveling inside of yarn has been established and derived using torus coordinate occurred on spinning machine. According to this theory, the relationship of yarn count number correlated by twist of yarn has been found.

In this research, the connection between twist and yarn count number has been established. According to Rohlena [11], the relationship between yarn count number and yarn twist is correspondence with the movement of fibre inside the yarn, furthermore it can be formulated on torus coordinate as we have established in this research. Based on this study, The higher of yarn count number, the higher is the twist of yarn. Besides yarn fineness or yarn count number in indirect system, twist also influences the strength of yarn. Lawrence.[7] proposed that the tenacity of rotor open end spinning yarn is influenced by the rate of twist and the relation can be shown as: the higher the strength of yarn per tex and vice versa the lower is the yarn twist. As stated by Some researchers, [1-13] yarn strength decreases when yarn twist and yarn count number increase.

the strength of yarn per tex is affected by the rate of yarn twist has been reported by Becker, et al [5]. Based on this study, the relation of yarn count number to twist has been predicted as Equation (25). Based on the Equation (25) as well as the data from experimental result, it can be established that the correlation of yarn twist to yarn number (i.e. yarn count number) can be formulated by using following equation:

$$T = \frac{0.7}{2\pi r_{yarn}} \frac{\sqrt{F_{take-off}}}{n_r \pi d} \sqrt{N_m}. \qquad (26)$$

## 4. Conclusion

Fiber movement on yarn has been analyzed to present and to predict the relationship of yarn count in metric to yarn twist. In this study, it was found that yarn twist is affected by yarn count number in metric on torus coordinate. It has been established that the formula to relate the relationship is $T = \frac{0.7}{2\pi r_{yarn}} \frac{\sqrt{F_{take-off}}}{n_r \pi d} \sqrt{N_m}$

## Acknowledgements

Our interest in yarn mechanics was helped by all our friends in STTT Bandung who supported this research. Their help and encouragement have been valued to us in pursuing this project.

## References

[1] Putra, V G V, Maruto G and Rosyid, M F., 2017, *Indian Journal of Fibre and Textile Research*, **42**. Pp. 359-363.
[2] Putra, V G V, Rosyid M F and Maruto, G., 2016, , *Global Journal of Pure and Applied Mathematics*, **12**.,No.1 pp. 405-412.
[3] Putra, V G V. &Rosyid, M F, 2015, *Journal of Applied Mathematics and Physics*,.**3**. Pp. 623-630.
[4] Putra, V G V,Rosyid, M F & Maruto, G., 2016, *Global Journal of Pure and Applied Mathematics*, **12**. pp. 4107-4114.
[5] Backer, Hearle & Grosberg, 1969, *Structural Mechanics of Fibres, Yarns and Fabrics* (New York: Wiley-Interscience)
[6] Furter R & Meier S., 2009, *Measurement and Significance of Yarn Twist* (Switzerland : Uster Technology AG,)
[7] Lawrence C A, 2003, *Fundamentals of Spun Yarn Technology* (New York: CRC Press).
[8] Musa & Ayse, 2006, *FIBRES & TEXTILES JOURNAL*, **5**, pp 84-87.
[9] Penava,, 1997, *Journal of the Textile Institute*, ***88***, pp. 21-32.
[10] Prendž*ova*,, 2000, *International Journal of Polymeric Materials*, , ***47***, pp. 701-707.
[11] Rohlena, V, 1975, *Open-End Spinning*, (New York :Elseiver Scientific Publishing Company).
[12] Trommer, G., 1995, *Rotor Spinning*, (Frankfurt :Deutscher fachverlag).
[13] Zeidman, Shawney dan Herington, 2003, *Indian Journal of Fibre and Textile Research*, **28**, pp. 123-133.
[14] Zhang, Xuefeng dan Yonglan , 2011, *FIBRES & TEXTILES JOURNAL*, **19**,pp. 17-20.
[15] Putra, VGV, Wijayono, A, Martina, T, Rosyidan, C., 2019, *Wahana Fisika*, **4**, pp. 27-34
[16] Wijayono, A., Putra VGV., 2018, *Arena Tekstil*, **33**, pp 85-92.

Journal of Physics: Conference Series **1381** (2019) 012052 doi:10.1088/1742-6596/1381/1/012052

# Modelling of Yarn Count and Speed of Delivery Roll to Yarn Strength in Spinning Machines Based On Analytical Mechanics

V G V Putra*, R Sahroni, A Wijayono, D Kusumaatmadja

Department of TeknikTekstil, Politeknik STTT Bandung, Jalan Jakarta No.31 Bandung, Jawa Barat, Indonesia.40272

*Corresponding author: valentinus@kemenperin.go.id

**Abstract**. In textile, the application of analytical mechanics can be found in a spinning process. The aim of this research is to determine a model as well as to predict the yarn strength of ring-spun yarn influenced by the yarn count number (in unit m/g) and also the delivery speed of yarn ($v_f$). In this article, the yarn count and the speed of delivery roll of ring-spun yarn were studied comprehensively, and based on this, a ring-spun yarn strength prediction model has been found and established by analytical mechanics.

## 1. Introduction

Yarn structure and its relationship with the properties have attracted many researchers and scientist in textile science. The importance of this subject has increased due to the need of yarn with best possible quality. Fibre migration or fibre movement inside of OE yarn is the distance change of a fibre (along its length) of the yarn axis, which occurs during the open end spinning yarn (OE yarn)production. The research of fibre movement and its effect of characteristics of yarn have been studied by many researchers both theoretically and experimentally [1-15]. According to Putra, et al [6,7,11} the fibre movement of yarn can be analyzed by classical mechanics in a certain coordinate. According to Lawrence[7], the property of spun yarn can be explained by the fibre movement and yarn structure. According to Lawrence[5]., the strength of yarn is affected by the rate of twist and the relation is shown as the lower of twist, the higher is the strength of yarn per tex and vice versa. Rohlena [8] stated that breakage rate is affected by the twist. The lower the twist, the lower is the breakage rate. According to some researcher [5-11], the yarn diameter affected the strength of yarn and yarn count number. Putra et al [7] reported that the wider yarn diameter , the stronger is the strength of yarn. According to some researchers [1-11] the application of physics, especially classical mechanics (analytic) in various fields of applied studies has been found, one of which is in the textiles field, namely in the case of yarn spinning. In the process of making staple yarn, there are several machines used in the process of making yarn, one of those machines is a Ring Spinning machine. In textile science, especially yarn spinning, yarn movement theory which discusses the influence of direct yarn count (Tex) in dimensions of long density or indirect yarn count in metrics having units ($N_m$) and delivery speed of yarn ($V_f$) against yarn strength, $T_w$ , in the Ring Spinning machine has not been yet studied in detail. Hongwei Zhang et al. [4]explained the influence of yarn count on the strength of Ring-Spinning yarn is inversely proportional, i.e. the greater the number of yarn ($N_m$) in units (m / g), the smaller is the yarn strength value, Andriansyah, et al. [1] explained the relationship between yarn strength as a function of yarn count is inversely proportional.

Some researchers who have attempted to model the movement of yarn in the spinning process are: Putra, V.G.V, Rosyid, M.F and Maruto, G [7] shows the modeling of yarn motion using analytical mechanics on solenoid coordinates, Putra et al [7] modeling yarn movements using mechanics classic and calculus tensor in torus coordinates, Putra and Iskandar [6] try to modeling the relationship between the effect of twisting shape on twist value, Zeidman, Shawney and Herington [10] explain the movement of fibers in yarns with twisting value using geometric modeling in cylindrical coordinates is defined as the ratio of the yarn rotation speed to the delivery speed of yarn. In Zeidman's, Shawney and Herington [10]model, the relationship between yarn counts, twisting value, and twisting angle andits effect on yarn strength in the spinning process did not been explained. Lawrence [5] describes the movement of yarns in the spinning process in Ring Spinning machines using analytical mechanics. Lawrence's [5]model do not elaborate in detail the influence of yarn counts and the yarn delivery speed on the strength of the yarn. Several other researchers such as Rohlena [8], Hearle and Gupta [2] modeled the equation form of fibers motion and yarn in the staple yarn spinning process through a twist review, Herawati et al. [3] modeled the equation of yarn motion in the rotor machine to find the relationship between the yarn strength to the rotational speed of the yarn on the rotor. Theoretical research and experimental validation of the spinning process is very necessary to get the equation of motion and form of modeling. In this study, the formulation of the yarn motion equation in the Ring-Spinning machine was examined to obtain the relationship between the yarn strength as a function of yarn speed and also the yarn counts.

## 2. Research Method

In this study the yarn motion equation in the Ring-Spinning machine will be modeled using classical mechanics analytic as well as nonlinear multivariate statistics with validation in the experiment on the ring spinning machine (Figure 1) and assuming the bobbin moves in circles together with the spindle and the yarn moves through the traveler towards bobbin.

**Figure 1.** Ring-Spinning Machine System

Bobbins rotate with angular speed,$n_b$, along with spindles that rotate with rotational speed, $\omega = n_b = n_s$, and yarns on traveler with traveler's rotational speed, $n_t \approx n_s$. The yarn moves with the delivery speed, $v_f$, passes through a lappet that has certain spinning tension, $T_s$. The yarn moves to the ring rail with a certain constant speed (there is no acceleration) and the strength of the initial yarn on the lappet is $T_o$ with the coefficient of friction between the yarn and lappet is μ, while the angles formed are $\Delta\theta$ as in Figure 2 and Figure 3.

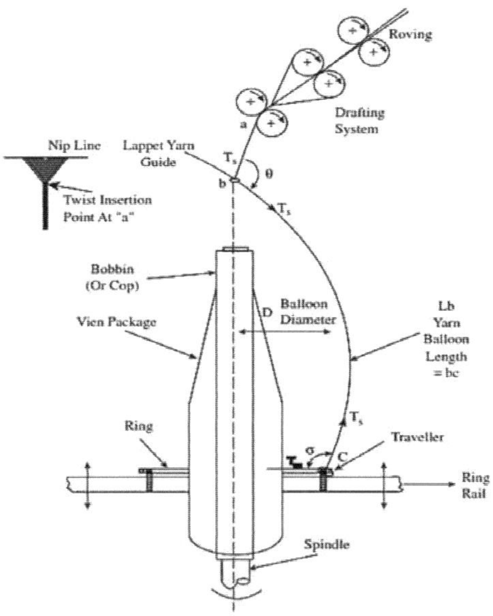


**Figure2.** Yarns movement on a ring-spinning machine

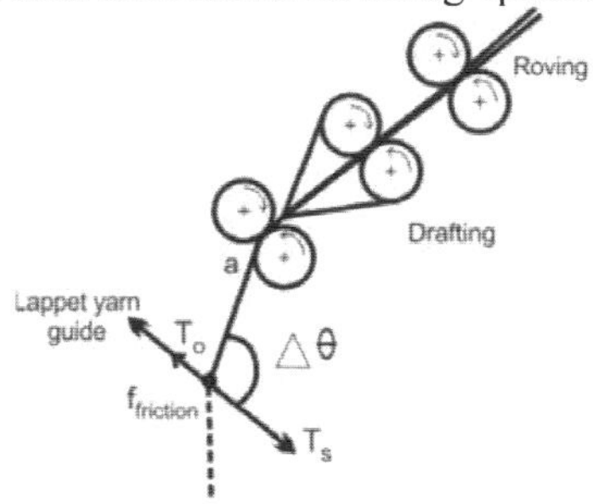


**Figure 3**. Yarn movement in the lappet area

The equation of yarn movement on lappet can be formulated as follows:

$$\sum F = \frac{dp}{dt} \tag{1}$$

$$\sum F = m\frac{dv}{dt} + v\frac{dm}{dt} = m\frac{dv}{dt} \tag{2}$$

When the speed is constant, then

$$\sum F = m\frac{dv}{dt} = 0 \tag{3}$$

Review the components on the x-axis and y-axis, so they can be described as follows

$$\sum F_x = 0 \tag{4}$$

$$(T_s - T_o)cos\left(\frac{\Delta\theta}{2}\right) - f_{friction} = 0 \tag{5}$$

For very small angle of $\frac{\Delta\theta}{2}$, then

$$(T_s - T_o) - f_{friction} \approx 0 \tag{6}$$

$$T_s - T_o \approx f_{friction} \tag{7}$$

$$\Delta T \approx f_{friction} \tag{8}$$

$$\Delta T \approx \mu N \tag{9}$$

For the force component on the y-axis it is obtained:

$$\sum F_y = 0 \tag{10}$$

$$(T_s + T_o)sin\left(\frac{\Delta\theta}{2}\right) - N = 0 \tag{11}$$

For very small angle of $\frac{\Delta\theta}{2}$, then

$$(T_s + T_o)\frac{\Delta\theta}{2} \approx N \tag{12}$$

If $T_s + T_o \approx 2T$, then

$$2T\frac{\Delta\theta}{2} \approx N \tag{13}$$

$$T\Delta\theta \approx N \tag{14}$$

By substituting Equation (9) in Equation (14), it is obtained:

$$T\Delta\theta \approx \frac{\Delta T}{\mu} \tag{15}$$

So that equation (16) is generated

$$\frac{\Delta T}{T} = \mu\Delta\theta \tag{16}$$

Forvery small $\Delta T$and $\Delta\theta$ and by doing integrals on both sides, it is obtained

$$\int \frac{dT}{T} = \int \mu\, d\theta \tag{17}$$

So that Equation (18) is generated

$$\ln\left(\frac{T_s}{T_o}\right) = \mu\Delta\theta \tag{18}$$

Or can be written

$$\frac{T_s}{T_o} = e^{\mu\Delta\theta} \tag{19}$$

$$T_s = T_o e^{\mu\Delta\theta} \tag{20}$$

The yarn which moves at a constant speed and the strength of $T_s$ from the lappet guide is then pulled towards the bobbin with a certain strength of $T_w$ as shown in Figure 4below

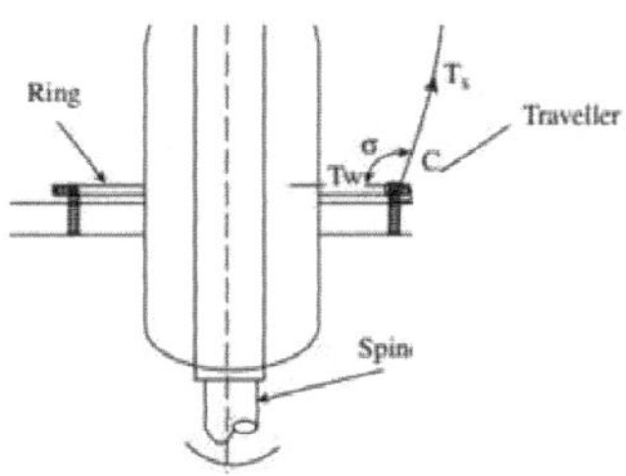


**Figure 4.**Movement in bobbins

The yarn moves constantly towards the bobbin with a certain speed which is affected by the traveler's coefficient of friction for the yarn of $\mu_t$. The strength of the yarn leading to the bobbin can be modeled as follows:

$$\sum F_x = 0 \tag{21}$$

$$(T_w - T_s)cos\left(\frac{\Delta\sigma}{2}\right) - f_{gesek} = 0 \tag{22}$$

For very small angle of $\frac{\Delta\sigma}{2}$, then

$$(T_w - T_s) - f_{friction} \approx 0 \quad (23)$$
$$T_w - T_s \approx f_{friction} \quad (24)$$
$$\Delta T \approx f_{friction} \quad (25)$$
$$\Delta T \approx \mu_t N \quad (26)$$

For the force component on the y-axis it is obtained:

$$\sum F_y = 0 \quad (27)$$
$$(T_w + T_s) sin\left(\frac{\Delta\sigma}{2}\right) - N = 0 \quad (28)$$

For very small angle of $\frac{\Delta\sigma}{2}$, then

$$(T_w + T_s)\frac{\Delta\sigma}{2} \approx N \quad (29)$$

If $T_w + T_s \approx 2T$, then

$$2T\frac{\Delta\sigma}{2} \approx N \quad (30)$$
$$T\Delta\sigma \approx N \quad (31)$$

By substituting Equation (31) in Equation (26), it is obtained:

$$T\Delta\sigma \approx \frac{\Delta T}{\mu_t} \quad (32)$$

So that Equation (33) is generated

$$\frac{\Delta T}{T} = \mu_t \Delta\sigma \quad (33)$$

For ΔT and Δσ are very small and by doing integrals on both sides, So that Equation (34) is generated

$$\int \frac{dT}{T} = \int \mu_t \, d\sigma \quad (34)$$

And then Equation (35) is:

$$\ln\left(\frac{T_w}{T_s}\right) = \mu_t \Delta\sigma \quad (35)$$

Or can be written in Equation (36)

$$T_w = T_s e^{\mu_t \Delta\sigma} \quad (36)$$

By substituting Equation (20) in Equation (36), it is obtained:

$$T_w = T_o e^{\mu\Delta\theta} e^{\mu_t \Delta\sigma} = T_o e^{(\mu\Delta\theta + \mu_t\Delta\sigma)} \quad (37)$$

If the value of $\mu\Delta\theta = \mu_t \Delta\sigma$, then can be written

$$T_w = T_o e^{2\mu\Delta\theta} \quad (38)$$
$$T_w - T_o = T_o e^{2\mu\Delta\theta} - T_o = T_o(e^{2\mu\Delta\theta} - 1) \quad (39)$$
$$\frac{T_w - T_o}{T_o} = \frac{\Delta T}{T_o} = e^{2\mu\Delta\theta} - 1 \quad (40)$$

The change in yarn strength ΔT is due to the strength of the yarn that leading to the bobbins and the strength of spinning along the area $y = R$ on the ring rail area can be formulated as follows:

$$\sum F = my\omega^2 \quad (41)$$
$$\Delta T = my\omega^2 \quad (42)$$
$$\Delta T = (\lambda dy)y\omega^2 \quad (43)$$

The direct yarn count $T_t$ in textiles is defined as the mass of yarn per unit length and has dimensions of long density in units of tex or g / km

$$\Delta T = (T_t dy)y\omega^2 \quad (44)$$

So that

$$\int \mathrm{d}T = \int (T_t\omega^2)\, y dy \tag{45}$$

$$\Delta T = \frac{1}{2} T_t \omega^2 (R^2 - y_o^2) \tag{46}$$

For condition of $y_o = 0$ (at the center of the bobbin), then

$$\Delta T = \frac{1}{2} T_t \omega^2 R^2 \tag{47}$$

Substitute Equation (47) into Equation (40) then get Equation (48)

$$\frac{T_t \omega^2 R^2}{2T_o} = e^{2\mu\Delta\theta} - 1 \tag{48}$$

The initial strength of the yarn in the lappet area can be written as Equation (49)

$$T_o = \frac{T_t \omega^2 R^2}{2(e^{2\mu\Delta\theta} - 1)} = \frac{1000\, n_b{}^2 R^2}{2N_m (e^{2\mu\Delta\theta} - 1)} \tag{49}$$

According to Rohlena[8], Lawrence [5] and also Trommer[9] regarding yarn numbering on textiles using a metric system (indirect yarn count) can be written as $N_m = \frac{1000}{T_t}$ in units (m/g). the speed of the delivery can be written as $v_f = n_b R$, so that for $\mu, \Delta\theta, R$, the value is constant so it can be written as follows

$$T_o = \left(\frac{1000}{2(e^{2\mu\Delta\theta} - 1)}\right) \frac{n_b{}^2 R^2}{N_m} = \left(\frac{1000}{2(e^{2\mu\Delta\theta} - 1)}\right) v_f{}^2 N_m{}^{-1} \tag{50}$$

Substitution of Equation (50) to Equation (38) shows that the final strength of the yarn is equal to

$$T_w = \left(\frac{500\, e^{(2\mu\Delta\theta)}}{(e^{2\mu\Delta\theta} - 1)} v_f{}^2\right) N_m{}^{-1} \tag{51.a}$$

$$T_w = a v_f{}^b N_e{}^{-1} \tag{51.b}$$

with values $a > 0$ and $b$=2 which are constant. Equation (51) is the equation for the ideal condition of yarn movements. In order to validate the results of the yarn, experiment can be used using the Toyoda RY 240 machine with a constant spindle rotational speed and front roller speed varies. For yarn with a certain counts in $N_e$ which is formulated $N_e = 0.59\ N_m$ (conversion of unit numbers of yarns in metrics) in units of $N_e$ (hank / lbs), with 1 hank is 840 yd and 1 lbs has a value of 453.6 g (Trommer [9]) and $N_e$ yarn count are 40 (hank / lbs), 50 (hank / lbs) and 60 (hank / lbs) with the delivery speed of yarn is 70.7 yd / min, 63 yd / min and 57.4 yd / min with yarn strength is 211 cN, 169 cN and 140 cN. To analyze the fiber movement inside yarn in the traveler, hence we assume that the fiber makes a torus coordinate as shown in Figure 5. The square of line element on torus coordinate can be shown as Equation (52)

$$ds^2 = (b + r\cos v)^2 du^2 + r^2 dv^2 + dr^2 \tag{52}$$

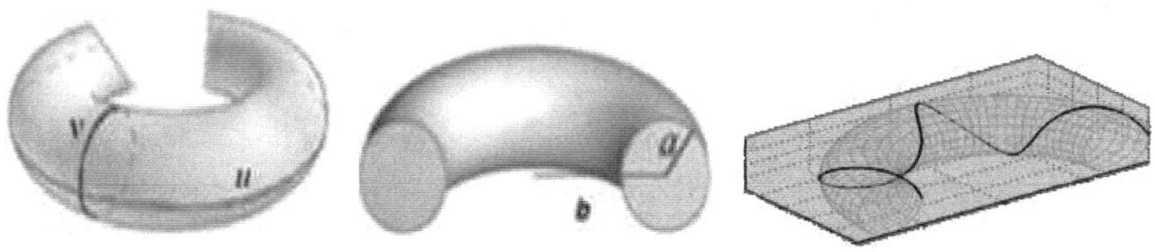


**Figure 5.**Fiber movement inside yarnin traveler area

The metric element can be written as below Equation (53)

$$g_{mn} = \begin{pmatrix} g_{11} & 0 & 0 \\ 0 & g_{22} & 0 \\ 0 & 0 & g_{33} \end{pmatrix} = \begin{pmatrix} (b+rcosv)^2 & 0 & 0 \\ 0 & r^2 & 0 \\ 0 & 0 & 1 \end{pmatrix} \tag{53}$$

The geodesic equation for metric element in Equation (53) can be written as below

$$\frac{d^2x^1}{dt^2} + \Gamma^1_{\alpha\beta}\frac{dx^\alpha}{dt}\frac{dx^\beta}{dt} = \frac{d^2x^1}{dt^2} + \Gamma^1_{11}\frac{dx^1}{dt}\frac{dx^1}{dt} + \Gamma^1_{12}\frac{dx^1}{dt}\frac{dx^2}{dt} + \Gamma^1_{13}\frac{dx^1}{dt}\frac{dx^3}{dt} + \Gamma^1_{21}\frac{dx^2}{dt}\frac{dx^1}{dt} + \Gamma^1_{22}\frac{dx^2}{dt}\frac{dx^2}{dt} + \Gamma^1_{23}\frac{dx^2}{dt}\frac{dx^3}{dt} + \Gamma^1_{31}\frac{dx^3}{dt}\frac{dx^1}{dt} + \Gamma^1_{32}\frac{dx^3}{dt}\frac{dx^2}{dt} + \Gamma^1_{33}\frac{dx^3}{dt}\frac{dx^3}{dt} = \frac{d^2u}{dt^2} - \frac{2rsinv}{b+rcosv}\frac{du}{dt}\frac{dv}{dt} + \frac{2cosv}{b+rcosv}\frac{du}{dt}\frac{dr}{dt} = a^u \tag{54}$$

$$\frac{d^2x^2}{dt^2} + \Gamma^2_{\alpha\beta}\frac{dx^\alpha}{dt}\frac{dx^\beta}{dt} = \frac{d^2x^1}{dt^2} + \Gamma^2_{11}\frac{dx^1}{dt}\frac{dx^1}{dt} + \Gamma^2_{12}\frac{dx^1}{dt}\frac{dx^2}{dt} + \Gamma^2_{13}\frac{dx^1}{dt}\frac{dx^3}{dt} + \Gamma^2_{21}\frac{dx^2}{dt}\frac{dx^1}{dt} + \Gamma^2_{22}\frac{dx^2}{dt}\frac{dx^2}{dt} + \Gamma^2_{23}\frac{dx^2}{dt}\frac{dx^3}{dt} + \Gamma^2_{31}\frac{dx^3}{dt}\frac{dx^1}{dt} + \Gamma^2_{32}\frac{dx^3}{dt}\frac{dx^2}{dt} + \Gamma^2_{33}\frac{dx^3}{dt}\frac{dx^3}{dt} = \frac{d^2v}{dt^2} + \frac{1}{r}sinv(b+rcosv)\frac{du}{dt}\frac{du}{dt} + \frac{2}{r}\frac{dv}{dt}\frac{dr}{dt} = a^v \tag{55}$$

$$\frac{d^2x^3}{dt^2} + \Gamma^3_{\alpha\beta}\frac{dx^\alpha}{dt}\frac{dx^\beta}{dt} = \frac{d^2r}{dt^2} + \cos v\,(b+rcosv)\frac{du}{dt}\frac{du}{dt} - r\frac{dv}{dt}\frac{dv}{dt} = a^r \tag{56}$$

Consider The yarn is assumed to be circular in cross section and the radius of yarn is constant $a$. The fiber is rotated with angular speed $\dot{v} = n_{spindle}$ in unit rpm and moves parallel to the yarn length by speed $\dot{u} = v_f = n_b R$ as the speed of the delivery, hence the metric element can be determined as

$$g_{mn} = \begin{pmatrix} g_{11} & g_{12} \\ g_{21} & g_{22} \end{pmatrix} = \begin{pmatrix} (b+acosv)^2 & 0 \\ 0 & a^2 \end{pmatrix} \tag{57}$$

The geodesic equation for metric element in Equation (57) of torus for $u$ and $v$can be obtained as

$$\ddot{u} - \frac{2asinv}{b+acosv}\dot{u}\dot{v} = 0 \tag{58}$$

$$\ddot{v} + \frac{1}{a}sinv(b+acosv)\dot{u}^2 = 0 \tag{59}$$

**Table 1**- Simulation of fibre movement inside yarn.

| | |
|---|---|
| Influence of angular velocity on fibre inside a yarn $\dot{v}$ =*10* $Twist = 3$ | 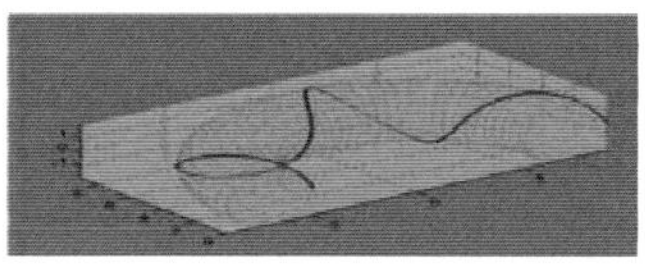 |
| Influence of angular velocity on fibre inside a yarn $\dot{v}$ =*20* $Twist = 5$ |  |

Using Equation (58) and Equation (59) we can establish the relationship of rotor angular speed and yarn count number. Putra, Rosyid, &Maruto [11] explained that the higher the yarn count number, the higher is the yarn twist. Putra, Rosyid, &Maruto [7] also said that the higher the yarn twist, the smaller is the strength of yarn. The prediction of fibre movement related by the yarn twist has been simulated and is shown in Table 1 . The simulation of fibre movement inside the OE yarn on torus coordinate was carried out using Equation (58), Equation (59) and the computer program. To determine the form of non-linear equations can be used the form of multivariate regression equation which is a form of static model with variable input that has more than one value, such as the relationship between the speed of delivery and the yarn count to the yarn strength. Such as modeled as follows:

$$\sum_{i=1}^{n} \hat{y}_i = a_o + a_1 \sum x_{i1} + a_2 \sum x_{i2} + \cdots + a_k \sum x_{ik}$$
$$\hat{y}_1 = a_o + a_1x_{11} + a_2x_{12} + +a_3x_{13} + \cdots + +a_kx_{1k}$$
$$\hat{y}_2 = a_o + a_1x_{21} + a_2x_{22} + +a_3x_{23} + \cdots + +a_kx_{2k}$$
$$\hat{y}_n = a_o + a_1x_{n1} + a_2x_{n2} + +a_3x_{n3} + \cdots + +a_kx_{nk} \tag{60}$$

$$\begin{pmatrix} \hat{y}_1 \\ \vdots \\ \hat{y}_n \end{pmatrix} = \begin{pmatrix} 1 & \dots & x_{1k} \\ 1 & \ddots & \vdots \\ 1 & \dots & x_{nk} \end{pmatrix} \begin{pmatrix} a_0 \\ \vdots \\ a_k \end{pmatrix} \tag{61}$$

$$\hat{y}_i = x_{ik}a_k \tag{62.a}$$
$$\hat{y} = Xa \tag{62.b}$$

The difference between experimental data and predictive modeling data is referred to as error $\epsilon$ and has a value of

$$\sum_{i=1}^{n} (y_i - \hat{y}_i) = \epsilon \tag{63}$$

If Equation (63) is squared, it will produce Equation (64) as shown below

$$\epsilon^T \epsilon = L \tag{64}$$

To find the value of $a$, optimization can be done through the squared differential Equation (56) with respect to $a$

$$\frac{dL}{da} = \frac{d}{da}(y^T y - y^T Xa - (y^T Xa)^T + a^T X^T Xa) = -2y^T X + 2a^T X^T X = 0 \tag{65}$$

$$2y^T X = 2a^T X^T X \tag{66}$$

$$a^T X^T X = y^T X \text{ with} a^T = (X^T X)^{-1} y^T X$$

$$(a^T X^T X)^T = (y^T X)^T \tag{67}$$
$$(X^T X)a = X^T y \tag{68}$$
$$a = (X^T X)^{-1} X^T y \tag{69}$$
$$\hat{y} = Xa = X(X^T X)^{-1} X^T y \tag{70}$$

Based on Equation (70) above with the matrix X weobtained Equation (71) below

$$T_w = a_o {N_e}^{a_1} {v_f}^{a_2} \tag{71}$$
$$\ln T_w = \ln a_o + a_1 lnN_e + a_1 lnv_f \tag{72.a}$$
$$Y = A_o + A_1X_1 + A_2X_2 \tag{72.b}$$

$$\begin{pmatrix} \hat{y}_1 \\ \hat{y}_2 \\ \hat{y}_3 \end{pmatrix} = \begin{pmatrix} 1 & \ln(40{,}1) & \ln(70.7) \\ 1 & \ln(50{,}1) & \ln(63{,}0) \\ 1 & \ln(60{,}1) & ln(57{,}4) \end{pmatrix} \begin{pmatrix} a_0 \\ a_1 \\ a_2 \end{pmatrix} \tag{73}$$

by using MATLAB programming, we get Equation (74) with $a_o = e^{Ao}$ and $a_1 = A_1$, $a_2 = A_2$

$$T_w = 451{,}73\, {v_f}^{0{,}5} {N_e}^{-0{,}8} \approx 451{,}73\, {v_f}^{0{,}5} {N_e}^{-1} \tag{74}$$

The results of theoretical predictions and experiments on the yarn strength can be shown as follows (Table 2):

**Table 2.** Multivariate Regression Modeling Results

| ***Ne*** *(hank/lbs)* | ***$V_f$*** ***(yd/min)*** | ***F experiment (cN)*** | ***F calculation (cN)*** |
|---|---|---|---|
| 40 | 70,7 | 211 | 199 |
| 50 | 63 | 169 | 157 |
| 60 | 57,4 | 140 | 129 |

The three-dimensional curve display results of the relationship between the speed of delivery and the yarn count to the yarn strength in the MATLAB software can be shown as follows (**Figure6**)

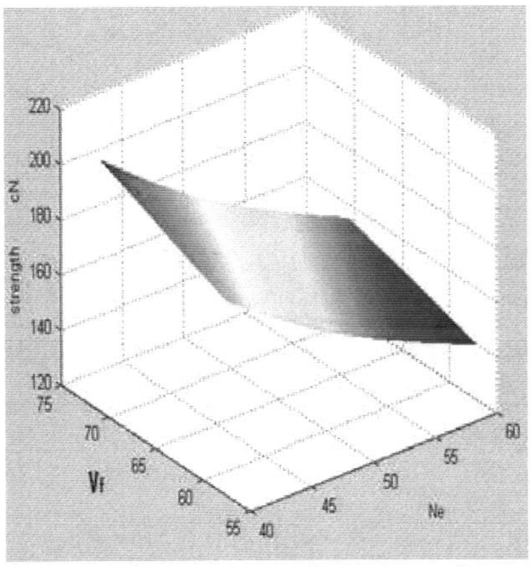


**Figure6.**Results of the yarn strength curve as a function of yarn number and introductory speed

## 3. Result and Discussion

Based on the results of calculations and experimental validation, the relationship of yarn strength as a delivery speed function and yarn count using classical mechanics can be formulated in Equations (75) and Equations (76) and the results of validation can be shown in Table 2. as follows:

$$T_w = \left(\frac{500\, e^{(2\mu\Delta\theta)}}{(e^{2\mu\Delta\theta} - 1)}\right) v_f{}^2 N_m{}^{-1} \quad (75)$$

$$T_w = konst\ v_f{}^2 N_m{}^{-1} = a\ v_f{}^2 N_e{}^{-1} \quad (76)$$

**Table 2**. Results of Classical Mechanics Modeling

| Ne (hank/lbs) | $V_f$ (yd/min) | F experiment (cN) | $F_{prediction}$ (cN) |
|---|---|---|---|
| 40 | 70,7 | 211 | 210 |
| 50 | 63 | 169 | 135 |
| 60 | 57,4 | 140 | 98 |

The results of modeling using classical mechanics show a fairly good value. Multivariate statistical modeling can be used to get the relationship between the yarnstrength to the delivery speed function and the yarn count, namely:

$$T_w = 451{,}73\ v_f{}^{0{,}5} N_e{}^{-0{,}8} \approx 451{,}73\ v_f{}^{0{,}5} N_e{}^{-1} \quad (77)$$

Based on Equation (75) and Equation (77) it can be concluded that the magnitude of the yarn delivery speed and the yarn count affect the strength of the yarn. The greater the yarn count, $N_e$, the yarn strength decreases, while the greater the yarn delivery speed, the greater is the yarn strength. In this study we also found a relationship between the yarn strength and the coefficient of yarn friction on the lappet guide and also the traveler is as in Equations (78) and Equation (79) as follows:

$$T_s = T_o e^{\mu\Delta\theta} \tag{78}$$

and also

$$T_w = T_o e^{(\mu\Delta\theta+\mu_t\Delta\sigma)} \tag{79}$$

The fiber movement inside yarn in traveler area can be written as

$$\ddot{u} - \frac{2asinv}{b + acosv}\dot{u}\dot{v} = 0 \tag{80}$$

$$\ddot{v} + \frac{1}{a}sinv(b + acosv)\dot{u}^2 = 0 \tag{81}$$

## 4. Conclusion

In this study the yarn strength equation has been successfully modeled as a function of the delivery speed and also the yarn count in the Ring-spinning process. Based on the results of calculations and experimental validation, the relationship of yarn strength as a delivery speed function and yarn count using classical mechanics can be formulated in Equations (82) and Equations (83) as follows:

$$T_w = \left(\frac{500\, e^{(2\mu\Delta\theta)}}{(e^{2\mu\Delta\theta} - 1)}\right) {v_f}^2 {N_m}^{-1} \tag{82}$$

$$T_w = 451{,}73\, {v_f}^{0{,}5} {N_e}^{-1} \tag{83}$$

The fiber movement inside yarn in traveler area can be written as

$$\ddot{u} - \frac{2asinv}{b + acosv}\dot{u}\dot{v} = 0 \tag{84}$$

$$\ddot{v} + \frac{1}{a}sinv(b + acosv)\dot{u}^2 = 0 \tag{85}$$

By considerate the fibre movement inside of yarn, the relationship among yarn strength, yarn count number, yarn speed and yarn twist was established and validated by experimental

## Acknowledgement

The author would like to thank the Department of Textile Engineering, STTT Polytechnic Bandung for supporting this research activity so that this research can be completed properly.

## References

[1] Andriansyah, 2017, *Experimental Modeling and Validation for Determining Actual Twist Type Z for Cotton Yarn Quality on a Ring Spinning Machine*, *Thesis* (Bandung : Politechnic STTT Bandung)
[2] Hearle, J.W.S. and Gupta, B.S., 1965, *Textile Research Journal*, **35** pp 788-795.
[3] Herawati RM Fauzi I. and Putra V G V., 2015, *Applied Mechanics and Materials,* **780**, pp. 69-74.
[4] Hongwei Zhang, Xuefeng and Yonglan , 2011, *FIBRES & TEXTILES JOURNAL,* **19,** pp. 17-20.
[5] Lawrence, CA 2003, *Fundamentals of Spun Yarn Technology*, (New York : CRC Press)
[6] Putra VGV, Dewanto A, Totong, 2016, *Global Journal of Pure and Applied Mathematics* **12**. pp. 4107-4114.
[7] Putra VGV, Maruto, G & Rosyid, M.F.,2017, *Indian Journal of Fibre and Textile Research,* **42**. pp. 359-363.
[8] Rohlena, V,1975, *Open-End Spinning*, (New York : Elseiver Scientific Publishing Company)

[9] Trommer, G., 1995, *Rotor Spinning*, (Frankfurt.: Deutscherfachverlag, )
[10] Zeidman, Shawney and Herington, 2003, *Indian Journal of Fibre and Textile Research*, **28**., pp. 123-133.
[11] Putra, V G V,Rosyid, M F. &Maruto, G.,2016, *Global Journal of Pure and Applied Mathematics*, **12**. pp. 405-412.
[12] Backer, Hearle&Grosberg, 1969, *Structural Mechanics of Fibres, Yarns and Fabrics*, (New York : Wiley-Interscience).
[13] Furter R & Meier S., 2009, *Measurement and Significance of Yarn Twist* (Switzerland : Uster Technology A).
[14] Putra, VGV, Wijayono, A, Martina, T, Rosyidan, C., 2019, *Wahana Fisika*, **4**, pp. 27-34.
[15] Wijayono, A., Putra VGV., 2018, *Arena Tekstil*, **33**, pp 85-92.

# Speed Control of Three Phase Induction Motor Using Universal Bridge and PID Controller

**H Hartono, R I Sudjoko*, P Iswahyudi**

Aviation Polytechnic of Surabaya, Indonesia

*rifdian.anto@gmail.com

**Abstract**. Speed control technique are generally essential in adjustable speed drive system. This system requires variable voltage and frequency supply which is obtained from a three phase voltage source inverter. This paper presents the speed control of induction motor fed by a three phase voltage source inverter using pulse width modulation method and universal bridge. To control the peak dc link voltage of voltage source inverter, a PID controller was designed. The performance analysis of open loop and closed loop speed control system is carried out in Matlab/Simulink. The simulation result shows that the speed controller has a good dynamic response and can control the induction motor successfully with a better performance.
**Keywords** : three phase induction motor, speed control, voltage source inverter, universal bridge, PID controller

## 1. Introduction

At present, asynchronous motors or induction motors are used in extensive industrial drives because of their simple, inexpensive or inexpensive engine construction and excellent reliability. This engine control is difficult because of the high motor complexity. Several control strategies have been presented to control motor engines. This technique includes PWM speed control. This control has a rapid expansion with the growth of power electronics [1]. They have succeeded in the application of inverter components which made it increasingly popular. In addition, PWM control, in particular, is regretful as one of the well-known techniques for controlling the high complexity of three-phase induction / asynchronous motor systems. This technique is done by adjusting the pulse width and the pulse duty ratio to set the average voltage. PWM technology is accompanied by the development of electronic power devices, has a good development and is now mature. By using the PWM technique, the waveform of the output signal of the inverter can be improved to minimize harmonic and ripple output torque. This technique reduces the design of the inverter, to speed up the level of adjustment and increase the dynamic response of the system. In the field of electric propulsion, the motor is very important to complete the variable speed motor [2].

### *1.1. Working principles of three phase induction motor*

The working principle of a three phase induction/asynchronous motor is slightly different from a single phase. When the stator holds a three-phase winding that is moved in space by 120 degrees, when a three-phase supply is fed to the stator winding, a rotating magnetic flux (rotating at synchronous speed) is created in the stator. [3]

Three-phase stators and rotors are considered as two basic parts of a three-phase AC induction motor. When the stator phase is energized by a three-phase AC power source, current flow is generated in the stator. The magnetic field synthesized by three-phase stator currents always rotates nonstop with variations in current. This rotating magnetic field cuts the rotor and the current produced in it interacts with the rotating magnetic field and thus produces a magnetic torque that makes the rotor spin (Figure 1). The rotational speed of the rotor n must be less than the rotating magnetic field n0. Reverse rotation of the rotor will be realized by the interchangeable three-phase resource position.

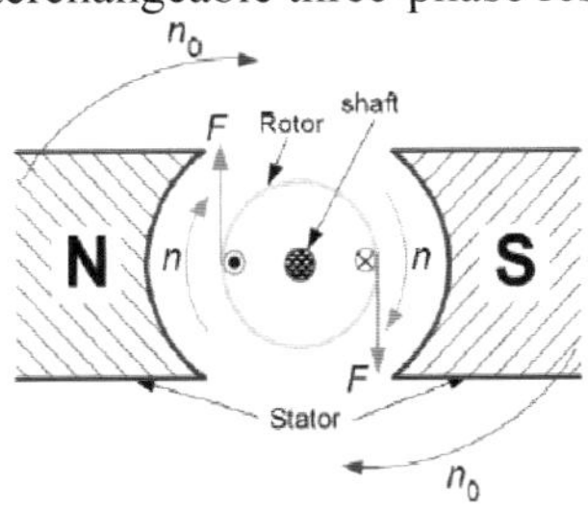


**Figure 1.** Rotating magnetic flux of asynchronous motor

When the filed is totating, the direction is the same with phase current and speed value is proportional to the frequency and inverse;y wiyh the polar pair number. Calculated per minute, the speed of the rotating magnetic field $n_0$ can be represented by this equation:[4]

$$n_0 = \frac{120\,x\,f}{P} \tag{1}$$

Where $n_0$ = synchronous rotational speed in rev/min
f = power supply frequency in Hz
p = number of motor poles

The speed which the stator flux rotate is called synchonous speed, depend on the number of poles of the motor and the power supply frequency. In the practical field, the speed value of rotor will be slower than synchronous speed. Induction motors are also called as asynchronous motors because the values of rotor speed is different with stator flux. The slip is differences between rotor speed and rotational of stator flux. The value of slip varies between 1% to about 6% of stator flux speed.

The value of slip is,

$$slip = s = \frac{(n_0 - n)}{n} \text{ p.u} \tag{2}$$

And the value of rotor speed is,

$$n = n_0(1 - s)\,rpm \tag{3}$$

Where $n_0$ = the speed of stator flux in rpm
n = The value of rotor speed in rpm
s = the value of slip in pu

When the AC power supplied the motor through stator windings,the stator flux rotate the same direction with the the supply .[5]

### *1.2. Induction motor drives*

Induction motors or asynchronous motor have two types of windings. They are stator and rotor winding. These winding are designed for three phase system. Induction motor or asynchronous motor usually operaated in variable speed drives. Three phase AC voltage supplied stator winding with balanced voltage. The stator induced the rotor with magnetic flux as transformer principal. The value of rpm and torque of induction motors/asynchronous motor can be adjusted by :

- Controlling the stator voltage
- Controlling the rotor voltage
- Controlling the frequency
- Controlling stator voltage and frequency
- Controlling the stator current

• Controlling the voltage, current and frequency

To get the rpm and torque, controlling the duty cycle, voltage, current and frequency are very important[5].

The purpose of this paper are controlling the rpm of an induction motor/asynchronous motor by adjusting the stator voltage. This method can be achieved by using sinusoidal pulse width modulation technique method through an universal bridge component and PID controller.

*1.3. Three phase dc to ac converter (inverter)*

In high power application, Three phase dc to ac converter (inverter) are widely used for adjustable frequency drive applications. A basic priciple of three phase dc to ac converter containing of three single phase switches which connected to one of the three load terminal ports..

Inverter is a circuit that is used to convert a DC voltage source into an AC voltage source. The power semiconductor components used can be in the form of SCRs, transistors, and MOSFETs that operate as switches and converters. Three phase inverter can be shown in fig.6

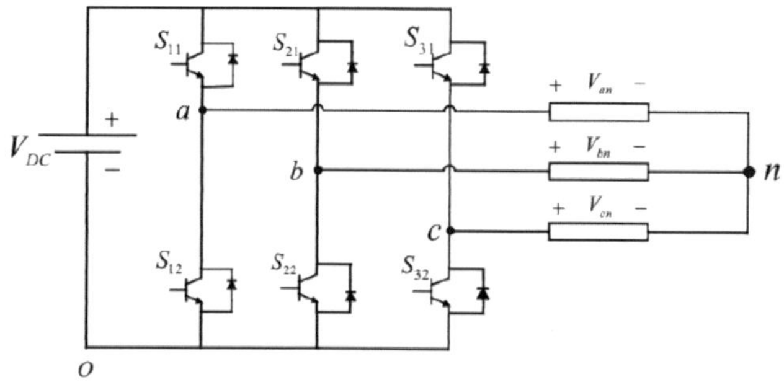


**Figure 2.** Three phase dc to ac converter cicruit

Judging from the conversion process, inverters can be divided into three types, namely inverters: series, parallel, and bridge. Bridge inverters can be divided into halfwave bridge inverters and fullwave bridges. The resulting output voltage can be in the form of one phase or three phases.[6]

*1.4. PID controller*

PID control systems are the most widely used in control systems industry. The success of the PID controller depends on its accuracy in determining PID constant (reinforcement). Practically the determination process PID constants are based on human expertise based on rules called rules of thumb. If the right result has been obtained, then this PID constant used for further control. This of course has weaknesses because this constant is the same for every error value that occurs and requires tuning reset if there are changes in plant parameters in the PID constant. For To overcome this, a method is needed to determine the PID constant exactly according to the plant. It is hoped that the performance of the PID control can be improved. In Figure 7 shows the priciple of PID controller.[6]

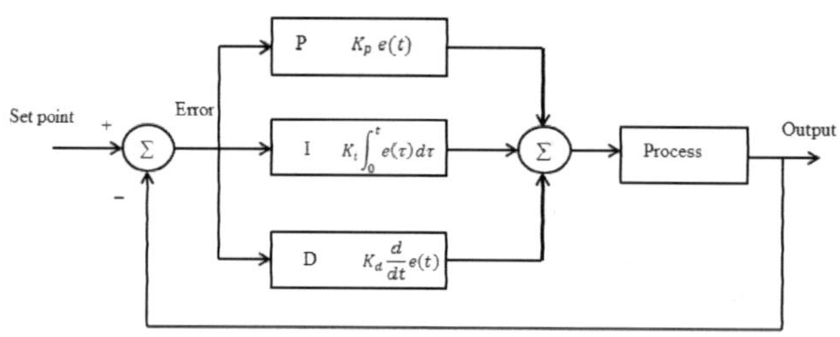


**Figure 3.** Diagram of PID control circuit [6]

Journal of Physics: Conference Series **1381** (2019) 012053 doi:10.1088/1742-6596/1381/1/012053

Common characteristics used in the regulation of an intermediate system others include stability, accuracy, response speed and sensitivity. In action proportional control, the output of the control system will be proportional to the input. The output signal is a reinforcement of the error signal with certain factors, This reinforcement factor is a proportional constant of the system, which is expressed by Kp, where Kp has a high / fast response. In integral control, the output always changes during deviations, and the speed at which the output changes is proportional to the deviation, a constant expressed as Ki, where Ki has a sensitivity which is high, i.e. by reducing the error generated from the feedback signal. The greater the value of Ki, the higher the sensitivity, but time needed to achieve stability more quickly, likewise the opposite. Whereas derivative control works by rate deviation changes, so this type of control is always used together with proportional and integral controllers. Constants expressed in Kd, where Kd is affecting the stability of the system, because of this control action able to reduce errors. By combining these PID control actions then expected to get a response that has a level of stability tall one.[6]

## 2. Methodology
There are tools which will be used for the effective implementation of this project, they include Matlab tool, Simulink tool and Simpowersystem tool.

### *2.1. Asynchronous machine/induction machine (squirrel cage)*
Induction/asynchronous machine block represent of 3 phase asynchronous engine such as wound rotor machine, squirrel cage machine. This machine will have two function, they are motor or generator.. These function is determined by the value of torque from the machine :

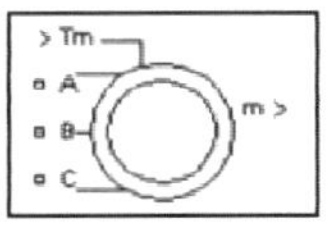


- the machine will acts as a motor when Tm value is positive
- the machine will acts as a generator when Tm value is negative

The instrument value of the machine component is measured in table .

**Table 1.** Asynchronous/Induction motor paramater

| | |
|---|---|
| Nominal power, voltage (line-line), frequency | 746 VA, 380 V, 50 Hz |
| Stator resistance & inductance | 0.009961 Ω , 0.000867 H |
| Rotor resistance & inductance | 0.005837 Ω , 0.000867 H |
| Mutual imductance | 0.03039 H |
| Inertia, friction factor, pole pairs | 0.4 , 0.02187 , 2 |

### *2.2. Universal bridge*

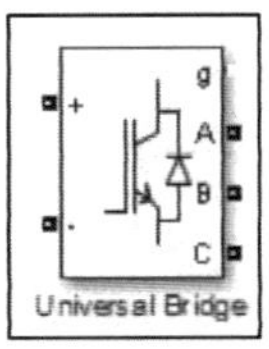


The universal bridge block component represents an universal voltage converter for 3 phase power system. This block contaian six switches which connected in a bridge arrangements. The universal bridge can be power electronic devices component [7]

The measurement value of this universal bridge can be described in table 2.

**Table 2.** Universal bridge componen block measurement value

| | |
|---|---|
| Power electronics devices | IGBT/Diodes |
| Number of bridge arms | 3 |
| Snubber resistance | 1e5 Ω |
| Snubber capacitance | inf |
| Ron | 1e-3 Ω |
| Tf , Tt | 1e-6 s, 1e-6 s |
| | |

*2.3. Pulse Width Generator*

Pulse Width Modulation (PWM) in general is a way of manipulating the signal width expressed by pulses in a period, to obtain a different average voltage. Some examples of PWM applications are data modulation for telecommunications, controlling power or voltage entering the load, voltage regulators, audio effects and amplification, as well as other applications. The simplest PWM signal generation is by comparing the sawtooth signal as a carrier voltage with reference voltage using an op-amp comparator circuit. The workings of this analog comparator are comparing sawtooth voltage waves with reference voltages. when the reference voltage is greater than the carrier voltage (sawtooth), the comparator output will be high. But when the reference voltage is less than the carrier voltage, the comparator output will be low. By utilizing the working principle of this comparator, to change the duty cycle of the output signal is enough to vary the reference voltage. [7]

*2.4. System overview*

The proposed circuit of the system had been designed and simulated using Matlab/Simpowersystem. Fig. 8 and Fig.9 shows the simulink circuit of the proposed model.

*2.4.1. Open loop system*

The open loop system consist of four main component. They are DC voltage supply, universal bridge, PWM generator and Asynchronous machine block. In open loop system, there are no feed back to control the speed. So the speed depend on the value of dc voltage supply. The open loop control system is traditionally control that used in several induction motor. Thi control is simple because contain a few component to implemented.(fig 8).

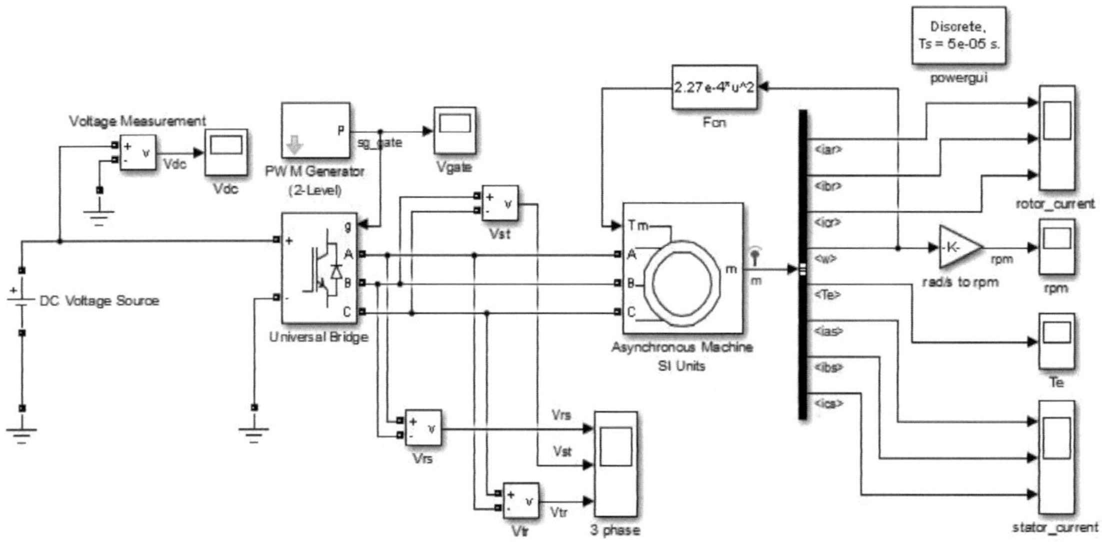


**Figure 4.** Simulink model of open loop system

*2.4.2. Closed loop system*

The improvement of open loop is closed loop system. The closed loop system consist of five main component. The component are contolled voltage source, PID controller block, universal bridge block, PWM generator block and asynchronous motor machine block.The circuit diagram of closed loop system shows in fig.8.

In closed loop system, speed sensor used as feed back signal to set point reference. The error is the differences between actual speed and rpm set point. The PID block process this error as signal control to controlled voltage sources.

Controlled voltage source block produce the varied dc voltage output as the input signal for universal bridge.

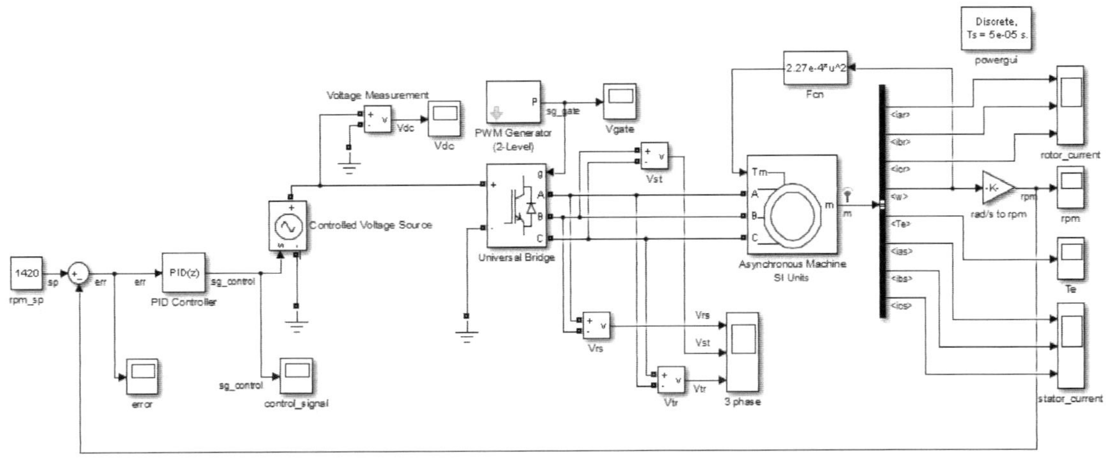


**Figure 5.** Simulink model of closed loop system

The universal bridge convert dc voltage to ac voltage from controlled voltage source. This bridge generate 3 phase voltage to supply asynchronous motor. The speed of asynchronous motor determined by the value of 3 phase voltage and frequency of the output universal bridge. This speed is adjusting to get the nearest value based on set poin rpm speed.

## 3. Results & Discussion

Simulation results are obatained from Matlab/Simulink software tools. The results of line voltage, phase voltage, current in the line are listed along with speed and torque results of induction motor in the following figures.

### *3.1 Results of open loop system*

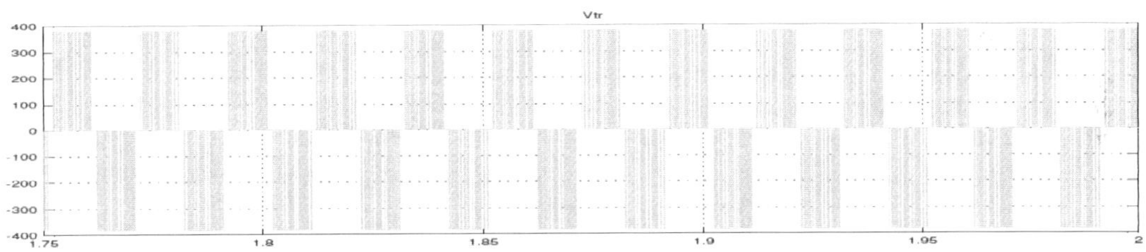


**Figure 6.** Line voltage output of universal bridge

From fig 6, the voltage value of inverter output are 380 Vac. Induction motor use this voltage for operation according to the nominal voltage rating.

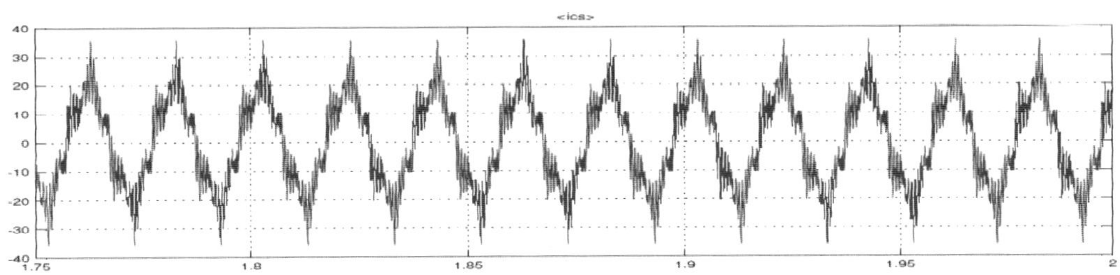


**Figure 7.** Stator current of induction motor

Induction motor produce 35 A for stator current. This current is equal to every phase because induction motor are a balanced load.

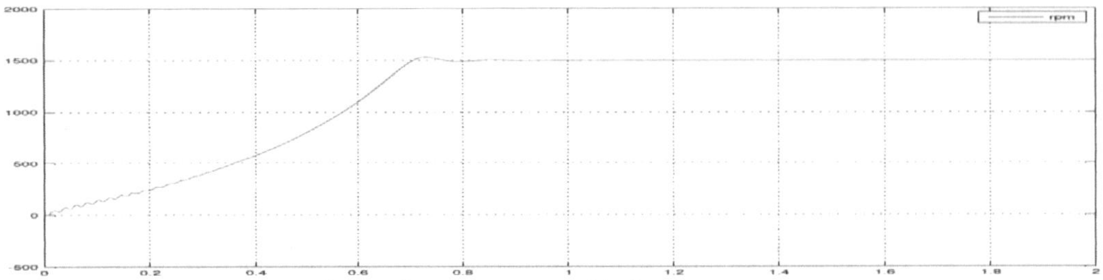


**Figure 8.** Rpm output of induction motor

From fig 8, the steady state speed of induction motor are 1500 rpm. The respons of open loop system can be shown from this figure. Open loop system respons are rise time (tr) = 0.7 s, delay time (td) = 0.55 s, peak time (tp) = 0.73 s and steady state time (ts) = 0.9 s.

*3.2 Results of closed loop system*

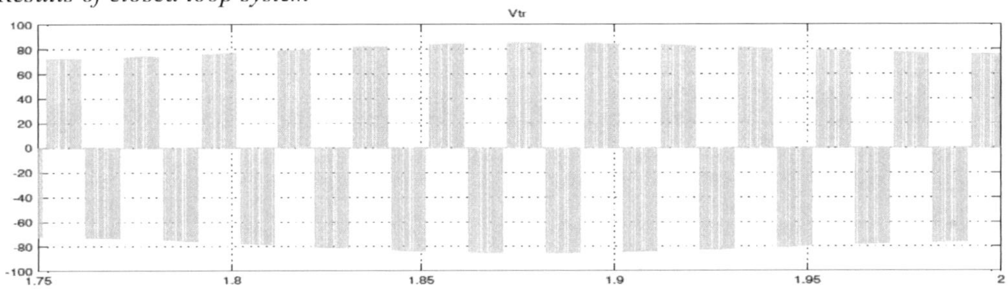


**Figure 9.** Line voltage output of universal bridge

From fig.9, the voltage value of inverter output are 85 Vac. Induction motor use this voltage for operation according to the nominal voltage rating.

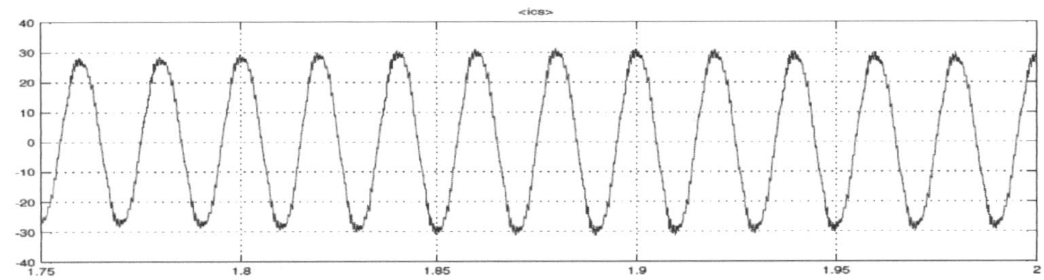


**Figure 10.** Stator current of induction motor

Induction motor produce 30 A for stator current. This current is equal to every phase because induction motor are a balanced load.

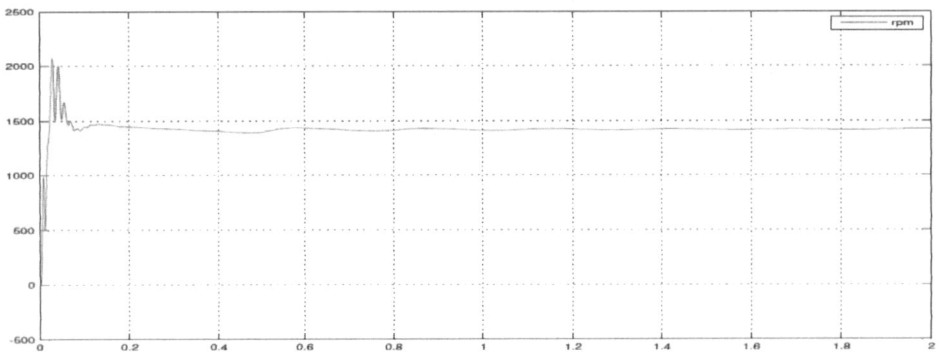


**Figure 11.** Rpm output of induction motor

From fig.11, the steady state speed of induction motor are 1420 rpm. The respons of closed loop system can be shown from this figure. Closed loop system respons are rise time (tr) = 0.02 s, delay time (td) = 0.015 s, peak time (tp) = 0.025 s and steady state time (ts) = 0.2 s.

## 4. Conclusion

Simulations have been carried out with Matlab / Simpowersystem. From the simulation results, it is obtained that the control of the induction motor using the PID controller and universal bridge produces a better response than without the PID control. In the open loop system, the time to get steady state speed is 0.9 seconds while in the closed loop system, the time to get steady state speed is 0.2 seconds. This shows that the PID controller obtained a faster response than without a controller. After all simulation we can conclude that the proposed system has a good ability to control the speed of the induction motor.

## References

[1] Bose Bimal K. 2001 Bimal K. Bose-Modern power electronics and AC drives -Prentice Hall PTR (2002).pdf
[2] Pwm T 2015 The PWM Control of the Three-phase Induction Motor Ping Wei, jinpeng Yu, Fatao Shi, Xiao Wei, Yan Wang, Quanwen Zhao 842–5
[3] Jatin A. Patel A P H H R 2015 Design of Sinusoidal Pulse Width Modulation Inverter *Int. J. Technol. Res. Eng.* **2** 1446–52
[4] H. Yakkundi P, Devadas K V and Suryasen K 2014 Design and Simulation of Space Vector PWM for Three-Phase Induction Motor *IOSR J. Electr. Electron. Eng.* **9** 01–8
[5] Emojong E E 2009 *Design of Inverter Drive for Ac Induction Motor*
[6] Yusoff S B M 2014 Pid Current Control Technique for Three Phase Induction
[7] Hingmire P K, Kumar S and Me R 2016 Development of a V / f Control scheme for controlling the Induction motor- both Open Loop and Closed Loop using MATLAB . *Int. J. Sci. Eng. Appl. Sci.* 403–7

Journal of Physics: Conference Series **1381** (2019) 012054 doi:10.1088/1742-6596/1381/1/012054

# Measurement of Wideband Indoor Radio Propagation Channel Using USRP

**B.B Harianto[1,2*], M. Ardiansyah[1], G. Hendrantoro[1], A. Mauludiyanto[1]**

[1]Institut Teknologi Sepuluh Nopember, Surabaya, Indonesia
[2]Aviation Polytechnic of Surabaya, Indonesia

bambangfarzardy@gmail.com*

*Abstract*—This paper describes the measurement of wideband radio channel sounding techniques for indoor radio propagation. The system consists of multiple frequency band-switching scheme, which is able to extend the overall sounding bandwidth without changing operation bandwidth of the hardware. It is used for low mobility and high data rate communications. The system of the channel sounding is developed on a universal software radio peripheral (USRP) that employs an orthogonal frequency division multiplexing signal at 2.190-2.290 GHz. In each sounding cycling, ten sub band frequency (10 MHz bandwidth) are used to reach an equivalent measurement over a 100-MHz bandwidth, corresponding to 10 ns time delay resolution. The validation test was perform using Vector Network Analyser (VNA). It will be showed variation result of channel impulse response consist of radio channel accuration and resolution between VNA and USRP. At the end of the chapter the proposed system will be used for measuring indoor channel propagation using five scenarios line of sight (LOS) and non-line of sight (NLOS).

## 1. Introduction

Wireless communication will increase rapidly because of people and device grow fast in the near future [1]. It will be need a 1000 times (1000x) system capacity for fifth generation (5G) wireless network to be achieved compared with the fourth generation (4G) network [2]. Wireless communication need propagation *channel parameters such as distance- path loss relation between transmitter and receiver, shadowing loss, small fading, power delay profile, delay spread, angular spread and soon are necessary in the design and implementation processes of the wireless network in any environment.* One of the approaches which can accurately determine the propagation channel is to conduct the measurement in a target environment by using channel sounder. The example of the simplest channel sounder is the system which consists of transmitter and receiver. To estimate the channel parameters, the transmitter (TX) part of channel sounder send the signal through the environment which consists of several scatters to the receiver (Rx). Then, the channel parameters can be estimated by performing the signal processing on the received and transmitted. The example of the channel information which can be easily obtained is the distance-path loss relation between transmitter and receiver by comparing the transmitted power with the received power versus their distance. In such case, the architecture of channel sounder is not complicated. However, to investigated the small fading, power delay profile and delay spread, the complicated synchronizations of time and frequency between transmitter and receiver are required. In case of indoor measurement, the vector network analyzer (VNA), which is the famous commercial equipment, usually utilized to be the channel sounder by many researchers e.g. [3] in order to obtain a precise synchronization between transmitter and receiver since both TX and Rx use the same local oscillator.

A more recent trend, there are several work which developed the channel sounder based on open-source software define radio platform called universal software define radio peripheral (USRP). The size of such channel sounder is small and it can be portable easier compared with the conventional channel sounder. Such characteristic is suitable for various indoor measurement. The hardware part of such open-source software define radio platform is usually connected to computer laptop, in which the data transfer between them is limited, resulting to the limited bandwidth of transmitter/receiver. For instance, by using the Nyquist sampling rate at two time of baseband signal, the maximum of bandwidth of channel sounder based on the famous low cost USRP-N210 model is 12.5 MHz . it means that such channel sounder can discriminate the delay time of signal (or delay resolution) only at $\frac{1}{12.5MHz} = 80\ ns$ which is not sufficient in some indoor measurement scenarios. Please note that the higher bandwidth can be obtained by using the higher price model such as USRP-X310/E310, but it requires more investment at least two time of low-cost USRP Models.

In this research, the portable channel is developed using LabVIEW software and the other platform of open source software define radio called universal software radio peripheral (USRP) which is much cheaper and smaller than conventional channel sounding model. This can be solved by adopting the multitoned-overlapping frequency stepping approach [4][5], the relative phase ($\Delta\emptyset$) between tones can be obtained and used to calculate the channel transfer function. Finally, the channel information can be extracted from the calculated channel transfer function. In this paper, by transmitting/receiving the two thousand and fifty-overlapping signal in each frequency stepping until to reach 2050 tones totally, the 100 MHz bandwidth of channel can be obtained. As for the synchronization, it is done by power detection algorithm.

The concept of multi tone over lapping frequency stepping approach and the synchronization are described in more detail in section 2. The validation result in laboratory environment is presented in section 2. The indoor channel propagation test conducted in laboratory are also presented in section 3. It shows the environments of the measurement site and how the proposed channel is used in the real measurement. Finally, the conclusion is made in section 4.

## 2. System Design and Calibration

System design and calibration need hardware setup and software design in order to make a wide band channel measurement. By using Vector Network Analyzer and LabVIEW the proposed system would be design.

### *2.1. Hardware Channel Sounding Desaign*

Measurements was arrenged using USRP N210, a type of software defined radio that able to sampling signals from dual ADC and DAC up to 100 MS/s and 400 MS/s. [6]. In this research, USRP N210 combine with SBX daughterboard that works at frequencies range from 400 to 4400 MHz. This daughterboard serves as a transceiver operate at the frequency of measurement proposed Its feature an output power up to 100 mW with a typical noise figure of 5 dB [7]. with a wide frequency band, the SBX daughterboard is suitable for any measurement research. The local oscillators for the receive and transmit can be arrange independently with dual operations [8].

Journal of Physics: Conference Series **1381** (2019) 012054 doi:10.1088/1742-6596/1381/1/012054

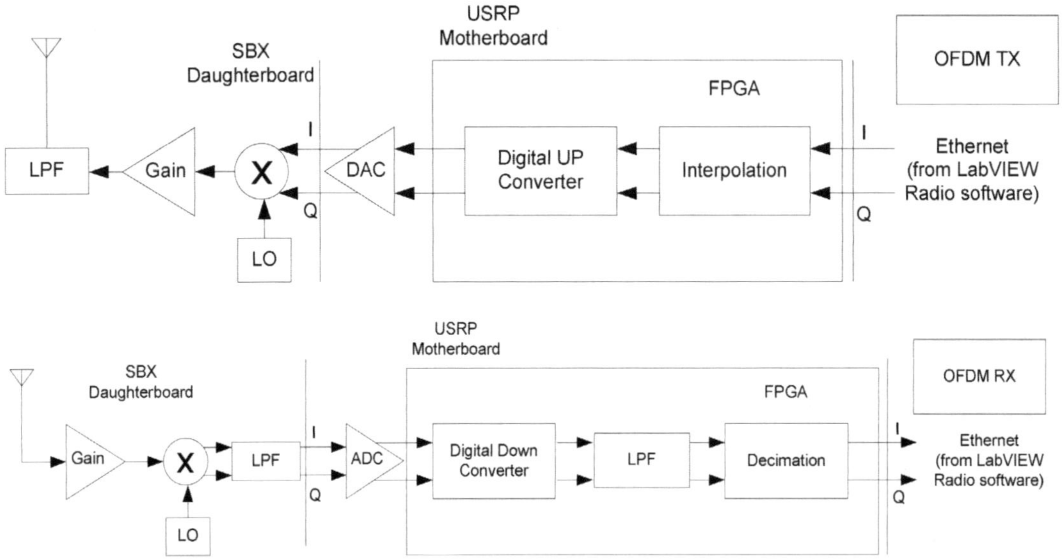


Figure 1. Blok diagram of USRP transmitter and receiver

### *2.2. Software Channel Sounding Design*

Wideband channel sounding using OFDM modulation to determine channel characteristics. In Figure 2 it can be explained that a number of 2050 bits are generated from data source, the output will be mapped and converted into symbol using 4QAM modulation with the same phase $X_1(f) = e^{j\pi/4}$ after passing serial to parallel zero padding will be added and modulated with 205 subcarriers. The output of zero padding is added with cyclic prefix and transmitted. The receiver symbol will be processed by eliminating cyclic prefix and zero padding to obtain the subcarrier value.

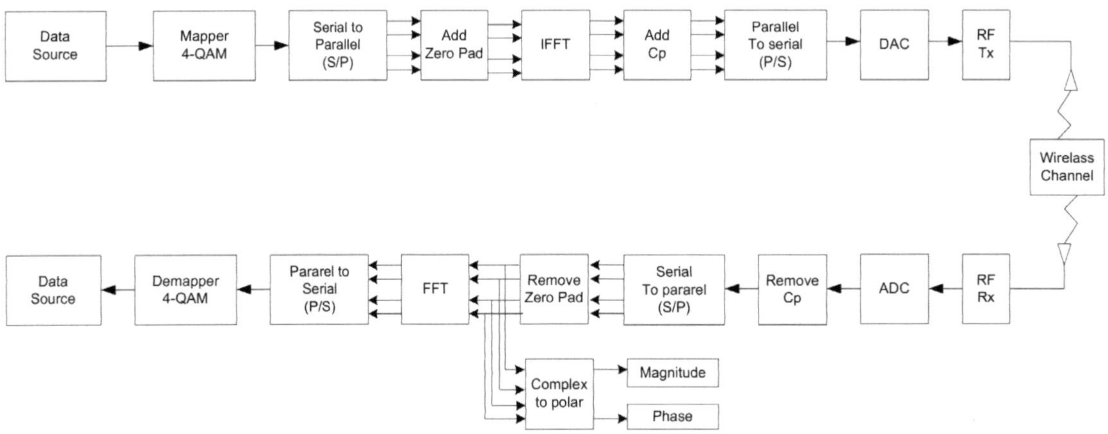


Figure 2. OFDM Software Architecture

measurement is done by sending a number of signals with 10 MHz and repeated ten times until reaching a bandwidth of 100 MHz the value of the sending from the subcarrier is 2050 after reaching the desired bandwidth. the output data is taken from the serial to parallel block after passing through the zero pad

Journal of Physics: Conference Series **1381** (2019) 012054 doi:10.1088/1742-6596/1381/1/012054

removal block. the data is still complex signal then separated into magnitude and phase. The raw data will be change into response frekwency using mathlab.

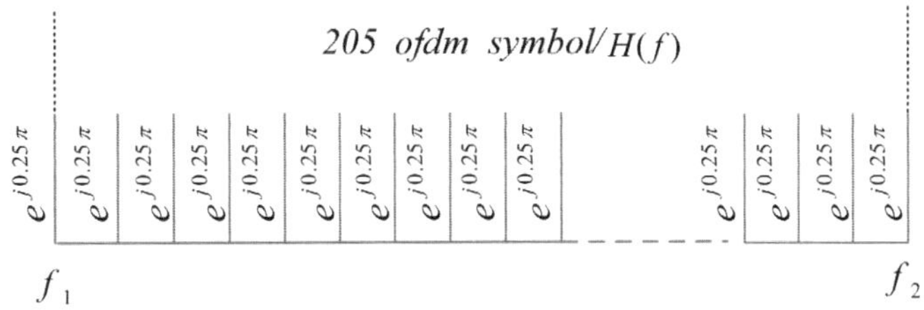


Figure 3. One frame of OFDM Symbol

Figure 3 shows an OFDM symbol consisting of 205 sub carrier frequencies with a phase difference of 45 degree. This OFDM symbol will be repeated 10 times when measure a wideband signal characteristic the desired number of subcarriers is adjusted using LabVIEW software before adding to the Zero Pad. the phase is made fixed with the aim of producing a stable frequency response.

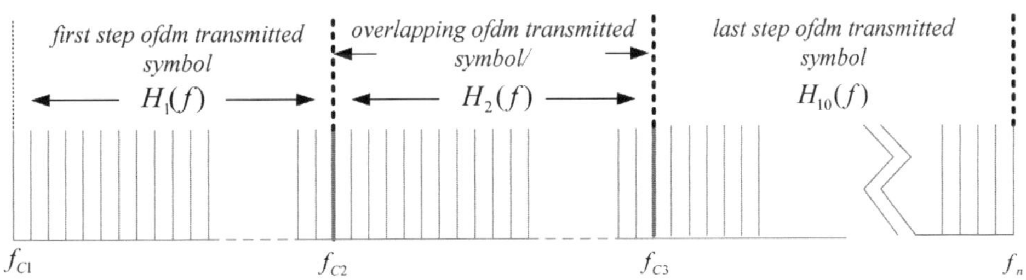


Figure 4. Multicarrier overlapping frequency sounding

Multiple overlapping frequency sounding is describe in the figure 4. All the signal will bw convert into, frequency response of the channel. By using *inverse fast Fourier transform* (IFFT) the frequency response convert into Channel Impulse Response $h[n] = h(nT_s)$, where $T_s = 1/(N\Delta f)$ is the sample period and $n$ is the discrete time index. Theoretically, the CIR $h(t)$is composed of infinite number of multipath component arrivals, which can be expressed by

$$h(\tau) = \sum_{i=1}^{\infty} a_i e^{j\emptyset_i \delta(\tau-\tau_i)} \quad (1)$$

Where,$\tau_1$ denotes the delay of the $i$th arrival in propagation between the transmitter and the receiver. The $a_i$ and $\emptyset_i \sim \upsilon[0,2\pi]$[11] represent, respectively, the amplitude and phase shift of the $i$th arrival. The multiband switching scheme, the receiver detects and store the probe signal frame and then switch to the next frequency band synchronously with the transmitter in a sounding cycle. In data post processing, synchronization is first performed to find the exact starting point of the received symbols. Then fast Fourier transform (FFT) is performed on both the transmitted and received symbols to acquire the $H_f$ on each frequency band. The channel transfer function of the sounder itself could be modeled as

$$H(f) = \frac{Y(f)}{X(f)} = A(f)e^{j\theta(f)} \quad (2)$$

Where $Y(f)$ is the received signal,$X(f)$ is transmit signal, $A(f)$ is amplitude response, $\theta(f)$ is the phase response, respectively

### 2.3. *Radio Channel Sounding Calibration*

To validate the performance of the proposed channel sounder, the measurement of radio channel sounding calibration as show in figure 5. The result of channel transfer function in term of amplitude (power in dB) and phase between two case compare between response channel and transmit signal [9]. Sounding calibration radio channel has a measurement distance of $l_1 = 3\ m$ and two cables that connect it to port1 and TX antennas, port 2 with Rx antenna cable length $l_2 = 4\ m$ and $l_3 = 3\ m$ . the radio channel that passes through the VNA using s12 parameter and the signal pass through $l_1, l_2, l_3$. by considering phase velocity of RF cable is relative to its material, i.e.,$v = \frac{1}{\sqrt{\mu\varepsilon}} = \frac{c}{\sqrt{\mu_r\varepsilon_r}}$, where c is the velocity of light in vacuum, $\varepsilon_r = \varepsilon/\varepsilon_0$ is the relative magnetic conductivity of the medium [9]. The dielectric of coaxial cable is polyethylene; whose nominal velocity ratio is 67% [10]. i.e., the phase velocity in cable is 0.67c. Therefore, the relative delay $(\tau)$ of the cable quite enough for influencing channel transfer function and should be compensated.

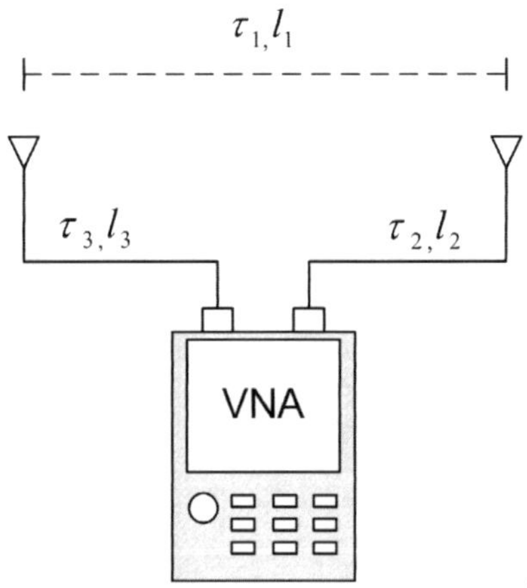


Figure 5. Radio channel sounding calibration

## 3. Indoor Channel Measurement

Measurements take place at the antenna and propagation laboratory Institut Teknologi Sepuluh Nopember Surabaya. The measurement is divided into five LOS and NLOS scenarios. The distance and delay resolution of each measurement are shown in table 1.

Tabel 1. Channel measurement distance and delay resolution calculation

| Tx | Rx | Delay (ns) |
|---|---|---|
| 1. | 3 m | 10 |
| 2. | 4.24 m | 14.13 |
| 3. | 3 m | 10 |
| 4. | 4 m | 13.33 |
| 5. | 5 m | 16.67 |

Indoor Channel measurement were performed using USRP N210 motherboard and SBX daughterboard. by using a frequency of 2,195 - 2,285 GHz with 10 MHz sub bandwidth, each step is consisting of 205 Sub carriers. Measurements using LOS propagation scenario and using the OFDM signal test, the mapper is used 4 QAM modulation with total delay resolution is 10 ns (100 MHz)

Journal of Physics: Conference Series **1381** (2019) 012054 doi:10.1088/1742-6596/1381/1/012054

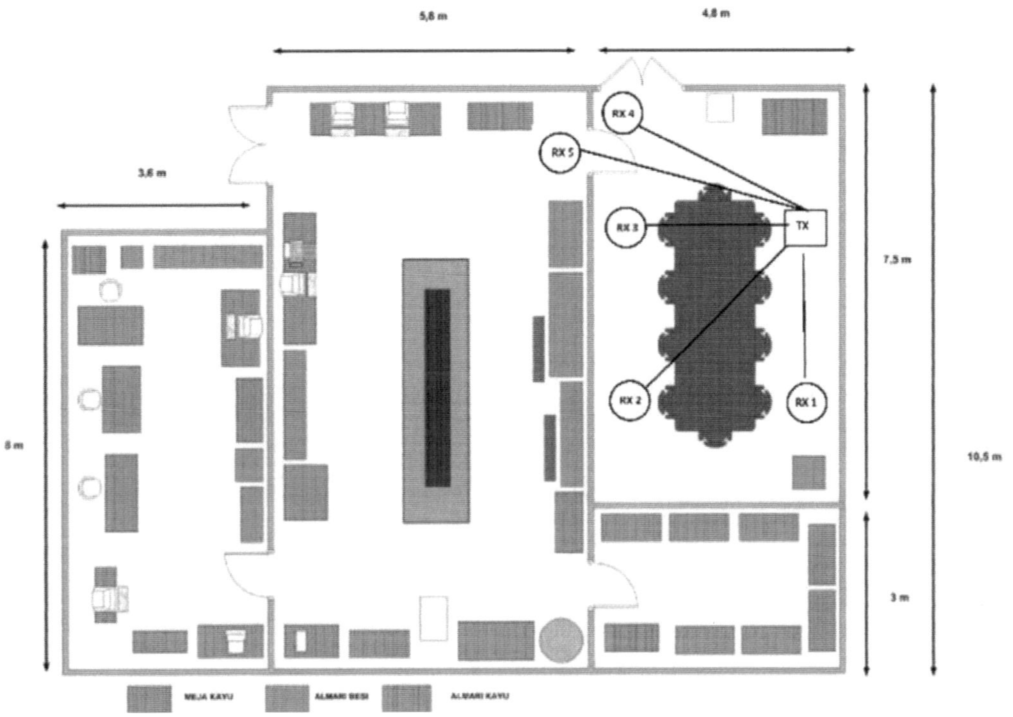


Figure 6. Floor plane of measurement site

Figure 7a-7e are measurement of USRP using data taken from the transmitter and receiver sides. from the measurement results we can find from the picture a channel impulse response in the form of magnitude and excess delay in ns. This data is used in data processing using one sample, namely the line of sight (LOS) condition. Figure 8a-8e are measuremen of USRP from the location setup non line of sight (NLOS)

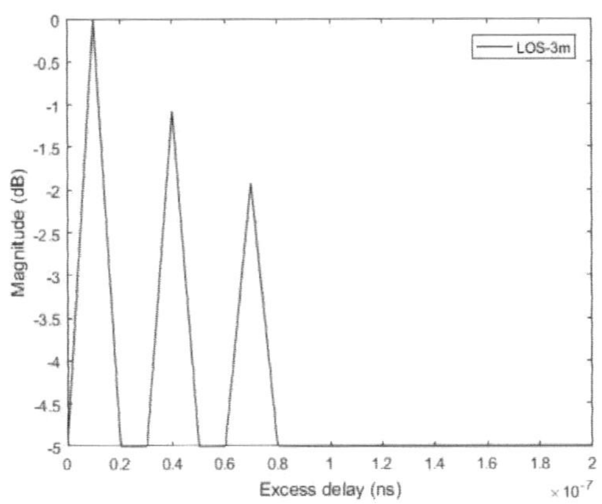


7a.

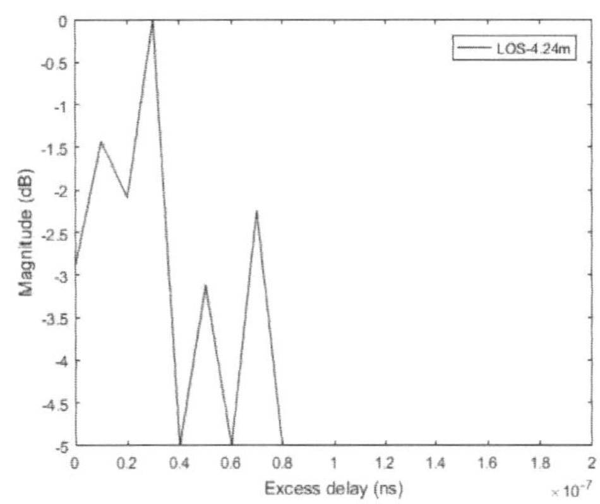


7b.

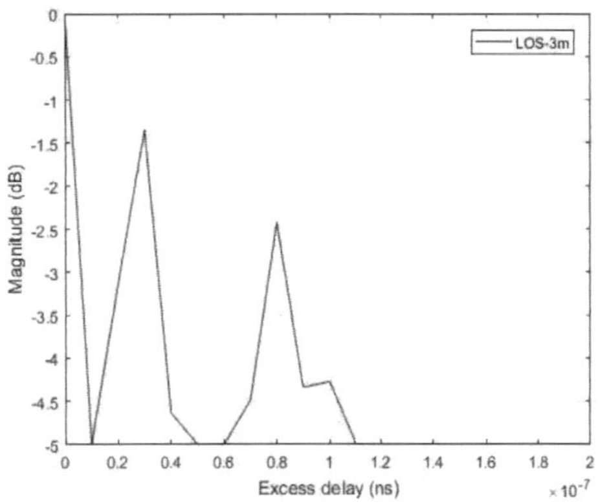


7c.

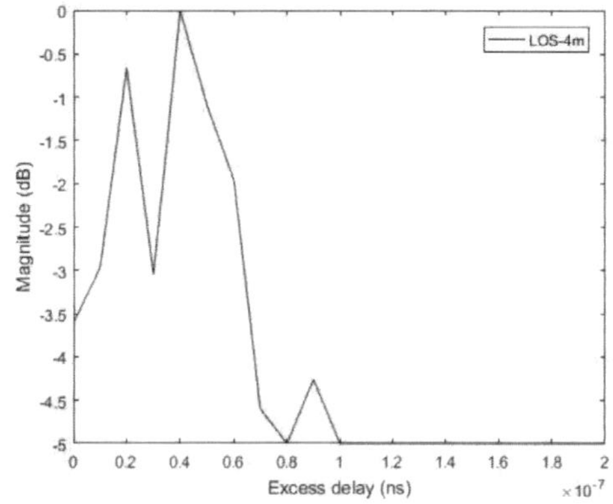


7d.

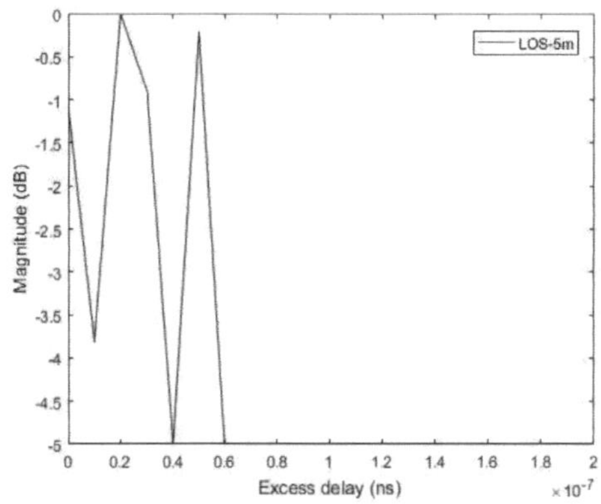


Figure 7e.

Figure 7. Typical impulse response LOS indoor propagation measurement figure 7a. Impulse response for RX 1, figure 7b. Impulse response for RX 1, figure 7c. Impulse response for RX 1, figure 7d. Impulse response for RX 1, figure 7e. Impulse response for RX 1

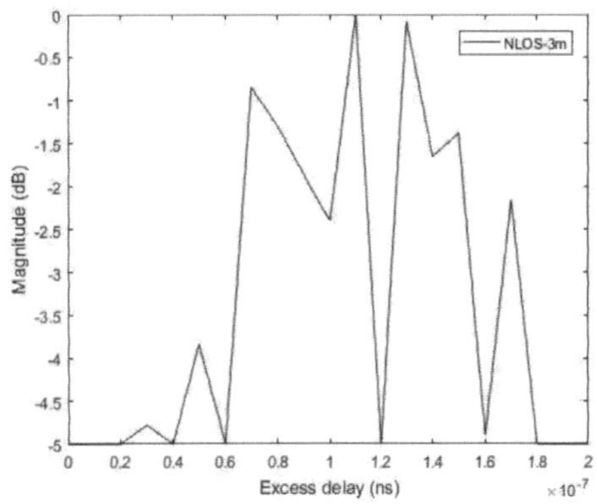


Figure 8a.

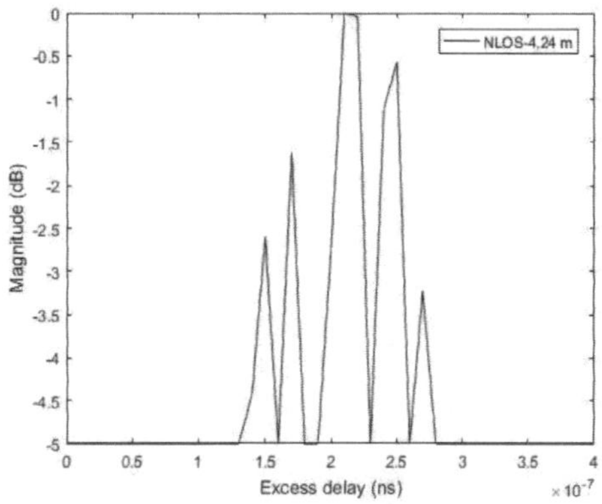


Figure 8b

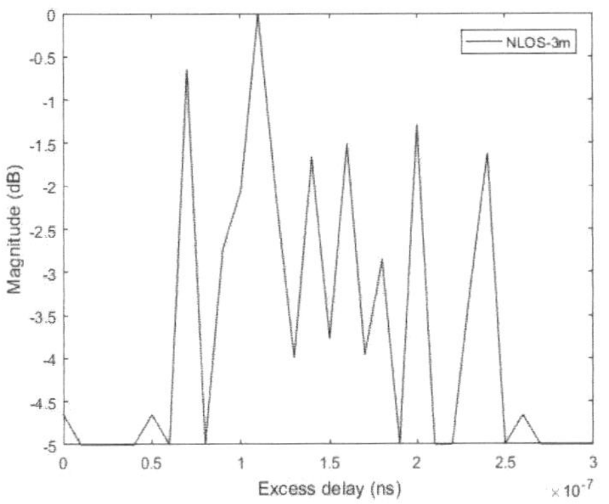


8c.

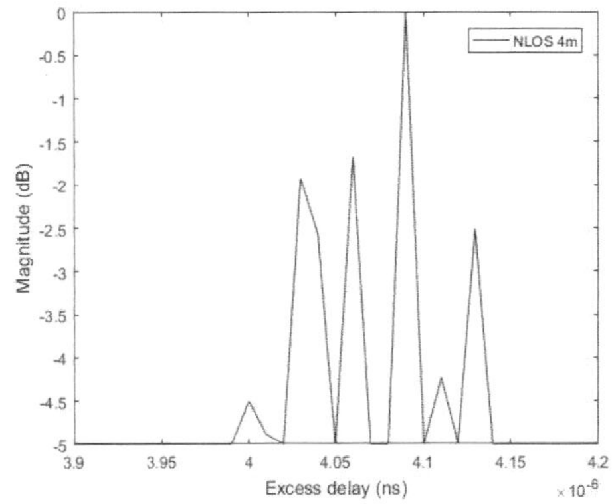


8d.

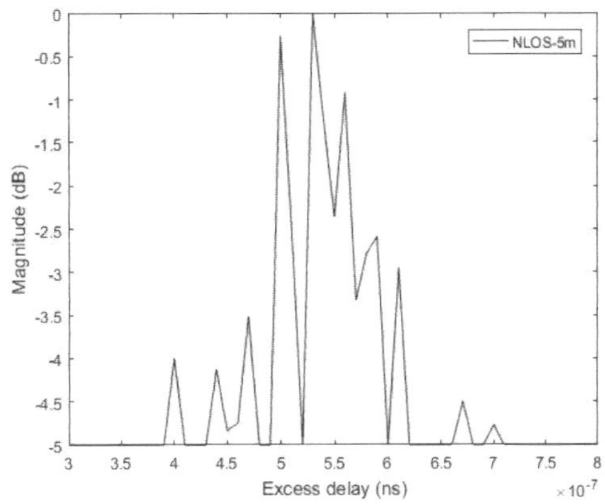


8d.

Figure 8. Typical impulse response NLOS indoor propagation measurement figure 8a. Impulse response for RX 1, figure 8b. Impulse response for RX 1, figure 8c. Impulse response for RX 1, figure 8d. Impulse response for RX 1, figure 8e. Impulse response for RX 1

## 4. Conclusion

This paper proposed the portable channel sounder which is suitable for indoor radio propagation channel by adopting LabVIEW software and low-cost commercially-available device called USRP. The multitoned-overlapping frequency stepping approach is implemented on USRP N210 device to obtain the wideband characteristic of propagation channel. The validation from VNA radio propagation is compared with proposed system.

## References

[1] Gupta A and Jha R K 2015 A Survey of 5G Network: Architecture and Emerging Technologies *IEEE Access*

[2] Wang C X, Haider F, Gao X, You X H, Yang Y, Yuan D, Aggoune H M, Haas H, Fletcher S and Hepsaydir E 2014 Cellular architecture and key technologies for 5G wireless communication networks *IEEE Commun. Mag.*

[3] Ghassemzadeh S S, Jana R, Rice C W, Turin W and Tarokh V 2004 Measurement and modeling of an ultra-wide bandwidth indoor channel *IEEE Trans. Commun.*

[4] Merwaday A, Rupasinghe N, Güvenç I, Saad W and Yuksel M 2014 USRP-based indoor channel sounding for D2D and multi-hop communications *2014 IEEE 15th Annual IEEE*

*Wireless and Microwave Technology Conference, WAMICON 2014*
[5] Srisooksai T, Takada J I and Saito K 2017 Portable wide-band channel sounder based software defined radio for studying the radio propagation in an outdoor environment *2017 International Symposium on Antennas and Propagation, ISAP 2017*
[6] Ettus Research, http:/www.ettus.com/product
[7] Ettus Research, http:/www.ettus.com/performance_data/sbx
[8] Zitouni R, Ataman S, Mathian M and George L 2015 Radio frequency measurements on a SBX daughter board using GNU radio and USRP N-210 *2015 IEEE International Workshop on Measurements and Networking, M and N 2015 - Proceedings*
[9] Li J, Zhao Y, Tao C and Ai B 2017 System design and calibration for wideband channel sounding with multiple frequency bands *IEEE Access*
[10] Febbo Enterprises,http://www.febo.com/reference/cable_data.html

Journal of Physics: Conference Series **1381** (2019) 012055 doi:10.1088/1742-6596/1381/1/012055

# Numerical Analysis of the Reynolds Number Effect on the Aerodynamic Performance Wing Airfoil Eppler 562 with Wingtip Fence

**S Hariyadi[1*], S Sutardi[2], W A Widodo[2], B J Pitoyo[1], N Pambudiyatno[1], and I Sonhaji[1]**
[1] Aviation Polytechnic of Surabaya, Jemur Andayani I/73 Wonocolo Surabaya, Indonesia
[2] Fluid Mechanic Laboratory, Mechanical Engineering Department, FTI, ITS, Surabaya, Indonesia

hudzaifahsetyo@gmail.com*

**Abstract.** Wings are a very important part of aircraft. In that section, most of the lift forces are generated on the airplanes. The aerodynamic performance produced by the wing greatly determines how optimal the cruising range of an aircraft. To improve the performance of the wings, researchers have been competing to make wing modifications of an aircraft. One modification that is used at this time is by adding end wall which is often referred to as a winglet. Winglets function as a barrier to fluid flow jumps from the lower surface to the upper surface. This fluid flow jumps is often called as a tip vortex. One type of winglet discussed in this study is the wingtip fence. This study took wing objects on unmanned aerial vehicle with numerical simulation using Ansys 19.0 software with turbulent model k-ω SST. The freestream flow rate to be used are 10 m/s ($Re$ = 2,34 x $10^4$) and 45 m/s ($Re$ = 1 x $10^5$). The angle of attack used are ($\alpha$) = 0°, 2°, 4°, 6°, 8°, 10°,12°,15°, 17°, and 19°. The wing model is an Eppler 562 (E562) airfoil with and without a winglet. From this study, it was found that wing aerodynamic performance with Eppler 562 (E562) airfoil was higher at $Re$ = 2,34 x $10^4$. Delay of the stall is more effective at $Re$ = 1 x $10^5$ compared to the $Re$ = 2,34 x $10^4$. But, the aerodynamic performance $Re$ = 2,34 x $10^4$ better than $Re$ = 1 x $10^5$ . Keywords: airfoil, *Eppler 562*, tip vortex, winglet, forward wingtip fence, rearward wingtip fence

## 1. Introduction

The times are always accompanied by technological developments. This encourages humans to create something that aims to provide more benefits to humans. In the industrial world, many applications have been found from the flow across the airfoil, for example on wings, propellers and others. In the case of engineering, especially in the field of fluid mechanics, there have been many studies on airfoil both in terms of simulations and experiments that aim to determine the flow characteristics of a fluid. In various literature, there has also been much discussion about the characteristics of flow and vortex shedding across the airfoil.

Several things make it possible to reduce drag and increase lift in the use of aircraft and UAV, including good wing designs, wingtip devices/winglets, boundary layer suction, and others. Winglets have advantages in terms of reducing the occurrence of tip vortex and reducing areas that are not effective because of the flow of flow from the lower surface to the upper surface. This is the concern of many researchers to continue to explore to provide space to research on the application of winglets that can result in a significant reduction in drag.

[1] analyzed the aerodynamic characteristics of multi-winglets that were applied to light aircraft. These winglets have demonstrated the potential for increasing aircraft aerodynamic efficiency by reducing induced drag. This experimental study focused on the half body wing model at $Re = 4 \times 10^5$ with six different multi-winglet configurations. The maximum aerodynamic efficiency improvement for the lift coefficient is 7.3%. Performance analysis was also carried out to reveal the potential for a 12% increase in the maximum climbing ratio. Winglets have led to a significant increase in performance parameters, with a gain of 7.3% in maximum aerodynamic efficiency or 11% in the best conditions. The maximum rate of climbing factor is also increasing by 12%. Analysis in this study also showed a significant reduction in the intensity of tip vortex.
[2]used the NACA 0012 airfoil and blended winglet with several cant angle variations. This numerical study uses the k-ω SST turbulence model, Courant Friedrichs Lewy Number 1.0, turbulent kinetic energy and an average specific dissipation of 0.8, $V_{inlet} = 55.5$ m/s with a turbulence intensity of 5%. Cant Angle selected are 0° (without winglets), 15°, 30°, 45°, and 60° and $\alpha = 0°, 2°, 4°, 6°$, and 8°. Cant angle 60° gives better $C_L$ but also higher $C_D$. Cant angle 15° provides a good $C_L/C_D$ distribution. It can happen because of the profile drag which is the effect of the viscosity and surface roughness. It is happens also in the pressure drag that occurs at the front of the airfoil but is not same with the rear of the wing.
[3] use NACA 4412 airfoil type sweept back wing with blended winglet on the normal wing, 30o winglet, and 90o winglet. This simulation study uses the Kω model, steady-state, and velocity inlet of 50 m/s. Winglet cant angle 30° gives better performance on $\alpha = 2°$ due to changes in overall flow and variations in the drag coefficient. This simulation study shows that by increasing the design of the winglet, with the correct cant angle variation, it can reduce the induced drag caused by vortex on the wingtip.
[4] researched a numerical simulation on the E420 airfoil on the ATLAS IV UAV wing with $Re = 10^5$ - $3 \times 10^5$. The consideration for choosing an Eppler in type airfoil is because it has Low Re airfoil, smooth trailing edge stall, and high lift characteristic and high thickness. In this study, it was found that the addition of winglets can reduce drag coefficient by 2.8% and increase aerodynamic performance by 80% in low angle of attack and 22% in operational regions.
This research will compare the minimum speed and maximum speed commonly used by UAVs so that their effects on aerodynamic performance are known. The maximum speed used refers to the research of [5].

## 2. Methodology

In this study numerical simulations of wings were used using Ansys 19.0 software. To produce data that is more accurate and does not depend on grid density an independence grid analysis is performed. Grid independence is a convergent solution determined from CFD calculations and is not influenced by the size of the cell meshing. In this study, the independence grid analysis was carried out using six different types of meshing with increasing density levels starting from meshing A to F meshing, which then obtained an error value from each meshing. In each meshing also seen the value of $y+$ where in this study the turbulence of the k-ω SST model which has the value of $y+ \leq 5$. For all values of drag coefficient ($C_D$) and $y+$ has been tabulated in Table 1.

The flow velocity to be used is 10 m/s ($Re = 2.34 \times 10^4$) and 45 m/s ($Re = 1 \times 10^5$) with $\alpha = 0°, 2°, 4°, 6°, 8°, 10°, 12°, 15°, 17°$ and 19°. Winglets that will be used are wingtip fence with forward and rearward variations. Reynolds Number is determined by chord. In this case, the length of the chord used is 20 cm. Figure 1 represents the domain simulation and the boundary conditions used in the simulation. The test specimen is in the form of an Eppler 562 type airfoil with and without a winglet attached to the tip in the form of a wingtip fence as shown in Figure 2 and the dimensions of the modeling dimensions in Figure 3 [6].

Journal of Physics: Conference Series **1381** (2019) 012055 doi:10.1088/1742-6596/1381/1/012055

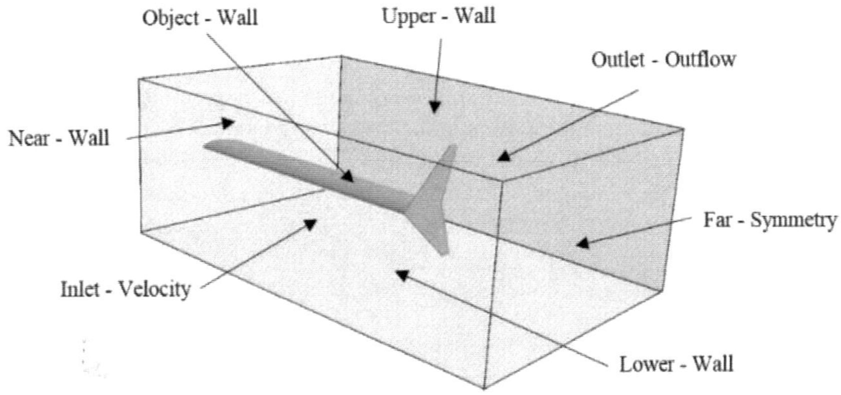


**Figure 1.** Simulation domain and boundary conditions

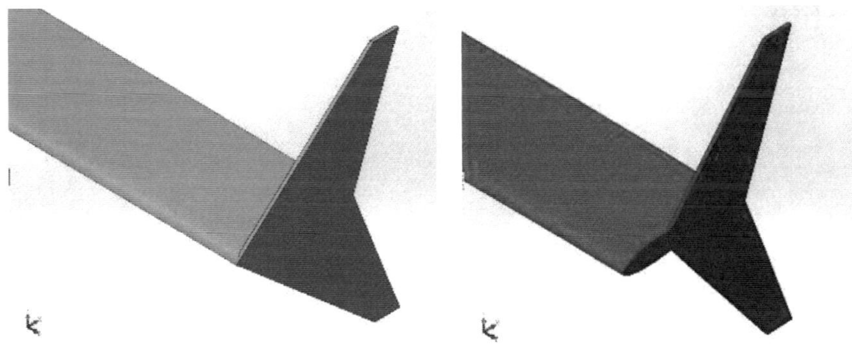

(a) Forward Wingtip Fence (b) Rearward Wingtip Fence

**Figure 2.** Specimen Model

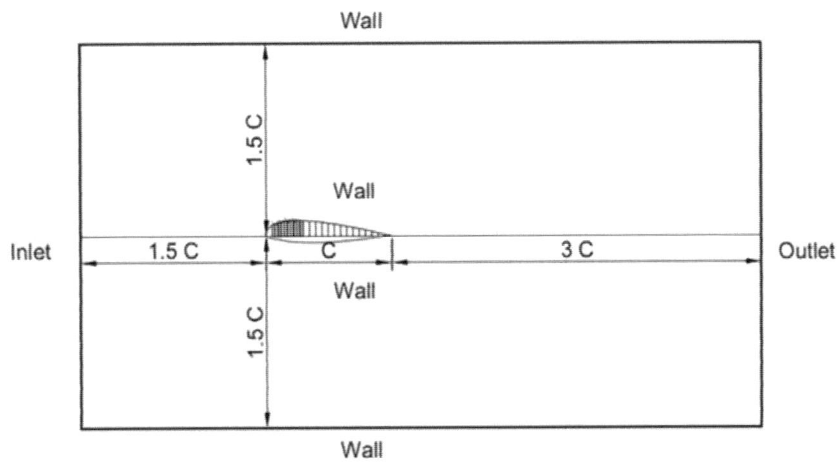


**Figure 3.** Modeling dimensions and background conditions [6]

The use of CFD application requires data accuracy in the post-processing and pre-processing steps. The grid independence step is needed to determine the level, the best and most efficient grid structure so that the modeling results are close to true. To get more complete information on the area around the wall, it requires the calculation of $y+$ on each meshing. The $y+$ calculation is based on the flat plate boundary layer theory calculations [7].

In addition to the calculation of y+, the inflation layer uses a quadrilateral meshing type so that information around the wall is more accurate. For the calculation of fewer cell numbers, the area far from the wall will use meshing type tetrahedrons.

Table 1. shows the meshing variations of the three dimensional independent test grid model at the 2.34 x $10^4$ Reynolds number. In this research, in order to get the good results, the *y*+ used *y*+ <1 as was done in the Kontogiannis research [4]. Based on table 1. The smallest $C_D$ value occurs in Meshing C. Besides, One of the considerations in conducting numerical simulations is the time and memory used, then the meshing used for subsequent simulations is Meshing C.

**Table 1.** Analysis of the E562 model-independent three dimensional grid without winglets

| *Meshing Type* | *Cells Number* | *Inflation Layer* | $C_D$ | $y^+$ | *Skewness Average* |
|---|---|---|---|---|---|
| *Meshing* A | 367075 | 40 | 0,86 | 2,1 | 0,346 |
| *Meshing* B | 469620 | 40 | 0,88 | 1,4 | 0,347 |
| *Meshing* C | 569233 | 40 | 0,90 | 0,8 | 0,343 |
| *Meshing* D | 685063 | 40 | 0,92 | 2,1 | 0,334 |
| *Meshing* E | 768003 | 40 | 0.90 | 0.8 | 0,351 |
| *Meshing* F | 875962 | 40 | 0.91 | 1.4 | 0,348 |

## 3. Result and Discussion

### *3.1. Drag Coefficients Analysis*

Winglets will increase the drag coefficient, especially the increase in drag pressure compared to viscous drag. Adding winglets to the wings increases skin friction drag. However, the role of drag pressure is far greater than viscous drag as research by [8]. In this study, it was seen that the drag coefficient produced by plain wing equipped with winglets was always higher than plain wing at 10 m/s (*Re* = 2.34 x $10^4$) and 45 m/s (*Re* = 1 x $10^5$). The effect of reynolds number is also seen in this study where an increase in reynolds number will increase the value of drag. This can be seen in Figure 4 where the reynolds number 1 x $10^5$ is higher than Re = 2.34 x $10^4$. However, the influence of geometry is seen at a high angle of attack both at 10 m/s (*Re* = 2.34 x $10^4$) and 45 m/s (*Re* = 1 x $10^5$). At the rearward wingtip fence, there is a visible increase in the drag coefficient at $\alpha = 15^o$ while in the forward wingtip fence there is a decrease at $\alpha = 17^o$. This is due to the influence of the geometry of the winglet shape. Allegedly induced drag affects the increase in total drag as well. Induced drag is part of the elevator, which means that as long as the lift is still there the Induced drag will remain. Induced drag cannot be ruled out, neither can an elevator. Lift and induced drag can be affected by the physical construction of aircraft wings, for example aspect ratio dimensions. The results of this study were compared with the [5] study. [5] uses *Re* = 2,358 x $10^6$ and *y*+ at intervals of 5 < *y*+ <30. The picture shows that the resulting drag coefficient is higher both on the plain wing and on the wingtip fence.

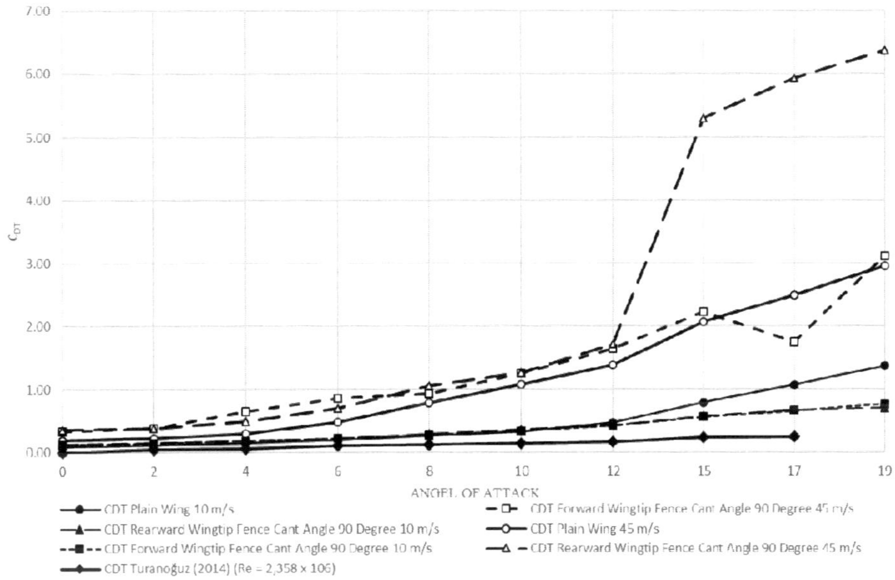

**Figure 4.** Comparison of the total drag coefficient ($C_{DT}$) to α on plain wing and winglet

*3.2. Lift Coefficients Analysis*

From these lift coefficient results, it is calculated to get the stall angle. Observation of the lift coefficient is done by 3D simulation for plain airfoil and airfoil using winglets. As an initial reference in analyzing the lift coefficient so that it can be seen influence of the Reynolds numbers variation and the addition of winglets, then a simulation of the plain airfoil is performed first. Figure 5 shows the graph of the relationship of $C_L$ to α for plain airfoil E562 with freestream speeds of 10 m/s ($Re = 2.34 \times 10^4$) and 45 m/s ($Re = 1 \times 10^5$).

In Figure 5 it appears that the lift coefficient increases with increasing angle of attack. It can be seen that using the wingtip fence can delay stalling. At the plain wing $Re = 2.34 \times 10^4$, a stall occurs at $\alpha = 12^\circ$ as in [5] while the addition of fence stall wingtip occurs at $\alpha = 15^\circ$ for rearward wingtip fence and $\alpha = 17^\circ$ for forward wingtip fence.

At plain wing $Re = 1 \times 10^5$, a stall occurs at $\alpha = 15^\circ$ while the addition of wingtip fence a stall occurs at $\alpha = 12^\circ$ for rearward wingtip fence, and $\alpha = 19^\circ$ for forward wingtip fence. The lift coefficient produced in this study is lower than that of [5]. However, it appears that the stall delay in this study resulted in a better angle of attack.

Journal of Physics: Conference Series **1381** (2019) 012055 doi:10.1088/1742-6596/1381/1/012055

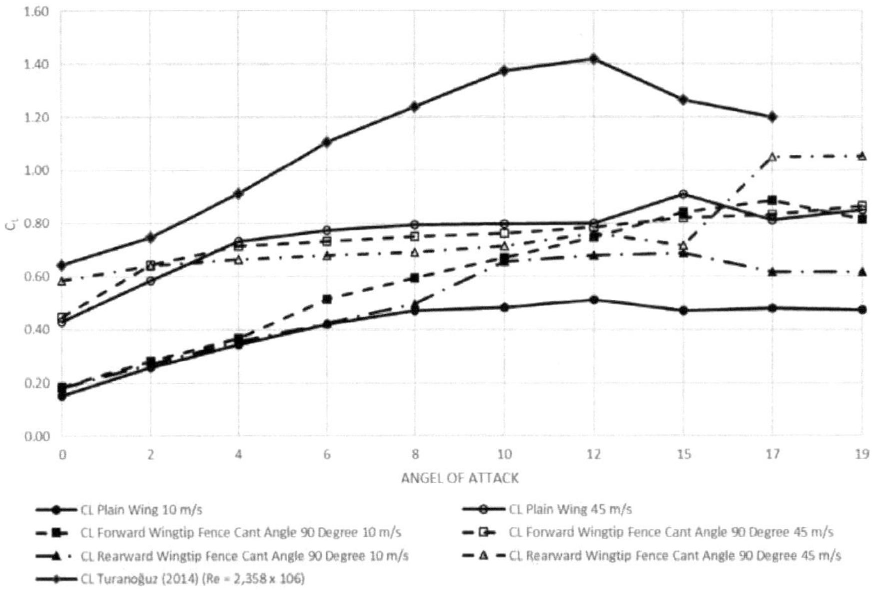


**Figure 5.** Comparison of the lift coefficient ($C_L$) to α on plain wing and winglet

### *3.3. Lift to Drag Ratio Analysis (CL / CD)*

Improved aerodynamic performance can be seen in the $C_L/C_D$ comparison image. Increased $C_L/C_D$ can occur among others due to the effective area of the wing better or reduced drag. Tip Vortex from the lower surface of the wing contribute to reducing effective area and also increase the induced drag of the wing. If the winglet can function effectively, it can hold the tip vortex.

Figure 6. shows the ratio of lift to drag ratio ($C_L/C_D$) on a plain wing and with a winglet at several angles of attack. It shown that the addition of winglets will increase the coefficient of lift and show a trend that increases with increasing angle of attack. Figure 6. also showed that the addition of winglets with freestream 10 m/s ($Re = 2.34 \times 10^4$) would improve effective wing performance at $\alpha = 6^o$ at the forward wingtip fence and $\alpha = 10^o$ at the rearward wingtip fence. At freestream 45 m/s ($Re = 1 \times 10^5$) it shows a decrease in aerodynamic performance both on the plain wing, forward wingtip fence and rearward wingtip fence.

On the forward wingtip fence freestream 10 m/s ($Re = 2.34 \times 10^4$) results in higher performance compared to other wing geometries at $\alpha = 6^o$ while the rearward wingtip fence produces a better performance at $\alpha = 10^o$. Aerodynamic performance on freestream 10 m/s ($Re = 2.34 \times 10^4$) is also better than [5] research on both the plain wing and wingtip fence.

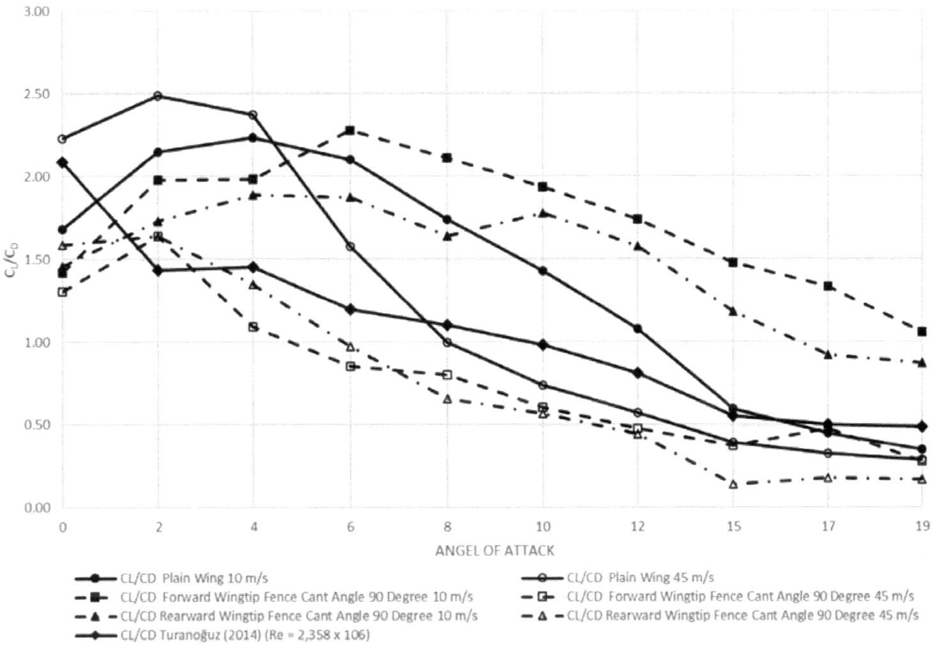


**Figure 6.** Comparison of the lift to drag ratio ($C_L/C_D$) to α on plain wing and winglet

## 4. Conclusion

From the results of numerical simulations that have been carried out about the effect of variations in Reynolds number and geometry on the characteristics of fluid flow through the E562 airfoil, it can be concluded that the use of winglets can improve wing performance but also the addition of winglets will increase drag as the angle of attack increases. From numerical studies it was found that the use of winglets can produce several flow characteristics, namely:

- With increasing Reynolds number and angle of attack will increase drag coefficient and lift coefficient. A higher Reynolds number increases a higher drag and a higher lift
- The use of winglets produces a drag coefficient and a lift coefficient higher than the plain wing. This applies to forward wingtip fence and rearward wingtip fence
- With the use of the wingtip fence, it can delay stalling. At plain wing $Re = 2.34 \times 10^4$, stall occurs at $\alpha = 12°$ while the addition of a fence stall wingtip occurs at $\alpha = 15°$ for rearward wingtip fence and $\alpha = 17°$ for forward wingtip fence.
- At the plain wing $Re = 1 \times 10^5$, the stall occurs at $\alpha = 15°$ while the addition of the wingtip fence stall occurs at $\alpha = 12°$ for the rearward wingtip fence and $\alpha = 19°$ for the forward wingtip fence.
- In the use of winglets, stall delays are better at freestream 45 m/s ($Re = 1 \times 10^5$) than freestream 10 m/s ($Re = 2.34 \times 10^4$)
- The aerodynamic performance produced by the addition of winglet geometry at freestream 45 m/s ($Re = 1 \times 10^5$) is lower than freestream 10 m/s ($Re = 2.34 \times 10^4$). At freestream 45 m/s ($Re = 1 \times 10^5$), the aerodynamic performance of plain wing is better than using winglets
- Adding winglets in freestream 10 m/s ($Re = 2.34 \times 10^4$) improves performance at $\alpha = 6°$ at the forward wingtip fence and $\alpha = 10°$ at the rearward wingtip fence. At freestream 45 m/s ($Re = 1 \times 10^5$) it shows a decrease in aerodynamic performance both on the plain wing, forward wingtip fence, and rearward wingtip fence. However, the forward wingtip fence improves performance at $\alpha = 15°$ compared to plain wing.

Journal of Physics: Conference Series **1381** (2019) 012055 doi:10.1088/1742-6596/1381/1/012055

## 5. References

[1] Cosin R and Catalano F 2009 Aerodynamic Analysis of the Use of Multi-Winglets in Light Aircrafts

[2] Dinesh M, V K M, Venkatesan D V, B S K, R S R and Kumar V R S 2014 Diagnostic Investigation of Aircraft Performance at Different Winglet Cant Angles *Int. J. Mech. Aerospace, Ind. Mechatronics Eng.* **8** 1987–96

[3] Myilsamy D and Thirumalai, Yokesh P P . 2015 Performance Investigation of an Aircraft Wing at Various Cant Angles of Winglets using CFD Simulation *Altair Technol. Conf.* 23

[4] Kontogiannis S G, Mazarakos D E and Kostopoulos V 2016 ATLAS IV wing aerodynamic design: From conceptual approach to detailed optimization *Aerosp. Sci. Technol.* **56** 135–47

[5] Turanoguz E and Alemdaroglu N 2015 Design of a medium range tactical UAV and improvement of its performance by using winglets *2015 Int. Conf. Unmanned Aircr. Syst. ICUAS 2015* 1074–83

[6] Mulvany N, Chen L, Tu J and Anderson B 2004 Steady-State Evaluation of Two-Equation RANS (Reynolds-Averaged Navier-Stokes) Turbulence Models for High-Reynolds Number Hydrodynamic Flow Simulations *Dep. Defence, Aust. Gov.* 1–54

[7] Hankins G A and Relf E F 1947 Fluid Mechanics *Proc. Inst. Mech. Eng.* **157** 277–81

[8] Hariyadi S S P, Sutardi and Widodo W A 2018 Drag reduction analysis of wing airfoil E562 with forward wingtip fence at cant angle variations of 75°and 90° *AIP Conf. Proc.* **2001**

Journal of Physics: Conference Series **1381** (2019) 012056 doi:10.1088/1742-6596/1381/1/012056

# Optimization of Blade Curve of Pump as Turbine for Piko Hydro Power Plants

**M Mustafa*, H Hantarum, S Suryono**

Universitas Merdeka Madiun, Serayu Street 79 Madiun, East Java, Indonesia, 63133

*mustafa@unmer-madiun.ac.id

**Abstract**. The use of the NS 50 water pump that functions as a turbine for the PikoHidro Power Plant was an idea in finding simple and easy alternative energy in its assembly. In this application a water pump was used as a turbine so the working principle was reversed, where water from a certain height will move the pump, so that the pump impeller rotates and continues to rotate the generator and will generate electrical energy. This study emphasizes the experimental pump functioned as a turbine (PAT) by modifying the same in and out angles on the PAT blade. An NS-50 pump water pump with a capacity of 563.22 liters / minute and a total head of 20 m was tested. The variables used were water level (H) was 3 meters and variations in the different entry and exit angles of the impeller in the 7 specimens used. The pump was then connected to a generator and the result was electricity. The results showed that the best variation was in variations in the exit and entry angles of 30/30, rotation was 810 rpm and the power produced was 90 watts. The experimental results show good results theoretically.

## 1. Introduction

Today the development of energy is directly proportional to the development of the age. Population growth, industrial development, and technological progress are the main reasons and evidence that energy is very important for life. The need for energy is unfortunately not in line with the resources that are used or used as the main source of energy production. In 2010, many countries have realized the importance of utilizing renewable energy sources as a substitute for non-renewable energy such as petroleum, coal and gas which has had very damaging effects on the earth. With the depletion of reserves of non-renewable energy sources will release carbon emissions into the atmosphere which is a major contributor to global warming.

Energy needs will always increase as a function of population growth. For conventional energy such as oil and gas, high demand if not matched by production capacity causes scarcity resulting in price increases and energy crisis. One of the efforts related to the energy policy is to develop and increase energy diversity including potential energy now and in the future. This shows the importance of developing the renewable energy sector.

Indonesia is targeting an electricity rate of 90% for the population with a current (predicted) number of more than 273 million. The average standard for electricity is 450 watts per house. Thus, it takes around 9000 MW until 2025. Hydroelectric Power Plants (HPP) in Indonesia has the potential to reach 70,000 MW. Thus the 9000 MW shortfall is predicted to be fulfilled through hydropower. Of the 70,000

Journal of Physics: Conference Series **1381** (2019) 012056 doi:10.1088/1742-6596/1381/1/012056

MW potential, only around 6% has been utilized. In remote areas, meeting electricity needs through hydropower can be assisted with Micro Hydro Power Plants (MHPP).

Indonesia's natural conditions are very rich in the potential of water that can be used as electricity generation. Therefore, it is necessary to optimize the utilization of water resources for energy fulfillment. The Pikohidro system as one of the new renewable energy sources that can provide great benefits to the community in meeting electrical energy without having to incur high costs for power transmission systems or environmental care in general because the implementation of the system is integrated with its utilization.

The basic principle of hydropower, if water can be channeled from a certain height to a lower level, then the resulting head of water can be used to do work. These uses can move mechanical components into rotational energy that is channeled to the shaft to drive a generator to generate electricity. The choice of a good turbine for each particular place depends on the characteristics of the place, head and discharge available. Turbine innovations that are modern, expensive are all obstacles for rural communities. However, if we use a pump that is mass produced and is reasonably priced, clean, green and sustainable is a solution for the electricity needs of rural communities. The generation scale targeted for household use ranges from 400W - 5000 W.

One economical alternative to building small-scale hydroelectric power plants is to use pumps as turbines. The field of science that specializes in operating pumps as turbines is often referred to as PAT, short for Pumps As Turbine. Several types of water pumps can be applied as water turbines, usually pumps are driven by electric motors to raise a certain amount of water to a certain height. In pump applications as turbines, the working principle of the pump is reversed, which is given a drop of water from a certain height to rotate the pump impeller. This impeller rotation will be continued to turn the generator so that electricity is produced.

The important thing in efforts to optimize this turbine is that the turbine's performance will change if there is a change in the curve and shape of the turbine blade. In this study will test experimentally to determine the performance of the output power generated by the turbine blade by modifying the entry and exit angles of the blade and blade shape using variation of BW1, BW2, BW3, RD4, FW5, FW6 and FW7 on the use of pumps as turbines. This study was held in Mechanical Engineering Department Laboratory at Merdeka University of Madiun.

### *1.1. Formulation of the problem*
The formulation of the problem in this research is how the influence of impeller blade curvature on the pump as a turbine on turbine mechanical power and turbine efficiency.

### *1.2. Scope of problem*
Limitation problems in this study are:
1. The height of the turbine with the upper reservoir is 2, 3 and 4 meters.
2. Opening valve regulating discharge 20%, 40%, 60%, 80% and 100%.
3. Turbine impeller curvature:
    - BW1 : radius curvature 15 cm with backward orientation
    - BW2 : radius curvature 13 cm with backward orientation
    - BW3 : radius curvature 11 cm with backward orientation
    - RD4 : straight blade with no radius curvature
    - FW5 : radius curvature 15 cm with forward orientation
    - FW6 : radius curvature 13 cm with forward orientation
    - FW7 : radius curvature 11 cm with forward orientation

## 2. Basic theory
Water turbine is a device to convert potential energy of water into round energy and then converted into electrical energy by generators. Before carrying out the design and selection of water turbines, a

feasibility test and analysis of the water resources that will be utilized for its potential energy is needed, especially the availability of head and water discharge from the water source for the designed load.

After knowing the potential head that exists at the source of the water flow, then determine the type of turbine and the planned load. Where the burden of the plan must not exceed the availability of potential energy from water sources, because it will result in not achieving maximum operational efficiency and economically detrimental.

### *2.1. Pico Hydro*
Pico-scale hydroelectric power plants are power plants that produce an electric power output of no more than 5 kW. Piko-scale hydroelectric power plants in principle utilize different heights and the amount of water discharges per second that exists in the flow of water from irrigation channels, rivers or waterfalls. This water flow will rotate the turbine shaft so that it produces mechanical energy. This energy then drives the generator and the generator generates electricity.
Some text.

### *2.2. Pump as Turbine*
Hydropower systems convert energy from water that falls into mechanical energy with turbines. In some cases, it may be more appropriate to replace the turbine with a centrifugal water pump, and run it in reverse. One economical alternative to building small-scale hydroelectric power plants is to use pumps as turbines.

Usually the pump is driven by an electric motor to raise a certain amount of water to a certain height. In pump applications as turbines, the working principle of the pump is reversed, which is given a drop of water from a certain height to rotate the pump impeller. This impeller rotation will be continued to turn the generator so that electricity is produced.

## 3. Research Methods
The test was carried out at laboratory of Mechanical Engineering Department, Merdeka University of Madiun. The equipment used includes: NS 50 pumps, centrifugal pumps and electric generators.

### *3.1. Installation testing on pumps as turbines*

Journal of Physics: Conference Series **1381** (2019) 012056 doi:10.1088/1742-6596/1381/1/012056

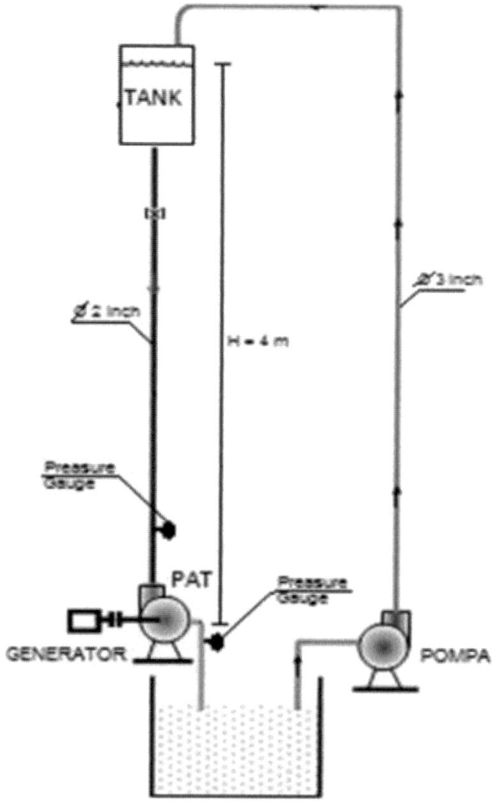


**Figure 1.** Installation testing on pumps as turbines

*3.2. Test Specimens*

In this test using 7 specimens consisting of: BW1, BW2, BW3, RD4, FW5, FW6, FW7. Specimen specimens are made from composite materials. This is intended to look for light objects that are light and resistant to pressure.

| | | |
|---|---|---|
| 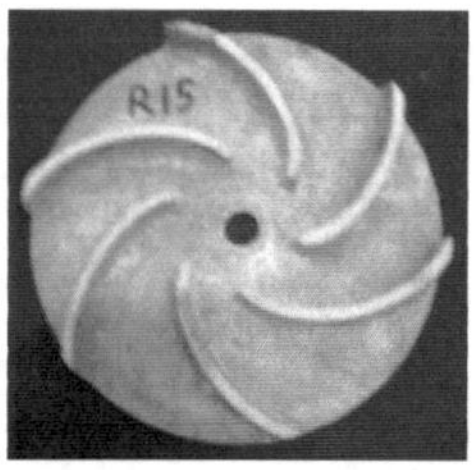 Specimen 1 (BW1) Curvature of the impeller radius of 15 cm (Backward) | 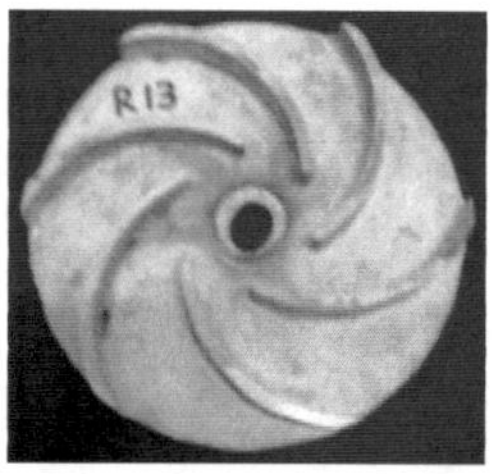 Specimen 2 (BW2) Curvature of the impeller radius of 13 cm (Backward) | 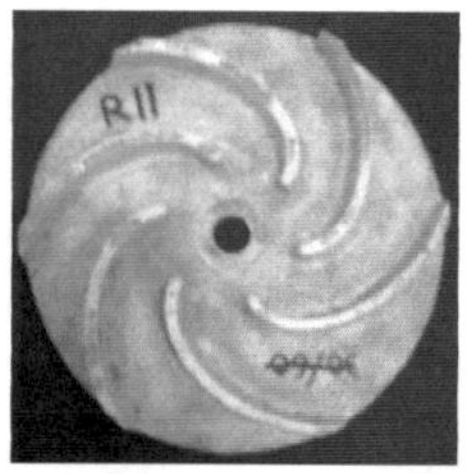 Specimen 3 (BW3) Curvature of the impeller radius of 11 cm (Backward) |
| | | |

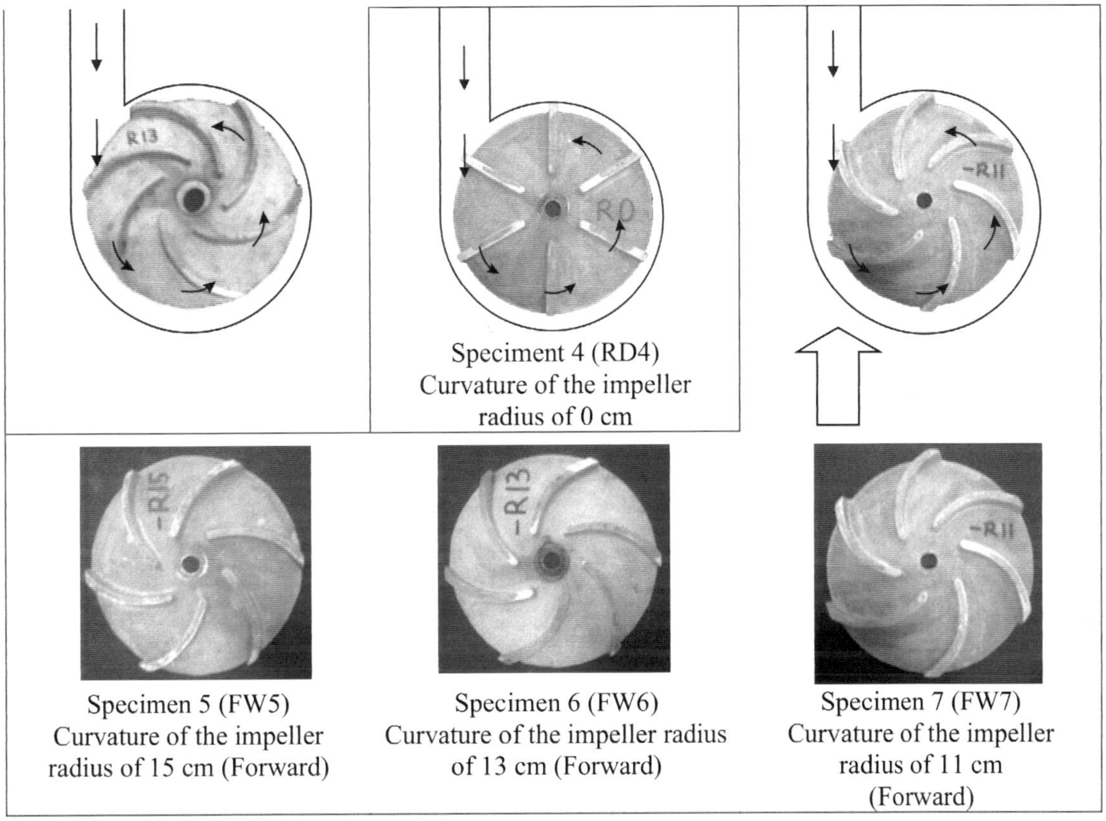


**Figure 2.** The relationship between mechanical power and turbine efficiency

## 4. Result and Discussion

Tests carried out on the pump as a turbine with a head surface height of 2, 3 and 4 meters while the variation of discharge at the valve openings 20%, 40%, 60%, 80% and 100%

### *4.1. Power of Fluid flow*

The power of the fluid flow (Ph) can be seen in Figure 2. below.

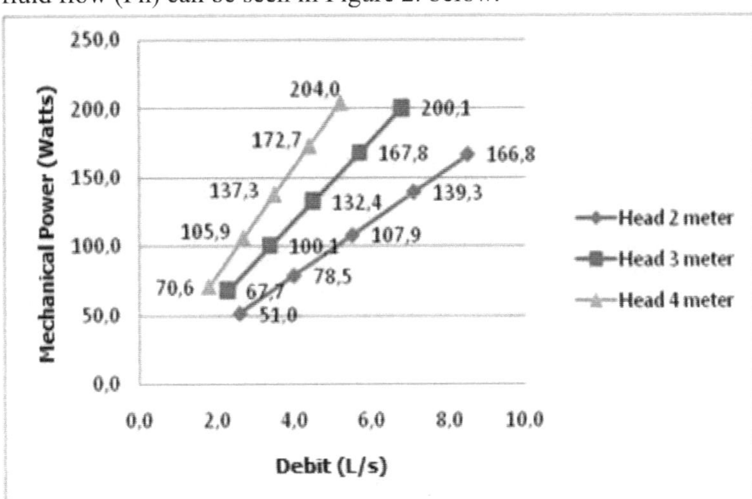


**Figure 3.** Relationship of discharge (l / s) to water power (watts)

### *4.2. Turbine Efficiency*
Turbine Efficiency can be seen in Figure 3. below, that describe the relationship between mechanical power and turbine efficiency .

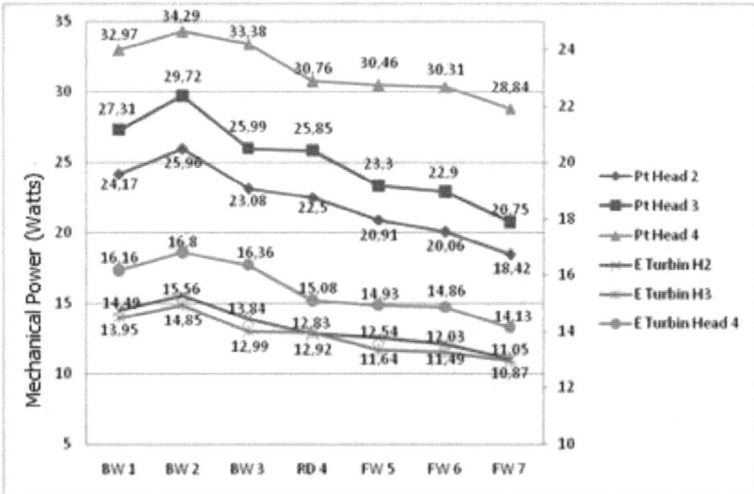


**Figure 4.** The relationship between mechanical power and turbine efficiency

## 5. Conclusion
From the test results it can be concluded as follows.

1. The highest mechanical power value is 34.29 watt at 4 meter head and BW2 specimens indicate that the ability of the turbine to rotate is obtained from the incoming water power and torque magnitude. The amount of shaft rotation and output power is influenced by the height of the falling water and the volume entering the turbine.
2. The highest output power value of 90 watts at a height of 4 meters in the BW2 specimen is due to the influence of the shaft rotation on the rotation of the generator as well as the ability of the blade angle to change the flow speed.
3. The highest efficiency at the 4 meter head is the highest magnitude at 29.92%. The two highest values were obtained in BW2 specimen variations.

## References
[1] Abidin, M.A. , 2015, *The influence of the large angle of the blade curvature on the performance of an overshot curved blade*; Undergraduate thesis; Mechanical Engineering Department Brawijaya University of Malang.

[2] Arriaga, M., 2010, *Pump as Turbine-A pico-hydro alternative in Lao People's Demcratic Republic*, Renewable Energy, 35.

[3] Derakhshan,S., Mohammadi,B., dan Noubakhsh, A.,2009. *Efficiency emprovement of centrifugal reverse pumps*, ASME Journal of Fluids Engineering p. 131.

[4] Dietzel, F., 2005, *Turbines, Pumps and Compressors*, Publisher : Erlangga, Jakarta.

[5] Himawanto, D. A., dan Danardono, D., 2015, *Effect of the Bucket and Nozzle Dimension on the Performance of a Pelton Water Turbine*. Modern Applied Science, 9(1), 25.

[6] Jain, S. V., Swarnkar, A., Motwani, K. H., dan Patel, R. N., 2015, *Effects of impeller diameter and rotational speed on performance of pump running in turbine mode*. *Energy Conversion and Management*, *89*, 808-824.

[7] Morris, A. L.,dan Harris, C. L.,2011, *A technique for probing sublexical representations*. Psychonomic Bulletin & Review, *8*(1), 118-126.

Journal of Physics: Conference Series **1381** (2019) 012056 doi:10.1088/1742-6596/1381/1/012056

[8] Nautiyal,H.,dan Kumar,A., (2010). *Reverse running pumps analytical, experimental and computational study:A review*, Renewable and Sustainable Energy Reviews 14.
[9] Singh, P.,danNestmann, F., (2010).*An optimizationroutineon aprediction andselectionmodelfor theturbine operationof centrifugalpumps*, ExperimentalThermaland Fluid Science34.
[10] Singh, P., danNestmann, F., (2011).*Internal hydraulic analysis of impeller rounding in centrifugal pumps as turbines*.Experimental Thermal and Fluid Science, 35(1), 121-134.
[11] Situmorang, dkk., 2014, Performance of Shimizu Type Ps-128 Bit Water Pump Works as a Water turbine; Undergaraduate thesis; Mechanical Engineering Department, Sam Ratulangi University.
[12] Yang, S. S.,danQu, X. Y. (2012). Effects *of impeller Trimming Influencing Pump As Turbine.Computers & Fluids*, *67*, 72-78.
[13] Zuo, Z., dan Wu, Y., (2015). *Pressure fluctuations in the vaneless space of high-head pump-turbines-areview.Renewable and Sustainable Energy Reviews*, 41, 965-974.

Journal of Physics: Conference Series **1381** (2019) 012057 doi:10.1088/1742-6596/1381/1/012057

# Design Tools for ControllingBrown Planthopper Pests Using Ultrasonic Waves

**S N Utama*, H Setyaningrum, B Sholeh, T Taufiqurrahman**

Universitas Darussalam Gontor, Siman Road Km 5, Siman, Ponorogo, East Java, Indonesia 63471

*shoffin@unida.gontor.ac.id

**Abstract.** The brown planthopper pest is a main pest of rice production in East Java and mostly Indonesia areas. The main controls of brown planthopper are mostly using chemical pesticides, which caused resistance on pest. Resistant varieties are the main choice for overcoming attacks of planthopper. However, resistant varieties had limitations because brown planthopper adapted to new biotypes. Conventional Integrated Pest Management (IPM) is reactive to pest attacks and often continues to rely on insecticides. One of the promising ecofriendly technologiesis ultrasonic waves. This study the aim was to design a planthopper pest control system in a room using the HC-SR04 ultrasonic sensor with a frequency of 40 KHz. The trial in the first stage was limited to testing the functionality to the system and testing the range of tools carried out to determine the area covered by the wave ultrasonic. The results of testofthe system are all system components are working properly. The ranges of the tools area maximum range is three meters in surrounding of the ultrasonic sensor.

## 1. Introduction

Planthopper,*Nilaparvatalugens*Stål is the most dangerous pest in tropical climates especially Southeast Asia [1]. The morphology of plant hopper small size 4 -5 mm with brown color, the femaleproducing egs 100-500 eggs in normal temperature (± 25°C). The adults of planthopper reach to 30 days lifetime[2], distributed widely in South, Southeast, and East Asia; the South Pacific islands; and Australia [3]. Various methods applied to overcome this pest attack, both using biological and chemical methods. Those both method has limited in point of view effectiveness and side effects caused such us environment and resistant issues. Chemical method of pest control has been found to be very effective but quite expensive to maintain. Also, these chemicals are highly poisonous and harsh both to human beings and pest. Air and water pollution is most important risk factor arising due to the use of chemical method. Human risk factors such as respiratory diseases, particularly for bronchial asthma and chronic obstructive pulmonary disease (COPD).In addition, it carries genetic mutations to offspring produced from internal pests that.According to those arising issues, the alternate methods of control increasingly[4] including the acoustic method with ultrasonic wave.

The method developed in this study is a physics method using ultrasonic waves. Ultrasonic waves are a strategic step, since the effects of waves that damage the body's tissues with no pollution impact. This is evident to several studies that have been conducted using ultrasonic waves, such as to overcome beetle pests, planthopper, and mice [5]. The type of ultrasonic wave used in this study is a longitudinal wave. Longitudinal waves are waves that have a vibration direction parallel to the direction of propagation which possible for back and for moved. When slinky is moved back and forth then slinky will form a density and stretch.

The ultrasound sensor is one sensor that produces analog quantities for using in tools. One type of sensor that is often used is parallax ultrasound, which has output as digital scale. In build a ultrasonic tool, many components are arranged including ultrasound wave generators and transmitters. The components are embedded in the board with integrated circuit with microcontroller. Other components that also used are registers, AT commands, and I / O systems. Recently, the microcontroller has been packaged in an active module kit such as the Arduino board [6] with many types based on capacity.

Several studies on control pest, especially insects are limited. Pramana et.al (2017) applied the research, used frequency of ultrasonic waves above 20 Khz and showed results afraidness of insect pest on apes.Similarly, Utama (2017) conducted a study to design a mosquito repeler system to use ultrasonic sensor with solar panel as energy sources, results showed tools triggeredmosquito movement up to 15-42.5 %. The results also indicate the mosquito reaction pattern movement which tends to stay away from ultrasonic wave sources. Yusianto and Sudibyo (2016) also carried out the brown planthopper pest control design for rice plants[7] with low effect on controlling planthopper.

Brown Planthopper as an insect group, able to listening sounds above the threshold of the average human hearing capacity, which is more than 20 kHz. Ultrasonic waves are ranged of 30-100 kHz expressed as very high-amplitude [8], which able to capture by the antenna of planthopper.When ultrasonic waves are emitted, the senses of the planthopper will recognize the sound caused by ultrasound waves as a threat, so that if the planthopper is continuously in the vortex of the wave, they will be uncomfortable and eventually stay away[9].

The aim of this research is to design a prototype of a planthopper repellent (call 'OUTWERE') in a room using the HC-SR04 ultrasonic sensor with a frequency of 40 KHz. The trial was carried out to determine the performance of ultrasound distance range. The system can provide information on areas covered by waves whose output is displayed on the LCD 16x2 cm.

## 2. Methods

### *2.1 The component of "Outwere" repellent*

In this study the structure of planthopper pest control system devices (Outwere) using 3 main components;arduinouno (Fig. 1), sensor ultrasonic HC-SR04 (Fig. 2),and LCD 16x2 cm (Fig. 3). The Arduino board (figure 1)type is a ready-to-use hardware with microcontroller ATMEGA328 modules and IDE software to program so that we can easily use them. This microcontroller consists of CPU, memory, and I / O that we can control by programming it. I / O is also often referred to as GPIO (General-Purpose Purpose Input Output Pins) which means, the pin that we can program as input or output as needed[10]. The information about coverage area ultrasonic will be displayed on a 16x2 LCD (figure 2).

**Figure1.**Arduino-ATmega 328 Microcontroller

**Figure 2.** HC-SR04 Ultrasonic.

**Figure 3.** 16x2 Liquid Christal Display.

Ultrasonic Sensor HC-SR04 (figure 2) is a sensor issued by Devantech Company which widely used in the world of robotics and aeronautics. This sensor serves to measure the distance between 3 cm to 3 meters. The HC-SR04 sensor has 2 main sensor parts namely transmitter (N1076) and receiver (N1081). The transmitter sensor emits an ultrasonic signal with a frequency of 40 KHz and the receiving sensor captures the reflected 40 KHz frequency [11].

### *2.2 The Design Of Ultrasonic Brown Planthopper Pest Repeller*

The design part of planthopper pest control systems is achieved by generation of Pulse Width Modulation (PWM) wave frequency by Arduino And HC-SR04 ( Figure4). The design briefly explains the details of a Mockup diagram of planthopper pest control systems with all components. Arduino requires 5-9 V regulated voltage to operate. HC-SR04 requires 5V regulated voltage to operate. The LCD is connected to the microcontroller to display the coverage area frequency ultrasonic. Whereby, the HC-SR04 data port is connected to microcontroller at PORT 4,5,6,7. The conductors were arranged in an isosceles triangle construction. The layout and dimensions to the model are shown in Figure 4.

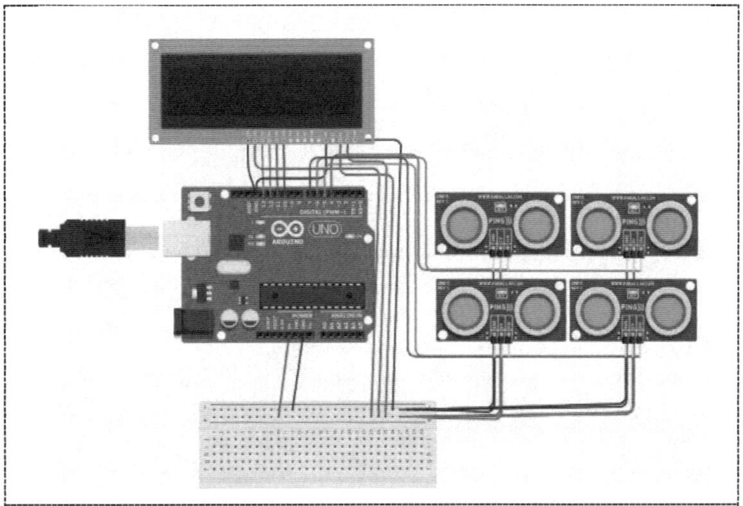

**Figure 4.** Complete design of ultrasonic waves to control brown planthopper pest

## 3. Results and Discussions

The results of the brown plant hopper pest control system (Outwere) were arranged by the Atmega 326 microcontroller with the Arduino Uno board as the control center. HC-SR04was used as actuators to produce Ultrasonic sound waves. For the wideness of range, four HC-SR04 sensors are gathered (Figure 5). All the components are wrapped with paperboard as packaging tools and displayed used a 1. 25 m single pillar (Figure 6). These designs to give comfortable and easy use of tools in test room.

**Figure 5.** 'OUTWERE' frontal site

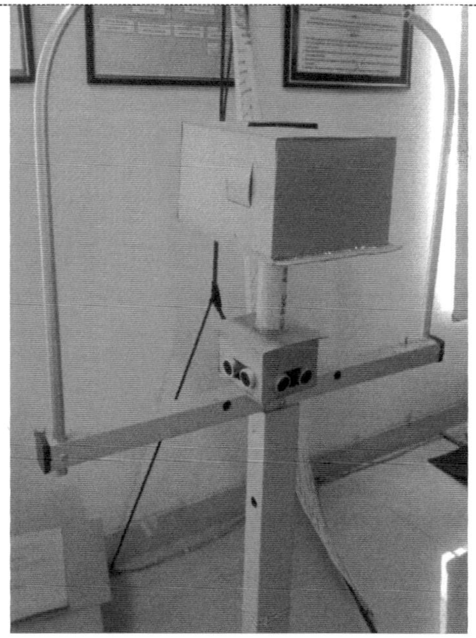

**Figure 6.** 'outwere' in staging

Test functionality by checking each device function, the results of the functional test can be seen in table 1

**Table 1.** Test result for functionality

| No | Device | Trials | Results |
|---|---|---|---|
| 1 | Microcontroller | Upload and run programs | Ok |
| 2 | LCD | Display information | Ok |
| 3 | HC-SR04 | Measure Distance | Ok |

The test results for each component in Outwere (table 1) are work well based on each results test. The microcontroller is tested by uploading the source sketch program, and the results are that the program runs well without errors. The LCD indicated works well by displaying welcome posts with a blue backlight. HC-SR04 given 10uS pulse is required at the trigger input and starts the ranging, then the module will send out an eight cycle burst of ultrasound at 40 kHz and raise its echo setting. The Echo is a distance object that is pulse width and the range in proportion. Then calculate the range through the time interval between sending trigger signal and receiving echoes signal. We suggest to use over the 60ms measurement cycle, in order to prevent trigger signal to the echo signal[11].

After the functionality, test is completed, it is continued with the ultrasonic wave range test using the principle of finding the distance through the ultrasonic waves that are emitted and received again. Ultrasonic sensors mounted on four sides can be used to determine the area that can be reached by ultrasonic waves. The formula used for the area calculation in this system is shown in the following formula (1)

$$L= (p1+p2+5)*(l1+l2+5) \quad (1)$$

p1, p2 are the length values of the measurements of two ultrasonic sensors on two opposite sides.*l*1, *l*2 are the width values of the measurements of the ultrasonic sensors on the 2 sides of the opposite. Because the position of 2 opposite sensors has a distance of 8 cm, the length and width are added to the value of 8cm.

The trial area calculation is done by comparing the area that has been determined by the area of the ultrasonic sensor calculation results. The results of the trial calculation of the area of the ultrasonic waves are seen in table 2.

**Table 2**. Measurement Area Ultrasonic Wave

| No | Actual area ($cm^2$) | Sensing Area ($cm^2$) | absolute error ($cm^2$) | Average absolute error ($cm^2$) | Average relative error (%) |
|---|---|---|---|---|---|
| | 400 | 420 | 20 | | |
| **1** | 400 | 400 | 0 | 13,4 | 3,4 |
| | 400 | 420 | 20 | | |
| | 2500 | 2652 | 152 | | |
| **2** | 2500 | 2756 | 256 | 204 | 8,16 |
| | 2500 | 2704 | 204 | | |
| | 4000 | 4264 | 264 | | |
| **3** | 4000 | 4335 | 335 | 396,3 | 9,9 |
| | 4000 | 4590 | 590 | | |
| | 6000 | 6448 | 448 | | |
| **4** | 6000 | 6875 | 875 | 556 | 9,27 |
| | 6000 | 6375 | 375 | | |

Table 2 is the first stage of a trial to find out wideness ultrasonic wave range of brown plant hopper pest control devices "Outwere." In this study, a test area was prepared with an area ranging from 400

$cm^2$ - 6000 $cm^2$. In testing in a room of 400 $cm^2$, the measurement results obtained by the instrument amounted to 420 $cm^2$,400cm2, 420 $cm^2$with an average error value of 34 where the average relative error was found to be 8.5%. The second test in a room with an area of 2500 $cm^2$ from the measurement results obtained an area of 2652 cm2, 2756 $cm^2$, 2704$cm^2$ so it is known that the average absolute error is 204 $cm^2$ and Average relative error is 8.16%. The third test in a room with an area of 4000 cm2 obtained the results from an area of 4264$cm^2$, 4335$cm^2$, 4590$cm^2$. The average absolute error from the third test was 396.3 $cm^2$ and the average relative error was 9.9%. For the 4th test, a room of 6000 $cm^2$ was used, and the results of the ultrasonic sensor were 6448 $cm^2$, 6875 $cm^2$, 6375 $cm^2$ with an average absolute error of 556 $cm^2$ and Average relative error of 9.27%. The results in Table 2 shows the ultrasonic waves from HC-SR04 can still reach an area of 6000 $cm^2$ even though the calculation results of ultrasonic sensors still show differences between the actual area and the sensing area with an average relative error between 8.16% - 9.27%.The range of error may have occurred caused the range of between two opposite sensors. The maximum effective ranges of the tools are three meters from each sensor. This resulting comforted with Mansyur, et. al (2009) [9].

The last trial was the exposure of ultrasonic waves to brown plant hopper who was carried out in a closed room with a size of 120x50x30 cm. in the initial stages of this study tried to see the behavior of brown plant hopper in a container given ultrasonic waves for 15 minutes. This exposure test room can be seen in Figure 7. In the test room, a sample of 80 brown plant hopper was inserted, then the 'outwere' tool was placed in the middle and turned on for 15 minutes. Observations showed that no brown plant hopper was perched at the center of the wave. During the wave exposure, brown plant hopper moves away from the ultrasonic wave center.

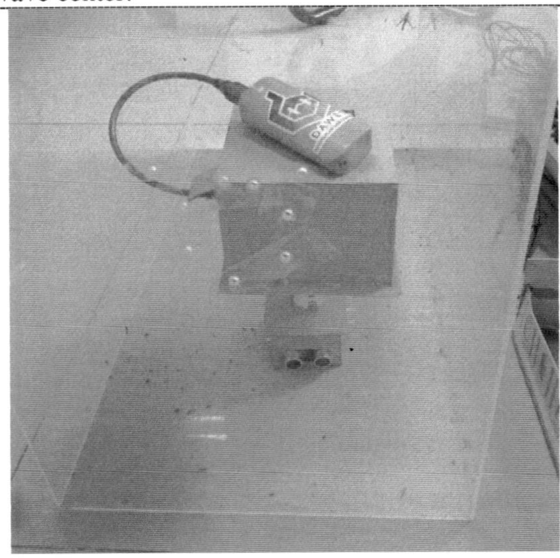

**Figure 7**. Test room of exposure ultrasonic waves to brown plant hopper

The results of this research showed the planthopper are triggered to move away from the sensors and the range. These current results agreed with some Mansyur, et. al (2009) [9] and Julian & K. Triyan (2017) [8] since the ultrasonic give an effect to behavior movement.

## 4. Conclusion

The tools of Outware reading need improvement since average reading relative error between 8.16% - 9.27% to get more precisely measurement. The effective ranges of tools are three meters from the sensors. We suggest to deeply studies in arrangement of components to get the wider effect of controlling pest, including for setting frequency used in the tool.

## References

[1] Horgan F G and Sydney T 2017 Integrated pest management for sustainable rice cultivation : a holistic approach *Burleigh Dodds Science Publishing Limited* pp 1–34

[2] Budiman A and Utomo P 2018 Pengembangan Website Informasi Hama Wereng *J. Pilar Teknol.* **3** 14–9

[3] Dyck V . and Thomas B 1979 *Brown planthopper : threat to rice production in Asia.* (Los Ba�os, Philippines: International Rice Research Institute)

[4] Kailas K N, S B P and Namdev K K 2015 Development Performance And Evaluation Of Ultrasonic Pest And Insect *Int. J. Eng. Res. Gen. Sci.* **3** 19–25

[5] Agusdian R and Rakhmadi F A 2012 Sistem Proteksi Tanaman Padi Dari Serangan Hama Wereng Menggunakan Gelombang Ultrasonik Dan Penunjuk Arah Angin *Seminar Nasional Fisika dan Pendidikan Fisika*

[6] Budiarso Z and Prihandono A 2015 Implementasi Sensor Ultrasonik Untuk Mengukur Panjang Gelombang Suara Berbasis Mikrokontroler *J. Teknol. Inf. Din. Vol. 20, No.2, Juli 2015 171-177* **20** 171–7

[7] Telaumbanua M, Anggraini R, Sasongko F I, Fitri A, Sari R F M and Waluyo S 2018 Control System Design for Rat Pest Repellent in the Rice Field Using a Modified ATMega328 Microcontroller Modified with Ultrasonic Sound Wave *Int. J. Eng. Invent.* **7** 22–8

[8] Julian T and Triyana K 2017 Pengujian Akuisisi Data sENSOR Ultrasonik HC-SR04 dengan Mikrokontroler Atmega 8535 ( Testing Data Acquisition of Ultrasonic Sensor HC-SR04 using Atmega 8535 Microcontroller ) *UNIERA* **6** 35–40

[9] Mansyur M, Rianti E D D and Setiawan H 2009 Optimasi Frekuensi dan Dosis Paparan Gelombang Ultrasonik Untuk Membunuh Jentik Nyamuk *J. Ilm. Kedokt. Wijaya Kusuma*

[10] Utama S N 2017 Perancangan System Pengusir Nyamuk Menggunaan Sensor Ultrasonik Dengan Panel Surya Sebagai Sumber Energi *Semin. Nas. Teknol. dan Rekayasa* 1–6

[11] Latha N A, Murthy B R and Kumar K B 2016 Distance Sensing with Ultrasonic Sensor and Arduino *Int. J. Adv. Res. Ideas Innov. Technol.* **2** 1–5

## Acknowledgments

This research was supported by Ministry of Research, Technology and Higher Education of the Republic of Indonesia. We thank our colleagues from University of Darussalam Gontor, who provided insight and expertise that greatly assisted the research, although they may not agree with all of the interpretations/conclusions of this paper.

Journal of Physics: Conference Series **1381** (2019) 012058 doi:10.1088/1742-6596/1381/1/012058

# Distance Measurement of ESP8266 for Control and Monitoring in Smart Home Application

**L Syafaah*, A E Minarno, F D S Sumadi, G W Mukti**

Informatics Department, Universitas Muhammadiyah Malang, Indonesia

*lailis_tsd@yahoo.co.id

**Abstract.** Along with the growth of Internet of Things (IoT) projects, the development of smart home application was also incrementally raised. The research trends tended to implement the smart home application for monitoring the house hold devices through automation and also for metering the power usage independently. Therefore, the proposed research which was developed in this paper focused for building a smart home application which could control and monitor based on the power usage of the electronic household devices that serially connected to the sensor node. This paper also proposed a combination of MQTT as the messaging protocol for conserving the bandwidth usage. In term of the project's analysis, the main aspect was directed to investigate the feasibility of ESP8266 as the transmission module in our application. The developed system could successfully monitor the power usage (in Watt and Ampere) and perform the specified action from the client node. The research result also concluded that the ESP8266 was feasible for this kind of project.

## 1. Introduction

IoT has became one of the most discussed topic among researcher. The definition itself separates IoT into the two meanings which can be classified as the device that connects to the cloud or internet as the extension of it and the derivative device on local network that performs either transmission or receiving data [1]. In fundamental term, the IoT can be defined as any objects that directly connected to the internet identified by using Internet Protocol (IP) address. One of the IoT sector that prominently gains concern is smart home. The implementation of automatic system based on the connected object available on the recent house technology may bring this topic to next level. One of the beneficial purpose of deploying the smart home application is the utilization process for metering the power usage of the household electronic devices for preventing the unessential utilization of power [2]. Therefore the amount of household expense can be minimized.

This paper proposed to build an IoT based smart home application for monitoring the watt and current usage of household electronic devices as well as performing control mechanism by either replace or insert the electricity using the MQTT messaging protocol. In term of the research gap analysis, this research was directed to perform in-depth analysis regarding to the feasibility of using ESP8266 as the transmission module for supporting the smart home application.

## 2. Related Works

Previously there were several related papers that concerned their topic to investigate some areas in smart home project. Several papers focused their research for generating various type of smart home application, such as smart metering application [2], household monitoring application [3],[4],[9], automation of household devices [5], simulation of smart home application [6]. Some of the mentioned paper had already used the ESP8266 for their transmission module installed on each of the sensor node [2],[3],[4],[5],[8],[11].

In addition, There were also several papers that has investigated the implementation of MQTT protocol in IoT application [4],[7],[9],[10]. All of them analyzed the impact of deploying the MQTT regarding to its functionality upon developed in IoT project and the differentiation of packet sending and power consumption rate compared to the other messaging protocol in IoT (HTTP and CoAP). The most related paper [4], developed a monitoring application for household devices as well as could perform action related to the user defined action deployed using MQTT infrastructure. This paper did not perform a smart metering project which was related to the power management system of household application. Therefore, the proposed research was directed for extending both smart metering project [2] and the household monitoring system [4] to be able to monitor the energy consumption of each the available electronic household devices as well as perform the related action (disconnect/connect the electricity if the usage trend above/below the standard utilization). The proposed system also implemented the MQTT messaging protocol for conserving the bandwidth usage on each of the link on the IoT system.

In term of the feasibility experiment for testing the ESP8266 as the standard module for smart home application was not being performed on the mentioned paper above. Therefore, the main topic for the proposed research was deployed for calculating the distance of ESP8266 for control and monitoring in smart home application.

## 3. Proposed System

The configuration of the proposed system is described on the Figure 1. There exist 4 sensor module and 1 AC relay. The available sensors include the 5A Non Invasive Current sensor and Current Transformer sensor with a maximum current of 5 amperes, and the temperature and humidity sensor (DHT11). This current sensors produce an analog AC voltage generated from the current reading process. There were also an AC relay for either suspend or restore the electricity flow through the household electronic devices which reside on 220 Volts and 10 Ampere in maximum values.

The implementation topology deployed by using the MQTT messaging protocol. The main reason why the research deployed the MQTT because it provided publisher/subscriber mechanism which could significantly reduce the bandwidth utilization during the data transmission process for both link between publisher-broker or subscriber-broker. There were 3 nodes available including the sensor node performed as publisher/subscriber, the broker node, and the client node functioned as the publisher/subscriber. The diagram of the MQTT is described in Figure 2. The broker node was deployed using Raspberry pi 3+ and cloud MQTT server. The MQTT software installed in broker node was Mosquitto (for local network). Both sensor node and the client node behaved as publisher/subscriber. The sensor node published the data regarding to the sensor reading and subscribed the command from the client node to perform action (release/restore the electricity on relay module), while the client node subscribed the sensor's data and published the command action based on the available data metric from sensor node. Another notable advantage upon implementing the MQTT was the sensor node and the client node did not have to online at the same time since the broker node could have a capability to store the incoming sensor's data into its local database.

As far as the distance measurement of the proposed system was concerned there were two

Journal of Physics: Conference Series **1381** (2019) 012058 doi:10.1088/1742-6596/1381/1/012058

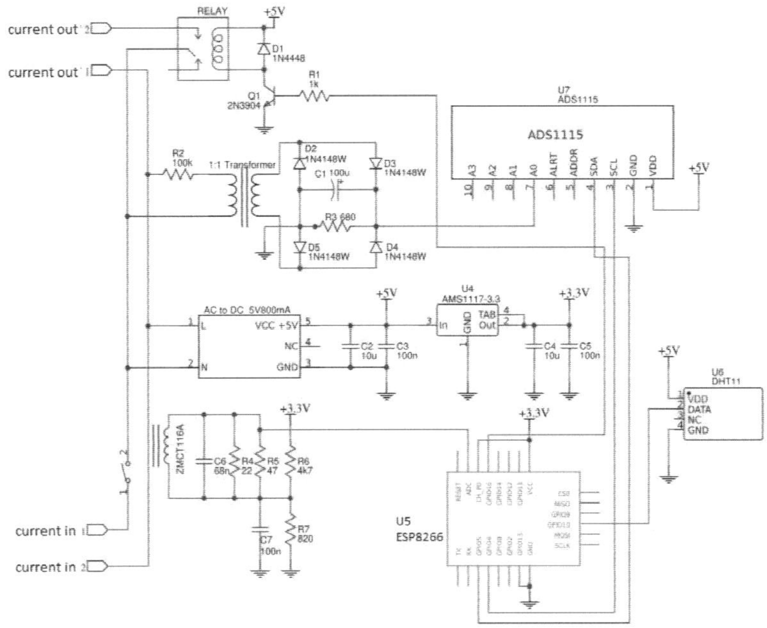


**Figure 1.** Block diagram of the sensor node

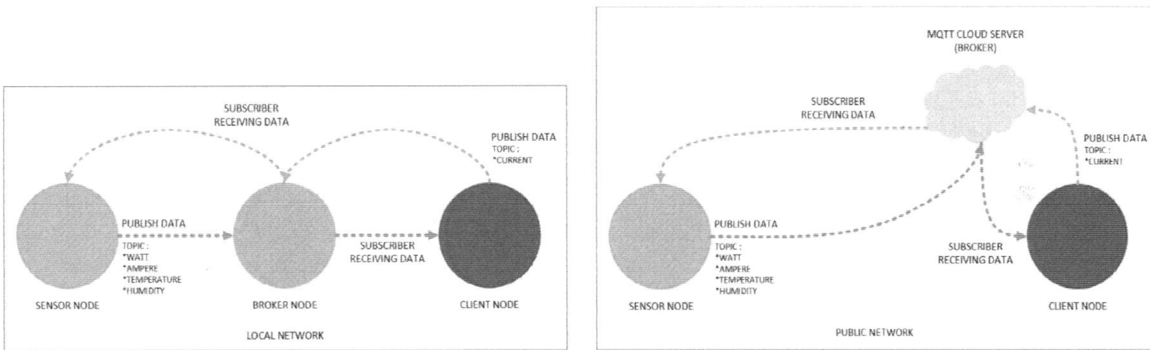


**Figure 2.** Topology of the smart home application

main experiment's scenario for evaluating the feasibility of ESP8266 including the No Line Of Sight (NLOS) and Line Of Sight (LOS). The map of the NLOS scheme and the possible location of physical interference are illustrated in Figure 3. There would be three different distance measurement in indoor side which varies in horizontal line. Each of the planned scenario in NLOS incremented the measurement distance within 2 metre interval which ultimately stopped on different final position scenario 1 after reaching 12 metre, scenario 2 after reaching 4 metre, and the scenario 3 after reaching 14 metre. The experiment only extracted the data from the link between sensor node and the broker node, since the link between the broker node and the client node was connected through the internet. In term, of the LOS scenario, the implementation of experiment was conducted by calculating the impact of increasing the distance measurement between sensor node and the broker node. The initial distance is 0 metre and the ultimate distance calculated after the packet loss hit 100%. The analysis was conducted by evaluating the Received Signal Strength Indicator (RSSI) in dBm and the packet loss percentage. In addition, this paper also investigated the comparison of power usage generated by the sensor node when implemented the MQTT and without implemented the MQTT.

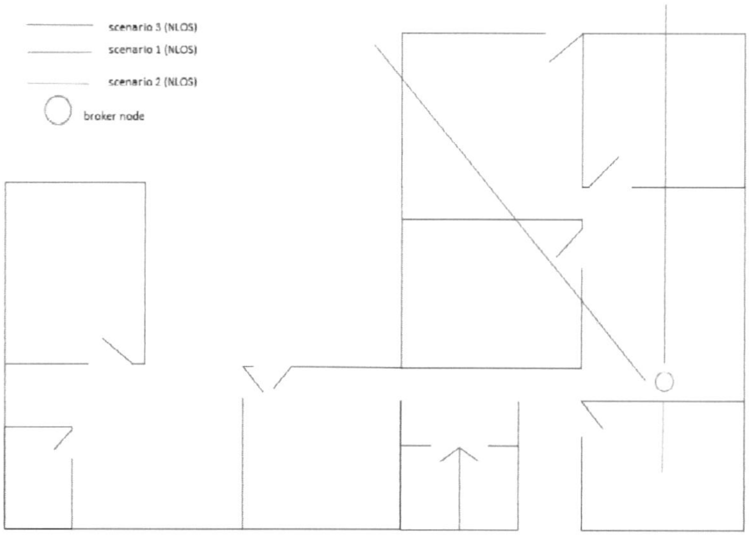


**Figure 3.** NLOS experiment scheme

## 4. Result and Analysis

In term of the research result and analysis, all of the proposed implementation scenario had been performed based on the given scenario. There were three implementation schemes including the the LOS approach, the NLOS approach, and the power consumption of sensor node in 10 hours period. Both in either LOS or NLOS, the experiment's variable that was used for analyzing the distance measurement of ESP8266 were packet loss ratio in percentage and the RSSI length in dBm. As can be seen form the Figure 4, 5, and 6, the measurement of packet loss variable through out the whole NLOS scenario still could produce healthy network environment by pointing almost 0 % in average even though there was an interval extension of the measurement distance. However on the first scenario, when the sensor node was shifted to 6 metre, there was a slight increase on the packet loss percentage, which could be caused by the similar signal interference during the experiment affecting some of the unreceived packet. Those trends in NLOS scenario indicated that ESP8266 could possibly become the alternative for either behave as the transceiver or receiver module. The implementation of MQTT transmission protocol during the experiment also did not causing the packet loss percentage to grow significantly which also demonstrated that MQTT was suitable for being deployed in IoT smart-home project. As far as the RSSI value is concerned, the Figure 7, 8, and 9, illustrates the received signal in dBm for the NLOS scenario. The graph shows that the RSSI value is increased along with the growth of the measurement distance. Under normal circumstances or the closest distance (0 metre) possible the normal value of RSSI was pointed at 48 dBm in average for all of the NLOS scenarios. There were no particular standard regarding to the RSSI value, therefore the authors could only assume that the vast decremented value of RSSI indicating that the signal strength was poor and it was not suitable for performing data transfer. Based on the experiment mapping stated on the section 3, the growth of distance which added by the physical interference (such as walls or house's furniture) has became the main aspect that brought the RSSI value to

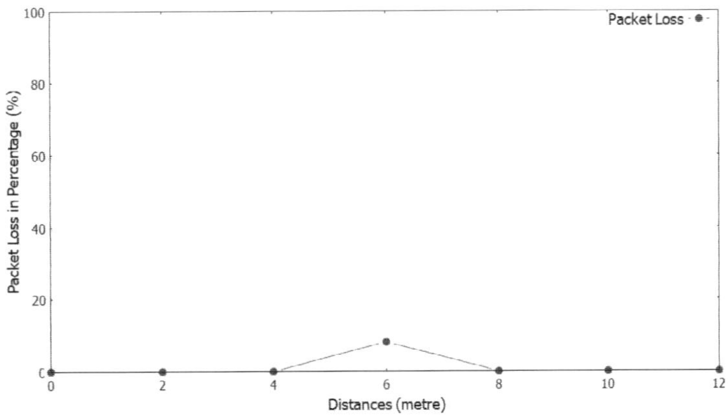


**Figure 4.** Packet loss ratio extracted from scenario 1

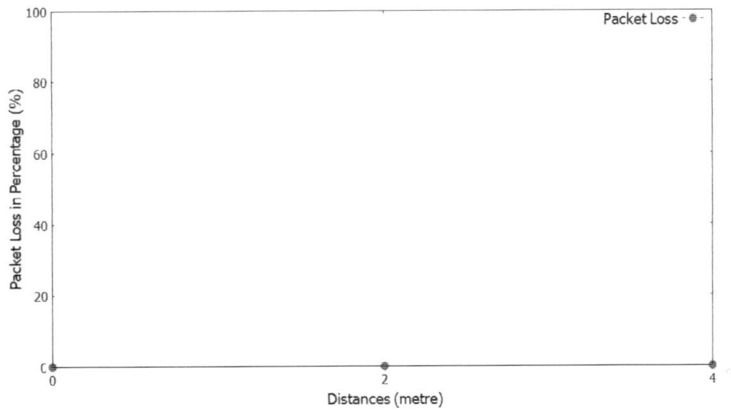


**Figure 5.** Packet loss ratio extracted from scenario 2

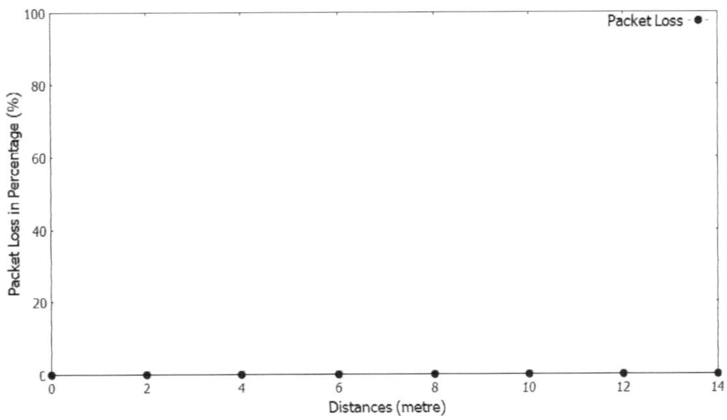


**Figure 6.** Packet loss ratio extracted from scenario 3

drop. Despite of the values of RSSI upon reaching 14 metre on NLOS's scenario 3, the packet loss variable still pointed in 0%, since the transmitted data was consisted of low bytes raw data generated by the publisher node. In term of the LOS scenario upon reaching more than 30 metre

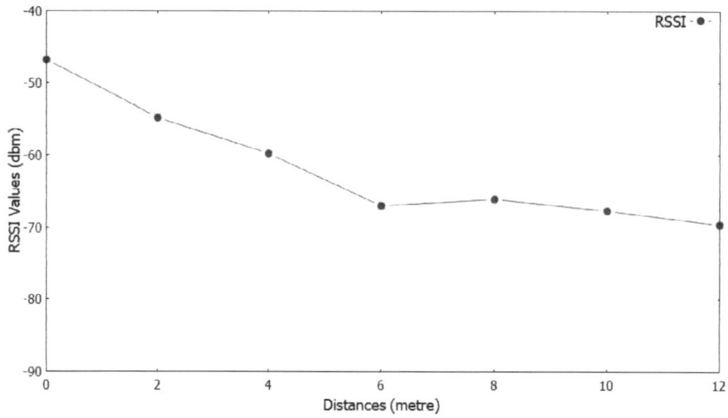


**Figure 7.** RSSI value extracted from scenario 1

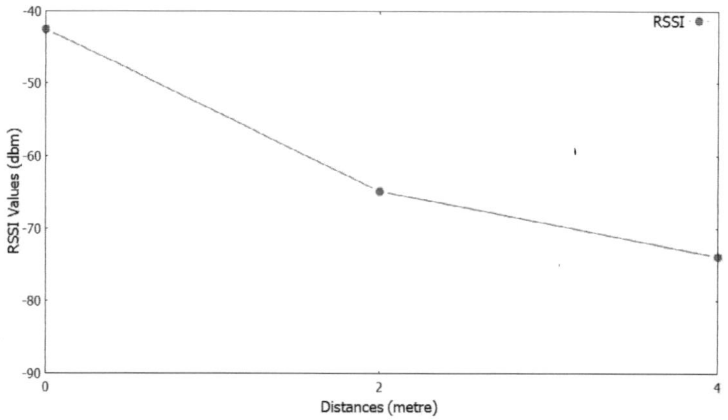


**Figure 8.** RSSI value extracted from scenario 2

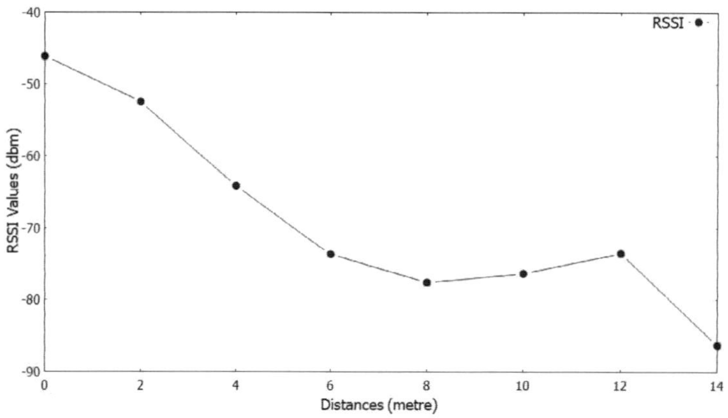


**Figure 9.** RSSI value extracted from scenario 3

distance, the value of packet loss percentage was increased. However, the packet loss percentage during 50 metre distance was monitored at 0%. This circumstance could be occurred due to the environmental aspects such as the wind that existed during the experiment illustrated in Figure

. The RSSI ratio also significantly rise along with growth of the distance. Since there were no physical interference, the RSSI value reached -95 at 75 metre. The distance measurement in LOS scenario were dependence several factors including the extension of distance measurement and natural phenomenon such as wind that could interfered the air which became the transmission media for ESP8266. The experiment's measured the transmission distance for ESP866 ranged within 190 metre. As far as the power consumption of sensor node is concerned, the Figure 12

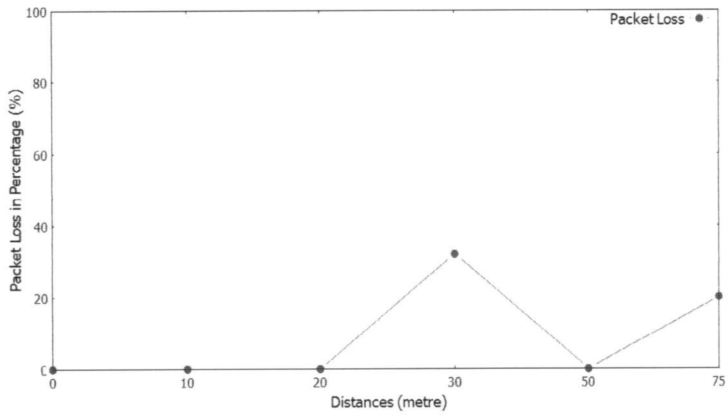


**Figure 10.** Packet loss value during LOS scenario

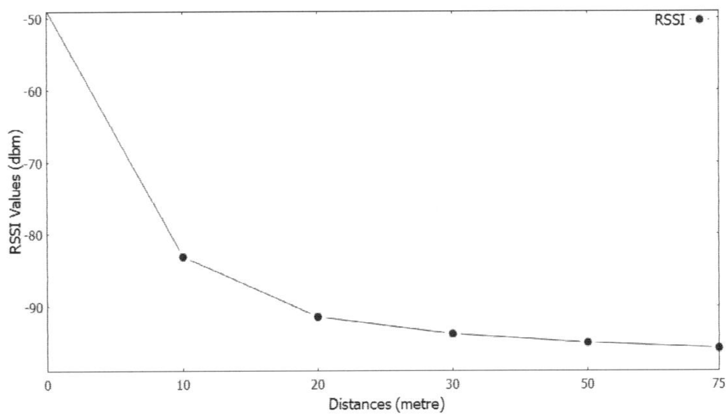


**Figure 11.** RSSI value extracted during LOS scenario

illustrated the power usage of sensor node while implements the MQTT and without implements it. The notable value for both schemes indicated that the implementation of MQTT did not consume more power even though there were a specific mechanism for encapsulating topic for each of the transmitted data.

## 5. Conclusion
In conclusion, ESP866 could become an alternative for being deployed as transmission module. However the developer should maintain the size of the transmitted data, therefore, the packet loss percentage can be minimized. The distance coverage based on the experiment reached up to 14 metre on indoor cases. The deployment of MQTT also did not interfere the transmission

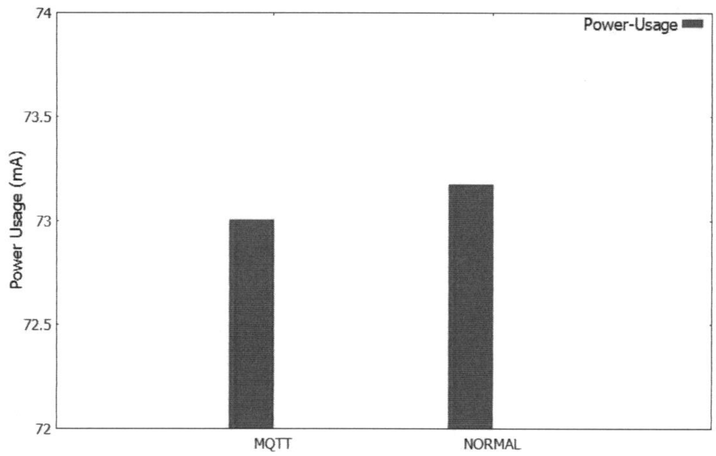


**Figure 12.** Power usage of sensor node within 10 hours period

process. The MQTT benefits the usage of the bandwidth during the data transmission since the subscriber only subscribes their concerned topic.

## References

[1] Li B, Lv S and Pan Q 2018 The Internet of things and Smart Grid IOP Conference Series: Earth and Environmental Science 113 012038

[2] Kaur M, Mathew L, Alokdeep and Kumar A 2018 Implementation of Smart Metering based on Internet of Things IOP Conference Series: Materials Science and Engineering 331 012015

[3] Hutabarat D P, Budijono S and Saleh R 2018 Development of home security system using ESP8266 and android smartphone as the monitoring tool IOP Conference Series: Earth and Environmental Science 195 012065

[4] Amri Y and Setiawan M A 2018 Improving Smart Home Concept with the Internet of Things Concept Using RaspberryPi and NodeMCU IOP Conference Series: Materials Science and Engineering 325 012021

[5] Uriawan W, Zulfikar W B, Sofa R M and Ramdhani M A 2018 Internet of things for automatic garage doors using ESP8266 module IOP Conference Series: Materials Science and Engineering 434 012057

[6] Sihombing O, Zendrato N, Laia Y, Nababan M, Sitanggang D, Purba W, Batubara D, Aisyah S, Indra E and Siregar S 2018 Smart home design for electronic devices monitoring based wireless gateway network using cisco packet tracer Journal of Physics: Conference Series 1007 012021

[7] Atmoko R A, Riantini R and Hasin M K 2017 IoT real time data acquisition using MQTT protocol Journal of Physics: Conference Series 853 012003

[8] Yuliansyah H, Corio D, Yunmar R A and Kahar Aziz M R 2019 Smart-Room Technology Implementation Based on Internet of Things Toward Smart Campus in Institut Teknologi Sumatera IOP Conference Series: Earth and Environmental Science 258 012053

[9] Masdani M V and Darlis D 2018 A comprehensive study on MQTT as a low power protocol for internet of things application IOP Conference Series: Materials Science and Engineering 434 012274

[10] Wardana A A, Rakhmatsyah A, Minarno A E and Anbiya D R 2019 Internet of Things Platform for Manage Multiple Messag Telemetry Transport Broker Server

Kinetik: Game Technology, Information System, Computer Network, Computing, Electronics, and Control; Vol 4, No 3, August 2019

[11] Dadah F S, Ore-Ofe A, Usman A D, Mshelia Y A and Babatunde M O 2018 An Election Energy Threshold Based Multi-Hop Routing Protocol in a Grid-Clustered Wireless Sensor Network Kinetik: Game Technology, Information System, Computer Network, Computing, Electronics, and Control; Vol 3, No 4, November 2018

Journal of Physics: Conference Series **1381** (2019) 012059 doi:10.1088/1742-6596/1381/1/012059

# Design of Portable Electric Hydraulic Jack for Improving The Productivity

**P Rachmawati*, I R Kurniawan**

Universitas Muhammadiyah Yogyakarta, Indonesia

*putri.rachmawati@vokasi.umy.ac.id

**Abstract**. The hydraulic system is a system that works based on character and potential that exists in liquid. Maintenance is a work done in sequence to maintain or to improve existing facilities to comply with the standards. Nonconformities and limitations of tools and design concepts used by humans while working are still found in everyday life. The purpose of this research is to design the hydraulic portable device to facilitate the mechanics in car maintenance and repair especially a part of the bottom machine. The data collection is done by directly measuring the performance test of the portable hydraulic jack in Autocar Andi Jaya compared with the original jack. The conclusion of this research is a tool that has benefits in supporting the car when car service is performed. This tool generates the total number of completed car maintenance and repair outputs of 30 minutes of normal service completion compared to the ordinary jack. Therefore, an electric portable hydraulic jack can speed up the performance of the mechanics and the results in improving productivity.

## 1. Introduction

Jack is a tool used to lift a car so that mechanics can check parts of the bottom of the car. Many types of jacks can be found in car workshops both large workshops and simple workshops supported by hydraulic jacks and car lifts. Hydraulic jacks are the most widely used jacks because they are lightweight and easy to use [1], affordable cost of applying access tools, easy maintenance, and portable. However the hydraulic jack can only lift one side of the car that is approved by the mechanic to service the underside of the car. Usually, the mechanics can solve it more than one side of the car using the help of Jack Stand / stand jack, but the use of Jack Stand is not effective because the new Jack Stand can be applied after the car is lifted using a hydraulic jack. Electric hydraulic jack is a tool used by technicians to improve the car using the working principle of hydraulics so that the technician does not automatically need to spend a lot of labor[2]. In order for electrical control to be applied to the manual control of the hydraulic pump, the author needs to align the manual hydraulic pump (driven) system with electrical control (which drives) using an electric motor. With the installation of an electric motor, this will help the mechanic to make car repairs [3] . The driving motor as a converter of electrical energy becomes a mechanical drive, the use of an electric drive motor on a manual hydraulic pump. From the problems that arise in the upper hydraulic jack, the author has the idea to install an electrical control system on the hydraulic pump that will be applied to a portable hydraulic jack. Portable Hydraulic Jack is a jack that can lift all sides of the car used by mechanics to facilitate mechanics in carrying out car repair and maintenance activities, so from the results of this study will be a jack that can lift all sides of the car and work automatically.

The design of a portable electric hydraulic jack is needed for car service aids so that it can be effective, efficient and make it easier for changing part while washing cars in the process of car engine maintenance in the workshop.

Research purpose is application design of a portable electric hydraulic jack in the workshop. Benefits of research are to simplify mechanics in carrying out the maintenance and repair process of the lower part of the car and speed up mechanics in service and car wash.

## 2. System Design

Designing or designing is a series of procedures to translate the results of analysis and a system into a programming language to describe in detail how the system components are implemented [4].

Hydraulic systems are systems that are often used and developed for industrial activities, from light industry to heavy industries such as the vehicle and mining industries as a means of driving on cutting machines, folding machines, press machines and lifting equipment with a capacity of hundreds of tons [4]. The parts or tools of this hydraulic system are quite simple, so that the operator or user obtains more secure safety and security, besides that the hydraulic system has the advantage of including an energy transfer system using fluid (hydraulic oil) so that it looks more flexible. The piston rod translation from the working cylinder caused by fluid pressure in the cylindrical space of the movement is used for forward and backward and up and down motion according to the installation ie vertical or horizontal direction [5]. Jack is a device that serves to help lift a vehicle not to support [6]. Ergonomics is a study of the relationship between people and their work environment.

## 3. Methodology and Result

The concept in this design is divided into the design of its own jack and hydraulic body. The principle of this research is divided into 2 types, namely: jacks and electrical systems in hydraulics. Body simulation jacks use inventor application, with the same concept as the carlift as a car service jack. Electrical systems use designs with AutoCAD [1]. Body design uses folding frame simulation. The method of designing this study with this study can be seen in a flow chart below:

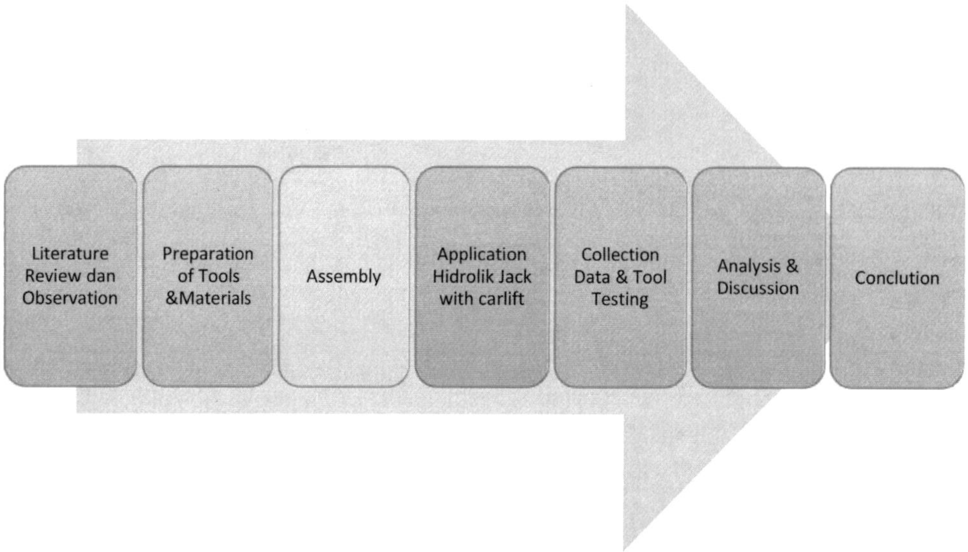


**Figure 1.** Process Flowchart

The folding frame design can be seen below;

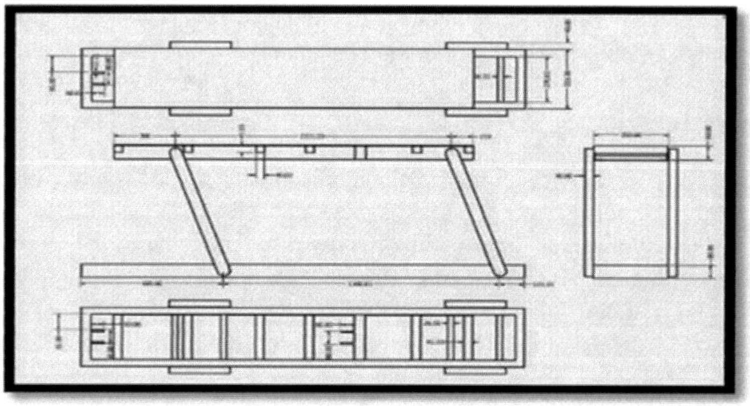

**Figure 2.** Folding Frame 2D

**Figure 3.** Folding Frame 3D

The result of the calculation obtained:

Table 1. Calculation of lift force of Portable Hidraulic Elektric Jack

| No | Hydraulic Tilt | Forlding Frame | Tilt Folding Frame Support style (kN) | Folding Single Frame Support Force (kN) | Hydraulic Lift Force (kN) | Total hydraulic Lift (kN) |
|---|---|---|---|---|---|---|
| 1 | 11° | 12° | 13,07 | 6,54 | 77,04 | 71,65 |
| 2 | 24° | 30° | 5,44 | 2,72 | 36,14 | 30,75 |
| 3 | 29° | 45° | 3,84 | 1,92 | 30,32 | 24,93 |
| 4 | 32° | 60° | 3,14 | 1,57 | 27,74 | 22,35 |
| 5 | 31° | 70° | 2,89 | 1,45 | 28,54 | 23,15 |

The data table 1. above shows the results of hydraulic lifting force and folding frame support force and folding frame support force derived from portable electric hidraulic jack. If viewed through the graph, you will see the ups and downs of lift force data and support forces in each slope of the degree passed by portable electric hidraulic jack. The following graph is from the lift table data above:

Note : a = 11° dan 12°, b = 24° dan 30°, c = 29° dan 45°, d = 32° dan 60°, e = 31° dan 70°

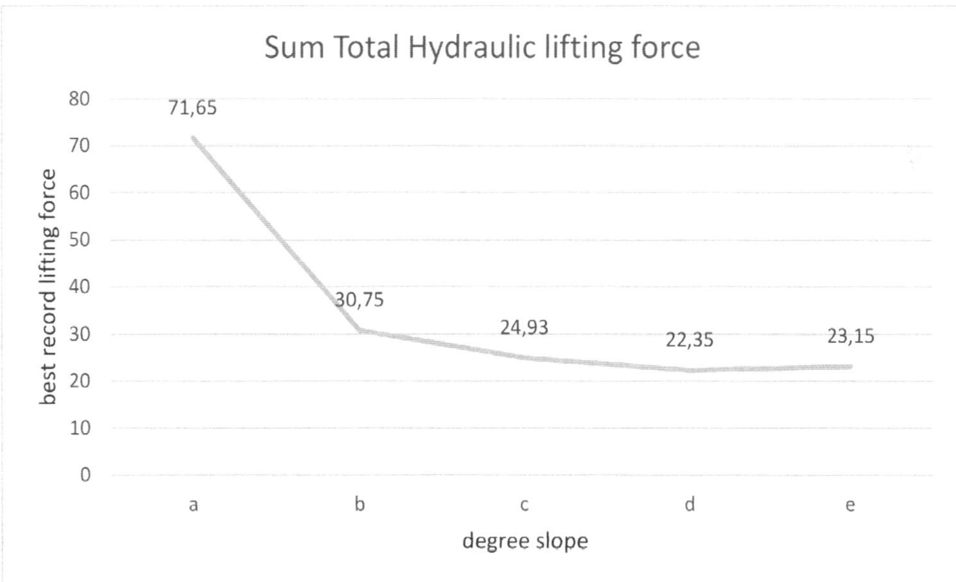


Figure 4. Sum Total Hydraulic Lifting Force

From the calculation data and graph above it can be concluded that the higher the value of the hydraulic tilt angle and the folding frame, the smaller the force produced. That is, both hydraulic and folding frame will be stronger in lifting and supporting the weight of the car being lifted. Thus the optimal lifting force of the folding frame comes from the slope of 70 ° and the hydraulic comes from the slope of 32 °. There are 2 calculations for power value (P), namely input and output power because the electric motor uses gearbox components. Input power using the formula P = V x I: t, that V = 180 Volts, current (I) = 0, 75 Amperes, then divided by time. The results of the input electric motor are 135 Watt. The output power at the electric motor power is not the same because there is a gearbox component used the formula P = T.n that T is the output torque = 577.5 Nm. n (the RPM coming out of the gearbox that has a value of 57 revolutions per minute), the result of the output power value is 32, 92 Watts. The power that came out was very large because of the gearbox components so that the rotation of the electric motor was very slow and very strong. The working principle of the installation of an electric control system on this portable hydraulic jack is when the switch is on then the electric motor will function to rotate the slab and on the slab, there is a lever so the lever will fluctuate pumping the hydraulic pump.

## 4. Conclusion

The results of replacing spare parts and total service with this portable electric hydraulic jack are compared with a crocodile jack of only 100 minutes while a crocodile jack of 130 minutes. The results obtained show that the design of this tool can increase mechanical productivity.

## 5. Acknowledgement

We are very grateful to LP3M Universitas Muhammadiyah Yogyakarta for helping to complete this research. We are thank you for Diploma program majoring the Mechanical Engineering Muhammadiyah University of Yogyakarta in supporting costs to attend the conference.

## References

[1] Sainath K 2014 Design of Mechanical Hydraulic Jack *IOSR J. Eng.* **4** 15–28
[2] Chin P E H 1985 United States patent *Geothermics* **14** 595–9
[3] Agrawal M, Thakur M G, Mahajan D and Chahar T S 2018 Inbuilt Hydraulic Jack System for Four Wheelers : -A new trend in automobile sector 1386–90
[4] Middlesworth M 2014 A Step-by-Step Guide Rapid Entire Body Assessment ( REBA ) *Ergon. Plus Inc* 1–11
[5] Frederick R C, Dixon B, Middlesworth M, Agrawal M, Thakur M G, Mahajan D, Chahar T S, Chin P E H, B-b S, A-a S and Sainath K 2014 Optimizing Performance and Fuel Efficiency for a Formula SAE Car *IOSR J. Eng.* **4** 15–28
[6] Plantard P, Shum H P H, Le Pierres A S and Multon F 2017 Validation of an ergonomic assessment method using Kinect data in real workplace conditions *Appl. Ergon.* **65** 562–9

Journal of Physics: Conference Series **1381** (2019) 012060 doi:10.1088/1742-6596/1381/1/012060

# Supply chain risk assessment of cotton shirt production uses the house of risk method

**A Nalhadi[1], A Kurniasari[1], N Djamal[1], S Suryani[2], S Supriyadi[1]**

[1]Department of Industrial Engineering, Faculty of Engineering, Universitas Serang Raya, Banten 42116, Indonesia

[2]Department of Management, Faculty of Management, Sekolah Tinggi Ilmu Ekonomi Al-Khairiyah, Banten 42441, Indonesia

E-mail: supriyadi@unsera.ac.id

**Abstract**. The purpose of this study is to mitigate risks in supply chain activities in the production of cotton shirts. The method used in this research is the House of Risk (HOR) approach based on the Supply Chain Operation Reference (SCOR) model. The House of Risk (HOR) offers a framework for effectively controlling risks in managing risks from risk agents. Aggregate Potential Risk (ARP) is used to determine the priority of risk sources that need to do in the design of mitigation strategies. The research results identify as many as 26 Risk Events and 21 Risk Agents. Three Risk Agents based on ARP values are Procedures for Delivery of Less Organized Raw Materials, limited and Sudden Product Demand from consumers Mitigation actions used to prevent the causes of risk from occurring are creating standard operating procedures for raw material delivery activities. Also, a contract system establishes between suppliers of raw materials and freight forwarding services. Migration in human resources is evaluating employee performance. Empowerment of the community around the company, subcontracting to other similar companies, reviewing raw material stock is an act of meditation in sudden product demand.

## 1. Introduction

In carrying out business processes, a company has a problem of uncertainty or unplanned events that disrupt the flow of production processes. This uncertainty has an impact on increasing the needs of data sources in anticipating it [1,2]. The determining factors and disorders cause the occurrence of risk[3]. Operational considerations are related to uncertainty in demand, the supply of raw materials and costs. While disruption factors such as natural disasters or economic crises[4].

House of risk (HOR) is one of the uncertainty measurement analysis methods in the supply chain developed by Pujawan, and Geraldin[5]. This model is a development of the quality function deployment (QFD) method, which uses the house of quality (HOQ) to prepare mitigation actions in the face of potential risks in the supply chain. House of risk aims to identify hazards in the supply chain and design prevention strategies to minimize the likelihood of agent risk[6,7]. The House of risk (HOR) divides strategy design into two stages, namely the risk identification and risk treatment phases.

Uncertainty in a make to order industry is very likely to occur. Some of these uncertainties include highly fluctuating demand [8,9], short product life cycle [10–12], outsourcing system when demand is high [13]and the quality of human resources. This research focuses on a cotton t-shirt convection business in Lampung, whose production system is still dependent on consumer demand. Some of the obstacles faced by business owners in running their enterprises include sudden desire and the quality of human resources. This study aims to identify and design a strategy of possible risks that might occur in the supply chain of the production of Katon shirts. The results of the risk mitigation undertaken are expected to be management input in the face of possible risks that occur in the future.

## 2. Method

This research was conducted based on the results of interviews with business owners regarding the constraints that occur while running their business. This research does not discuss related to quantitative data, which is a personal data analysis of company data. The research step guide by the concept of the house of risk[5], which consists of the risk identification phase and the risk treatment phase.

In the concept of the house of risk, risk assessment uses Failure Mode and Effect Analysis (FMEA). The occurrence assessment is carried out to determine the probability level of the risk agent and severity associated with the risk event, which is to assess the severity of the risk event. The steps of the risk identification phase are as follows:

1. Identify risk events that can occur in every business process. This risk can finish through mapping the supply chain of cotton t-shirt production based on the SCOR model (plan, source, make, deliver, and return).
2. Estimating the impact of severity risk events (1-10).
3. Identify sources of risk and assess the likelihood of each risk source occurrence (1-10).
4. Develop matrix relationships. The relationship between each source of risk and each risk event, Rij (0, 1, 3, 9) where 0 indicates no correlation and 1, 3, 9 show a successively low, medium and high relationship.
5. Calculate the aggregate risk potential (agent = j = ARPj) determined as a result of the likelihood of events occurring at the source of risk j and the aggregate of causal effects of each risk event caused by the source of risk j.
   $$ARPj = Oj \sum Si.Rij \qquad (1)$$
6. Ranking risk sources based on a collection of potential risks in descending order (from large to lowest value).

The steps in the risk treatment phase are:

1. Select some risk sources with high priority ranking that might use the Pareto analysis of the ARPj.
2. Identify the consideration of actions relevant to the prevention of sources of risk.
3. Determine the relationship between each precautionary measure and each source of risk, Ejk. Values (0, 1, 3, 9)which indicate that there are no correlations, low, medium, and the long relationship between actions k and source j, respectively.
4. Calculate the total effectivenessof each action.
   $$TEk = \sum ARPj \ x \ Ejk(2)$$
5. Estimating the degree of difficulty in performing each action. The value scale used is weight 3 for mitigation actions that are easy to implement. Weight 4 for mitigation actions that are somewhat difficult to implement and weight 5 for mitigation actions that are difficult to implement.
6. Put those values in a row in the bottom row of the valid total.
   $$ETDk = \frac{Tek}{Dk} \qquad (3)$$
7. Determine the priority ranking of each action (Rk), where ranking 1 gives the meaning of the work with the highest ETDk.

## 3. Result

The mapping of the company's business activities in the production of cotton shirts does by a discussion with the owner of the company. Identification of company supply chain activities based on the SCOR model that occurs in the process of plan, source, make, delivery, and return. Risk event identification is used to determine the risks that may arise in the process of the supply chain of cotton t-shirts with a total of 26 risk events identified (Table 1). Identification of risk agents in every possible risk event that arises made by observation and direct interviews with business owners with the results of 21 risk agents (table 2).

The determination of severity level (table 1) and assessment of occurrence level assessment (Table 2) uses a scale of 1-10 based on the results of the business owner. They use the results as a basis for calculating the development of the matrix relationship and the Aggregate Risk Potential of an agent (ARP). It determines as a consequence of the likelihood of events occurring from the source of risk and the collection of impacts due to risk events caused by sources of uncertainty.The results of the identified risk agents that have the most considerable ARP value determined by the Pareto diagram will be input to HOR2. The priority risk agent to mitigate (Fig. 1) with the priority of mitigation based on the ARP value of A7 (Procedure for sending raw materials less organized) ARP 3344, A10 (Limited Human Resources) ARP 3258, and A4 (Sudden requests from consumers) ARP 3024.

**Table 1.** Risk event and severity

| Process | Sub Process | Risk Event | ( Ei ) | Severity |
|---|---|---|---|---|
| Plan | Product demand | Sudden changes in production plans | E1 | 6 |
| | Control of raw material inventory | Inadequate inventory parameters | E2 | 3 |
| Source | Procurement process | Delay in raw materials from suppliers | E3 | 6 |
| | | Disruption of raw material supply | E4 | 6 |
| | | Raw materials sent are not inspected by the recipient of the goods | E5 | 3 |
| | | Changes in the quality of raw materials | E6 | 8 |
| | | Error choosing supplier | E7 | 3 |
| | | Error item posted by the supplier | E8 | 3 |
| | | Lack of communication with suppliers | E9 | 2 |
| | | Lack of expertise and qualifications for human resources | E10 | 6 |
| Make | Execution | Defect products | E11 | 9 |
| | | Inventory cannot use | E12 | 1 |
| | Production scheduling | Delay in production | E13 | 6 |
| | | Inefficient process | E14 | 3 |
| | Packing process | Engine failure (downtime) | E15 | 9 |
| | | Lack of engine maintenance | E16 | 9 |
| | | The packaging is damaged or torn | E17 | 6 |
| | Conduct product inspection | Delay in completion of production | E18 | 6 |
| | | Do not carry out production activities | E19 | 9 |
| | | Lack of employee ability | E20 | 3 |
| | | Lack of quality checking products | E21 | 8 |
| Deliver | delivery process | The product sent to the wrong destination | E22 | 9 |
| | Warehousing | Stacking of finished products | E23 | 3 |
| | | Damage to raw materials during storage | E24 | 3 |
| | | Late delivery of products to consumers | E25 | 9 |
| Return | Return of the defected product | Delay in the return process from the management industry to consumers | E26 | 9 |

**Table 2.** Identify causes of occurrence risks and value

| Risk Agent | Code | Occurrence |
|---|---|---|
| Significant increase in demand | A1 | 6 |
| Seasonal factors demand | A2 | 6 |
| Request for sudden purchase of raw materials | A3 | 7 |
| Immediate product demand from consumers | A4 | 7 |
| Direct engine damage | A5 | 4 |
| The recipient of raw materials is not precise | A6 | 4 |
| Procedure for sending raw materials is unorganized | A7 | 8 |
| Lack of communication and information | A8 | 6 |
| Depends on one supplier | A9 | 6 |
| Limited human resources | A10 | 6 |
| The efficiency factor during the production process takes | A11 | 4 |
| Unused raw materials | A12 | 5 |
| Lack of preparation when production carry out | A13 | 4 |
| Lack of maintenance management | A14 | 3 |
| Lack of packaging quality | A15 | 5 |
| Product quality inspection is not thorough | A16 | 6 |
| A sudden product design change from consumers | A17 | 6 |
| Lack of storage space/storage material | A18 | 4 |
| Lack of space/storage for finished products | A19 | 4 |
| Interference during trip | A20 | 1 |
| Distance and communication factors with consumers | A21 | 2 |

Unorganized shipping procedures due to only one supplier in the business. A single supplier is very influential on the stability of the raw material flow. The selection of suppliers of raw materials should consider supplier reliability [14]so that it does not interfere with the production process. In the supply chain principle, flexible supplier sources are more profitable for the company in minimizing inventory constraints or risks related to raw material delays [15].

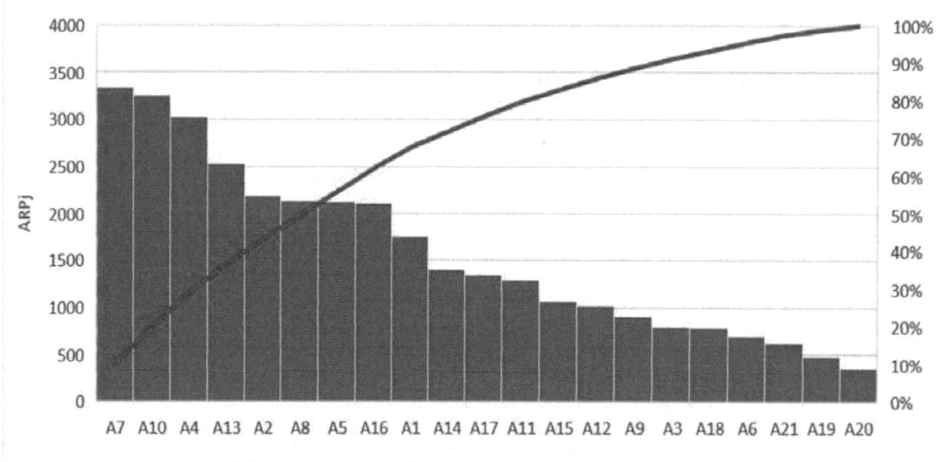


**Figure 1.** ARP risk agent pareto diagram

**Table 3.** Matrix of mitigation strategy (HORII)

| Risk Agent | Making the standard operational procedure activities PA1 | Making a supplier contract system with delivery service PA2 | Evaluate employee performance PA3 | Empowerment of the surrounding community company PA4 | Subcontracting with similar companies PA5 | Review of raw materials PA6 | ARP |
|---|---|---|---|---|---|---|---|
| A7 Procedure for sending raw | 9 | 3 | 1 | 0 | 1 | 0 | 3344 |
| A10 Limited human | 0 | 0 | 9 | 9 | 3 | 0 | 3258 |
| A4 Sudden product demand | 0 | 0 | 0 | 1 | 9 | 9 | 3024 |
| Total Effectiveness(TEk) | 30096 | 10032 | 32666 | 32346 | 40334 | 27216 | |
| Degree of Difficulty (Dk) | 4 | 5 | 4 | 5 | 3 | 5 | |
| Difficulty ratio (ETD) | 7524 | 2006.4 | 8166.5 | 6469.2 | 13444.7 | 5443.2 | |
| Ranking | 3 | 6 | 2 | 4 | 1 | 5 | |

The results of the identification of mitigation actions from three Priority Risk Agents produce six mitigation actions with notation PA1 (making operational standard procedures for the delivery of raw materials), PA2 (creating a contract system between suppliers of raw materials and freight forwarding services), PA3 (evaluating employee performance) , PA4 (community empowerment around the company), PA5 (subcontracting with other similar companies), and PA6 (review of raw material stock). After identifying mitigation actions, a correlation value (Ejk) is measured between the mitigation actions with priority risk agents and Total Effectiveness (TEk) by adding up the results of multiplying the correlation values (Ejk) with the ARP of each priority risk agent (AJ). The next step is to measure the degree of difficulty until the stage 2 HOR model is obtained (Table 3).

In the supply chain system, collaboration with other companies is the key to success in running a business. Subcontracting with other companies is one way to increase operating efficiency [16–18]with established quality standards. Trust maintenance and excellent communication minimize the issue of outsourcing. Outsourcing system produces product quality and time faster than adding human resources whose condition has not tested. Production capacity constrains the selection of this system when sudden or high order.

Evaluate employee performance by making precise employee administration data, then target data for making the right products that can be achieved by employees. Compensation [19]and skill improvement are ways to improve employee performance and retain skilled employees. They are making standard operational procedures for the delivery of raw materials to ensure the availability of raw materials in supporting the production process so that the product arrives at consumers on time.

## 4. Conclusions

Based on the analysis of data that has finished, obtained six mitigation actions in handling three Priority Risk Agents. Three priority action mitigations are procedures for sending raw materials that are less organized, limited human resources, and sudden product demand from consumers. Six mitigation actions are prioritized to be realized by carrying out sub-contracts or cooperation with similar companies, evaluating employee performance, making standard operating procedures for raw material delivery activities, empowering communities around the company, reviewing raw materials, and establishing supplier contract systems with shipping services. Cooperation with similar companies in certain conditions has a significant influence in maintaining the sustainability of the company. This research can be forwarded to a lean supply chain to get a more effective and efficient supply chain system.

**References**

[1] Vanany I, Zailani S and Pujawan N 2009 Supply Chain Risk Management *Int. J. Inf. Syst. Supply Chain Manag.* **2** 16–33
[2] Kusmantini T, Guritno A D and Rustamaji H C 2015 Mapping of supply chain risk in industrial furniture base on house of risk framework *Eur. J. Bus. Manag* **14** 104–15
[3] Tang C S 2006 Perspectives in supply chain risk management *Int. J. Prod. Econ.* **103** 451–88
[4] Chopra S and Sodhi M S 2004 Supply-chain breakdown *MIT Sloan Manag. Rev.* **46** 53–61
[5] Nyoman Pujawan I and Geraldin L H 2009 House of risk: a model for proactive supply chain risk management *Bus. Process Manag. J.* **15** 953–67
[6] Anggrahini D, Karningsih P D and Sulistiyono M 2015 Managing quality risk in a frozen shrimp supply chain: a case study *Procedia Manuf.* **4** 252–60
[7] Parinduri S K, Sinulingga S and Sembiring M T 2019 Design the Supply chain risk mitigation with supply chain risk management approach in spring bed factory *IOP Conference Series: Materials Science and Engineering* vol 505 (IOP Publishing) pp 1–7
[8] Wang X, Chan H K, Yee R W Y and Diaz-Rainey I 2012 A two-stage fuzzy-AHP model for risk assessment of implementing green initiatives in the fashion supply chain *Int. J. Prod. Econ.* **135** 595–606
[9] Shen B, Choi T-M, Wang Y and Lo C K Y 2013 The coordination of fashion supply chains with a risk-averse supplier under the markdown money policy *IEEE Trans. Syst. Man, Cybern. Syst.* **43** 266–76
[10] Xie C, Anumba C J, Lee T, Tummala R and Schoenherr T 2011 Assessing and managing risks using the supply chain risk management process (SCRMP) *Supply Chain Manag. An Int. J.* **17** 323–36
[11] Chiu C-H and Choi T-M 2016 Supply chain risk analysis with mean-variance models: A technical review *Ann. Oper. Res.* **240** 489–507
[12] Choi T-M and Chiu C-H 2012 Mean-downside-risk and mean-variance newsvendor models: Implications for sustainable fashion retailing *Int. J. Prod. Econ.* **135** 552–60
[13] Xiao T and Jin J 2011 Coordination of a fashion apparel supply chain under lead-time-dependent demand uncertainty *Prod. Plan. Control* **22** 257–68
[14] Levary R R 2007 Ranking foreign suppliers based on supply risk *Supply Chain Manag. An Int. J.* **12** 392–4
[15] Tang O and Musa S N 2011 Identifying risk issues and research advancements in supply chain risk management *Int. J. Prod. Econ.* **133** 25–34
[16] Olson D L and Wu D 2011 Risk management models for supply chain: a scenario analysis of outsourcing to China *Supply Chain Manag. An Int. J.* **16** 401–8
[17] Gonzalez R, Gasco J and Llopis J 2005 Information systems outsourcing reasons in the largest Spanish firms *Int. J. Inf. Manage.* **25** 117–36
[18] Lockamy III A and McCormack K 2010 Analysing risks in supply networks to facilitate outsourcing decisions *Int. J. Prod. Res.* **48** 593–611
[19] Distelhorst G, Hainmueller J and Locke R M 2016 Does lean improve labor standards? Management and social performance in the Nike supply chain *Manage. Sci.* **63** 707–28

# The effect of temperature on measurement accuracy

**H A Khoiri[1*], W Isnaini[1] and T A Edison[1]**

[1] Industrial Engineering Department, Universitas PGRI Madiun, Madiun, Indonesia

*Email: halwaannisa@unipma.ac.id (corresponding author)*

**Abstract.** The work environment gives some effects on the employee. One of these is the temperature. Inappropriate working temperatures can interfere with employee work performance. This study aims to proof the effect of temperature on the work efficiency by using the student as a respondent. The student must finish the task in two type temperature condition: with Air conditioner (AC) and no air conditioner (Non-AC). The task was to measure twenty-five workpieces (length and diameter) repeatedly using five types of measuring instruments. Different environmental temperatures are thought to affect student concentration during repeated measurements. The effect of temperature on the measurement results was tested using Two Way ANOVA. Other factors that are taken are the measuring instruments. The results obtained are the temperature conditioned using AC and Non-AC affect the measurement results with p-value<0.05. In addition, the temperature and measuring instruments together influence the measurement results with p-value <0.05.

## 1. Introduction

Employee performance is an important factor for the progress of a company. One of the things that supports employee performance is the working environment. A conducive working environment is closely related to how an employee can complete a task or job [1].

In the working environment, temperature becomes an important component because it is related to how the body can survive in different temperatures. Improper temperature can interfere employee concentration in completing their works [2]. In industry, one type of task or job that requires concentration is quality control, one of which is workpiece measurement. In measuring workpieces, especially measurements made repeatedly, an employee must maintain the stability of his concentration so the quality control can be completed properly.

In previous studies, the temperature can affect employee performance. According to [3], inappropriate temperatures make that the results of the work are not optimal. Another study was conducted by [4] who examined the effect of temperature and noise on the workload of the production operator's cardiovascular system. The temperature used in the experiment used two levels, and the results obtained that the temperature significantly affected the cardiovascular system and contributed 17.556%. This paper analyse how temperature affects the results of repeated measurements on a workpiece. These measurements are carried out in the laboratory so that after this analysis can be used as suggestions in laboratory settings to optimize the student practicum.

## 2. Literature Study

The literatures used in this paper is as follows.

### *2.1. Ergonomic*

Ergonomics is a branch of science that studies human traits, abilities, and limitations that are useful for designing working systems so that humans can work comfortably [5]. The purpose of ergonomics as follows.

a. reduce injuries and illnesses that occur at work,
b. improve social welfare through improving the quality of social contacts, managing and coordinating work appropriately
c. creating a rational balance between aspects, namely technical, economic, anthropological, and cultural aspects of each work system carried out to create high quality work.

Another definition of ergonomics according to Manuaba (1999) is the ability to apply information based on human character, capacity, and limitations on the work system so as to create a comfortable working environment.

### *2.2. Working Environment*

Humans and the environment are interdependent entities. Humans must be able to adapt to the environment, including the work environment. The condition of the working environment affects the physical and psychological conditions of workers and directly impacts the performance results. In the work system, humans are as controllers, planners, as well as interacting with the system. In creating a comfortable working environment, several aspects must be considered, namely differences in physical, psychological conditions, as well as the strengths and weaknesses of people who are in the work system. According to [6], the strengths and weaknesses of human physical condition are indeed the main things, but not only factors that can affect work productivity.

There are several things that must be considered in the work environment, including noise, light intensity, temperature, and others. Employees can work optimally if the working environment is suitable. Conversely, improper environmental conditions will impact on employees within a certain period of time. A further impact is the occurrence of work accidents [7].

One part of the work environment that affects workers is temperature. High temperatures are dangerous in the work environment [8]. Other research states that high temperatures can reduce performance because it affects the human psychological system [2].

### *2.3. Two-Way Analysis of Variance (Two-Way ANOVA)*

The experimental design is used to obtain an appropriate sample for the population in the study conducted, then the corresponding conclusions are obtained [9]. The analysis used in experimental design is Analysis of Variance (ANOVA) which examines the influence or contribution of factors to the response. One type of ANOVA is the Two-Way ANOVA which is used if the experiment involves two factors, and a combination of levels between factors [10] is carried out. The mathematical model used in the design of this experiment is as follows.

$$Y_{ijk} = \mu + \tau_{ij} + \varepsilon_{k(ij)} \quad (1)$$

where:

$i$ : level of Factor A
$j$ : level of Factor B
$\tau_{ij}$ : effects caused by the treatment
$\varepsilon_{k(ij)}$ : error in each treatment

The hypothesis used in ANOVA is

$H_0$ : $\mu_1 = \mu_2 = \cdots = \mu_n$
$H\_1$ : there is at least one different $\mu_i$

The decision to reject H0 if the value $F > F_{table}$, and if the decision use p.value then $H_0$ is rejected where $p.value < \alpha$.

Journal of Physics: Conference Series **1381** (2019) 012061 doi:10.1088/1742-6596/1381/1/012061

## 3. Data and Methods

In this study, the experimental unit was a student who measured 25 workpieces. The response variable is the result of measuring the length and diameter of the workpiece parts (six parts) as Figure 1.

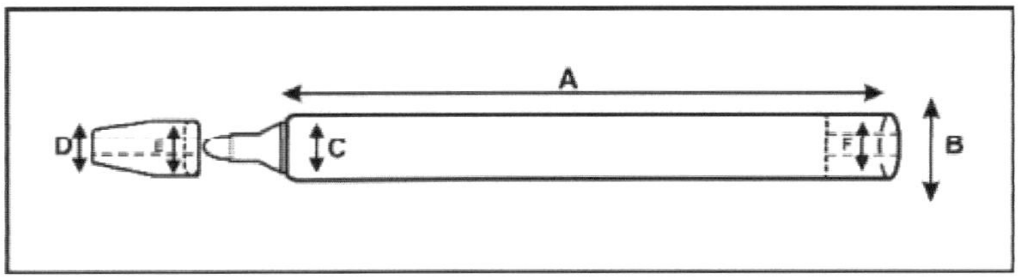


**Figure 1.** Measured workpiece

Part A is the length of workpiece, then part B to F are the diameter of workpiece. The independent variable or factors that influence the response variable are temperature and the measuring instrument used by student in measuring workpieces. The temperature factor has 2 levels and the measuring instruments has 5 levels. The level for each factor as follows.

**Table 1**. Factor and level for experiment

| Factor | Symbol | Level | Symbol |
|---|---|---|---|
| Temperature | Q | AC | Q1 |
| | | Non-AC | Q2 |
| Measurement Instruments | P | Vernier Calliper | P1 |
| | | Vernier Calliper Digital | P2 |
| | | Mistar | P3 |
| | | Micrometer Screw | P4 |
| | | Micrometer Screw Digital | P5 |

The experimental design used was Two Way Analysis of Variance (Two-Way ANOVA) involving two factor and its interaction. The data structure as follows.

**Table 2**. Data Structure

| Temperature | P1 | P2 | P3 | P4 | P5 |
|---|---|---|---|---|---|
| Q1 | $x_{111}$ | $x_{112}$ | $x_{113}$ | $x_{114}$ | $x_{115}$ |
| | $x_{211}$ | $x_{212}$ | $x_{213}$ | $x_{214}$ | $x_{215}$ |
| | $x_{311}$ | $x_{312}$ | $x_{313}$ | $x_{314}$ | $x_{315}$ |
| | $\vdots$ | $\vdots$ | $\vdots$ | $\vdots$ | $\vdots$ |
| | $x_{2511}$ | $x_{2512}$ | $x_{2513}$ | $x_{2514}$ | $x_{2515}$ |
| Q2 | $x_{121}$ | $x_{122}$ | $x_{123}$ | $x_{124}$ | $x_{125}$ |
| | $x_{221}$ | $x_{222}$ | $x_{223}$ | $x_{224}$ | $x_{225}$ |
| | $x_{321}$ | $x_{322}$ | $x_{323}$ | $x_{324}$ | $x_{325}$ |
| | $\vdots$ | $\vdots$ | $\vdots$ | $\vdots$ | $\vdots$ |
| | $x_{2521}$ | $x_{2522}$ | $x_{2523}$ | $x_{2524}$ | $x_{2525}$ |

The method carried out in this paper is as follows.

- do data retrieval by measuring 25 workpieces repeatedly using five measuring instruments,

- workpiece measurements are carried out in two different conditions, the first condition use Air Conditioner (AC) and the second condition does not use,
- the parts of workpiece are measured according to Figure 1,
- after the data is collected, an analysis is conducted using Two Way ANOVA to see whether differences in the use of measuring instruments and AC affect the measurement results,

## 4. Results and Discussion

In this paper, the statistical method used Two-Way ANOVA. The hypotheses are tested based on factors are thought to influence the measurement results are as follows.

$H0_1$ : Temperature does not significantly influence the measurement results of the workpiece
$H0_2$ : Measurement instrument does not significantly influence the measurement results of the workpiece
$H0_3$ : Interaction between temperature and measurement instrument does not significantly influence the measurement results of the workpiece
$H1_1$ : Temperature does significantly influence the measurement results of the workpiece
$H1_2$ : Measurement instrument does significantly influence the measurement results of the workpiece
$H1_3$ : Interaction between temperature and measurement instrument does significantly influence the measurement results of the workpiece

The results of data analysis are shown as follows.

**Table 3.** The results of statistical test for each part of workpiece

| Part | df | Sum of Sq. | Mean Sq. | F-ratio | P-value |
|---|---|---|---|---|---|
| **A (length)** | | | | | |
| Q | 1 | 334.75 | 334.75 | 3774.75 | 0.00 |
| P | 4 | 400.27 | 100.07 | 1128.38 | 0.00 |
| Q*P | 4 | 374.52 | 93.63 | 1055.79 | 0.00 |
| Error | 240 | 21.28 | 0.09 | | |
| Total | 249 | 1130.82 | | | |
| **B (diameter)** | | | | | |
| Q | 1 | 4.38 | 4.38 | 27.42 | 0.00 |
| P | 4 | 48.50 | 12.12 | 75.95 | 0.00 |
| Q*P | 4 | 38.32 | 9.58 | 60.02 | 0.00 |
| Error | 240 | 38.31 | 0.16 | | |
| Total | 249 | 129.51 | | | |
| **C(diameter)** | | | | | |
| Q | 1 | 0.37 | 0.37 | 17.08 | 0.00 |
| P | 4 | 1.97 | 0.49 | 23.00 | 0.00 |
| Q*P | 4 | 0.86 | 0.21 | 10.03 | 0.00 |
| Error | 240 | 5.14 | 0.02 | | |
| Total | 249 | 8.34 | | | |
| **D(diameter)** | | | | | |
| Q | 1 | 368375 | 368375 | 1 | 0.32 |
| P | 4 | 1476599 | 369150 | 1 | 0.41 |
| Q*P | 4 | 1473576 | 368394 | 1 | 0.41 |
| Error | 240 | 88443951 | 368516 | | |

| Part | df | Sum of Sq. | Mean Sq. | F-ratio | P-value |
|---|---|---|---|---|---|
| Total | 249 | 91762501 | | | |
| | | | | | |
| **E(diameter)** | | | | | |
| Q | 1 | 2.21 | 2.21 | 88.14 | 0.00 |
| P | 4 | 8.64 | 2.16 | 86.12 | 0.00 |
| Q*P | 4 | 4.38 | 1.10 | 43.65 | 0.00 |
| Error | 240 | 6.02 | 0.03 | | |
| Total | 249 | 21.25 | | | |
| **F(diameter)** | | | | | |
| Q | 1 | 652450 | 652450 | 1.01 | 0.316 |
| P | 4 | 2592270 | 648067 | 1 | 0.406 |
| Q*P | 4 | 2571957 | 642989 | 1 | 0.41 |
| Error | 240 | 154772911 | 644887 | | |
| Total | 249 | 160589588 | | | |

The decision to reject or accept H0 is based on F value or the significance of p-value. Hypotheseis H0 is rejected if the $F > F_{tabel}$, where the determination of $F$ dan $F_{tabel}$ is based on degrees of freedom of factor 1 and factor 2. If using the p-value significance, H0 will be rejected when the $p - value < \alpha$, in this case use $\alpha$ 5%. Based on Table 3, all parts of the workpiece measured show temperature, measuring devices, and their interactions significantly affect the measurement results, except for measurements of diameters D and F, because in both parts p-value> α. If you look further, the SS error value in D and F is very large compared to other parts, thus affecting the significance test of these two factors. In addition to factor 1 and factor 2, there are also significant interactions affecting differences in measurement results, then it means that the combination of temperature and measurement instruments used affects simultaneously.

After the ANOVA test, the significant factors in each part of the workpiece are known. Furthermore, the factors calculated contribution value (% contribution) to the measurement results in all parts of the workpiece, except parts D and F. The results of the percentage contribution for each factor are as follows.

**Table 4** The contribution value for each part of workpiece

| Part | df | SS | MS | SS' | % kontribusi | $R^2$ (%) |
|---|---|---|---|---|---|---|
| A (length) | | | | | | |
| Q | 1 | 334.75 | 334.75 | 334.66 | 29.59 | 98.05 |
| P | 4 | 400.27 | 100.07 | 399.91 | 35.36 | |
| Q*P | 4 | 374.52 | 93.63 | 374.16 | 33.09 | |
| Error | 240 | 21.28 | 0.09 | | | |
| Total | 249 | 1130.82 | | | | |
| B (diameter) | | | | | | |
| Q | 1 | 4.38 | 4.38 | 4.22 | 3.26 | 69.31 |
| P | 4 | 48.5 | 12.12 | 47.86 | 36.95 | |
| Q*P | 4 | 38.32 | 9.58 | 37.68 | 29.09 | |
| Error | 240 | 38.31 | 0.16 | | | |
| Total | 249 | 129.51 | | | | |

| | | | | | | |
|---|---|---|---|---|---|---|
| C(diameter) | | | | | | |
| Q | 1 | 0.37 | 0.37 | 0.35 | 4.20 | 36.21 |
| P | 4 | 1.97 | 0.49 | 1.89 | 22.66 | |
| Q*P | 4 | 0.86 | 0.21 | 0.78 | 9.35 | |
| Error | 240 | 5.14 | 0.02 | | | |
| Total | 249 | 8.34 | | | | |
| E(diameter) | | | | | | |
| Q | 1 | 2.21 | 2.21 | 2.18 | 10.26 | 70.40 |
| P | 4 | 8.64 | 2.16 | 8.52 | 40.09 | |
| Q*P | 4 | 4.38 | 1.1 | 4.26 | 20.05 | |
| Error | 240 | 6.02 | 0.03 | | | |
| Total | 249 | 21.25 | | | | |

Based on Table 4, it can be seen each factor and the interaction between factors contribute differently to each part is measured. And the result is the difference factor of the measuring instrument used gives the highest contribution to the difference in measurement results. Temperature contributes an average of 11.83% to differences in measurement results, measuring instruments contribute an average of 33.77% and interactions between temperature and measuring instruments contribute 22.90%. The smaller percentage of temperature contribution from the measuring instrument is caused by the number of treatments used in the experiment. The temperature used only distinguishes measurements in the AC and Non-AC rooms without the room temperature being measured, so the measurement results obtained do not provide a large difference. The second factor, namely measuring instruments contribute above 30% to the measurement results because the types of measuring instruments used are different so that the accuracy provided is also different.

Temperature and type of measuring instrument used have a significant effect on the measurement results, which on a larger scale affect to quality control. This quality control determines the products received or rejected so that high accuracy is needed because it is related to production results.
Based on the analysis of the measurement results, the largest R ^ 2 values are part A almost 100%, part B and E are below 75%, and part C is below 50%. If the value of R ^ 2 is small, it means that there are other factors outside of the observation that have a large influence on the measurement results. For example in part C, the value of R ^ 2 is only 36.21%, then it means temperature, measuring instruments, and their interactions only contribute 36.21% to the measurement results, while 63.79% is influenced by factors other than these two factors, for example the physical conditions who take measurements.

## 5. Conclusion

The temperature affects the measurement results, but the contribution percentage is smaller than the type of measuring instrument used. It is caused by temperature settings that only use and do not use AC (not specifically regulated at certain temperatures). In the next research can be developed by specifically setting the temperature (using more than two levels) so that the effect given to the measurement results is clearer and reduces errors.

## References

[1] Ali A S, Chua S J L and Lim M E L 2015 The effect of physical environment comfort on employees' performance in office buildings: A case study of three public universities in Malaysia *Struct. Surv.*

[2] Badayai A R A 2012 A Theoretical Framework and Analytical Discussion on Uncongenial Physical Workplace Environment and Job Performance among Workers in Industrial Sectors *Procedia - Soc. Behav. Sci.* **42** 486–95

[3] Seppänen O, Fisk W and Lei Q 2006 Effect of Temperature on Task Performance in Office Environment *Lawrence Berkeley Natl. Lab.* 11

[4] Febriyanti Y 2007 *ANALISIS PENGARUH TEMPERATUR DAN KEBISINGAN TERHADAP KERJA SISTEM CARDIOVASCULAR OPERATOR PRODUKSI (Studi Kasus PT General Electric Lighting Indonesia)* (Universitas Sebelas Maret Surakarta)

[5] Wignjosoebroto 2012 Ergonomi, studi gerak dan waktu : teknik analisis untuk peningkatan produktivitas kerja *Univ. Sumatera Utara Libr.*

[6] Kroemer A D and Kroemer K H E 2016 *Office ergonomics: Ease and efficiency at work: Second edition* (New Jersey: Prentice International Hall, Inc)

[7] Yildizel S A, Kaplan G, Arslan Y, Yildirim and Ozturk A U 2015 A study on the effects of weather conditions on the worker health and performance in a construction site **4** 291–5

[8] Monazzam M R, Golbabaei F, Hematjo R, Hosseini M and Dehghan S F 2006 The Assessment of Heat Stress and Heat Strain in Pardis Petrochemical Complex, Tehran, Iran **5** 6–11

[9] Walpole R, Myers R, Myers S and Ye K 2012 *Probability & Statistics for Engineers & Scientists* (London: Pearson Prentice Hall)

[10] Montgomery D C 2012 *Design and Analysis of Experiments Eighth Edition* (New York: John Wiley & Sons Inc.)

# Phytoremediation as a Sustainable Way for Land Rehabilitation of Heavy Metal Contamination

**P Alvernia[1], T E B Soesilo [2*]**

[1]Magister Program of Environmental Science, School of Environmental Science, Universitas Indonesia, Salemba, Indonesia.
[2]School of Environmental Science, Universitas Indonesia, Salemba, Indonesia.

*Corresponding author e-mail: soesilo@indo.net.id

**Abstract**. Land contamination from heavy metals leads to a decrease in soil functions. Chemical contamination on agriculture land will cause negative impacts especially affecting human health. The factor that causes heavy metals to be included in the group of pollutants is the nature of heavy metals to be non-degradable. Several physicochemical techniques have been utilized to remove the pollutants from the contaminated environment. However, such techniques require high costs, intensive labor, irreversible change of the nature of the soil which cannot be recovered and other contribution towards pofllution. One of the methods of toxicity reduction or applicable heavy metal mobility through in situ or ex situ as well as being environment-friendly is phytoremediation. Phytoremediation technology includes phytostabilization, phytostimulation, phytotransformation, phytofiltration, and phytoextraction. This paper aims to study phytoremediation frorem the sustainable indicators, namely environmental, social and economic aspects. Phytoremediation is one of the sustainable ways to recover contaminated lands by offering the lowest cost compared to other methods. Healthy and fertile agriculture land is the key to food security so that by utilizing phytoremediation it may recover the polluted land with low fertility through sustainability aspects.

## 1. Introduction
Land contamination from heavy metals are one of the dangerous forms of contamination for living organisms. Heavy metal-contaminated lands will cause a decrease of soil function. This may give an adverse effect especially towards human health and ecosystem, land productivity as well as social and economic welfare [1]. Based on such things, there needs to be a form of human intervention through the existing technology in order to overcome the pollution. Several ways may be used, one of them is with physicochemical technique. One of the forms of contamination control with physicochemical technique is by giving chemical substance to the contaminated land and could also be done by replacing the contaminated land through the removal of land [2]. Removal of land means adding a significant number of clean soil to cover the contaminated soil surface. In addition to the high costs, this way is also deemed to remove the original nature of the soil that cannot be recovered. Conventional remediation technique is deemed to be incompatible to be conducted especially in widespread area [3].

One of the alternatives to overcome heavy metal-contaminated lands which is environmental friendly with affordable cost, namely phytoremediation. Phytoremediation is a technology utilizing specific plants to remove the pollutants from the soil and reduce the toxic effects from the contaminants in the surrounding environment [4]. Phytoremediation technology could be implemented to remove contaminants from the contaminated soil with affordable cost, effective and environmental friendly [5]. Phytoremediation is the use of plants associated with microorganisms to remove pollutants from contaminated soils and waters. Phytoremediation has the potential to be implemented on several types of substance including severe environment pollutants and could recover soil functions [6], [7]. Phytoremediation could include soil fertility allowing it to follow up the growth of plants with economic added value [8]. However, the use of this technology could not be fully implemented by farmers with contaminated agriculture lands. The lack of knowledge by the farmers is assumed to cause the lack of such implementation in heavy metal-contaminated lands. The importance of the farmer's knowledge towards the utilization of phytoremediation is also considered to be the key to success for contaminated land rehabilitation. Based on such things, this paper will study the phytoremediation technology of the heavy metal contaminated soil in a sustainable matter.

## 2. Materials and Methods

This research utilizes literature study method about phytoremediation technology as way of heavy metal-contaminated lands. The data used in this paper comprises of secondary data from previous researches. The source of research data utilized namely regarding phytoremediation technology as the solution to heavy metal-contaminated lands in a sustainable manner. The research utilized in this literature study is namely a research conducted in the last ten years.

## 3. Result and Discussion

### *3.1 Result*

Potential danger of heavy metal towards the environment could be determined from the total heavy metal concentration on the area [1]. The accumulation of heavy metal on the agriculture land is caused by the industrialization and the agricultural activity such as fertilizers, pesticides and irrigation [2]. Heavy metal contamination could give an effect which is carcinogenic and mutagenic towards the living organisms [3].

Table 1. Impact of Heavy Metal on Human Health

| No | Source of pollutants | Type of heavy metals | Effects on humans | References |
|---|---|---|---|---|
| 1 | Air, food, water, soil, dust, biosolids, metal mining | Lead (Pb) | Damage to the kidney, production of Cardiovascular system (CVD) | [4,5] |
| 2 | Biosolids and manure, air | Cadmium (Cd) | Lung cancer, liver cancer, bone damage | [3,6] |
| 3 | Pesticides, food (sea fish) | Mercury (Hg) | Damage to the fetus and baby's brains | [7,8] |
| 4 | Biosolids, irrigation, food, water, air | Arsenic (As) | Cardiovascular, skin, digestion, respiratory, neurological, endocrinological , cancer | [9] |
| 5 | Fertilizers, pesticides, water, food, biosolids | Copper (Cu) | Nausea, vomit, stomachache, diarrhea | [3,8] |
| 6 | Fertilizers, pesticides, food, industrial waste, dust, biosolids, metal mining and milling | Zinc (Zn) | Epigastric pain, diarrhea, increasing risk of prostate cancer, respiratory problems, lethargy, microcytosis, neutropenia | [8,10,11] |

| | | | | |
|---|---|---|---|---|
| 7 | Industrial waste, irrigation, food, beverage, biosolids | Chromium (Cr) | Asthma, pneumonia, skin allergy | [12,13] |
| 8 | Industrial waste, fossil fuel combustion, nickel mining and metal coating, food, beverages as well as air | Ni (Nickel) | Lung cancer, laryngeal cancer, prostate cancer, skin allergy | [8,14,15] |

**Phytoremediation Mechanism**

Phytoremediation mechanism consists of phytoextraction, phytofiltration, phytostabilization, phytovolatilization and phytodegradation. Phytoextraction is an absorption of pollutants by the plant roots and accumulate to the parts of the roots, stems and leaves. Phytofiltration is the utilization of plant roots to absorb, precipitate, accumulate heavy metals from the waste stream. Phytovolatilization occurs when the plan absorbs the heavy metals and releases it through the leaves to the air and occasionally the heavy metals experience degradation first before being released through the leaves. Phytodegradation is the metabolism of the heavy metals in the plant tissue by enzymes such as dehalogenase, reductase and oxygenase [16].

Table 2. Plants' Ability to Absorb Heavy Metals in soil

| **No** | **Plant species** | **Family** | **Metal** | **Part (Accumulation)** | **Duration of treatment** | **Translocation Factor** | **References** |
|---|---|---|---|---|---|---|---|
| 1 | Sunflower (*Helianthus annuus* L.) | Asteraceae | Pb | Shoot<br>Root<br>Seed | After 126 days | TF > 1 | [17] |
| 2 | Flax (*Linum usitatissimum* L.) | Linaceae | Pb<br><br>Cd<br><br>Zn | Shoot<br>Capsule<br>Seed | 155 days (harvesting stage) | Pb, TF > 1 (shoot and capsule), Pb, TF < 1 (seed)<br><br>Cd, TF <1<br><br>Zn, TF >1 (capsule and seed) Zn, TF <1 (shoots) | [18] |
| 3 | *Solanum nigrum L.* | Solanaceae | Cu,<br>Zn<br>Pb<br>Cd<br>Mn<br>Hg | Leaves<br>Root<br>Shoot | 7 months | Cu, TF = 0,62<br>Zn, TF = 0,30<br>Pb, TF= 0,58<br>Cd, TF = 1,47<br>Mn, TF = 0,39<br>Hg, TF = 0,33 | [19] |
| 4 | Vetiver (*Chrysopogon zizanioides*) | Poaceae | Pb<br>Cr<br>Ni<br>Co | Root<br>Shoot | 1 year | Pb, TF= 0.5<br>Cr, TF=1.1<br>Ni, TF= 0.6<br>Co, TF=0.6 | [20] |
| 5 | *Erodium glaucophyllum* L. | Geraniaceae | Pb<br>Cd<br>Zn<br>Fe<br>Mn<br>Cu | Root<br>Shoot | 2 months | Pb, TF>1<br>Cd, TF>1<br>Zn, TF>1<br>Fe, TF>1<br>Mn,TF<1<br>Cu, TF<1 | [21] |

*3.2 Discussion*

Sustainable development is a form of economic growth that meets the current necessity and desire without compromising the fulfilment of future necessities in which the concept consists of three aspects, namely environmental, social and economic [22–24]. Those three aspects are interrelated and inseparable from one another. This paper discusses the use of phytoremediation reviewed from the sustainable indicators, namely environmental, social and economic aspects. Sustainable agriculture is a harmonious and balanced agricultural system in accordance with environmental governance. The importance of sustainable agriculture theory based on the awareness will maintain environmental health and preservation that has become a world trend which is internationally developed through global regulations requiring agricultural products to be safe to be consumed, environmental-friendly and with high nutrients [25].

Sustainable agriculture thrives to support farmers to implement sustainable agricultural practices in the farmer's agricultural land [26]. One of the examples, namely 'sustainable intensification' as an alternative which is more environmental-friendly without utilizing additional lands or causing yield losses [27]. Sustainable agricultural system is a growth system from traditional agriculture with an objective to meet the maximum targets that have been planned, overcome global economic problems and maximize the fast and instant necessities. Based on its managements, sustainable agricultural system that is environmental-friendly performed through the use of natural resources in an optimal, sustainable and profitable manner as well as be utilized sustainably for the benefit of the present and future generations.

Based on the sustainable agriculture theory, phythoremediation is one of the sustainable ways to recover contaminated lands. Phytoremediation could increase the soil fertility that could be follow up the plant growth with economic value added [28]. Healthy and fertile agricultural land is the key to food security thus by using phytoremediation is expected to be able to recover contaminated lands with low fertility rate. Indonesia is one of the countries that has agreed to the implementation goals of Sustainable Development committed to support the implementation of SDGs. One of the principles of SDGs, namely integration carried out in an integrated and interrelated manner on the dimensions of environment, social and economy. The research utilized by the researcher is one of the efforts to achieve sustainable development goals. Subject to Sustainable Development Goals 2030, on the goal number three, namely "Good Health and Well-Being" especially with the target to alleviate mortality rate and diseases caused by hazardous chemicals and also pollution and air, water and soil contamination. Moreover, the indication of this objective, namely mortality rate caused by unsafe water, unsafe sanitation and lack of hygiene.

Chemical contaminations especially on water and soil in agricultural land will cause adverse effect especially on human health. The polluted agricultural land becomes a serious a public health concern. The health impact due to heavy metals, namely physiology effects, effects on the skin, effects on respiration, effects on the kidney, effects on the liver, carcinogenic, growth and reproduction. Based on such things, there needs to be recovery efforts on the heavy metal contaminated agricultural land to achieve the third objective. Phytoremediation as a sustainable way is studied from three main aspects, namely, natural, social and economic aspects.

The environmental aspects on this research, namely the land contamination by heavy metals caused the decrease of soil function leading to the disruption to the surrounding environment as well as towards humans. Heavy metals are part of pollutant groups due its non-degradable nature and be easily absorbed. Based on such things, there needs to be land improvement efforts, one of the methods of toxicity decrease or the applicative heavy metal mobility as well as being environmental-friendly is phytoremediation. All plants are capable to absorb metals in various amounts, but several plants accumulate certain metal elements with high level of concentration. The result in Table 2 shows that each type of plant can absorb heavy metals respectively. Translocation factor shows the ability of plants to translocate metals from roots through shoots and leaves of plants. The plants can be used as bio accumulators if the translocation factors higher than one.

Based on the economic aspects, the utilization of phytoremediation to absorb heavy metals is deemed to be efficient as it does not require expensive costs. According to Mahar *et al.* [28], the purpose of phytoremediation to be studied from the economic aspects is the sustainable land management that could raise the soil fertility so as to enable plant growth with economic added values. In addition to phytoremediation, there are a number of ways to remove pollutants from the contaminated environment, one of which is by physicochemical methods [29]. However, with such ways, it is required in a significant amount depending on the contaminated area thus potentially involving expensive costs. The conventional remediation technique is deemed to be incompatible to be performed in a widespread area [30]. However, according to Zehra *et al.* [17] the utilization of phytoremediation on an agricultural land that is not balanced with economic income is not feasible for farmers. Plants that become phytoremediator agents must also give additional benefits for the economic life of farmers. Based on such purposes, the researcher believes that phytoremediation is an environmental-friendly technology with relatively cheap costs. The research outcome from Mwegoha [31] shows that the use of phytoremediation is 60-80% cheaper than the physicochemical techniques.

Apart from the environmental and economic aspects, the social aspects are also an important case of how the farmers are able to apply phytoremediation on their own agricultural lands. Based on such things, this becomes important for the farmer's knowledge on the choice of plant types to optimally reduce heavy metals. The farmer's knowledge on phytoremediation is still very limited and has not been widely implemented. Based on such things, there needs to be a study on the farmer's knowledge so that the farmers are able to utilize this technology in accordance with necessary steps. One of the ways that could be used to increase the farmer's knowledge is by convening agricultural extension. Agricultural extension is an activity to disseminate information with the view of the farmers to be well-informed, capable and be able to overcome agricultural problems to improve welfare. As opposed to the other techniques, the use phytoremediation depends on the location of the contaminated land. Based on such thigs, therefore this becomes impractical if field trials of the phytoremediation were conducted on every region. The land recovery efforts with conventional techniques are more feasible compared to the phytoremediation technology. The result of the physicochemical process is more predictable and controllable than the physiological parameter [32]. In addition, the conventional technique is easier to be applied in the field compared to the phythoremediation that also requires longer time. Based on such things, there needs to be a development on phytoremediation regarding the environmental and commercial matters. This is to ensure that the phytoremediation could become a solution to the heavy metal-contaminated lands.

## 4. Conclusion

Land contamination by heavy metals could be alleviated with phytoremediation technology. The advantages of phytoremediation from other technologies are being environmental-friendly with affordable cost compared to other methods. Moreover, the use of appropriate plant not only removes soil pollutants but also could be utilized as the farmer's income. The farmer's knowledge on phytoremediation needs to be further studied. The choice of appropriate plants as phytoremediation agent on the polluted land area becomes the key to success in this phytoremediation technology.

## Acknowledgments

This paper is supported and funded by the International Indexed Publication Grant for Student Final Project (PITMA) of the University of Indonesia 2019.

## Reference

[1] Zhang W, Chen L, Zhang R and Lin K 2015 Effects of decabromodiphenyl ether on lead mobility and microbial toxicity in soil *Chemosphere* **122** 99–104
[2] Zhang X, Zhong T, Liu L and Ouyang X 2015 Impact of soil heavy metal pollution on food safety in China *PLoS One* **10** 1–14
[3] Šukalić A, Mačkić S, Leto A, Džubur A, Ahmetović N and Antunović B 2018 Human health risk assessment of heavy metals from the agricultural soil in South Herzegovina *Agric. Conspec. Sci.* **83** 45–50
[4] Iqbal M P 2012 Lead pollution - A risk factor for cardiovascular disease in Asian developing countries *Pak. J. Pharm. Sci.* **25** 289–94
[5] Zhou J, Du B, Hu Y, Liang J, Liu H, Fan X, Zhang L, Cui H, Liu X and Zhou J 2019 A new criterion for the health risk assessment of Se and Pb exposure to residents near a smelter *Environ. Pollut.* 218–27
[6] Chen C, Wang X and Wang J 2019 Phytoremediation of cadmium-contaminated soil by Sorghum bicolor and the variation of microbial community *Chemosphere* **235** 985–94
[7] Ou L, Wang H, Chen C, Chen L, Zhang W and Wang X 2018 Physiologically based pharmacokinetic (PBPK) modeling of human lactational transfer of methylmercury in China *Environ. Int.* **115** 180–7
[8] Wuana R A and Okieimen F E 2014 Heavy metals in contaminated soils: A review of sources, chemistry, risks, and best available strategies for remediation *Heavy Met. Contam. Water Soil Anal. Assessment, Remediat. Strateg.* **2011** 1–50
[9] Chakraborti D 2011 *Arsenic: Occurrence in Groundwater* (Elsevier Inc.)
[10] Deshpande J, Joshi M and Giri P 2013 Zinc: The trace element of major importance in human nutrition and health *Int. J. Med. Sci. Public Heal.* **2** 1
[11] Plum L M, Rink L and Hajo H 2010 The essential toxin: Impact of zinc on human health *Int. J. Environ. Res. Public Health* **7** 1342–65
[12] Kirti Sekhawat, Sreemoyee Chatterjee B J 2015 Chromium Toxicity and its Health Hazards *Int. J. Adv. Res.* **3**
[13] Sun H, Brocato J and Costa M 2015 Oral Chromium Exposure and Toxicity *Curr. Environ. Heal. reports* **2** 295–303
[14] Zambelli B, Uversky V N and Ciurli S 2016 Nickel impact on human health: An intrinsic disorder perspective *Biochim. Biophys. Acta - Proteins Proteomics* **1864** 1714–31
[15] Mahurpawar M 2015 Effects of heavy metals on human health *Int. J. Reseacrh-Granthaalayah* **2350** 2394–3629
[16] Yao Z, Li J, Xie H and Yu C 2012 Review on remediation technologies of soil contaminated by heavy metals **16** 722–9
[17] Zehra A, Sahito Z A, Tong W, Tang L, Hamid Y, Khan M B, Ali Z, Naqvi B and Yang X 2020 Assessment of sunflower germplasm for phytoremediation of lead-polluted soil and production of seed oil and seed meal for human and animal consumption *J. Environ. Sci. (China)* **87** 24–38
[18] Hosman M E, Soliman El-Feky S, Elshahawy M I and Shaker E M 2017 Mechanism of Phytoremediation Potential of Flax (Linum usitatissimum L.) to Pb, Cd and Zn *Pelagia Res. Libr. Asian J. Plant Sci. Res.* **7** 30–40
[19] Li X, Zhang X, Wang X and Cui Z 2019 Phytoremediation of multi-metal contaminated mine tailings with Solanum nigrum L. and biochar/attapulgite amendments *Ecotoxicol. Environ. Saf.* **180** 517–25
[20] Vijayan V D and Sushama P K 2018 Phytoremediation as an effective technology for the removal of heavy metals from dump yard soils *Nat. Environ. Pollut. Technol.* **17** 1353–8
[21] Jeddi K and Chaieb M 2018 Evaluation of the potential of Erodium glaucophyllum L. for phytoremediation of metal-polluted arid soils *Environ. Sci. Pollut. Res.* **25** 36636–44
[22] Tanguay G A, Rajaonson J, Lefebvre J F and Lanoie P 2010 Measuring the sustainability of cities: An analysis of the use of local indicators *Ecol. Indic.* **10** 407–18
[23] Wang B, Xie H L, Ren H Y, Li X, Chen L and Wu B C 2019 Application of AHP, TOPSIS, and

TFNs to plant selection for phytoremediation of petroleum-contaminated soils in shale gas and oil fields *J. Clean. Prod.* **233** 13–22

[24] Yang B, Xu T and Shi L 2017 Analysis on sustainable urban development levels and trends in China's cities *J. Clean. Prod.* **141** 868–80

[25] Lestari Y K and Suryana A T 2013 Sustainability of Organic Rice Farming in Indonesia *Sustain. Org. Rice Farming Indones.* 55–60

[26] Pretty J and Bharucha Z P 2014 Sustainable intensification in agricultural systems *Ann. Bot.* **114** 1571–96

[27] Teschner N, Orenstein D E, Shapira I and Keasar T 2017 Socio-ecological research and the transition toward sustainable agriculture *Int. J. Agric. Sustain.* **15** 99–101

[28] Mahar A, Wang P, Ali A, Awasthi M K, Lahori A H, Wang Q, Li R and Zhang Z 2016 Challenges and opportunities in the phytoremediation of heavy metals contaminated soils: A review *Ecotoxicol. Environ. Saf.* **126** 111–21

[29] Wuana R A, Okieimen F E and Imborvungu J A 2010 Removal of heavy metals from a contaminated soil using organic chelating acids **7** 485–96

[30] Witters N, van Slycken S, Ruttens A, Adriaensen K, Meers E, Meiresonne L, Tack F M G, Thewys T, Laes E and Vangronsveld J 2009 Short-rotation coppice of willow for phytoremediation of a metal-contaminated agricultural area: A sustainability assessment *Bioenergy Res.* **2** 144–52

[31] Mwegoha W J S 2008 The use of phytoremediation technology for abatement soil and groundwater pollution in Tanzania: opportunities and challenges *J. Sustain. Dev. Africa* **10** 140–56

[32] Conesa H M, Evangelou M W H, Robinson B H and Schulin R 2012 A critical view of current state of phytotechnologies to remediate soils: Still a promising tool? *ScientificWorldJournal.* **2012**

Journal of Physics: Conference Series **1381** (2019) 012063 doi:10.1088/1742-6596/1381/1/012063

# Estimated leachate rate affected by climate change in landfill for radioactive contaminated hazardous waste from petroleum industries

**C A W Dwipayana[1, 2], S S Moersidik[1], M A Pratama[1]**
[1] Department of Civil Engineering-University of Indonesia, Kampus UI, Depok 16424, Indonesia
[2] Nuclear Energy Regulatory Agency, Jl.Gadjah mada No. 8, Jakarta 10120, Indonesia

Corresponding author: ssarwanto@eng.ui.ac.id;

**Abstract.** Naturally occurring radionuclides are present in the beneath the earth at different levels and can be concentrated and increased by operations related with the oil and gas industries. NORM disposal must provide adequate protection to both human health and the environment. Leachate is the major danger posed by contemporary landfills, which is affected by weather. In this paper a leach rate simulation has been carried out using the required design of class I landfill to see the effect of climate change scenarios. Simulations are carried out using the US EPA Hydrologic Evaluation of Landfill Performance software. The inputs are 60 years generated weather data from the LARS-WG program using 30 year historical data from Bogor Climatology Station. Climate change modeling uses GCM HadGEM3-ES with RCP scenarios 2.6, 4.5 and 8.5. The annual average rainfall for RCP 2.6, 4.5 and 8.5 climate change scenarios are, 3154.27mm, 3434.47mm, and 3316.33mm respectively, lower than the site base simulation 3854.24mm. The monthly average temperature of RCP scenario is higher than site base scenario, 3.78%, 4.60%, and 6.51% respectively for RCP 2.6, 4.5 and 8.5. But the leachate rate from the barrier soil layer has no difference for each climate change scenario.

## 1. Introduction

Naturally occurring radionuclides are present in the beneath the earth at different levels and can be concentrated and increased by operations related with the oil and gas industries. Examples of materials that can contain high concentrations of NORM are sludge, drilling mud and pipe scales, and radioactive materials can be transferred from site to site when machinery and materials are reused [1].

Even though the risk of radiation exposures arising from NORM contaminated waste streams can be generally accepted by both workers and the general public, the extent of this danger and the consequent necessity to regulate NORM have also been discussed. In the past few years, the petroleum industry has adopted methods for managing and disposing of NORM-contaminated wastes and equipment that are more restrictive than past practices and are likely to provide greater isolation of the radioactivity. Presently acceptable techniques of NORM management and disposal are including (1) burial in authorized NORM or low-level radioactive waste disposal facilities, (2) dilution and burial in authorized installations, (3) downhole encapsulation within plugged and discontinued well by license only, (4) underground injection into underground forming by license only, and (5) decontamination and reuse of controlled devices [2].

The Minister of Environment and Forestry Regulation Number P.63/Menlhk/Setjen/KUM.1/ 7/2016 of Requirements and Procedures for Hazardous and Toxic Waste Disposal in the Landfill Facilities, regulate that radioactive contaminated hazardous waste must be disposed of in class I or class II landfills.

The difference in the requirement of the two landfills is the number of geomebranes. Class I landfills have 2 geomembrane layers while class II landfills only have 1 geomembrane layer under the waste layer. While the required capping design for the two types of landfills is the same, they have 1 layer of geomembrane

Landfilling is a managed technique for land disposal of solid waste with the double goal of eliminating public health and environmental risk while reducing nuisance, with no surface water or groundwater resources contaminated. Leachate is the major danger posed by contemporary landfills. Leachates are compounds of water and water-soluble waste that accumulate as the water passes through the site. This water could be produced by precipitation or by the waste [3]. Climate (precipitation), landscape (run-on/runoff), waste cap, plants and waste type are important factors that influence leachate generation [4].

We need to consider climatic variations, look at particular methods and investigate site development strategies if we want to effectively manage a shift from open dumping disposal sites to engineered landfill in the Asian community [5]. Climate change has become a major problem in the future where water availability may be questioned in the coming years [6]. An embedded assessment is therefore necessary, which can measure the effects of climate change on the waste disposal process. General Circulation Models (GCMs) are used to predict future climate change. GCMs are an Earth system-based computer model and are numerically used to simulate the current and future project climate strengthened by greenhouse gas emissions and particulate matter. Due to its coarse spatial resolution, the GCM output can not be utilized for hydrological evaluation [7]. One of stochastic weather generator intended for climate change effect research is Long Ashton Research Station Weather Generator (LARS-WG). It has been evaluated and better than some other generators in a variety of environments. The latest research has tested LARS-WG at various locations worldwide and demonstrated their capacity to model extreme rainfall with sensible skills [8].

In this paper a leach rate simulation has been carried out using the required design of class I landfill to see the effect of climate change scenarios. Simulations are carried out using the US EPA Hydrologic Evaluation of Landfill Performance software. Weather information from LARS-WG is produced over 60 years using 30 years of Bogor Climatology Station's historical information.

## 2. Methodology

### *2.1. Estimation Future Climate Change*

LARS-WG is a stochastic weather generator and is used in present and future circumstances to simulate weather information at one location. LARS-WG uses recorded daily weather information for a particular location to calculate a set of weather variables distribution probability parameters and their correlation by randomly chooses a value from the suitable distributions to create the simulated weather time series of arbitrary lengths [8].

In this paper, 30 years (1984-2013) historical data from Bogor Climatology Station will be used to generate future daily weather data for 60 year. As general circulation models (GCMs), HadCM3 with A2 scenario will be used. This model has been developed in the Hadley Centre of the UK national meteorological services and is a combined atmosphere-ocean GCMs. HadCM3 was selected as it is commonly used in numerous researches on the effect of climate change. It can also simulate for a century and demonstrates a little change in surface climatic conditions [9]. Representative Concentration Pathway (RCP) that be used in this study are RCP 2.6, 4.5 and 8.5. RCPs are based on multi-gas emission scenarios, the four RCPs cover future levels ranging from very elevated (RCP8.5) to very low (RCP2.6). The RCP numerical values (2.6, 4.5, 6.0 and 8.5) are the same as the 2100 concentrations [10].

### *2.2. Landfill Design*

Landfill design to be used is those required in the Minister of Environment and Forestry Regulation Number P.63/Menlhk/Setjen/KUM.1/7/2016 [10]. Landfill design parameters not included in the Regulation are taken from hazardous landfill of the PPLI Cileungsi. PPLI Cileungsi has 1 landfill cell which has been closed with an area of around 6.3 ha. The waste thickness used in this simulation is

12.5m, this value is obtained from the practical approach of landfill volume data that has been closed with the final cover at PPLI Cileungsi which is 767,761 $m^3$, with a landfill area of 62,693 $m^2$. The thickness of the landfill layer used in this study is as in table 1.

**Table 1.** Landfill Layer Thickness

| No | Lapisan | Thickness ( m) |
|---|---|---|
| 1. | Top Soil-Fine Sandy Loam | 0.6 |
| 2. | Drainage Layer -Sand | 0.3 |
| 3. | Geomembrane-HDPE-2mm | 0.002 |
| 4. | Capping Barrier-Barrier Soil | 0.6 |
| 5. | Temporary Cover-Fine Sandy Loam | 0.15 |
| 6. | Waste | 12.5 |
| 7. | Protective Layer-Sandy Clay Loam | 0.3 |
| 8. | Leachate Collection Layer-Gravel | 0.3 |
| 9. | First Geomembrane-HDPE-2mm | 0.002 |
| 10. | Barrier Layer-Barrier Soil | 0.3 |
| 11. | Leakage Detection Layer-Gravel | 0.3 |
| 12. | Leakage Detection Layer -Drainage Net (0.6cm) | 0.006 |
| 13. | Second Geomembrane-HDPE-2mm | 0.002 |
| 14 | Base Layer-Barrier Soil | 1.0 |
| 15. | Local Soil | 4.0 |

**Tabel 2**. HELP Input Parameters

| Parameters | Value |
|---|---|
| Evapotranspiration Depth | 20cm |
| Maximum Leaf Area Index | 3.5 |
| Relative Humidity | |
| • Quarter 1 | 87.4 % |
| • Quarter 2 | 85.1 % |
| • Quarter 3 | 80.1 % |
| • Quarter 4 | 84.5 % |
| Average Wind Speed | 1 km/hour |
| Landfill Area | 6.3 Ha |
| Percentage of area where runoff is possible | 100% |

*2.3. Hydrology Modeling*

The leachate quantification of the landfill is generally modelled on the basis of principles of water balance by estimating the water percolation in the landfill and by degradation or evaporation of water [11]. Various software has been created to simulate leachate generation in landfill sites [12]. In this paper, the Hydrologic Evaluation of Landfill Performance (HELP) software will be used. The HELP system (version 3.07) for predicting landfill leachate volume and water balance analysis in landfill systems and capping systems is the most commonly used model in the United States Environmental Protection Agency [13].

Generated daily weather data of 60 years for each RCP and site base scenario are used to simulate water balance in landfill. Weather data are include daily rainfall data, daily minimum and maximum temperatures, and also solar radiation. The humidity and wind velocity parameters are taken from the average weather climatology station in 1984-2013. Other input parameters can be seen in table 2. Leachate rate from hydrological modelling will be an input for radiological modelling.

## 3. Result and Discussion

From LARS-WG simulation, daily future weather data of 60 years are gained. The daily data are processed into average monthly data for each parameter, as shown in figure 1 to figure 4.

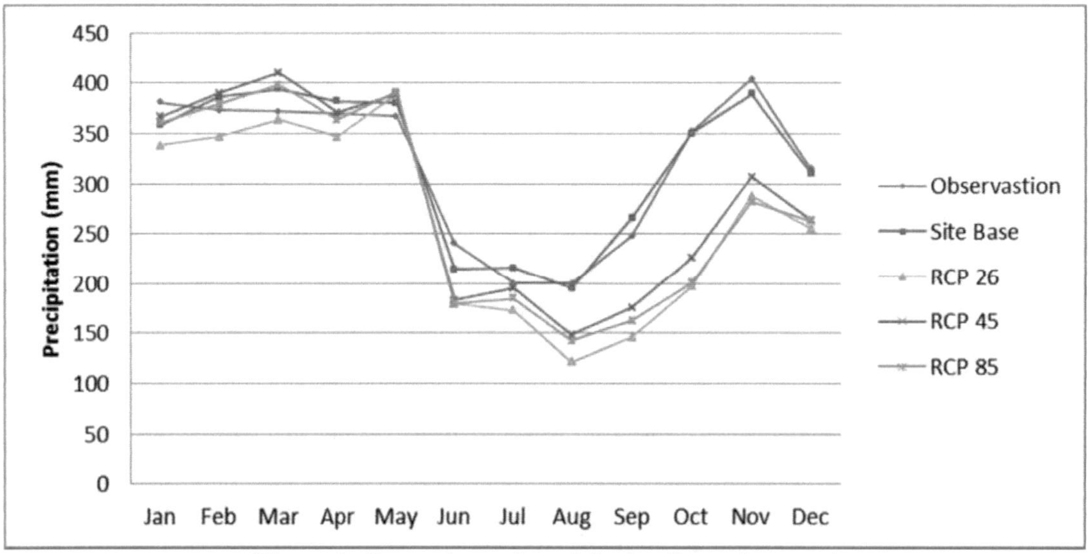


**Figure 1**. Comparison of precipitation for each scenario

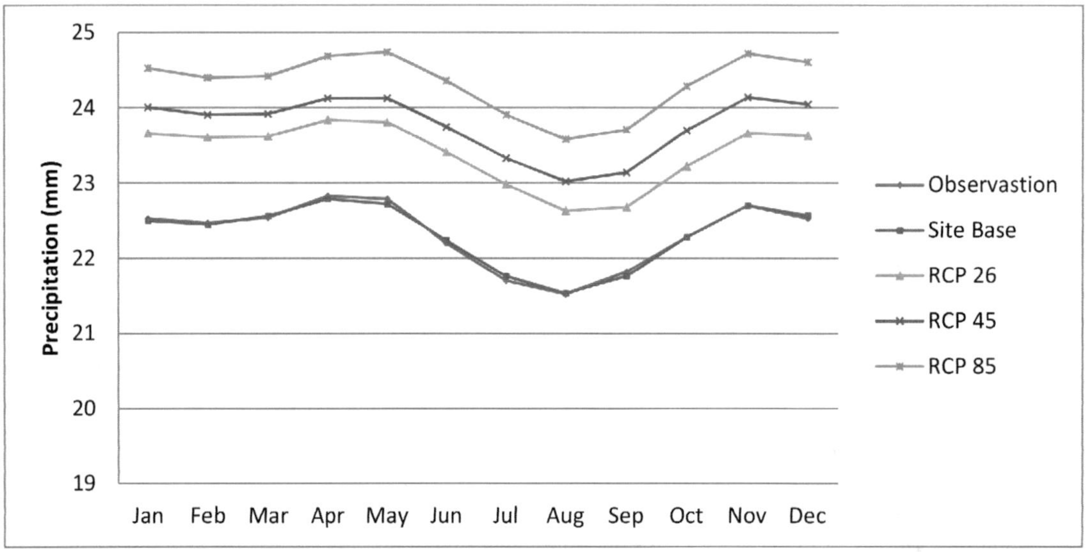


**Figure 2**. Comparison of minimum temperature for each scenario

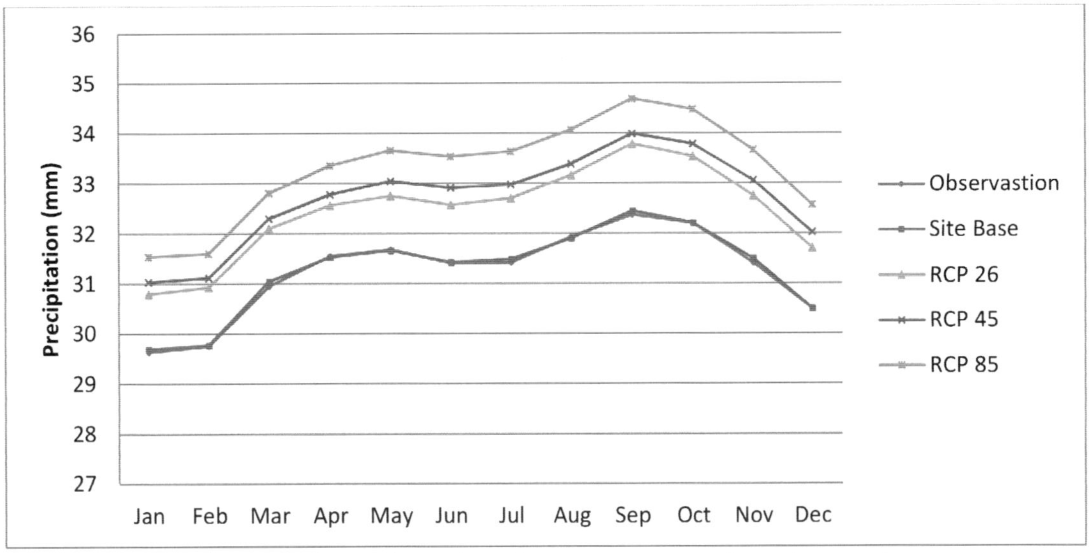


**Figure 3**. Comparison of maximum temperature for each scenario

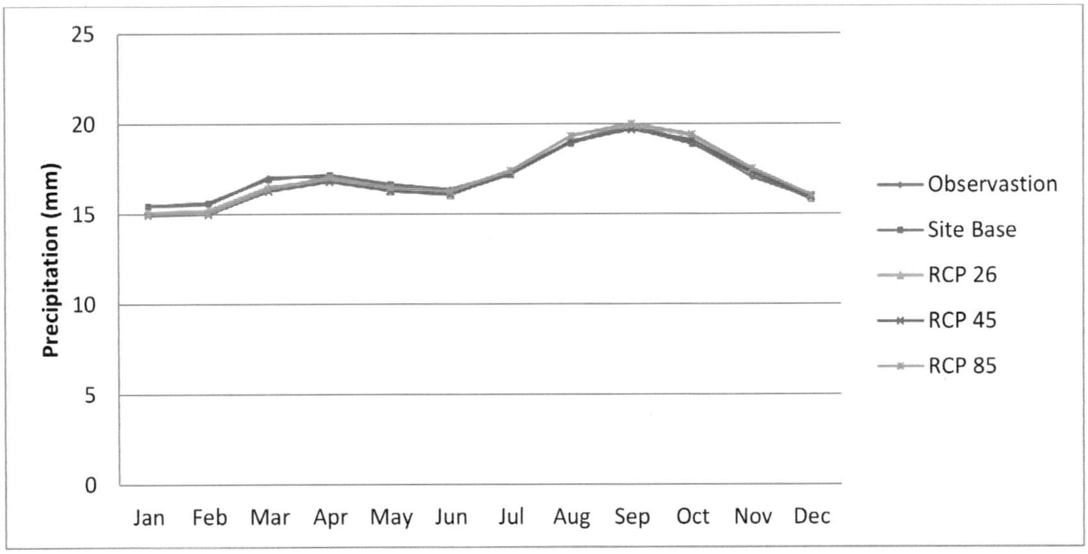


**Figure 4**. Comparison of solar radiation for each scenario

Visual examination is done for each average monthly data to know the difference of each RCP Scenario. From figure 1 to 4 it can be seen that LARS WG is able to produce good predictions, which can be seen that the site base curve always coincides with the observation curve.

From figure 1, precipitation for future is slightly lower than historical weather. The annual average rainfall for RCP 2.6, 4.5 and 8.5 climate change scenarios are, 3154.27mm, 3434.47mm, and 3316.33mm respectively, the site base simulation 3854.24mm.

Temperature in future is predicted increase for maximum and minimum temperature. The monthly average temperature of RCP scenario is higher than site base scenario, 3.78%, 4.60%, and 6.51% respectively for RCP 2.6, 4.5 and 8.5 (figure 2 and 3). There is no difference for solar radiation data of historical and future climate (figure 4).

Predicted daily weather data of each RCP Scenario is used to simulate leaching rate of designed landfill. HELP Software produce leach rate and volume for several layers. Comparison of leach rate for each climate scenario is shown in table 3 and figure 5. Difference value of precipitation make difference value of runoff, however the water that unfiltered to the waste has same rate value. It is caused by appearance of geomembrane in capping barrier. Geomembrane reject almost all water that try to infiltrate to waste layer. From this fact it can be concluded that the presence of geomembranes is very important in preventing contamination by reducing leachate rate.

**Table 3**. Comparison of HELP output for each scenario

| | Site Base | RCP 2.6 | RCP 4.5 | RCP 8.5 |
|---|---|---|---|---|
| precipitation | 3854.24 | 3154.27 | 3434.47 | 3316.33 |
| runoff | 2092.808 | 1443.83 | 1713.684 | 1599.634 |
| evapotranspiration | 1220.172 | 1181.51 | 1185.787 | 1186.34 |
| percolation through capping drainage layer | 540.45687 | 528.0892 | 534.1679 | 529.5237 |
| percolation through capping barrier | 0.89467 | 0.85743 | 0.87764 | 0.86124 |
| percolation through protective layer | 0.89467 | 0.85743 | 0.87764 | 0.86124 |
| leachate collection system | 0.89456 | 0.85733 | 0.87754 | 0.86113 |
| leachate through barrier layer | 0.00011 | 0.00007 | 0.00011 | 0.00011 |
| leachate through base layer | 0.00007 | 0.00006 | 0.00007 | 0.00007 |

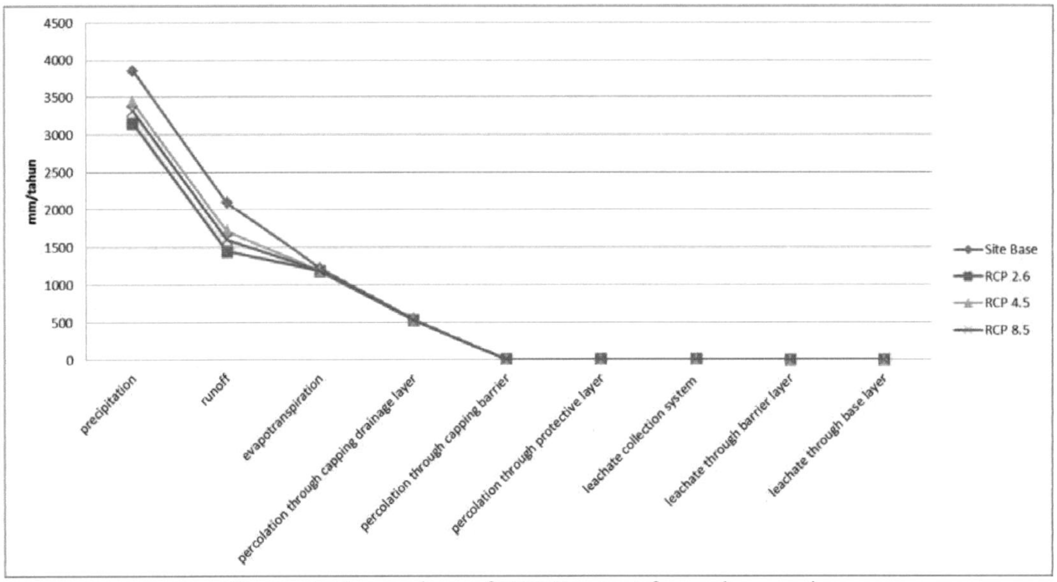


**Figure 5**. Comparison of HELP output for each scenario

## 4. Conclusion

LARS WG is able to produce good predictions of future daily weather data. Precipitation for future is slightly lower than historical weather. Temperature in future is predicted increase for maximum and minimum temperature. The monthly average temperature of RCP scenario is higher than site base scenario, 3.78%, 4.60%, and 6.51% respectively for RCP 2.6, 4.5 and 8.5. There is no difference for solar radiation data of historical and future climate. Leachate rate from the barrier soil layer has no difference for each climate change scenario.

**Acknowledgments**

The Hibah PITTA B (Hibah Publikasi Internasional Terindeks Tugas Akhir/International Indexed Publication for Final Project) funding scheme, with contract number NKB-0785/UN2.R3.1/HKP.05.00/2019, endorsed this research by Universitas Indonesia.

## References

[1] OGP 2016 *Guidelines for the Management of Naturally Occurring Radioactive Material (NORM) in the Oil and Gas Industry.*

[2] Smith K P Blunt D L Williams G P and Tebes C L 1995 Radiological Dose Assessment Related to Management of Naturally Occurring Radioactive Materials Generated by the Petroleum Industry in *SPE/EPA Exploration and Production Environmental Conference.*
[3] Alslaibi T M Abustan I Mogheir Y K and Afifi S 2013 Quantification of leachate discharged to groundwater using the water balance method and the Hydrologic Evaluation of Landfill Performance (HELP) model *Waste Manag. Res.* **31(1)** 50–59.
[4] Nasar A M Qrenawi L Jaber A and Jazar M A 2007 *Assessment of Solid Waste Dumpsites in Gaza Strip* The Japan International Cooperation Agency (JICA)
[5] Visvanthan C Trankler J Kuruparan O and Xiaoning Q 2002 Influence of Landfill Operation and Waste Composition on Leachate Control : Lysimeter Experiments Under Tropical *Second Asian Pacific Landfill Symp. Landfill Gas Leachate Manag. Sepetember 25-28, Seoul Korea Sess. 4* p. 7.
[6] Hassan Z and Harun S 2013 Impact of Climate Change on Rainfall over Kerian, Malaysia with Long Ashton Research Station Weather Generator (LARS-WG) *Malaysian J. Civ. Eng.* **25(1)** 33–44.
[7] Karamouz M Fallahi M Nazif S and Rahimi M 2009 Long Lead Rainfall Prediction Using Statistical Downscaling and Artificial Neural Network Modeling *Sci. Iranica* **16** 165-17
[8] Osman Y Al-Ansari N Abdellatif M Aljawad S B and Knutsson S 2014 Expected Future Precipitation in Central Iraq Using LARS-WG Stochastic Weather Generator *Engineering* **06(13)** 948–959.
[9] McCarthy J Canziani O Leary N Dokken D and White K 2001 *Impacts, Adaptation and Vulnerability* Cambridge University Press, New York
[10] Al-Ghafri A Gunawardhana L and Al-Rawas G 2014 An Assessment of Temperature and Precipitation Change Projections in Muscat, Oman from Recent Global Climate Model Simulations *Int. J. Students Res. Technol. Manag.* 2**(03)** 109–112.
[11] Kjeldsen P and Beaven R 2011 Landfilling: Hydrology in *Solid Waste Technology and Management* **2(10.3)** 709-733 Wiley Chichester West Sussex UK
[12] Beck-Broichsitter S Gerke H H and Horn R 2018 Assessment of leachate production from a municipal solid-waste landfill through water-balance modeling *Geosci.* **8(10).**
[13] Yong R N Mulligan C N Fukue M 2015 *Sustainable Practices in Geoenvironmental Engineering* 2nd ed. Boca Raton, FL: CRC Press.

# Analysis study of nonwoven pineapple leaf fibre, nonwoven pineapple layered double weave and tricot knitting fabric as absorber material

**N I Khasanah[1] and A I Makki[1,2]**

[1]Politeknik STTT Bandung, Jl Jakarta No 31 Bandung
[1]Graduate of Textile Engineering, Politeknik STTT Bandung
[2]Assistant Professor at Textile Engineering, Politeknik STTT Bandung

Corresponding : ibrahimmakki@sttttekstil.ac.id;

**Abstract**. Sound pollution can be minimized by noise intensity using sound absorbing material. Nonwoven fabrics are the material most widely used for acoustic applications. This research investigated about absorption coefficient of nonwoven pineapple leaf fiber (PNW), and its layered combination with double weave (PDW) and tricot knitting fabric (PTK). The variation of fabric will be tested for sound absorption coefficient using an impedance tube. Absorption coefficient test results showed there was an effect in absorption coefficient on nonwoven with layered.

## 1. Introduction

Sound pollution (noise) can be defined as the excessive noise that can cause an imbalance level of the humans or animals. Excessive noise will impact badly on physiological and psychological effects on humans that can cause permanent hearing damage, increasing efficiency, reducing stress in the workplace, the interference pattern sleep and disorders in communicating[1]

The noise happens around us can come from a variety of sources. The source of this distinguished sources of static and moving resources. Examples of static is the factory and machinery construction. The example moving resource such as motor vehicles, trains and airplanes[2]. Noise in Office buildings, schools, hospitals, etc. are sourced from the activity surrounding the building. Sound absorption is the ability of a material to muffle the sound of coming, calculated in percent or fractional value $0 \leq \alpha \leq 1$. A value of 0 means there is no attenuation sound (the whole sound comes reflected perfectly) whereas a value of 1 means the sound came completely absorbed (Nothing reflected back)[3].

Porous materials absorbed the most commonly used as sound absorbent fibrous material is usually in the form of boards (composite), foam, fabric, rugs, pillows, etc.[4]. According to Padhye and Nayak[1], structure of textile fibrous materials is a great example of a porous material used for sound absorption. There are a number of textile material which can be used for sound absorbent. The material can be classified as a nonwoven fabric, woven and knitted.

Soltani and Zarrebini[5] are conducting research on sound absorption coefficients and characteristics on a woven fabric with a variation of the plain weave structure, twill 2/1, 3/1 2/2, ribs and satin. Tang

Journal of Physics: Conference Series **1381** (2019) 012064 doi:10.1088/1742-6596/1381/1/012064

et al.[6] analyzes woven fabric corduroy against the absorption of sound by performing experiments and simulations. Mankondi and Mistry[7] conducts research on the combination of jacquard woven fabric with nonwoven fabric against the absorption of sound. Monaragala[8] and Ozturk, et.al.[9] using fabric knit spacer for sound absorption.

The use of natural fibers as sound absorber has been done, including hemp, coconut, and abaca. natural fibers are made in the form of composites[10]. Pineapple leaf fiber as a textile material has also been previously studied[11]. Pineapple leaf extract has been studied as a sound absorber in a composite form[12]. In this study, pineapple leaf fiber will be made nonwoven as sound absorber material.

## 2. Method research

This research uses experimental methods. The study begins by make a sample pineapple leaf fiber nonwoven fabric, double weave fabric[13], and tricot knit fabric[14]. Sample fabric that has been created is then tested with absorption coefficient of a tube impedance[15]. Absorption coefficient testing performed on sample of nonwoven fabric, nonwoven layered of double weave, nonwoven layered of tricot knitted fabric. Thickness, weight per meter square and density are calculated to support the results of the study[16][17].

## 3. Result and Discussion

Nonwoven fabric made using a thermal bonding method with low melt polyester as a fastener by a ratio of 75% comparison of pineapple leaves fiber and 25% low melt polyester. Type double weave made that kind of a double weave with a bond, the composition of warp and weft threads 1a1.Top side weave is twill 1/3 and bottom weave is twill 3/1. Double weave is made from 100% cotton. Figure of top and bottom weave can see in figure 1, and weaving plans of double weave can see in figure 2.

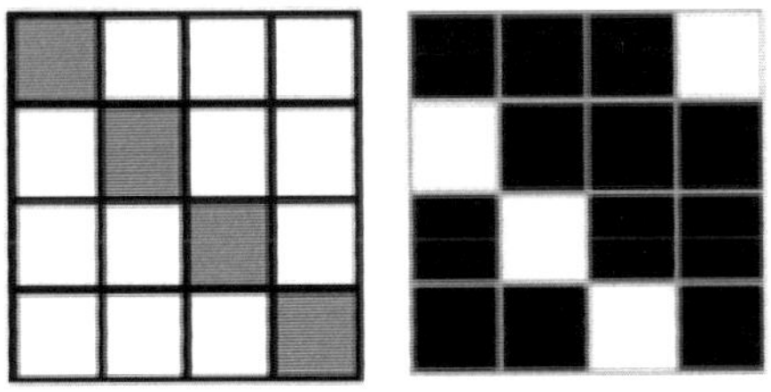

(a) (b)

**Figure 1.** (a) top side weave is twill 1/3, (b) bottom side weave is twill 3/1

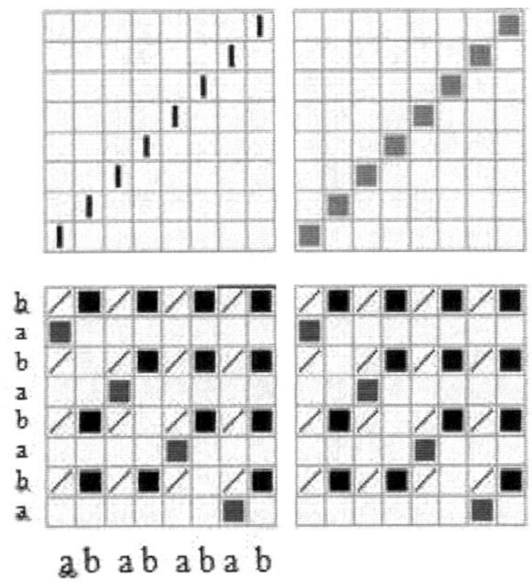


**Figure 2.** Weaving plans of double weave

Tricot knit fabric made of 100% polyester yarn using a double needle machine with 22E gauge. Guide bars are used as many as four of the total guide bar that is installed seven pieces. Lapping diagram of the tricot fabric used is shown in figure 3. Guide bar used is guide bar numbers 1,2,4 and 5. Lapping diagram of each bar is shown in figure 4.

Journal of Physics: Conference Series **1381** (2019) 012064 doi:10.1088/1742-6596/1381/1/012064

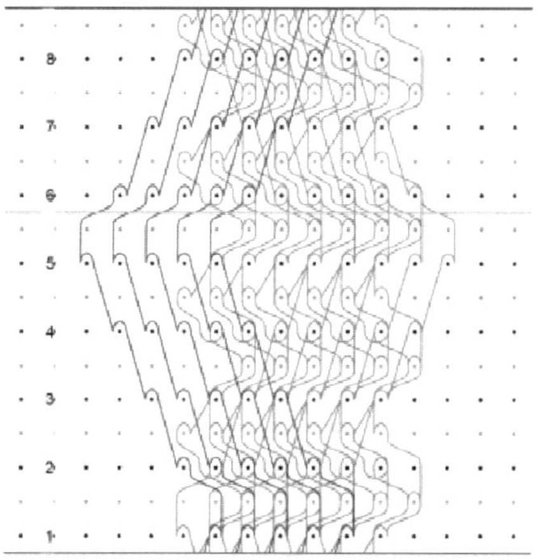

Figure 3 Lapping diagram Tricot knit fabric

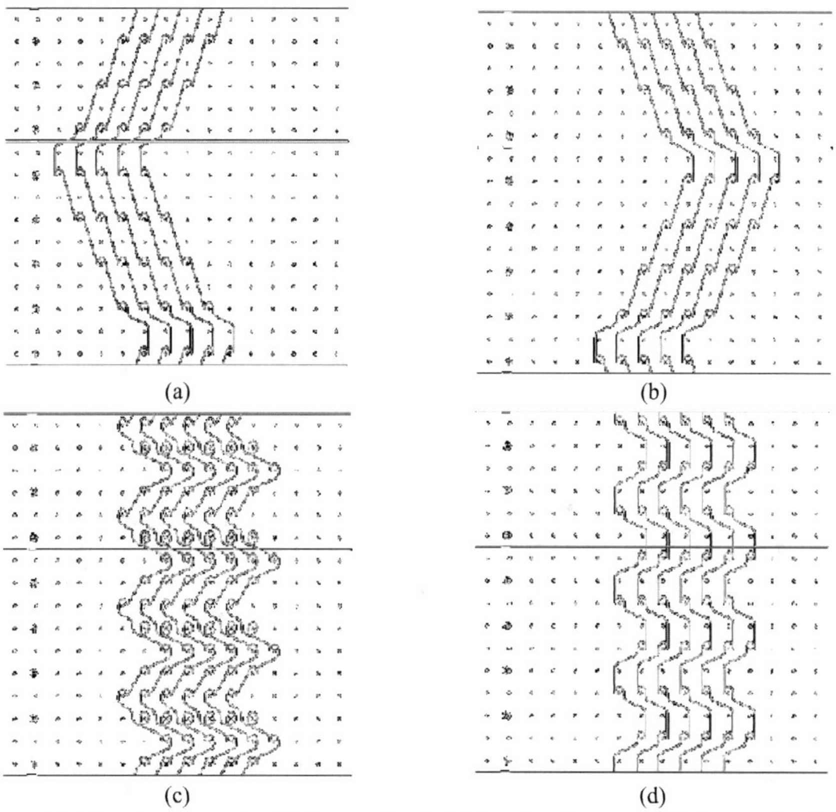


**Figure** 4 (a) Lapping diagram guide bar number 1, (b) Lapping diagram guide bar number 2, (c) Lapping diagram guide bar number 4, (d) Lapping diagram guide bar number 5

The results of the absorption coefficient test of the fabric variations made are shown in table 1. The absorption coefficient test is carried out twice for each sample. Tests are carried out at low, normal and moderate frequencies at frequencies 100 - 5000 Hz. In figure 5 the frequency ranges are listed according to ISO 11564[18]. Based on these graphs all materials used can absorb sounds with different absorption coefficients.

Absorption coefficients in all samples increase at frequencies of 1000-5000 Hz. At low frequencies the absorption coefficient is relatively low. This is because porous material can absorb sound at high frequencies. Sound absorption in fabrics is caused by the morphology of the fabrics that have cavity so that when sound passes through the material, the sound will be absorbed by the porosity in the fabrics.

The addition of layer to nonwoven fabric has an effect on sound absorption. Especially at high frequencies. This might be due to the addition of sample thickness which affects the density. At low frequency the nonwoven sample of pineapple fiber layered with double weave absorbs less sound so that more noise is transmitted. Conversely at high frequencies the nonwoven layered double weave absorbs the most sound. This is possible because of the influence of the porosity by the woven structure on the top and bottom for double weave fabric. Tricot upholstery has less influence on sound absorption compared to double weave, this is because the characteristics of the tricot that are made that have many open voids compared to the double weave so that they cannot absorb much sound.

The highest sound absorption coefficient at the frequency of 5000 Hz based on the graph above is equal to 0.70760 in the nonwoven layered double weave, while the lowest absorption coefficient on the double fabric material is 0.05252 on the same fabric combination at the frequency of 100 Hz.

**Table 1** Result of absorption coefficient

| Frequency (Hz) | Nonwoven pineapple leaf fiber | Nonwoven Pineapple + Double weave | Nonwoven Pineapple + Tricot Knitted Fabric |
|---|---|---|---|
| 100 | 0,1407 | 0.05252 | 0,099389 |
| 124 | 0,1672 | 0.07188 | 0,112073 |
| 160 | 0,1334 | 0.07233 | 0,121352 |
| 200 | 0,1307 | 0.06346 | 0,140659 |
| 250 | 0,1330 | 0.08702 | 0,121732 |
| 314 | 0,1230 | 0.10024 | 0,12389 |
| 400 | 0,1649 | 0.09920 | 0,112567 |
| 500 | 0,0638 | 0.10674 | 0,066939 |
| 630 | 0,1474 | 0.12513 | 0,149091 |
| 800 | 0,1510 | 0.16461 | 0,135188 |
| 1000 | 0,1697 | 0.18907 | 0,132164 |
| 1250 | 0,2282 | 0.21595 | 0,160132 |
| 1600 | 0,2102 | 0.27801 | 0,229578 |
| 2000 | 0,2388 | 0.29391 | 0,248456 |
| 2500 | 0,2590 | 0.35791 | 0,237869 |
| 3150 | 0,3062 | 0.45641 | 0,330113 |
| 4000 | 0,3254 | 0.57728 | 0,441594 |
| 5000 | 0,4589 | 0.70860 | 0,481594 |

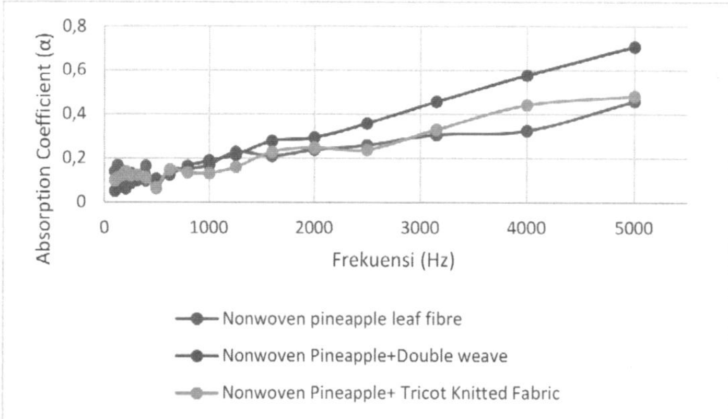


**Figure 5.** Sound frequency graph of sound absorption coefficient

Factors that affect sound absorption are fabric density. The higher the fabric density, the higher the sound absorption. Material that has a high fabric density has good sound absorption at high frequencies. Data density of each sample are shown in table 2. The relationship between fabric density and absorption coefficient is shown in Figure 6.

Tabel 2 density of sample

| | Nonwoven pineapple leaf fiber | Nonwoven Pineapple + Double weave | Nonwoven Pineapple + Tricot Knitted Fabric |
|---|---|---|---|
| Density (g/cm3) | 0,514 | 0,674 | 0,554 |

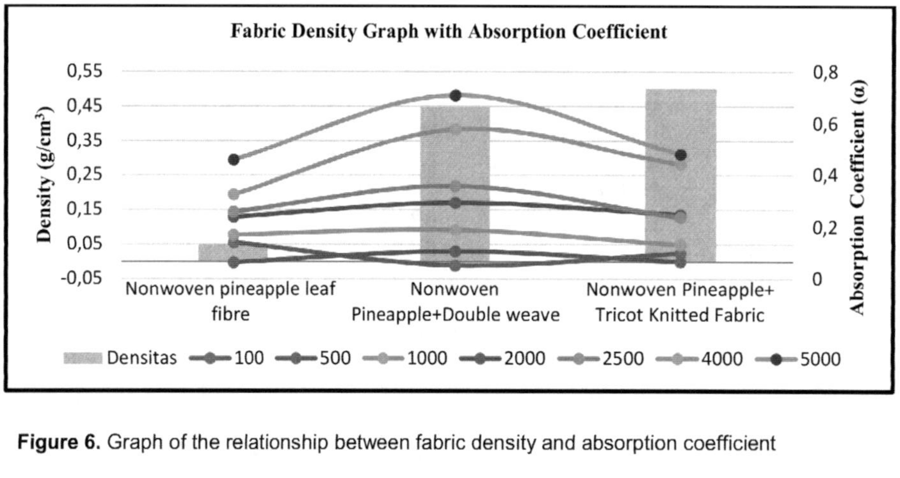


**Figure 6.** Graph of the relationship between fabric density and absorption coefficient

The density of pineapple leaf fiber nonwoven fabric layered with tricot knit fabric is higher than the nonwoven fabric layered with double weave. This is because the tricot fabric has the highest thickness compared to the double weave fabric. So that when the nonwoven fabric is layered with tricot, the thickness exceeds the thickness of the nonwoven fabric layered double weave.

## 5. Conclusion

Cavities in the fiber are able to absorb sound at low and high frequencies. Thickness and weight in square affect the fabric density which can increase sound absorption. In this study, nonwoven layered with double weave fabric and tricot fabric is increases sound absorption at high frequencies. Morphological analysis can be done next to see more clearly the influence of porosity. Nonwoven layered using double weave fabric and tricot fabric can be tried to determine the effect on sound absorption and become a solution to add aesthetics.

## References

[1] Padhye R and Nayak R (edt). 2016 *Textile Science and Clothing Technology* (Singapore: Springer)
[2] Mediastika Christina 2005 *Akustika Bangunan: Prinsip-prinsip dan Penerapannya di Indonesia* (Jakarta: Erlangga)
[3] Satwiko Prasasto 2009 *Fisika Bangunan* (Yogyakarta: C. V Andi Offset)
[4] Everest F Alton and Pohlmann Ken C 2009 *Master Handbook of Acoustic Fifth Edition* (United States of America: McGraw-Hill)
[5] Soltani Parham and Zarrebini Mohammad 2013 *The Journal of The Textile Institute* (London: Taylor & Francis)
[6] Tang Xiaoning, Zhang Xiansheng, Zhuang Xingmin, Zhang Huiping, and Yan Xiong 2017 *Journal of Industrial Textile* 0(00) 1-20
[7] Mankodi Hireni R and Mistry Purvi 2014 *International Journal of Industrial Engineering & Technology (IJIET)* **4** 19-26
[8] Monaragala R M 2011 *Advances in Knitting Technology* (United Kingdom: Woodhead Publishing Limited)
[9] Ozturk M K, Nergis B U, Candan C 2010 *7th International Conference* – TEXSCI
[10] Eriningsih Rifaida, Widodo Mukti, Marlina Rini 2014 *Arena Tekstil* **29** 1-8
[11] Hidayat Praktino 2008 *Teknoin Jurnal Teknologi Industri* **13** 31-5
[12] Putra Azma, Or Khai Hee, Selamat Mohd Zulkefli, Noor Mohd Jailani M, Hassan Muhamad Haziq and Prasetiyo Iwan 2018 *Applied Acoustic* **136** 9-15
[13] Widodo Sugeng and Wardiningsih Wiah 2005 *Disain Tekstil. Bandung* (Indonesia: Sekolah Tinggi Teknologi Tekstil Bandung)
[14] David J Spencer 2001 *Knitting Teknology Third Edition* (United Kingdom: Woodhead Publishing)
[15] ___ 1998 *ISO10534-2: Acoustic–Determination of absorption coefficient and impedance in impedance tubes part 1 :method using standing wave ratio*
[16] ___ 2010 *SNI ISO 5084:Cara Uji Tebal Tekstil dan Produk Tekstil*
[17] ___ 2010 *SNI ISO 3801:Cara Uji Berat Kain per Satuan Panjang dan Berat Kain per Satuan Luas*
*[18]* ___ 1997 *ISO 11654: Acoustic – Sound Absorption for use in building – rating of sound absorption*

Journal of Physics: Conference Series **1381** (2019) 012065 doi:10.1088/1742-6596/1381/1/012065

# Acoustic absorptive properties of Kapok fiber, Kapok fiber layered tricot fabric and Kapok fiber layered double weave fabric

**A I Makki, E Oktariani***

Politeknik STTT Bandung, Jalan Jakarta No. 31 Bandung

Corresponding author : ekaoktariani90@gmail.com

**Abstract.** Kapok fiber was one of the natural cellulosic fiber that can be found widely in Asia's forest especially in Indonesia. Kapok fiber was short and smooth surface causing poor inter-fiber cohesion that could not be processed in spinning. This fiber could be utilized to be accoustic sound proofing material. This research investigated about absorptive properties of non woven kapok fiber (SK) and its layered combination with tricot knitted fabric (TC) and double weave fabric (DWK). Combination of various kind of fabric's type improved thickness and air cavity that impact absorptive properties of fabric. Impedance tube were used to characterization sound absorptive properties at low until normal frequency on 100 until 5000 Hz. The optimum sound absorptive were showed by combination of non woven kapok fiber with double weave fabric with thicknes 11,25 mm. It shows thickness impact to air cavity and gave significant effect to sound absorptive properties.

## 1. Introduction

Prolonged exposure to noise pollution has harmful effect for human health. Noise exposure could cause cognitive decline and health problem such as high stress level, sleep disturbance, hearing impairment and cardiovascular effect. According to WHO, more than 30% of the population in EU countries is exposed to noise exposure at night. The growing awareness towards engineering and research to obtain sound absorbing material as noise control. Sound absorbing material are generally made from synthetic fiber that consumes high energy level on production process. In contrast, sustainable material have good potency as sound absorbing materials [1]. The previous research shows that natural kenaf fiber [1] and Pineapple leaf fiber [2] gave good results as sound absorber material. Kapok fiber is one of natural fiber that have good potency as sound absorber material. kapok fiber have good potency as sound absorbing material due to its special properties that increase the chance of friction between sound wave and fibers [3]. Kapok fiber is a buoyancy material, combining kapok fiber with low melting fiber can improve the bouyancy charactheristic of kapok fiber assemblies [3]. The previous research shows kapok fiber with smaller bulk density compared with glass wool gave similiar absorbancy coefficient [3]. The recent research shows that combining kapok fiber with polypropilene fiber provide very good sound absorption behaviour in 250 – 2000 Hz [4]. Production of kapok fiber as sound absorber material required less energy and less chemical compound.

Textile plays important role in sound absorber material. There was some factor that give significant impact on sound absorbing material such as bulk density, thickness and arrangement of fiber [1]. Knitted and woven are potential for sound absorbing material, altough, both of them have micro-perforated acoustic absorption panel, they have different acoustic absorption mechanism[5]. Sound absorption properties of knitted fabric are highly influenced by fabric structure [6]. In the other hand,

Journal of Physics: Conference Series **1381** (2019) 012065 doi:10.1088/1742-6596/1381/1/012065

Non woven fabric contains micron and submicron pores making them preferred as sound absorbing material [7]. There is many factors that can improve sound absorptive properties of non woven fabric, such as fiber parameters, process parameter and physical parametes [8]. This paper focuses on assesment kapok fiber and its layered combination to absorbance properties.

## 2. Methods Research

This research were used three types of fabric : tricot knitted fabric , double weave fabric, and non woven fabric from kapok fiber. All fabric were made in Polytechnic STTT Bandung. Tricot knitted fabric were produced on Double Needle Knitted Fabric RD7/2-12 EN/119740 (Karl Meyer, Germany) using polyester fiber 100%. Structure of tricot knitted fabric shows on Fig. 1

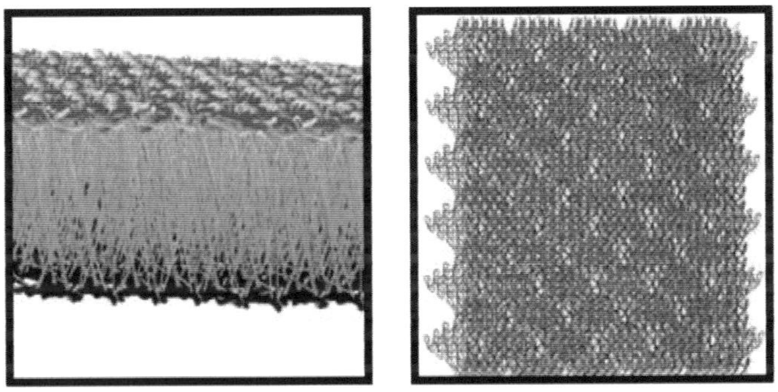

**Figure 1.** Structure of Tricot Fabric

Double weave fabric were produced on Toyoda Machinery using 1/3 twill effect on top surface and 3/1 twill effect for bottom surface as seen on Fig. 2. Cotton fiber are used as materials on double weave fabric.

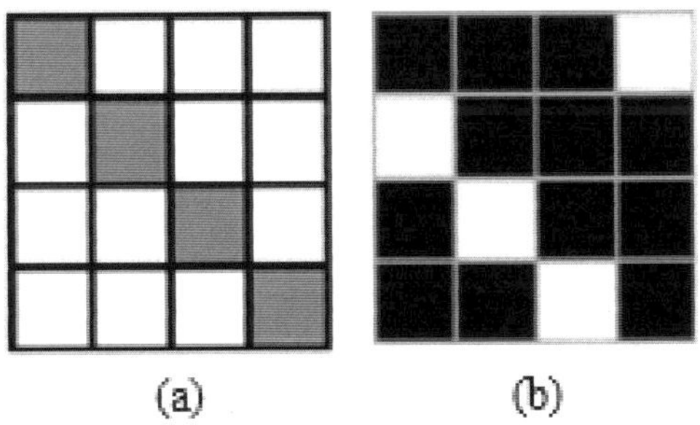


**Figure 2**. Weaving Plan for Double Weave Fabric

Non woven fabric from kapok fiber were processed by mixing 25% low melt polyester binder with kapok fiber using manual thermal bounding machine at temperature 130°C. Each non-woven kapok fabric then layered by tricot knitted fabric and double weave fabric. Characterization of sound absorption properties was measured by impedance tube in the frequency range 100 – 5000 Hz [9,10]. GSM (Gram Square Metre) was measured using SNI ISO 3801 Method and thickness of fabric was observed using SNI ISO 5084 Method to discover characteristic properties of fabrics and relation to sound absorption properties [11,12].

## 3. Result and Discussion

### *3.1. Morphological Structure*

Sound absorption properties is highly influenced by morphological structure of fiber. The cross sectional of kapok fiber as shown in figure 3, denote irregular oval to round shape. The longitudinal view shows a hollow morphological structure, a special structure on natural fiber that play role on sound absorption properties.

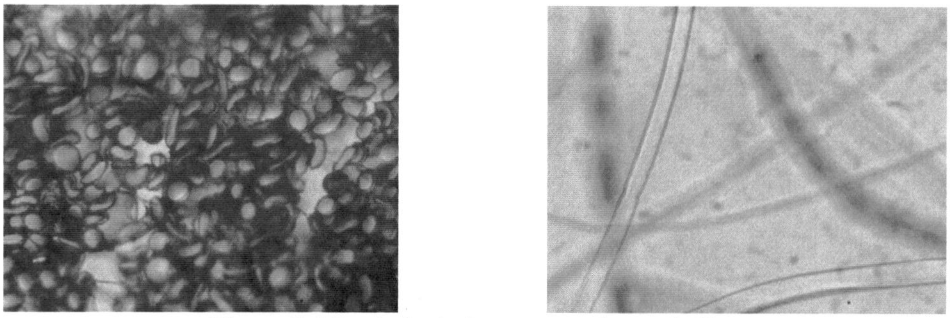

**Figure 3**. Morphological structure of Kapok fiber

Buoyancy behaviour on kapok fiber are affected by hollow stucture. These charachteristic makes kapok fiber has high porosity resulting improvement on sound absorption properties. Large diameter of cross sectional kapok fiber increases friction resistance between air and fibre, resulting dissipation of energy [13]. This unique structure enabled the air of sound wave diffuse between and into the fibre.

*3.2. Thickness*

Thickness was observed in each sample as shows in Figure 4. Data in figure 4, indicates thicknes on this research was not give significant impact. Altough TC Fabric was the highest thickness, DWK fabric absorb sound more effective than other fabric. This phenomena conducted due to Tricot fabric and non woven kapok fiber have high thickness because of spacer apart the fabric, but the space area was highly loose and open structure. The sound energy could through the fabric easily, caused sound energy transmitted. This phenomenon are suitable with previous research, that explain the fabric with loose open structure has lowest sound absorption properties [14].

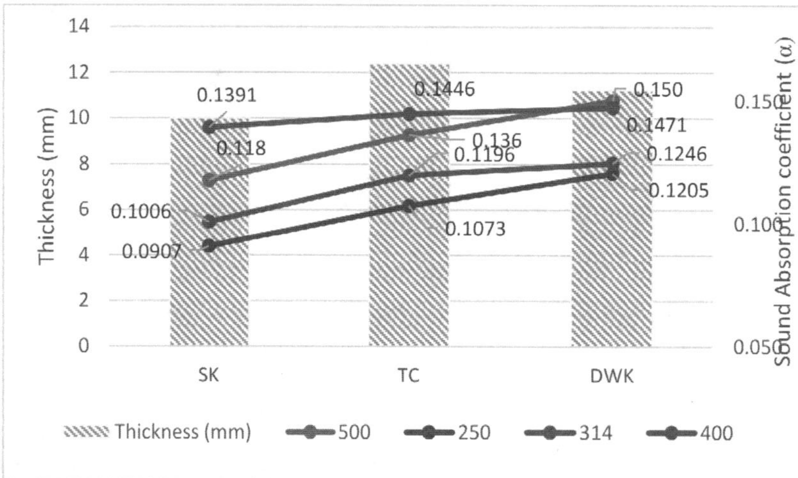


**Figure 4.** Relation between Thickness and Sound Absorption Coefficient

*3.3. Density*

Sound Absorption coeffiecient on various sample was observed. Graphic in Figure 5 shows all sound absorber material give similiar coefficient absorbancy, 0,98, at High frequency 5000Hz. Otherwise non woven kapok fiber layered by double weave fabric (DWK) give the best results at frequency between 1600 - 3160Hz compared with single non-woven kapok fiber (SK) and kapok fiber layered by tricot fabric (TC), there was 0.95, 0.89, 0.90 perspectively. This phenomenon due to complex pores structure on DWK cause the sound wave trapped on the fabric and absorb to the fabric. Twill effect on double layer woven fabric obstruct sound energy to release. In the other hand, Tricot fabric have perforated structure with big pores it cause the energy are released easily.

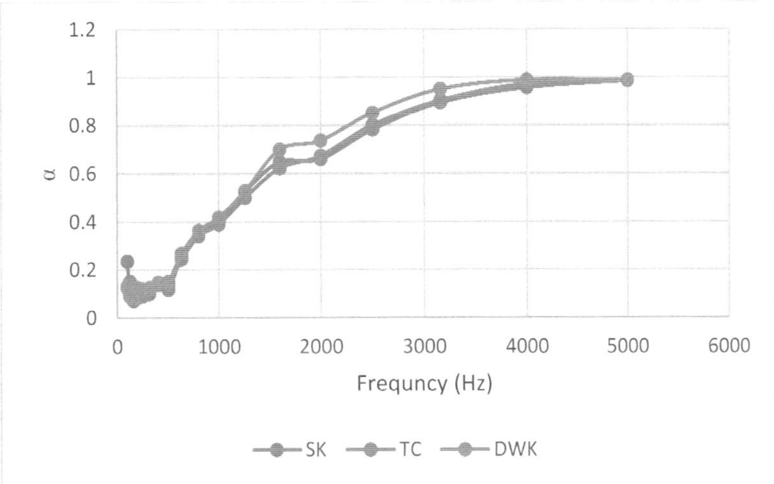


**Figure 4**. Sound Absorption Coeficient on Various Fabric

Density is the most important factor on sound absorption properties. Dense properties of sound absorber were affected by material and web formation of material, as shown in Figure 6.

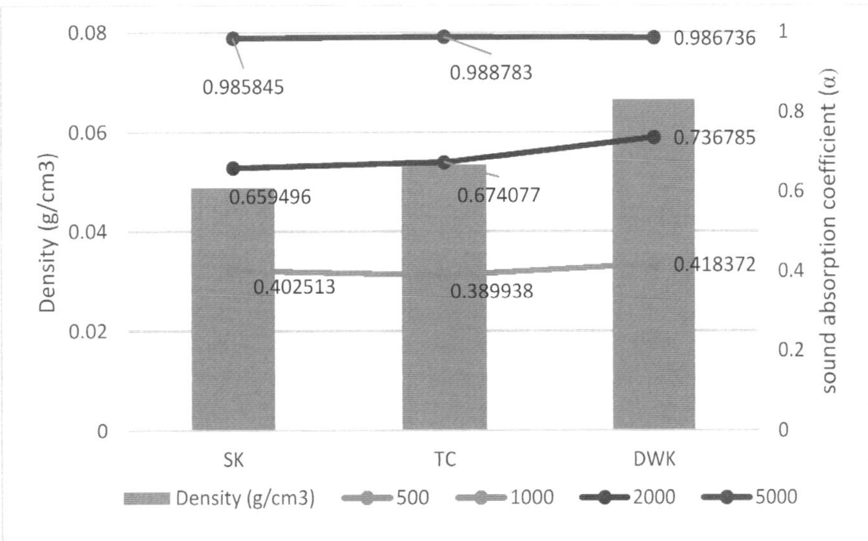


**Figure 5**. Relation Density of Fabric to Sound Absorption Coefficient

Figure 5 shows relation between density of various fabric to sound absorption coefficient on the range frequency 500 – 5000 Hz. DWK fabric have high density however the density is still can absorb the sound energy. Double weave fabric has open structure forming flexible complex pores between warp and weft.This structure impact to high friction and resistance between air and fiber. DWK were composed by cotton and kapok fiber. Both of this fiber have good absorption properties resulting good absorption properties on DWK fabric.

## 4. Conclusion

Textile has potential properties as sound absorber material. Woven, knitted and non woven fabric are most commonly textile product that could be used as sound absorber material. This material could absorb the sound energy and convert into heat energy. Fiber material, thickness, and density of sound absorber material are major factors that impact significantly to sound absorption properties. Non woven kapok fabric layered by double weave fabric give good result in absorbing sound at range frequency 100 – 5000 Hz.

**References**

[1] Asdrubali F, Schiavoni S, Horoshenkov, K.V. 2012. *A review of Sustainable Materials for Acoustic Applications*. Building Acoustis. Vol 19. P. 283-312.
[2] Putra, Azma dkk. 2018. *Sound Absorption of Extracted Pineapple-Leaf Fibres*. Applied Acoustics
[3] Zheng, Yian dkk. 2015. *Research and Application of Kapok Fiber as an Absorbing Material : Mini Review*. Journal of Environmental Science.
[4] Veerakumar, A., Selvakumar, N. 2012. *A Preliminary Investigation on Kapok/Polypropylene Nonwoven Composite for Sound Absorption*. Indian Journal of Fiber & Textile Research. Vol 37. p 385-388.
[5] Tang, X., Yan, X. 2017. *Multi-layer Fibrous Structures for Noise Reduction.* The Journal of Textile Institute.
[6] Monaragala, R.M. 2011. *Knitted Structures for Sound Absorption.* Woodhead Publishing Limited. United Kingdom.
[7] Shahani, Fereshteh. dkk. 2014. *The Analysis of Acoustic Charactheristics and Sound Absorption Coefficient of Needle Punched Nonwoven Fabrics*. Journal of Engineered Fibers and Fabrics.
[8] Puranik, P.R., Parmar, R.R., Rana, P.P. 2014. *Nonwoven Acoustic Textiles – Review.* International Journal of Advanced Research in Engineering and Technology (IJARET)
[9] _____. Acoustic Sound Absorption, ISO 11564.
[10] _____. 1998. ISO 10534-2: Acoustic –Determination of absorption coefficient and impedance tubes
[11] _____. 2010. SNI ISO 3801: Cara Uji Berat Kain per Satuan Panjang dan Berat Kain per Satuan Luas.
[12] _____. 2010. SNI ISO 5084: Cara Uji Tebal Tekstil dan Produk Tekstil
[13] Xiang, H., Wang, D., Liu, H. 2013. *Investigation on Sound Absorption Properties of Kapok Fibers*. Chinese Journal of Polymer Science (English Edition).
[14] Mankodi, Hireni dan Purvi Mistry. 2014. *Woven Fabrics Combination For Acoustic of Building Interior*. University of Baroda, India

# Experimental Study on Variation of Tilted Angles Toward Acoustic Power of Thermoacoustic Engine

**R A Anugrah**

Department of Mechanical Engineering, Faculty of Vocational Program, Universitas Muhammadiyah Yogyakarta,Jl. Brawijaya,Tamantirto, Kasihan, Bantul, Yogyakarta 55183 Indonesia

Corresponding author: rinasaanugrah@umy.ac.id

**Abstract.** Tilted angles have to be researched for determining the good position of thermoacoustic engine. The research aims to find the best acoustic power with a tilted angle of the position, therefore the performance of Standing-Wave Thermoacoustic Engine (SWTE) is going to be increased. The performance is affected by the tilted angles. The natural convection which occurs in the core of Themoacoustic Engine will be different every tilted angle. The heat transfer characteristics within it were investigated under three tilted angles including, -90°, 0°, and 90°, with the resonator length 780 mm. This study was for getting the acoustic power which used the two sensors method with Data Acquisition (DAQ) NI-DAQmx 15.5 and two pressure transducers. It was calculated by Matlab R2013a. In this research is found that the acoustic power which is the best value is resulted on 90° of tilted angle, while the smallest value is found on 0°. Thes matter can be occured because phase difference of acoustic wave on the 90° is the biggest value, while the 0° is the smallest value. Further, have to study how phase difference has big effect toward the horizontal position which is resulted on the worse value compared the vertical position.

## 1. Introduction

In converting from heat energy to become another energy, acoustic power of thermoacoustic engines has been tried to contend with internal combustion engines for last several years ago. One of the examples is converting to electric energy. The four types of acoustic to electric conversions are such as electromagnetic devices, piezoelectric devices, magneto hydrodynamic devices, and bidirectional turbines [1]. Research has applied one of the types, piezoelectric devices in a study that is thermoacoustic piezo system called Rijke-Zhao system and has explained the energy conversion process and harvesting heat energy by the type [2]. Before converting heat energy to electric energy, heat energy first is converted to acoustic waves, then to be converted to electrical energy. For generating acoustic waves, thermoacoustic engine needs waste heat sources that contain abundant heat sources.

Furthermore, the thermoacoustic engine has also been applied for cooling system. Research used the waste heat from the inside of microelectronic components to generate acoustic waves has been finished [3]. The sound was later used as a cooling driver of thermoacoustic to be cold the surface of the computer processor. The capability of cooling of the device could produce begin 0.1 up to 0.6 Watt per $cm^2$ in cross-sectional area of the resonator. The other study [4] of thermoacoustic refrigerator driven by reused waste heat from exhaust gases has applied to refrigeration system of vehicles. The system can substitute the conventional refrigeration system that has applied before.

The one of parameters which affect the acoustic power of Standing-Wave Thermoacoustic Engine (SWTE) is tilted angles of the device. Namely by varying position placement on SWTE device with

various angles, that is tilted angles such as horizontal position 0°, slope position -45°, and 45°, and vertical (-90°) and (90°). There was a research which was conducted by varying the tilted angles at range of -90° to 90°, and the result was comparing any others working fluid except air on some parameters analysis [5]. From now on, there have conducted same research [6], in the research, it was just explained about one parameter, onset temperature that was affected by tilted angles it was onset and also the damping behaviors of the onset temperature. When a position of the tilted angle called -90°, that the condition of the direction of the temperature gradient is similiar as the direction of gravity, thus further it can be reinforcement the natural convection that can be occurring. So the onset temperature of SWTE, which was the greatest is when the device in (-90°) of this tilted angle. Furthermore, there is research that has conducted same research with variation with different aim [7] and any other research with the same variation also like tilted angles [8].
There have not been researches that published that have presented and have analyzed the connection of different tilted angles toward acoustic power of SWTE with common air as working fluid. This research aims for the acoustic power could be greater because it could be interpreted that the performance of SWTE is in good way. So in this research, it uses variations of tilted angles, there are three tilted angles including (-90°), (0°), and (90°). Therefore, research needs to be done with the variable of the variants that were mentioned above. The aim of this research is to investigate the effect of tilted angles with the working gas,which just uses air at 1 atm of pressure and 25°C until 27°C room temperature toward acoustic power. The results of the experiment are essential for an effective position based on tilted angles, then, this is obtained the best performance or describe and illustrate the scientific reason why this phenomenon occurs in SWTE.

## 2. Experimental Apparatus and Measurement

### 2.1. Component of Standing-Wave Thermoacoustic Engine (SWTE)

Fig. 1. below illustrates the components of SWTE. Names of each components are below the figure.

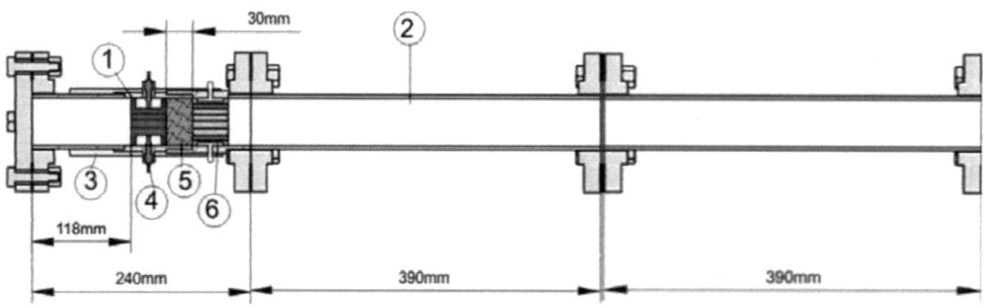


**FIGURE 1.** Components of Standing-Wave Thermoacoustic Engine

Information:

1. Hot Heat Exchanger (HHX)
2. Resonator
3. *Thermal insulation*
4. *Glow plug*
5. *Wire mesh*
6. Cold Heat Exchanger (CHX)

### 2.2. Positioning Thermoacoustic Engine Device with Variation of Tilted Angles

To get result of performance of SWTE with variation tilted angle, the apparatus of experimental research can be changed every different tilted angles. In process of experiment, the thermoacoustic engine was put on the floor, except on horizontal position, the device was propped up iron cantilever. The each experiment of thermoacoustic engine was conducted for each tilted angle. The angle was defined as a positive value in case the HHX was above the CHX (tilted angle on 90°). It was defined as a negative value if the HHX was below the HHX (tilted angle on -90°) as shown as in Fig. 2.

(a) (b)

**FIGURE 2.** The position of tilted angles of thermoacoustic engine
(a) -90° and (b) 90°

The research were performed for every 90° from -90° to 90°. The position of 0° is shown in Fig. 3. that the resonator lengths was defined 780 mm. Hence, three tilted angles including -90°, 0° and 90° have been tested in total. The thermoacoustic engine was arranged of two heat exchangers (hot heat exchanger and cold heat exchanger), thermoacoustic core, and resonator tube. The thermoacoustic core which is between two heat exchangers was the main component of thermoacoustic engine which is called stack.

**FIGURE 3.** The position of tilted angle on 0°

The geometry of the component of thermoacoustic engine are presented by Table 1. There are sizes of components such as diameter of HHX, Stack, CHX, and Resonator Length. Also, the length for each components.

**Table 1.** Size of component of SWTE.

| | HHX | Stack | CHX | Resonator |
|---|---|---|---|---|
| Diameter (mm) | 52 | 52 | 52 | 52 |
| Length (mm) | 40 | 30 | 40 | 780 |

Journal of Physics: Conference Series **1381** (2019) 012066 doi:10.1088/1742-6596/1381/1/012066

### 2.3. Experimental Steps

In the research, the temperature of thermoacoustic core which consist of HHX, Stack, and CHX were measured every variation of tilted angles. The steps of experimental research were similar to each position of angle of tilted angles. The corresponding of experimental set up with several steps, there are:

1. HHX of thermoacoustic engine apply an electrical heater that is called Battery Charger Transformer (BCT) which the six glow plugs are stacked around a cylinder copper in the duct. A voltage regulator was used to control the heat temperature. The BCT was stabilized at 230 Watts. It would be heated until to maximum temperature at 320ºC in this experiment. The CHX was cooled by circulate common water which the temperature was steady at 25º – 28ºC. The temperature of HHX and CHX were measured by thermocouples which the type is K-type which were completed both of them. The thermocouples T1 and T2 which are the thermocouples in Fig. 4. were located between left and right side of the stack. The schematic diagram of this experimental research is shown at Fig. 4.

2. The data of parameter of acoustic wave were taken down and conjoint by Data Acquisition (DAQ) with series NI-DAQmx 15.5 and NI LabVIEW 2015 of the software program. This parameters of the acoustic wave such as pressure amplitude with time domain.

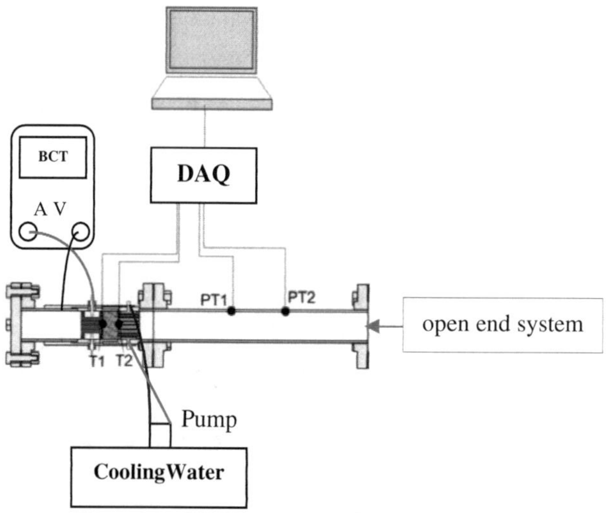


**FIGURE 4.** The diagram of schematic design of experimental research

3. Result of raw data were times to pressure amplitude, then, the time domain was converted to become to frequency domain by Fast Fourier Transform (FFT) formula. Data Acquisition (DAQ) and software program, that is NI LabVIEW 2015 were used to record the process of taking raw data from pressure transducer to obtain the values such as frequency, pressure amplitude, and phase of acoustic wave with the FFT method. The values of them which were taken to caculate acoustic power by subtitute them inside an equation were presented in the Equation (1) below. PT1 and PT2 are $P_A$ and $P_B$ in the equation which were pressure transducers with 50 mm of distance between them on the center of the resonator.

4. The stack was chosen with type of stack wire mesh screen with the material of stainless steel and onset temperatures were in interval 200ºC – 250ºC. Oscillation of acoustic wave in the resonator occurs in open end system.

The module in Data Acquisition (DAQ) National Instrument NI-DAQmx 15.5 which is presented in Fig. 5. with code NI9213 was connected to thermocouples. It was used to measure temperature in HHX and CHX. While the other one as code NI9234 was connected to pressure transducers for measure the pressure amplitude with domain of time and could be converted with FFT to domain of frequency.

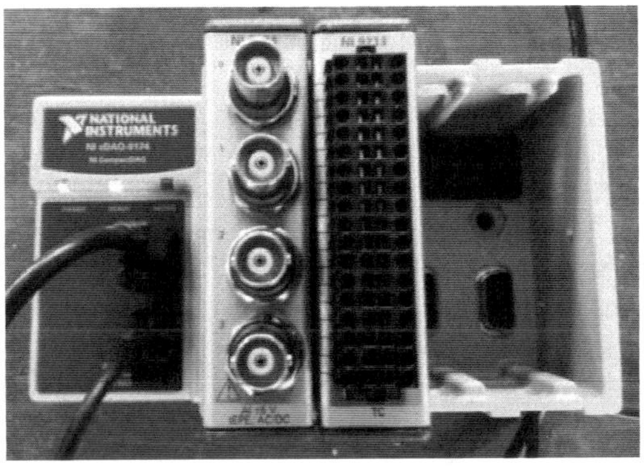


**FIGURE 5.** Data Acquisition (DAQ) NI-DAQmx 15.5 with 2 module (NI9213 and NI9234)

Pressure transducer is presented in Fig. 6. below with type PCB 102 which accurately measured parameters of acoustic wave in resonator. There are two pressure transducers that could be used to measure pressure amplitudes based on two sensor method.

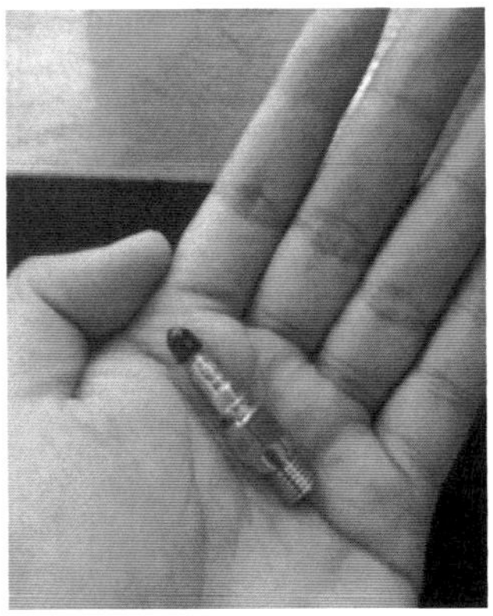

**FIGURE 6.** Pressure Transducer (PT) PCB 102 to measure pressure amplitude

The layout of LabVIEW when recording data in real-time condition is presented in Fig. 7.On the left side is the appearance of acoustic waves with time domain, while, right side is frequency domain.

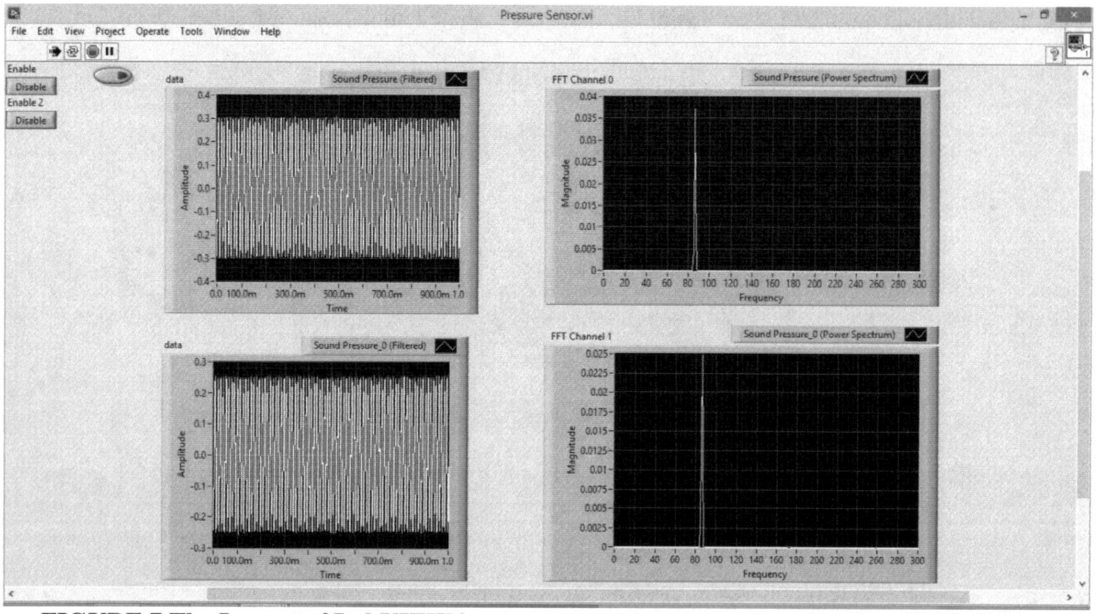


**FIGURE 7.The Layout of LabVIEW in measurements pressure amplitude with FFT (Fast Fourier Transform)**

Then, the magnitude of pressure amplitude was substituted in formulation to calculate acoustic intensity. Because it is the function of acoustic intensity, it can be resulted the data of acoustic power.

## 3. Method For Calculating Acoustic Power

In the beginning, calculating the acoustic intensity (*I*) before finding acoustic wave was determined to use the two pressure sensor method [9]:

$$I = \frac{1}{8\omega\rho}\{Im[H](|p_A|^2 - |p_B|^2) + 2\ Re[H]|p_A||p_B|\sin\theta^o\} \quad (1)$$

Equation (1) calculates the acoustic intensity which is between the two sensors (in this research called pressure transducer) which is used to measure $P_A$ and $P_B$. Where $P_A$ and $P_B$ are the pressure amplitude of acoustic waves measured at two adjacent pressure sensor positions along the resonator, and $\theta^o$= arg[$P_A/P_B$] is the phase difference of pressure amplitude of acoustic wave $P_A$ relative to $P_B$. Then $\rho$ is represented the density of the working gas, Re[ ] and Im[ ] represent the real and imaginary components.

Defining *H* with:

$$H = \frac{kF}{\cos(k\Delta x/2)\sin(k\Delta x/2)} \quad (2)$$

*k* constitutes the complex wave number and *F* constitutes a complex factor which are calculated by:

$$F = 1 - \frac{2J_1(i^{3/2}\sqrt{2}r_0/\delta_v)}{i^{3/2}(\sqrt{2}r_0/\delta_v)J_0(i^{3/2}\sqrt{2}r_0/\delta_v)} \quad (3)$$

and $$k = -ik_0\sqrt{\frac{J_0(i^{3/2}\sqrt{2}r_0/\delta_v)}{J_2(i^{3/2}\sqrt{2}r_0/\delta_v)}}\sqrt{\gamma + (\gamma - 1)\frac{J_0(i^{3/2}\sqrt{2Pr}r_0/\delta_v)}{J_2(i^{3/2}\sqrt{2Pr}r_0/\delta_v)}} \quad (4)$$

where $J_n$ is the $n_{th}$ order complex of Bessel function, and $\sigma$ is the Prandtl number and $\gamma$ is the specific heat ratio respectively.

So, acoustic power ($\dot{E}$) is able to be calculated by dividing acoustic intensity ($I$) with cross-sectional area ($A$) of thermoacoustic core which is the same and connected by resonator.

$$\dot{E} = I.A \quad (5)$$

The calculating to acoustic power above was conducted with using software program as Matlab R2013a in order to be easier and fast to result end data, which is the acoustic power.

## 4. Result and Discussion

Effect of Different Tilted Angles toward Acoustic Power

The plotting data of acoustic power which are obtained from the experiments with variation of different tilted angles are presented in Fig. 8. We are able to know effect of different tilted angles toward Acoustic Power. The researcher found phenomenon that is the best position of this device at 90° of the tilted angle.The acoustic power at tilted angle of 90° was found of 10.7 Watts. While at 0°, it was found of 0.5 Watts. Besides,at -90°, it was found of 6.7 Watts. This is presented in Fig. 8. that can be inferred as we put a position of thermoacoustic engine with the core below the resonator on position (-90°). The data of acoustic power is on the best position because of natural convection occurring in the core. While the experimental investigation has been implemented at different tilted angles with range begin -90° to 90°. The more the tilted angle decreases, the more the working fluid (gas fluid) and stack of heat transfer coefficient of the natural convection are going to increase [10]. So that, this tilted angles affect it in using less energy than others positions. It is only appropriate for using heat energy sources. However, this statement is not consistent to acoustic power.

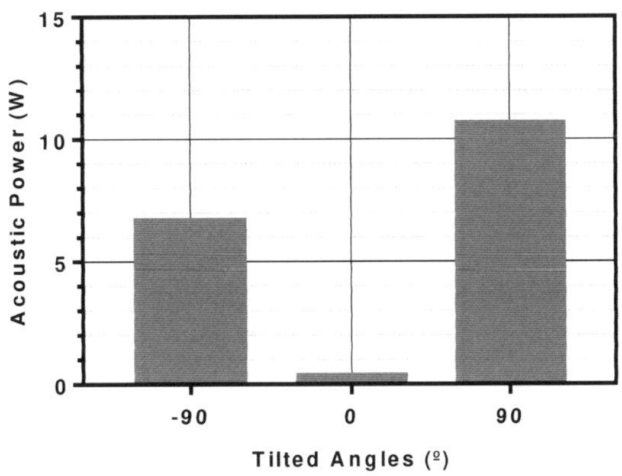


**FIGURE 8.** Effect of Tilted Angles toward Acoustic Power

Beside the statements above, the phase of wave precisely affect a result in power. Because of the difference of two waves' phase which is each measured by PT1 and PT2, it is called also phase difference becomes the function in calculating acoustic power. Table 1 presents the data of phase measured by PT1

($\theta_1{}^{\circ}$), phase measure by PT2 ($\theta_2{}^{\circ}$), and phase difference ($\theta^{\circ}$). Based on equation (1), phase difference ($\theta^{\circ} = |\theta_1{}^{\circ} - \theta_2{}^{\circ}|$) which was found at 0° of tilted angle is the smallest value among the others, so that it cause the experiment result the smallest value of acoustic power. The best acoustic power is found at 90° same case with the best value of the phase difference too.This research have not explained scientifically the phenomenon how it can be happening. So, it have to conduct the next research about the analysis of how phase difference can givea big effect toward acoustic power, which causes the horizontal position has worse value compared vertical position.

Table 2. Phase and phase difference of each tilted angles

| Tilted Angles | phase measured by PT1 ($\theta_1{}^{\circ}$) | phase measure by PT2 ($\theta_2{}^{\circ}$) | phase difference ($\theta^{\circ}$) |
|---|---|---|---|
| -90° | -2.2479 | -1.7334 | 0.5145 |
| 0° | 0.4184 | 0.3919 | 0.0265 |
| 90° | -2.8473 | -1.3577 | 1.4896 |

## Conclusion

According to the result, it is able to be concluded that the acoustic power has the greatest position at 90° of tilted angle which the value is 10.7 Watts. Also, the smallest value of acoustic power found at 0° which the value is 0.5 Watts. It can be inferred that the big effect is caused by phase difference. The next researcher has to conduct the next research about the analysis of how phase difference has big effect toward the horizontal position which is resulted on the worse value compared the vertical position.

## ACKNOWLEDGMENTS

This study was supported by Laboratory of Universitas Gadjah Mada. And big thanks to Universitas Muhammadiyah Yogyakarta to support administration of this publication.

## REFERENCES

[1] M. A. G. Timmer, K. de Blok, and T. H. van der Meer, "Review on the conversion of thermoacoustic power into electricity," *The Journal of the Acoustical Society of America*, vol. 143, no. 2, pp. 841–857, Feb. 2018.

[2] D. Zhao, "Waste thermal energy harvesting from a convection-driven Rijke–Zhao thermo-acoustic-piezo system," *Energy Conversion and Management*, vol. 66, pp. 87–97, Feb. 2013.

[3] O. G. Symko, E. Abdel-Rahman, Y. S. Kwon, M. Emmi, and R. Behunin, "Design and development of high-frequency thermoacoustic engines for thermal management in microelectronics," *Microelectronics Journal*, vol. 35, no. 2, pp. 185–191, Feb. 2004.

[4] D. L. Gardner and C. Q. Howard, "Waste-Heat-Driven Thermoacoustic Engine and Refrigerator," p. 4, 2009.

[5] C. Shen, Y. He, Y. Li, H. Ke, D. Zhang, and Y. Liu, "Performance of solar powered thermoacoustic engine at different tilted angles," *Applied Thermal Engineering*, vol. 29, no. 13, pp. 2745–2756, Sep. 2009.

[6] Y.-L. He, H.-B. Ke, F.-Q. Cui, and W.-Q. Tao, "Explanations on the onset and damping behaviors in a standing-wave thermoacoustic engine," *Applied Thermal Engineering*, vol. 58, no. 1–2, pp. 298–304, Sep. 2013.

[7] C. Shen, H.-X. Li, D.-W. Zhang, P. Yu, N. Pan, and S.-F. Wang, "Study on the heat transfer characteristic of solar powered thermoacoustic prime mover at different tilted angles," *Applied Thermal Engineering*, vol. 103, pp. 1126–1134, Jun. 2016.

[8] N. Pan, S. Wang, and C. Shen, "A fundamental study on characteristic of thermoacoustic engine with different tilt angles," *International Journal of Heat and Mass Transfer*, vol. 74, pp. 228–237, Jul. 2014.

[9] T. Biwa, Y. Tashiro, H. Nomura, Y. Ueda, and T. Yazaki, "Experimental verification of a two-sensor acoustic intensity measurement in lossy ducts," *The Journal of the Acoustical Society of America*, vol. 124, no. 3, pp. 1584–1590, Sep. 2008.

[10] R. A. Anugrah, A. Widyaparaga, I. M. Miasa, J. Waluyo, Sugiyanto, and S. Kamal, "Experimental study on performance of standing-wave thermoacoustic engine at different tilted angles and resonator length," presented at the ADVANCED MATERIALS: Proceedings of the International Workshop on Advanced Materials (IWAM-2017), Berhampur, Odisha, India, 2018, p. 070005.

# Investigation of resistivity for delineation aquifer layers and subsurface structures

**A Jufriadi*** **and H D Ayu**

Physics Education Study Program, Universitas Kanjuruhan Malang, Jl. S. Supriadi No 48, Malang 65148, Indonesia

akhmadjufriadi@unikama.ac.id

**Abstract**. Delineation of aquifer layers and subsurface structures using investigation of resistivity was accomplished. The results of geoelectric resistivity investigations was described of the detailed subsurface structures conditions and clear aquifer layers. The resistivity geoelectric and well logging methods were used in this study because both of these methods give correct and efficient results, due to the strong correlation between the resistivity values of rock layers and the subsurface structure conditions. Result of investigations show research areas have potential aquifers with wide-spread. The potential of aquifers consists of surface aquifers and deep aquifers, with surface aquifers at a depth of 41 to 50 meters with resistivity values of 30.5-145 ohm-meters, and deep aquifers at depths between 140 to 170 meters with a resistivity of 14.7-46.8 ohm-meter. While the subsurface structure was dominated by clay, sandy clay, sand.

## 1. Introduction

One of the problems of the coastal communities is getting fresh water from the underground aquifer layer, because some aquifers are saltwater aquifers. Thus, a detailed description of the subsurface structure conditions was needed to find the layer of fresh water aquifer.

Geophysical methods are often used to find the potential zone of underground water [1], and one of the best techniques for geophysical methods in underground water exploration and subsurface structural conditions is geoelectric resistivity [2]. The geoelectric method shows optimal results in determining rock resistivity [3], and is an important method in underground water investigations because the method is simple, inexpensive, does not damage the environment, is capable of detecting subsurface rock layers [4], and there is a strong relationship between resistivity and rock layer water content [5].

However, the description of subsurface structures is a complex problem, cause of irregular topography, surficial unconsolidated sediments in some areas [6]. So this research combines the resistivity method and the well logging data to decide the underground water aquifer layer and describe the subsurface structure well and in detail. In particular, the purpose of this study is to delineation the underground water aquifer layer and its thickness, and describe the subsurface structure based on its constituent sedimentary layer.

## 2. Location and geological setting
The research area is Sidoarjo District, East Java with a position of 112.5°-112.9° East Longitude and 7.3°-7.5° South Latitude (**Figure 1**). In general, the constituent rocks of Sidoarjo consist of passive tuffs, alluvium deposits, middle quarter volcanic rocks, and upper quarter volcanic rocks . Based on the geological map, the study area is composed of alluvial deposits consisting of sand, gravel, clusters, and clays (Qa), Quaternary volcanic deposits with wide distribution, and active volcanic deposits. The hydrogeological condition is a medium productive aquifer area with wide distribution[7].

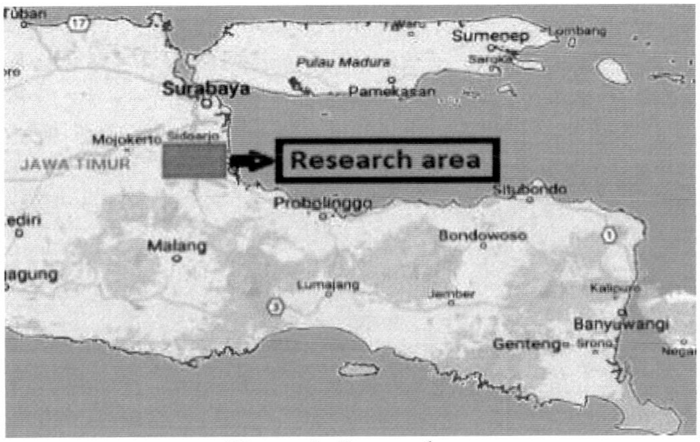


**Figure 1.** Research area

## 3. Materials and Methods
The application of the geoelectric method consists of three important stages, namely the stages of data acquisition, processing and interpretation. Data acquisition is done using a Wenner configuration (**Figure 4**) with the distance between electrodes ***a*** the same, namely C1-P1 = P1-P2 = P2-C2.

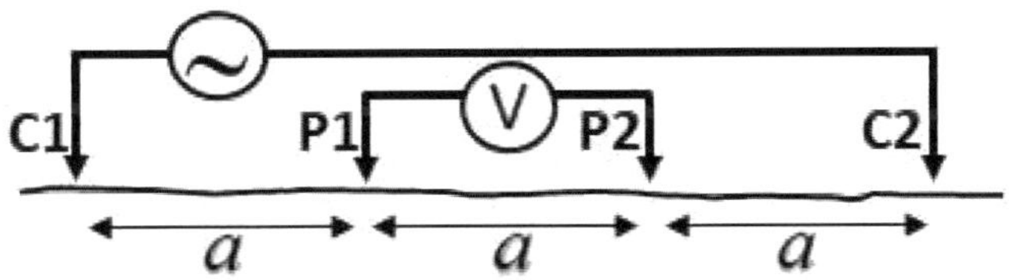


**Figure 2**. Wenner configuration

C1 and C2 are current electrodes 1 and 2, P1 and P2 are potential electrodes 1 and 2. While the field acquisition design used is a 2-dimensional design (**Figure 3**).

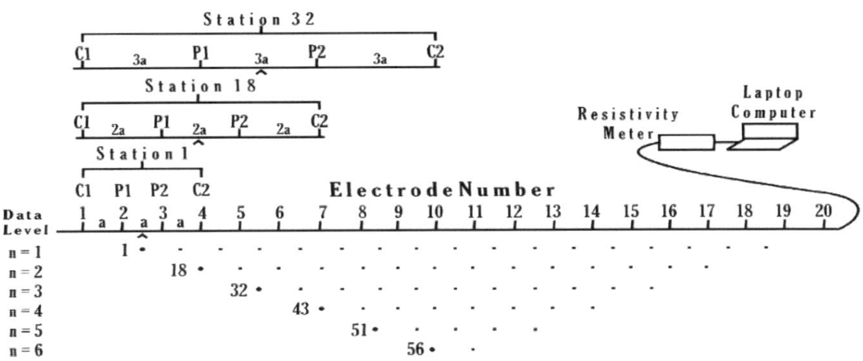


**Figure 3**. 2D survey design with Wenner configuration [8]

Data acquisition produces a potential difference value (ΔV) and the amount of current (I), so that it can be calculated by the apparent type of resistance, with the equation:

$$\rho = k\frac{\Delta V}{I}$$

$k$ is a geometry factor, which is expressed by the equation:

$$k = 2\pi a$$

2D survey design with a Wenner configuration starts with the position of electrodes C1, P1, P2, C2 respectively in positions 1, 2, 3 and 4 for the first measurement. The second measurement, the position of the electrode C1, P1, P2, C2 are shifted sequentially in positions 2, 3, 4 and 5. The measurements are continued until the end of the landscape with fixed spaces.

Processing and analysis of resistivity data using the RES2DINV program. This program uses inversion of quasi-Newton is based on an analytical calculation of Jacobian matrix for a homogeneous halfspace for the first iteration [9]

## 4. Results and Discussion

Delineation of aquifer layers and subsurface structures was carried out by 2D model interpretation, and well-logging data interpretation. The RES2DINV program is used to process resistivity data into a 2D model, thus producing clear and detailed descriptions of subsurface structures [10].

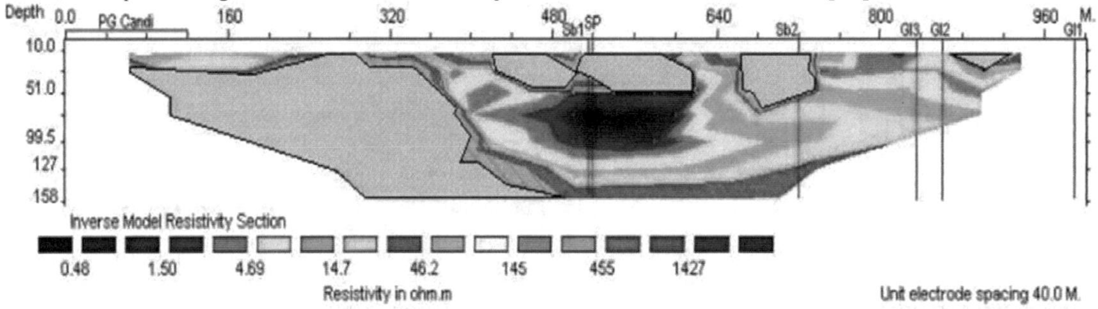


**Figure 4**. 2D model for aquifer layers

2D model for aquifer layers (**Figure 4**), illustrates the large distribution of resistivity, thickness, and depth of the aquifer layers indicated by green color with a resistivity around 14.7-46.8 ohm-meters. The resistivity range of the aquifer layer is almost the same as the resistivity range described in previous studies [2], [11], [1]. Aquifer layers are at measurement points 60 to 260, points 290 to 420, points 460 to 510, points 600 to 660, points 700 to 760. They are shallow aquifers with resistivity values of around 30.5-46.2 ohm-meters and depths of about 10 to 30 meters from ground level. Whereas deep aquifers have resistivity values of around 14.7 to 46.8 ohm-meters at the measurement point 370 to 880 with a depth of about 140-170. meters from the surface.

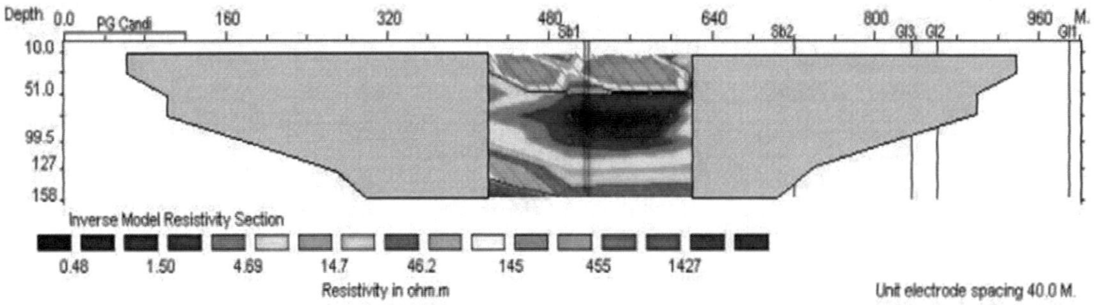


**Figure 5**. 2D model for subsurface structure at point 420 to 620

2D models for the conditions of subsurface structures (Figure 5), especially at point 420 to 620 show that the subsurface layer can be divided into 2 layers. The first layer at a depth of 0-40 meters, is a layer of sandstone and sand with resistivity values of 95.6-145 ohm-meters. The second layer with a depth of 40-170 meters, is a layer of clay and sand. But the second layer can be divided into several layers. In detail, the subsurface characteristics based on the 2D model for subsurface characteristic at point 420 to 620 is shown in **Table 1**.

**Table 1**. Subsurface structure at point point 420 to 620

| Depth (m) | Resistivity (ohm-meter) | Rock layer |
|---|---|---|
| 0-40 | 95,6-145 | Sand stone, sand |
| 40-50 | 30,5-46,2 | Clay, sand (shallow aquifer) |
| 50-110 | 0,48-3,1 | Clay, sand |
| 110-140 | 4,69-9,7 | Clay, sand |
| 140-170 | 14,7-46,8 | Clay, sand (deep aquifer) |

The 2D pseudo section model shows that aquifer potential is widespread and evenly distributed in the study area. The aquifer potential was indicated by the subsurface in the form of clay and sand. This result agrees with the hydrogeological map of the study area, which shows that the study area is a medium productive aquifer area with wide distribution. This also agrees with other studies, that the clay and sand material layers are rocks that makeup aquifers [12].

The results of well-logging (**Table 2**) show that the subsurface layer was divided into clay layers, sandy clay, sand (0-40 meters deep); inserts of clay, inserts of sand (41-50 meters depth); inserts of clay, sand (50-133 meters deep); and deep aquifers (133-168 meters deep).

**Table 2.** Subsurface characteristics based on well logging

| **Depth (m)** | **Rock layer** |
|---|---|
| 0-40 | Clay, sandy clay, sand |
| 41-50 | Clay inserts, sand inserts |
| 50-133 | Clay inserts, sand |
| 133-168 | Deep aquifer |

By comparation of the results investigation of resistivity and well-logging data, the subsurface structure consists of layers of sandy clay, sand, clay inserts, and sand inserts. While the potential of shallow aquifer layers is at a depth of about 41-50 meters and deep aquifers at a depth of 140-170 meters with wide-spread

## 5. Conclusion

The research area is an area that has a potential of shallow aquifers (about 41-50 meters deep) and deep aquifers (depth of about 140-170 meters), with a subsurface structure generally consisting of clay layers, sandy clay, sand, clay inserts, and inserts sand.

## References

[1] Kitila G, Lemessa G, Gebrekidan H and Alamirew T 2013 Identification of Groundwater Level by Using Geoelectrical Resistivity Method at Fincha'a Sugar Estate, Blue Nile Basin, Western Ethiopia *STAR J.* **2** 114–25

[2] Mohamaden M I I and Ehab D 2017 Application of electrical resistivity for groundwater exploration in Wadi Rahaba, Shalateen, Egypt *NRIAG J. Astron. Geophys.* **6** 201–9

[3] Wiyono, Lailatin N and Jufriadi A 2019 Interpretation Patterns for The Distribution of The Turbidity of Soil Water Around TPA Supit Urang ( Final Removal of Garbage ) Using Geoelectric Resistivity Methods Interpretation Patterns for The Distribution of The Turbidity of Soil Water Around TPA Su *9th Annu. Basic Sci. Int. Conf. 2019 (BaSIC 2019)*

[4] Jufriadi A and Ayu H D 2014 Investigation of Landslide Potential by Using Geolistrik Resistivitas Method in Ampelgading Malang Regency *FOTON, J. Fis. dan Pembelajarannya* **18** 13–8

[5] Eke K T and Igboekwe M U 2011 Geoelectric Investigation of Groundwater in Some Villages in Ohafia Locality, Abia State,Nigeria *Br. J. Appl. Sci. Technol.* **1** 190–203

[6] Benjumea B, Macau A, Gabàs A and Figueras S 2016 Characterization of a complex near-surface structure using well logging and passive seismic measurements *Solid Earth* **7** 685–701

[7] Budiono K, Handoko, Hernawan u and Godwin 2010 Penafsiran Struktur Geologi Bawah Permukaan di Kawasan Semburan Lumpur Sidoarjo , Berdasarkan Penampang Ground Penetrating Radar ( GPR ) *J. Geol. Indones.* **5** 187–95

[8] Loke M H 2013 *Tutorial : 2-D and 3-D electrical imaging surveys*

[9] Dahlin T and Loke M H 1998 Resolution of 2D Wenner resistivity imaging as assessed by numerical modelling *J. Appl. Geophys.* **38** 237–249

[10] Dahlin T and Bing Z 2004 a Numerical Comparison of 2D Resistivity Imaging With Ten *Geophys. Prospect.* **52** 379–398

[11] Basheer A A, Mansour K Q and Abdalla M A 2014 Geophysical investigation to reveal the groundwater condition at new Borg El-Arab industrial city , Egypt *NRIAG J. Astron. Geophys.* **3** 117–29

[12] Maria R, Rusydi A F, Lestiana H and Wibawa S 2018 Hidrogeologi dan potensi cadangan airtanah di dataran rendah indramayu *Ris. Geol. dan Pertamb.* **28** 181–92

# Requirement analysis of e-library application using Mandatory Desirable Inessential (MDI) and Technical Operational Economic (TOE) method

**AR Putera*, S Riyanto, M Arianto**
University of PGRI Madiun, Jl. Setiabudi No. 85 Madiun 63118, Indonesia

Email: andirahmanputera@unipma.ac.id

**Abstract:** The purpose of this study was to find out the requirement analysis to build e-library applications that will be managed by the Informatics Engineering Study Program at the University of PGRI Madiun. The focus of this study was requirement analysis using the Mandatory Desirable Inessential (MDI) method and Technical Operational Economic (TOE) methods. The analysis of online library information system requirements in this study was based on functional and non-functional needs analysis. This requirement analysis is based on functional requirements analysis that resulted 23 strategic information items to be done technically and operationally and resulted 4 non-functional items. Analysis of functional requirements based on feasibility resulted 12 technical items with middle or capable risks, and 9 operational items with middle or capable risks and 2 operational items at low risk or easy to do.

## 1. Introduction

Information technology that is growing rapidly must be used to support educational and learning activities in educational institutions. One of the needs of the education filed is the existence of a library or collection of books and research journals. The procurement of books and journals collection in fiction requires considerable costs and space, so a breakthrough is needed to provide a collection of books and journals effectively and efficiently that are used by users quickly and easily. To realize the existence of an effective and efficient library, it can utilize information technology in the form of e-library applications that can be accessed quickly and easily by its users. The use of e-library applications is a breakthrough to save the cost of procuring books and journals, where a collection of books and journals in e-libary can be in the form of e-book and e-journal[1].

The e-library application will make it easier for users to find and access the references needed. Users can download ebooks or e-journals in e-library applications as reference material for learning activities or scientific article writing. Users are also facilitated to find references quickly and easily by using keywords. In addition, this e-library application will also be connected to other e-library and e-journal applications, so that the available library availability is more complete and facilitate user[2].

The existence of e-library application also provides positive benefits in maintaining environmental balance, because the use of references is done by paper less. The use of paper to print reference books can be reduced because it is enough to use electronic media as a reference support. Reducing paper usage can be done significantly because e-library administrator and users do not have to print referenced books or journals.

E-library application is the future solution to solve problems related to the provision of references in the learning process from elementary school to college level[3]. The existence of a well-managed e-library will make it easier for users to access the ebook needed, so users can learn easily through their laptops and cellphones without having to visit a conventional library. The use of information technology for e-library activities is important to be carried out and well managed and also responds to the needs of users for learning references and scientific writing.

The results of the study related to information system requirements analysishave been carried out by many previous researchers. For instance research from Yasin, *et al.,*They stated that in the requirement analysis using PIECES analysis framework, analysis of Causality and Root problems with solutions using Fishbone analysis[4]. Research by Setiawan, *et al.,* This requirement analysis used an experimental method with reference to the System Development Life Cycle.This research started from preliminary studies, literature studies, problem formulation, and system requirements analysis[5]. Gumelar *et al.*Used Ripple Methodology for analysis and design that will produce definitions of requirements and design artifacts, then design testing process using Consistency Analysis to determine the consistency of the design and Litmus Test Service to find out the compliance of SOA criteria by the design[6].

The objective of this study was to find out the needs analysis to build e-library applications using the Mandatory Desirable Inessential (MDI) and Technical Operational Economic (TOE) methods that will be managed by the Informatics Engineering Study Program at the University of PGRI Madiun.

## 2. Literature

### Concept of Requirement

Need (*requirement*) is the characteristics of the system or product that will be developed in accordance with the wishes of customers. Requirement is classified into functional requirement and nonfunctional requirement.

a. Functional requirement is needs that explain the interactions between system and environmentthat are separate from implementation. The system is a set of elements that are interrelated and influence each other in carrying out activities to achieve a goal.
b. Non functional requirement is user aspects that can be seen on a system that is not directly related to functional behavior. Response time must be less than one second and accuracy

must be within a second. This requirement is forced by the client or the environment in which the system will operate[7].

**Elicitation**

Elicitation is a design made based on a new system desired by management and is carried out by researchers to be executed. Elicitation is obtained by interview method and carried out in three stages[7] as follows:

a. Elicitation Stage 1
At this stage, the elicitation contains all new system designs proposed by the management proposed through the interview process.

b. Elicitation Stage 2
This stage is the result of class 1 classification based on the Mandatory Desirable Inessential (MDI) method. The MDI method aims to separate the important system design and must be presented in a new system of designs that are capable of being executed by researchers. The MDI method includes three things, namely:
1) Mandatory the requirement must be exist and cannot be removed when creating new system.
2) Desirable, the requirement is not too important and can be removed but if the requirement is used in the form system, the system will be more perfect.
3) Inessential ,the requirement is not part of discussed system (outside of the system)

c. Elicitation Stage 3
This stage of this elicitation is the result of shrinkage from elicitation stage 2 that is carried out by eliminating all needs with the innessential option on the MDI method. Then all the remaining need is re-classified through the Technical Operational Economic (TOE) method, as follows:
1) Technical, how to create requirement proposed in the system ?
2) Operational, how to use need in the system to be developed ?
3) Economic (ekonomis), how much the cost needed to build the need in new system.
TOE method is divided into some options:
1) High (H), difficult to do because the manufacturing techniques and usage are difficult, the cost is expensive and these needs must be eliminated.
2) Middle (M), able to do
3) Low (L), easy to do

d. Final draft of elicitation, is the final result achieved from the eelicitation process and can be used as the basis for making a system to be developed[7].

## 3. Methodology

The discussion in this study was only emphasized on requirement analysis, where the analysis of this need will be the material or step of further research in building and designing system, implementing system, integrating and testing the systems created and applying and maintaining the system that has been running[8].

Requirement analysis in this study was carried out by some stagesand the system requirements analysis stage is the stage of determining all components needed by the system to be built. The researcher determines the system requirement that is expected to meet user need and determine the limits of the software. Information needed in conducting a requirement analysis was carried out byobservation and interview to competent stakeholders[9].This study was conducted at Informatics Engineering Study Program at the University of PGRI Madiun, Campus 3 Floor 3 Jl. Auri No.14 Madiun.

## 4. Results

The following is the elicitation of the Requirement of the online library information system that will be managed by the Informatics Engineering Study Program at the University of PGRI Madiun, the elicitation is based on interviews with stakeholders:

1. Elicitation Stage 1 and Stage 2

**Table 1.** Elicitation Stage 1 and Stage 2

| | **Functional**<br>**Strategic Information**<br>This system is expected to : | M | D | I |
|---|---|---|---|---|
| 1 | Integrated well | √ | | |
| 2 | Show master home menu | √ | | |
| 3 | Show master profile menu | √ | | |
| 4 | Show master catalog menu | √ | | |
| 5 | Show virtual reading room (RBV) menu | √ | | |
| 6 | Show master menu of IT discussion room | √ | | |
| 7 | Show e-resource master menu | √ | | |
| 8 | Show IT News | | √ | |
| 9 | Show announcement | | √ | |
| 10 | Show Television Informatics | | √ | |
| 11 | Show login facilities for admin | √ | | |
| 12 | Show admin menu | √ | | |
| 13 | Show edit home menu for admin | √ | | |
| 14 | Show edit profile menu for admin | √ | | |
| 15 | Show catalog edit menu for admin | √ | | |
| 16 | Show RBV edit menu for admin | √ | | |
| 17 | Show IT discussion room edit menu for admin | √ | | |
| 18 | Show e-resource edit menu for admin | √ | | |
| 19 | Show IT news edu for admin | | √ | |
| 20 | Show announcement edit for admin | | √ | |
| 21 | Show TV informatics TV for admin | | √ | |
| 22 | Show visitor data | | √ | |
| 23 | Show complete address for online library | | √ | |

2. Elicitation Stage 3

**Table 2.** Elicitation Stage 3

| | **Feasibility** | T | | | O | | | E | | |
|---|---|---|---|---|---|---|---|---|---|---|
| | **Risk** | L | M | H | L | M | H | L | M | H |
| | **Requirement Analysis**<br>The system is expected to : | | | | | | | | | |
| 1 | Integrated well | | √ | | | | | | | |
| 2 | Show master home menu | | √ | | | | | | | |
| 3 | Show master profile menu | | √ | | | | | | | |
| 4 | Show master catalog menu | | √ | | | | | | | |
| 5 | Show virtual reading room (RBV) menu | | √ | | | | | | | |
| 6 | Show master menu of IT discussion room | | √ | | | | | | | |
| 7 | Show e-resource master menu | | √ | | | | | | | |
| 8 | Show IT News | | √ | | | | | | | |
| 9 | Show announcement | | √ | | | | | | | |
| 10 | Show Television Informatics | | √ | | | | | | | |
| 11 | Show login facilities for admin | | √ | | | | | | | |
| 12 | Show admin menu | | √ | | | | | | | |
| 13 | Show edit home menu for admin | | | | | √ | | | | |
| 14 | Show edit profile menu for admin | | | | | √ | | | | |
| 15 | Show catalog edit menu for admin | | | | | √ | | | | |
| 16 | Show RBV edit menu for admin | | | | | √ | | | | |
| 17 | Show IT discussion room edit menu for admin | | | | | √ | | | | |
| 18 | Show e-resource edit menu for admin | | | | | √ | | | | |
| 19 | Show IT news edu for admin | | | | | √ | | | | |
| 20 | Show announcement edit for admin | | | | | √ | | | | |
| 21 | Show TV informatics TV for admin | | | | | √ | | | | |
| 22 | Show visitor data | | | | √ | | | | | |
| 23 | Show complete address for online library | | | | √ | | | | | |

3. Final Elicitation Requirement Draft

**Table 3.** Final Elicitation Requirement Draft

| **Functional** | |
|---|---|
| | **Requirement Analysis** |
| | **This system is expected to :** |
| 1 | Integrated well |
| 2 | Show master home menu |
| 3 | Show master profile menu |
| 4 | Show master catalog menu |
| 5 | Show virtual reading room (RBV) menu |
| 6 | Show master menu of IT discussion room |
| 7 | Show e-resource master menu |
| 8 | Show IT News |
| 9 | Show announcement |

| | |
|---|---|
| **Functional** | |
| | **Requirement Analysis** |
| | **This system is expected to :** |
| 10 | Show Television Informatics |
| 11 | Show login facilities for admin |
| 12 | Show admin menu |
| 13 | Show edit home menu for admin |
| 14 | Show edit profile menu for admin |
| 15 | Show catalog edit menu for admin |
| 16 | Show RBV edit menu for admin |
| 17 | Show IT discussion room edit menu for admin |
| 18 | Show e-resource edit menu for admin |
| 19 | Show IT news edu for admin |
| 20 | Show announcement edit for admin |
| 21 | Show TV informatics TV for admin |
| 22 | Show visitor data |
| 23 | Show complete address for online library |
| **Non functional** | |
| | **This system is expected :** |
| 1 | To Have interested display and user friendly (easy to understand by user). |
| 2 | To Have accuracy and speed in displaying data wanted by user |
| 3 | Available of site visitor buttons as feedback |
| 4 | Registered application in creative commons license. |

## 5. Conclusion

The requirement analysis of the online library information system that will be named informaticslibrary.com. This is an online library that will be managed by the Informatics Engineering Study Program. This requirement analysis is based on functional requirements analysis that resulted 23 strategic information items to be done technically and operationally and resulted 4 non-functional items. Analysis of functional requirement based on feasibility resulted 12 technical items with middle or capable risks, and 9 operational items with middle or capable risks and 2 operational items with low risk or easy to work.

## 6. References

[1] Nasrul N, Afrizal Y 2007 Sistem e- Library Berbasis Web di Politeknik Negeri Padang. *Jurnal TeknikA*. Vol. 2 No 27, pp 18-24.

[2] Putra, MS 2011 Perancangan dan Implementasi Digital Library Pada Warung Baca Dodi. *Jurnal Teknik Informatika,* Vol 1, No 1, pp. 1-16

[3] Listyorini T, Kudus UM, Iqbal MA 2015 Perancangan Pengembangan Digital Library Berbasis Web. *Jurnal SIMETRIS*, Vol 6 No 1. pp. 69-76.

[4] Alimudin Yasin, Yumarlin MZ, Fitriyadi T 2015 Analisis Kebutuhan Sistem Informasi di LPK RJ-Comp Yogyakarta. *Seminar Nasional Informatika 2015*. pp. 111-116.

[5] Setiawan A, Wibawa FM, Burhanudin A 2016 Analisis Kebutuhan Sistem Informasi Perwalian Mahasiswa Berbasis *Website* Studi Kasus Teknik Informatika UM Magelang. *KNSI 2016*. pp. 684-687.

[6] Gumelar AS, Saputra MC, Wardani NH 2017 Analisa Kebutuhan dan Perancangan Sistem Informasi Produksi Dinas Peternakan dan Kesehatan Hewan Kabupaten Malang Berbasis Teknologi *Service Oriented Architecture. Jurnal Pengembangan Teknologi Informasi dan Ilmu Komputer*. Vol. 1, No. 11, November 2017. pp. 1215-1225.

[7] Sudaryono 2015 *Metodologi Riset di Bidang TI*. (Yogjakarta: Andi Yogyakarta).

[8] Pressman RS 2010 *Software Engineering : a practitioner's approach*. (New York: McGraw-Hill)

[9] Riyanto S, Nugrahanti F. Pengembangan Pembelajaran Statistika Berbasis Praktikum Aplikasi Software SPSS dengan Bantuan Multimedia untuk Mempermudah Pemahaman Mahasiswa terhadap Ilmu Statistika. *DoubleClick: Journal of Computer and Information Technology*. Vol.1, No. 2, Febuary 2018, pp 62-67.

Journal of Physics: Conference Series **1381** (2019) 012069 doi:10.1088/1742-6596/1381/1/012069

# The development of the practicum based environmental pollution module for guided inquiry which collaborates video to improve student learning outcomes

**P A Nugroho[1], Y D Puspitasari[1]**
[1] Science Education, STKIP PGRI Nganjuk, Indonesia.
Email: purwoadinugroho@stkipnganjuk.ac.id

**Abstract.** The research aims to find out the improvement of student learning outcomes by applying the development of guided inquiry-based environmental pollution practicum modules that collaborate on videos. The syntax of guided inquiry, namely: identification problems, formulating hypotheses, collecting data, interpreting data, and conclusions. All are packaged in an environment-collaborating video pollution practicum module, where the learning outcomes used are from the implementation of the guided inquiry syntax listed in the module and student cognitive problems. The type of research is modified R&D (Research & Development) Borg and Gall (2003). The research sample is in the form of initial product trial sample, limited field trial sample, and operational field trial sample. The instruments used were a questionnaire, observation, and test. Based on the results of research when conducting water, soil, and air pollution practicum's that have been carried out using the application of video collaborating modules, the results are obtained based on the syntax implementation (70.8%, 77.2%, 85.4%) and based on the results cognitive learning (79.87%, 83%, 86.40%). The results of the assessment of student learning outcomes at each meeting are increasing, so that the application of guided inquiry-based environmental pollution modules that collaborate videos can improve student learning outcomes.

## 1. INTRODUCTION

Human life is very dependent on the state of the surrounding environment, where we need each other in various needs. Our activities and behaviors sometimes cause various kinds of environmental problems, such as pollution and environmental damage, whether intentionally or unintentionally. In everyday life, humans are faced with a variety of behaviors that can benefit the surrounding environment and that can damage the stability of the environment. To support environmentally friendly behavior, it is necessary to be given an understanding of the importance of preserving the environment [1].

Environmental pollution can be categorized into 3, namely: 1) Soil pollution, 2) Water pollution, and 3) Air pollution. Air pollution in Indonesia is caused by the increasing volume of motorized vehicles which reaches 72, 41% [1]. Water pollution occurs due to disposal of residual detergents, textile or batik factories, whose waste is directly discharged into the water flow. While the pollution of the land due to littering by 64.85% [1].

The rapid development of science and technology in the current era of globalization has a profound impact on all aspects of life. In addition to positive impacts, many also have negative impacts. The impact is very influential in the world of education today, such as the learning process, teaching materials used, and learning strategies.

The learning module is a very complete learning unit that can stand alone and inside there is a whole series of learning activities that are organized to help student learning activities to achieve a number of objectives in learning [2]. Learning modules are learning tools or tools that contain material, learning methods or models used, evaluation methods that are designed systematically and attractively to achieve optimal learning outcomes. In the module, it has been adjusted to 1) the appropriateness of the content, which is in accordance with the needs of students, in accordance with teaching material, and the addition of insight about pollution, 2) the feasibility of the language used clearly and in accordance with Indonesian language rules, 3) the feasibility of the presentation which includes indicators that will include achieved, and 4) the appropriateness of the graph where the letter is clear, there are photos, diagrams, views and a good layout [3].

In learning biology, activities cannot be separated from the role of lecturers as motivators and students as active learning objects in all their activities, so that they need appropriate and effective learning models that use guided inquiry. Guided inquiry in its application there is still the role of the lecturer therein, namely as a motivator while students actively engage in learning activities or processes. Guided inquiry-based learning where students are trained to improve thinking skills according to their syntax which aims to make learning activities more meaningful and structured [4].

Modules will be more effective if accompanied by learning videos about environmental pollution material and become more encouraging students to be active in learning activities by applying guided inquiry models. The use of learning videos which will attract the attention of students in the absorption of learning material that will make students more enthusiastic in learning [5]. From this background, a study was conducted under the title "Development of Guided Inquiry-Based Environmental Pollution Practicum Module that Collaborates on Videos to Improve Student Learning Outcomes".

## 2. METHOD

Research methods in the development of learning products in the form of modules that use the method and design of research development using a modified R&D (Research and Development) design from [6], with steps, such as: 1) research and information gathering, 2) planning, 3) product development initial, 4) initial product trials, 5) product revision I, 6) limited field trials, 7) operational field trials. The procedure for developing the Guided Inquiry-based environmental pollution practicum module using the modified [6] method is presented in **Figure 1.**

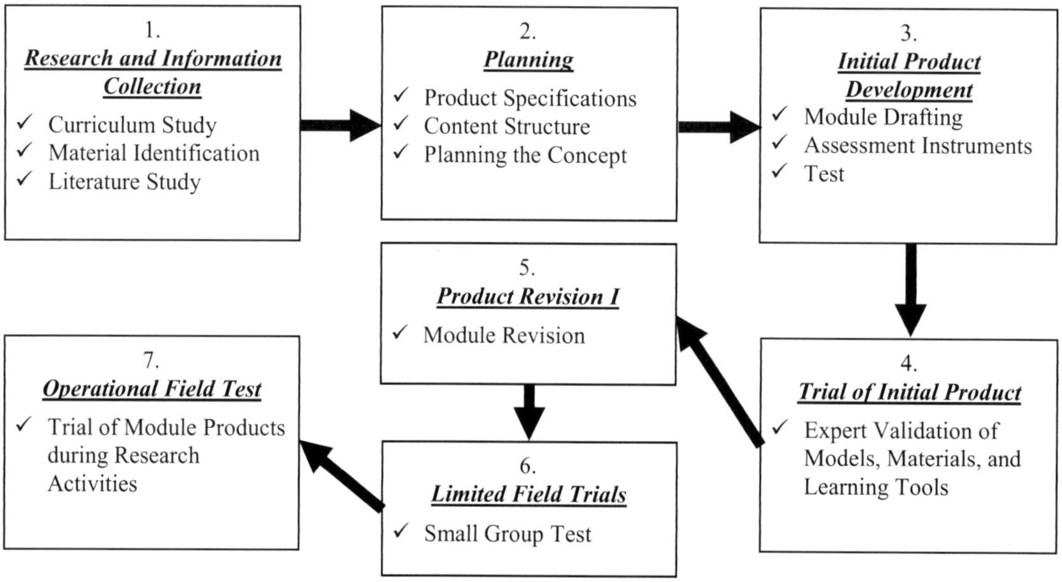


**Figure 1.** Product Development Procedure

Based on the explanation above, the data collection method can be explained as follows:

*2.1 Research and Information Collection*

a. Reviewing the curriculum to be used is an adjustment to the curriculum used at the STKIP PGRI Nganjuk.
b. Identifying the material to be used, of course, which can be collaborated with the guided inquiry model and the use of appropriate videos, and appropriately included in the planned module.
c. Literature study, where the studies used come from various sources, both books, journals, and all studies that can be accessed internet.

*2.2 Planning*

The planning is to prepare a draft module that is integrated with the syntax of guided inquiry, namely: a) determine the learning objectives based on indicators, b) determine the sub topics, c) determine the format and visualization of the module to be developed [7], d) determine the procedure research ranging from validation, testing, problem development.

*2.3 Initial Product Floating*

At this stage, a draft module is produced that is ready to be submitted for evaluation or validation. Module development is determined by the development of bases in accordance with the syntax of guided inquiry on video pollution environmental materials collaborating used for practicums.

*2.4 Trial of Initial Products*

The intended test is an evaluation or validation by the module design expert, the material, and the learning tools used. Where there is a scale 4 reference assessment [8] in the form of score conversion in **Table 1.**

**Table 1.** Converting Actual Scores to Scores of Four

| Value Range | Category |
|---|---|
| 3,51 - 4,00 | Very good |
| 2,51 - 3,50 | Well |
| 1,51 - 2,50 | Pretty good |
| 1,00 - 1,50 | Not good |

*2.5 Product Revision I*

Product revision I was based on validation conducted by the validators which could be used as input to improve the module content.

*2.6 Limited Field Trials*

The trial was conducted on STKIP PGRI Nganjuk upper semester science education students in order to get input or there are still errors in typing or sentences that are difficult to understand.

*2.7 Operational Field Test*

At this stage the effectiveness test is carried out on the experimental class which will be used as research. In the operational field test, it will be assessed, namely the implementation of the guided inquiry syntax already contained in the module. The values obtained were then analyzed using descriptive analysis with the value criteria according to [9], can be seen in **Table 2.**

$$\text{Value of Syntability of Syntax} = \frac{\sum \text{score acquisition}}{\sum \text{maximum score}} x100\%$$

**Table 2.** Implementation of the Syntax of Guided Inquiry Learning

| Criteria | Conversion Value | |
|---|---|---|
| | Number | Calculation |
| Very good | 4 | 81 - 100 |
| Well | 3 | 61 - 80 |
| Enough | 2 | 41 - 60 |
| Less | 1 | 20 - 40 |

## 3. RESULTS AND DISCUSSION

The development of guided inquiry-based practicum modules in collaborating videos is applied to students of the STKIP PGRI Nganjuk in science education program the fourth semester. The study was conducted three times, namely in the matter of soil pollution, water pollution, and air pollution, where the whole material is summarized into modules.

From the results of the study obtained several data that can be used as an assessment, including:

*3.1 Research and Information Collection*

The results of the literature study found that the use of the curriculum is in accordance with the curriculum used by STKIP PGRI Nganjuk with the use of RPS that is adapted to practicum material, namely environmental pollution. In the literature study it was also found that there was no complete practical module with the incorporation of learning models such as guided inquiry with the addition of videos.
Integrating guided inquiry models in biology practicum modules because the model used can make students build their own knowledge, so students will feel interested and motivated so that learning becomes easier [10].

### *3.2 Planning*
At this stage the learning objectives developed are to measure student learning outcomes. Besides that the material developed is in the form of practicum modules that are systematically packaged so that students can use them independently or in groups where students are directed to conduct experiments in accordance with the syntax of guided inquiry.

The components in the module such as: a) competencies to be achieved contained in indicators, b) discourse and pictures, c) observation and identification activities through discourse and pictures, d) defining and classifying, e) identifying, formulating problems and hypotheses, f) design experiments and conduct experiments, g) measure and collect data to be entered into the observation table, h) analyze, conclude and communicate the results of experiments.

### *3.3 Initial Product Floating*
Product development that will produce a draft module on environmental pollution starting from the cover, contents, and cover, namely:
a) The module cover, includes the title, the STKIP PGRI Nganjuk logo, and the pollution pictures
b) Preface, list of contents, introduction
c) Chapter I, Chapter II, and Chapter III.
d) Closing
e) Back cover of the module

### *3.4 Trial of Initial Products*
Initial product trials are stages of validation by experts, starting from design experts, materials, and learning tools. The results can be seen in this table below. The results of the validation by the module design expert can be seen in **Table 3.**

**Table 3.** Validation Results for Module Design Experts

| No. | Aspect | Average | Category |
|---|---|---|---|
| 1. | Quality of content / material | 3,3 | Good |
| 2. | The relevance and credibility of the source book | 3 | Good |
| 3. | The suitability of the guided inquiry base | 3,4 | Good |
| 4. | The quality of the presentation method | 3,4 | Good |
| 5. | Presentation of learning | 3 | Good |
| 6. | Use of language | 3,4 | Good |
| 7. | Use of illustrations | 3 | Good |
| 8. | Quality and completeness of supporting materials | 3 | Good |

| | | | |
|---|---|---|---|
| 9. | Finesse | 3,5 | Good |
| 10. | General appearance | 3,5 | Good |
| | **Average of all aspects** | **3,25** | **Good** |

In Table 3. Obtained the results of the assessment or validation from the module design experts obtained an average value of all aspects of 3.25 which is categorized as good but still needs to be revised or improved.

The results of the validation by material experts can be seen in **Table 4.**

**Table 4.** Results of Expert Material Validation

| No. | Aspect | Average | Category |
|---|---|---|---|
| 1. | The basic concept of the material | 3,67 | Very Good |
| 2. | The concept of sub material discussion | 3 | Good |
| 3. | Concept pictures | 4 | Very Good |
| 4. | Presentation of images | 3,67 | Very Good |
| 5. | Systematic delivery of material | 3 | Good |
| 6. | Relevance to everyday life | 3 | Good |
| 7. | Relevance and credibility of the source book | 3,33 | Good |
| 8. | Use of language in the material | 3,33 | Good |
| | **Average of all aspects** | **3,38** | **Good** |

In Table 4. Obtained the results of the assessment or validation from the material experts obtained an average value of all aspects of 3.38 which are categorized well but still needs to be improved or the addition of some material from other sources.

The results of the validation by the module design expert can be seen in **Table 5.**

**Table 5.** Results of Expert Validation of Learning Devices

| No. | Aspect | Average | Category |
|---|---|---|---|
| 1. | Formulation of learning objectives | 3 | Good |
| 2. | Selection and organization of teaching materials | 3,33 | Good |
| 3. | The choice of learning resources or learning media | 3 | Good |
| 4. | Learning method | 3,33 | Good |
| 5. | Assessment of learning outcomes | 3,67 | Very Good |
| 6. | Evaluation questions | 3,67 | Very Good |
| | **Average of all aspects** | **3,33** | **Good** |

In Table 5. Obtained the results of the assessment or validation from the experts of the device obtained a value of all aspects of 3.33 which are categorized as good but still need improvement.

### *3.5 Product Revision I*

Based on the validation that has been done, there needs to be an improvement, that is, from the design that is less bright in the color of the cover and the addition of images that have more visible pollution. In terms of material, some references were added from other books, and learning tools were added by several essay questions to increase student thinking.

Journal of Physics: Conference Series **1381** (2019) 012069 doi:10.1088/1742-6596/1381/1/012069

### *3.6 Limited Field Trials*
Based on the trial results there is no need for improvement, because the module has covered all the material and there is a guided inquiry stage in it.

### *3.7 Operational Field Test*
In the operational field test, student learning outcomes are assessed in two ways, namely from the implementation of the guided inquiry syntax in the module and the cognitive learning outcomes of students through tests, such as:

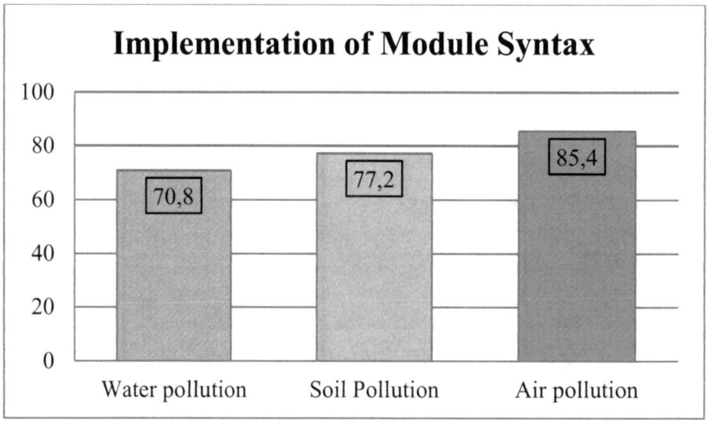


**Figure 2.** Results of Guided Inquiry Syntax Based on Modules

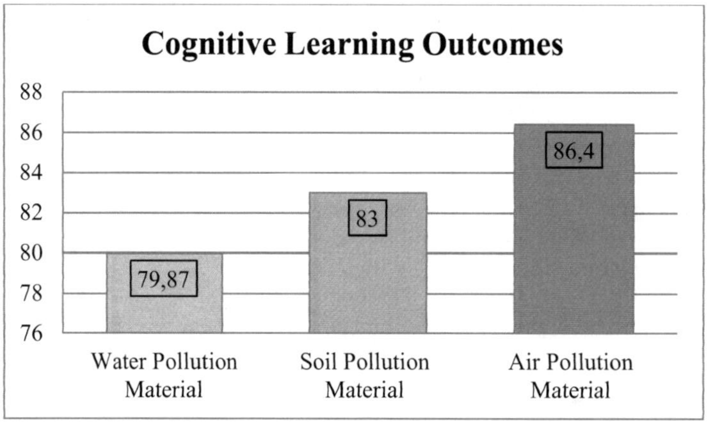


**Figure 3.** Cognitive Learning Outcomes

Journal of Physics: Conference Series **1381** (2019) 012069 doi:10.1088/1742-6596/1381/1/012069

## 4. CONCLUSIONS

With the application of the environmental pollution practicum module that collaborates on videos can improve student learning outcomes. This can be seen from the increase in each stage of the study, where the assessment is seen from the implementation of the syntax and test questions. Where the results obtained are based on the syntax implementation (70.8%, 77.2%, and 85.4%) and based on cognitive learning outcomes (79.87%, 83%, and 86.40%).

Suggestions for researchers who want to use almost the same method are more innovative in making modules so that they are more attractive to students because each teacher has different styles and creativity, of course to require extra time and energy.

## 5. REFERENCES


[1] Badan Pusat Statistik 2014 *Indikator Perilaku Peduli Lingkungan Hidup 2014* Jakarta BPS

[2] Nasution S 2003 *Berbagai Pendekatan dalam Proses Belajar dan Mengajar* Jakarta Bumu Aksara

[3] Susilo A, Siswandari, dan Bandi 2016 Pengembangan Modul Berbasis Pembelajaran Saintifik Untuk Peningkatan Kemampuan Mencipta Siswa Dalam Proses Pembelajaran Akuntansi Siswa Kelas XII SMA N I Slogohimo 2014 *Jurnal Pendidikan Ilmu Sosial Vol 26 No 1 Juni 2016 ISSN 1412-3835*

[4] Putri P N, Subandi, dan Munzil 2018 Pengaruh Strategi Inkuiri Terbimbing dan Kolb's Learning Style terhadap Kemampuan Berpikir Tingkat Tinggi *Jurnal Pendidikan Vol 3 No 12 Bulan Desember 2018 (1664-1671) EISSN 2502-471X*

[5] Wati L A 2013 Pemanfaatan Media Video Pembelajaran Untuk Meningkatkan Hasil Belajar IPS Pada Siswa Kelas IV SDN Babatan I/456 Surabaya *Jurnal Penelitian Pendidikan Guru Sekolah Dasar Vol 1 No 01 2013*

[6] Gall M D, Borg W R, and Gall J P 2003 *Educational research: An introduction (7th ed.)* White Plains New York Longman

[7] Mulyasa E 2006 *Implementasi Kurikulum 2004 Panduan Pembelajaran KBK* Bandung PT Remaja Rosdakarya

[8] Millah E S, Budipramana L S, dan Isnawati 2012 Pengembangan Buku Ajar Materi Bioteknologi di Kelas XII SMA IPIEMS Surabaya Berorientasi Sains, Teknologi, Lingkungan, dan Masyarakat (SETS) *Jurnal Bio Edu 1(1) 19-24*

[9] Suwandi S 2009 *Model Assesment dalam Pembelajaran* Surakarta Panitia Sertifikasi Guru Rayon 13 FKIP UNS

[10] Wahyudin S, dan A Isa 2010 Keefektifan Pembelajaran Berbantuan Multimedia Menggunakan Metode Inkuiri Terbimbing Untuk Meningkatkan Minat Dan Pemahaman Siswa *Jurnal Pendidikan Fisika FMIPA UNNES Semarang 58-62*

Journal of Physics: Conference Series **1381** (2019) 012070 doi:10.1088/1742-6596/1381/1/012070

# The selection of Calcium Milk Products that are appropriate for advanced age using PROMETHEE II Algorithm

**Khairunnissa Fanny Irnanda**[1] , **Fatimah Nur Arifah**[2], **Mokhamad Ramdhani Raharjo**[3], **Ardian Arifin**[4], **Agus Perdana Windarto**[5]
[1,5]STIKOM Tunas Bangsa, Pematangsiantar, North Sumatera - Indonesia
[2]STMIK Bina Patria, Indonesia
[3]Universitas Islam Kalimantan Muhammad Al Banjari Banjarmasin, Indonesia
[4]IKIP PGRI Pontianak, Indonesia

e-mail: agus.perdana@amiktunasbangsa.ac.id

**Abstract.** The selection of Calcium Dairy Products that are suitable for old age uses the PROMETHEE II Algorithm. This study aims to recommend calcium milk products that are appropriate for elderly people based on consumer selection. This research was conducted in Simalungun district using interview techniques, observation and questionnaires to 350 elderly respondents randomly. This study also uses a quantitative data approach, which is testing four calcium milk products (alternatives) that are considered appropriate for use in old age. The four dairy products are widely used for the elderly, namely: Anlene Gold (A1), Entrasol (A2), Prolansia (A3) and Appeton 60+ (A4). The assessment criteria used in the selection of calcium milk recommendations appropriate for old age are: price (C1), content (C2), side effects (C3) and taste variants (A4). The solution used is to use a decision support system with PROMETHEE II algorithm. The results of PROMETHEE II calculation show that Anlene (A1) is obtained as the first recommendation of the right calcium milk with a value (net flow 0.5) and Appeton 60+ (A4) as the second recommendation with a value (net flow 0.484).

## 1. Introduction

Milk is an almost perfect natural food. Most nutrients are in milk such as protein, calcium, phosphorus, vitamin A, and vitamin B1. Milk is the best source of calcium, because high calcium levels and lactose in milk help the digestive tract [1]. Fresh milk can be processed into various products that are popular and have a relatively long shelf life. Examples of processed products in the milk processing industry such as homogenized milk, skim milk and cream, butter, sweetened condensed milk, powdered milk, yogurt, sterilized milk, cheese, ice cream, caramel or confectionery, dodol milk, tofu milk and milk crackers [2]. Elderly is an aging process with increasing age of an individual which is characterized by a decrease in the function of organs such as the brain, heart, liver and kidneys and an increase in the loss of the body's active tissue in the form of body muscles. Decreasing the function of organs is a result of the reduction in the number and ability of the body's cells, so that the ability of the body's tissues to function normally disappears, so it cannot survive infection and repair the damage suffered. Astawan and Wahyuni (1988) stated that common changes experienced by the elderly such as (1) decreased fluid in the body, (2) increased fat levels in the body, (3) increased lime levels in the brain and blood vessels which resulted in decreased tissue bone, decreased hormone activity, decreased enzyme activity, and other physical changes [3]. The

1 Jenderal Sudirman Blok A No.1/2/3 Pematangsiantar, North Sumatera - Indonesia
2 Raden Saleh No.2, Potrobangsan, Kec. Magelang Utara, Kota Magelang, Jawa Tengah 56116- Indonesia
3 Adhyaksa, Jl. Kayu Tangi 1 Jalur 2 No.2, Sungai Miai, Kec. Banjarmasin Utara, Kota Banjarmasin, Kalimantan Selatan 70123 - Indonesia
4 Ampera No.88, Sungai Jawi, Kec. Pontianak Kota, Kota Pontianak, Kalimantan Barat 78116 - Indonesia
5 Jenderal Sudirman Blok A No.1/2/3 Pematangsiantar, North Sumatera - Indonesia

population of elderly aged = 60 years is 10% and is expected to increase in 2050 in a world where the elderly aged = 85 years increase by 0.25% (Holdsworth, 2014). According to the Badan Pusat Statistik (2015) report, the number of elderly people in the world, including Indonesia, is increasing every year. In 2012 the percentage of the population aged 60 years and above was 7.58%, while in 2013 it increased to 8%, in 2014 it increased to 8.2% and in 2015 it increased to 8.5%. Calcium is very important in bone growth, and teeth. Childhood, calcium and food absorption can reach 75%, then shrink to 20-40% as they reach adulthood. Peak bone mass formation occurs at the age of 25 years. Public awareness needs to be raised about the dangers of osteoporosis and people need to be invited to stay active in consuming foods with balanced nutrition, especially those with high calcium from an early age, so that they can ultimately improve the quality of life [4]. Efforts to maintain nutritional status or health status of the elderly are the impact of increasing Life Expectancy. The health condition of the elderly is determined by the intake of food and beverages both in terms of quantity and nutritional value.

The number of calcium milk products currently makes many choices for the elderly in consuming calcium milk products because the elderly are strongly advised to consume milk in order to meet the nutritional needs of the body. Not only for fulfilling nutrition, milk is also very much needed by the elderly to maintain their health. Based on these problems, the purpose of this study is to recommend the best calcium milk products for the elderly based on consumer choice. There are many branches of computer science (artificial intelligence) that can solve complex problems that are related, including datamining [4]–[6], decision support systems [7], artificial neural networks [8], [9][10], [11] and others [12]. Based on these problems, researchers used the PROMETHEE II algorithm which is part of a decision support system. Related research is carried out by [13] where the results of this study are able to recommend one social media to the formation of brand attachments using the PROMETHEE II method. It is expected that the results of the study can provide recommendations on the selection of calcium milk products that are appropriate for old age..

## 2. Methodology

The research data was obtained by conducting observations and interviews with consumers (elderly) by taking 150 samples randomly in Simalungun district. The following assessment criteria obtained from interviews and observations in determining the best calcium milk products based on consumer choice as shown in table 1 and alternative calcium milk products used as shown in table 2 below:

**Table 1**. Kriteria Pemilihan Produk Susu Lansia

| Criteria | Information | Initials |
|---|---|---|
| Criteria 1 | Price | C1 |
| Criteria 2 | Content | C2 |
| Criteria 3 | Side effects | C3 |
| Criteria 4 | Taste Variant | C4 |

**Table 2**. Alternatif Produk susu kalsium

| Product | Information |
|---|---|
| Anlene | A1 |
| Entrasol | A2 |
| Prolansia | A3 |
| Appeton 60+ | A4 |

The following is the suitability rating data from each alternative that has been processed based on the results of the questionnaire on 150 consumers, where the mentat data processing is assisted by using Microsoft Excel software as shown in table 3 below:

**Table 3.** Compatibility Rating of Each Alternative at Each Criteria

| Criteria | Min/Maks | Dairy products | | | |
|---|---|---|---|---|---|
| | | **A1** | **A2** | **A3** | **A4** |
| C1 | Maks | 0.72 | 0.7 | 0.7 | 0.6 |
| C2 | Maks | 0.7 | 0.6 | 0.63 | 0.7 |
| C3 | Maks | 0.72 | 0.73 | 0.7 | 0.6 |
| C4 | Maks | 0.78 | 0.8 | 0.63 | 0.2 |

## 3. Results and Discussion

Based on the explanation in tables 1, 2, 3, the following is how to complete the PROMETHEE II method in the case of selecting calcium milk products that are good for the elderly.
**STEP I.** Calculate Preference Value:
The existing criteria will then be calculated for the value and index of preference as follows:

**1. Price Criteria Value (C1)**

C1(A1,A2) : d = C1(A1) – C1(A2)
d = 0.72 – 0.7
d = 0.02
d > 0 then H|d| = 1

C1(A2,A1) : d = C1(A2) – C1(A1)
d = 0.7 – 0.72
d = -0,02
d > 0 then H|d| = 0

C1(A1,A3) : d = C1(A1) – C1(A3)
d = 0.72 – 0.7
d = 0,02
d > 0 then H|d| = 1

C1(A3,A1) : d = C1(A3) – C1(A1)
d = 0.7 - 0.72
d = -0.02
d < 0 then H|d| = 0

C1(A1,A4) : d = C1(A1) – C1(A4)
d = 0.72 – 0.6
d = 0.12
d > 0 then H|d| = 1

C1(A4,A1) : d = C1(A4) – C1(A1)
d = 0.6 - 0.72
d = -0.12
d < 0 then H|d| = 0

C1(A2,A3) : d = C1(A2) – C1(A3)
d = 0.7 – 0.7
d = 0
d = 0 then H|d| = 0

C1(A3,A2) : d = C1(A3) – C1(A2)
d = 0.7 - 0.7
d = 0
d = 0 then H|d| = 0

C1(A2,A4) : d = C1(A2) – C1(A4)
d = 0.7 – 06
d = 0.1
d > 0 then H|d| = 1

C1(A4,A2) : d = C1(A4) – C1(A2)
d = 0.6 - 0.7
d = -0.1
d < 0 then H|d| = 0

C1(A3,A4) : d = C1(A3) – C1(A4)
d = 0.7 – 0.6
d = 0,1
d > 0 then H|d| = 1

C1(A4,A3) : d = C1(A4) – C1(A3)
d = 0.6 – 0.7
d = -0,1
d < 0 then H|d| = 0

**2. Content Criteria Value (C2)**

C2(A1,A2) : d = C2(A1) – C2(A2)
d = 0.7 – 0.6
d = 0.1
d > 0 then H|d| = 1

C2(A2,A1) : d = C2(A2) – C2(A1)
d = 0,6 – 0.7
d = -0.1
d < 0 then H|d| = 0

C2(A1,A3) : d = C2(A1) – C2(A3)
d = 0.7 – 0.63
d = 0.07
d > 0 then H|d| = 1

C2(A3,A1) : d = C2(A3) – C2(A1)
d = 0.63 - 0.7
d = -0,07
d < 0 then H|d| = 0

C2(A1,A4) : d = C2(A1) – C2(A4)
d = 0.7 – 0.7
d = 0
d = 0 then H|d| = 0

C2(A4,A1) : d = C2(A4) – C2(A1)
d = 0.7 – 0.7
d = 0

d = 0 then H|d| = 0
C2(A2,A3) : d = C2(A2) – C2(A3)
d = 0.6 – 0.63
d = -0,03
d < 0 then H|d| = 0
C2(A3,A2) : d = C2(A3) – C2(A2)
d = 0.63 - 0.6
d = 0,03
d > 0 then H|d| = 1
C2(A2,A4) : d = C2(A2) – C2(A4)
d = 0.6 – 0.7
d = -0.1
d < 0 then H|d| = 0
C2(A4,A2) : d = C2(A4) – C2(A2)
d = 0.7 – 0.6
d = 0,1
d > 0 then H|d| = 1
C2(A3,A4) : d = C2(A3) – C2(A4)
d = 0.63 - 0.7
d = -0,07
d < 0 then H|d| = 0
C2(A4,A3) : d = C2(A4) – C2(A3)
d = 0.7 – 0.63
d = 0.07
d > 0 then H|d| = 1

**3. Side effects Criteria Value (C3)**

C3(A1,A2) : d = C3(A1) – C3(A2)
d = 0.72 – 0.73
d = -0.01
d < 0 then H|d| = 0
C3(A2,A1) : d = C3(A2) – C3(A1)
d = 0.73 – 0.72
d = 0.01
d > 0 then H|d| = 1
C3(A1,A3) : d = C3(A1) – C3(A3)
d = 0.72 – 0.7
d = 0.02
d > 0 then H|d| = 1
C3(A3,A1) : d = C3(A3) – C3(A1)
d = 0.7 - 0.72
d = -0.02
d < 0 then H|d| = 0
C3(A1,A4) : d = C3(A1) – C3(A4)
d = 0.72 – 0.6
d = 0.12
d > 0 then H|d| = 1
C3(A4,A1) : d = C3(A4) – C3(A1)
d = 0.6 – 0.72
d = -0.12
d < 0 then H|d| = 0
C3(A2,A3) : d = C3(A2) – C3(A3)
d = 0.73 – 0.7
d = 0.03
d > 0 then H|d| = 1
C3(A3,A2) : d = C3(A3) – C3(A2)
d = 0.7 - 0.73
d = -0.03
d < 0 then H|d| = 0
C3(A2,A4) : d = C3(A2) – C3(A4)
d = 0.73 – 0.6
d = 0.13
d > 0 then H|d| = 1
C3(A4,A2) : d = C3(A4) – C3(A2)
d = 0.6 – 0.73
d = -0.13
d < 0 then H|d| = 0
C3(A3,A4) : d = C3(A3) – C3(A4)
d = 0.7 - 0.6
d = 0.1
d > 0 then H|d| = 1
C3(A4,A3) : d = C3(A4) – C3(A3)
d = 0.6 – 0.7
d = -0.1
d < 0 then H|d| = 0

**4. Taste Variant Criteria Value (C4)**

C4(A1,A2) : d = C4(A1) – C4(A2)
d = 0.78 - 0.8
d = -0.02
d < 0 then H|d| = 0
C4(A2,A1) : d = C4(A2) – C4(A1)
d = 0.8 – 0.78
d = 0.02
d > 0 then H|d| = 1
C4(A1,A3) : d = C4(A1) – C4(A3)
d = 0.78 - 0.63
d = 0.15
d > 0 then H|d| = 1
C4(A3,A1) : d = C4(A3) – C4(A1)
d = 0.63 – 0.78
d = -0.15
d < 0 then H|d| = 0
C4(A1,A4) : d = C4(A1) – C4(A4)
d = 0.78 – 0.2
d = 0.58
d > 0 then H|d| = 1
C4(A4,A1) : d = C4(A4) – C4(A1)
d = 0.2 – 0.78
d = -0.58
d < 0 then H|d| = 1
C4(A2,A3) : d = C4(A2) – C4(A3)

d = 0.8 – 0.63
d = 0.17
d > 0 then H|d| = 1
C4(A3,A2) : d = C4(A3) – C4(A2)
d = 0.63 – 0.8
d = -0.17
d < 0 then H|d| = 0
C4(A2,A4) : d = C4(A2) – C4(A4)
d = 0.8 – 0.2
d = 0.6
d > 0 then H|d| = 1
C4(A4,A2) : d = C4(A4) – C4(A2)
d = 0.2 - 0.8
d = -0.6
d < 0 then H|d| = 0
C4(A3,A4) : d = C4(A3) – C4(A4)
d = 0.63 – 0.2
d = 0.43
d > 0 then H|d| = 1
C4(A4,A3) : d = C4(A4) – C4(A3)
d = 0.2 - 0.63
d = -0.43
d < 0 then H|d| = 0

**STEP II:** Calculate the Preference Index Alternative partner (A1,A2)
(A1,A2) = (1 + 1 + 0 + 0) = 2/4 = 0.5
Alternative partner (A2,A1)
(A2,A1) = (0 + 0 + 1 + 1) = 2/4 = 0.5
Alternative partner (A1,A3)
(A1,A3) = (1 + 1 + 1 + 1) = 4/4 = 1
Alternative partner (A3,A1)
(A3,A1) = (0 + 0 + 0 + 0) = 0/4 = 0
Alternative partner (A1,A4)
(A1,A4) = (1 + 0 + 1 + 1) = 3/4 = 0.75
Alternative partner (A4,A1)
(A4,A1) = (0 + 0 + 0 + 1) = 1/4 = 0.25
Alternative partner (A2,A3)
(A2,A3) = (0 + 0 + 1 + 1) = 2/4 =0.5
Alternative partner (A3,A2)
(A3,A2) = (0 + 1 + 0 + 0) = 1/4 =0.25
Alternative partner (A2,A4)
(A2,A4) = (1 + 0 + 1 + 1) = 3/4 = 0.75
Alternative partner (A4,A2)
(A4,A2) = (0 + 1 + 0 + 0) = 1/4 = 0.25
Alternative partner (A3,A4)
(A3,A4) = (1 + 0 + 1 + 1) = 3/4 = 0.75
Alternative partner (A4,A3)
(A4,A3) = (0 + 1 + 0 + 0) = 1/4 = 0.25

The following are the complete results of the Preference Index calculation as shown in the following table:

**Table 4.** Preference Index Value

| | **A1** | **A2** | **A3** | **A4** |
|---|---|---|---|---|
| **A1** | - | 0.5 | 1 | 0.75 |
| **A2** | 0.5 | - | 0.5 | 0.75 |
| **A3** | 0 | 0.25 | - | 0.75 |
| **A4** | 0.25 | 0.4 | 0.25 | - |

**STEP III:** Calculate PROMETHEE I

1. Calculating Leaving Flow
   Then obtained:
   A1 = 1 / (4-1) (0.5 + 1 + 0.75) = 0.749
   A2 = 1 / (4-1) (0.5 + 0.5 + 0.75) = 0.582
   A3 = 1 / (4-1) (0 + 0.25 + 0.75) = 0.333
   A4 = 1 / (4-1) (0.25 + 0.4 + 0.25) = 1.233
2. Calculating *Entering Flow*
   Then obtained:
   A1 = 1 / (4-1) (0.5 + 0 + 0.25) = 0.249
   A2 = 1 / (4-1) (0.5 + 0.25 + 0.4) = 0.382
   A3 = 1 / (4-1) (1 + 0.5 + 0.25) = 0.582
   A4 = 1 / (4-1) (0,75 + 0.75 + 0.75) = 0.749

**STEP IV:** Calculate PROMETHEE II:
Obtained:
A = 0.749 – 0.249 = 0.5
B = 0.582 – 0.382 = 0.2
C = 0.333 – 0.582 = -0.249
D = 1.233 – 0.749 = 0.4

Based on step 3, the process of step 4 is obtained to calculate PROMETHEE II so that the calculation results in step 4 are shown in the following table:

**Table 5.** the calculation results of PROMETHEE II

| | Net Flow | Ranking |
|---|---|---|
| Anlene (A1) | 0.5 | 1 |
| Entrasol (A2) | 0.3 | 3 |
| Prolansia (A3) | -0.249 | 4 |
| Appeton 60+ (A4) | 0.484 | 2 |

Based on table 5, the process of calculating Net Flow (PROMETHEE II) is obtained from the difference between Leaving Flow and Entering Flow (PROMETHEE I). The biggest results obtained are the best alternatives. In this case, calcium milk products that are recommended for the elderly based on the calculation of PROMETHEE II are Anlene (A1).

## 4. Conclussion

Based on the results of the study it can be concluded that the PROMETHE II method can be applied to the selection of calcium milk products for the elderly based on consumer choice. The results of the study mention of the 5 alternatives used and 4 assessment criteria obtained by Anlene (A1) as the best recommendation (net Flow: 0.5) and Appeton 60+ (A5) as the second recommendation (Net Flow: 0.484).

## References

[1] I. Utami, "Hubungan antara literatur," pp. 8–31, 2009.
[2] S. U. dan Abu bakar, *Teknologi Pengolahan Susu*, vol. 4, no. 2. 2009.
[3] D. Yudianti, "Analisis Asupan Air Dan Mutu Gizi Asupan Pangan Pada Lansia Di Indonesia," 2011.
[4] Sudirman, A. P. Windarto, and A. Wanto, "Data mining tools | rapidminer: K-means method on clustering of rice crops by province as efforts to stabilize food crops in Indonesia," *IOP Conf. Ser. Mater. Sci. Eng.*, vol. 420, p. 12089, 2018.
[5] H. Siahaan, H. Mawengkang, S. Efendi, A. Wanto, and A. P. Windarto, "Application of Classification Method C4 . 5 on Selection of Exemplary Teachers," in *IOP Conference Series*, 2018, pp. 1–6.
[6] B. Supriyadi, A. P. Windarto, T. Soemartono, and Mungad, "Classification of natural disaster prone areas in Indonesia using K-means," *Int. J. Grid Distrib. Comput.*, vol. 11, no. 8, pp. 87–98, 2018.
[7] Ermatita, Sri Hartati, R. Wardoyo, and A. Harjoko, "Electre Methods in Solving Group Decision Support System Bioinformatics on Gene Mutation Detection Simulation," *Int. J. Comput. Sci. Inf. Technol.*, vol. 3, no. 1, pp. 40–52, 2011.
[8] T. Budiharjo, Soemartono, T., Windarto, A.P., Herawan, "Predicting school participation in indonesia using back-propagation algorithm model," *Int. J. Control Autom.*, 2018.
[9] T. Budiharjo, Soemartono, T., Windarto, A.P., Herawan, "Predicting tuition fee payment problem using backpropagation neural network model," *Int. J. Adv. Sci. Technol.*, 2018.
[10] Sumijan, A. P. Windarto, A. Muhammad, and Budiharjo, "Implementation of Neural Networks

in Predicting the Understanding Level of Students Subject," *Int. J. Softw. Eng. Its Appl.*, vol. 10, no. 10, pp. 189–204, 2016.

[11] A. P. Windarto, L. S. Dewi, and D. Hartama, "Implementation of Artificial Intelligence in Predicting the Value of Indonesian Oil and Gas Exports With BP Algorithm," *Int. J. Recent Trends Eng. Res.*, vol. 3, no. 10, pp. 1–12, 2017.

[12] R. Rahim *et al.*, "Enhanced pixel value differencing with cryptography algorithm," in *MATEC Web of Conferences 197*, 2018, vol. 3011, pp. 1–5.

[13] F. Adelia, D. Wahyuli, T. Imanda, and A. P. Windato, "Analisis Promethee II Pada Faktor Penyebab Mahasiswa Sulit Menemukan Judul Artikel Ilmiah," *Jurnal Ilmiah KOMPUTASI*, vol. 17, no. 2, pp. 131–135, 2018.

# AUTHOR INDEX

# AUTHOR INDEX

# AUTHOR INDEX

**Institute of Physics**
Dirac House, Temple Back
Bristol BS1 6BE UK

ISSN: 1742-6588
ISBN 978-1-7138-1137-4

9 781713 811374